# Animals in Traditional Folk Medicine

Rômulo Romeu Nóbrega Alves
Ierecê Lucena Rosa
Editors

# Animals in Traditional Folk Medicine

## Implications for Conservation

Springer

*Editors*
Rômulo Romeu Nóbrega Alves
CCBS—Departamento de Biologia
Universidade Estadual da Paraiba
Campina Grande-PB
Brazil

Ierecê Lucena Rosa
CCEN/Depto. de Sistemático
Universidade Federal da Paraíba
João Pessoa-PB
Brazil

ISBN 978-3-642-29025-1 ISBN 978-3-642-29026-8 (eBook)
DOI 10.1007/978-3-642-29026-8
Springer Heidelberg New York Dordrecht London

Library of Congress Control Number: 2012940466



Printed on acid-free paper

Springer is part of Springer Science+Business Media (www.springer.com)

# Preface

This book deals with the use of animals for medicinal purposes—one of the oldest forms of biodiversity use known to humankind, which is still practiced throughout the world. Ingredients derived from wild animals are not only widely used in traditional remedies, but are also increasingly valued as raw materials in the preparation of modern medicines.

Hence, conserving the resource base upon which traditional medical systems rely is critical both from an ecological and a cultural perspective. Following this recognition, "Animals in traditional folk medicine: implications for conservation" provides a compilation of current articles discussing key issues on the use of animal products, their trade and conservation, highlighting the importance of inter- and cross-disciplinary approaches to increase our comprehension of the interfaces between traditional medicines and the conservation of biodiversity and ecosystem processes.

The book has its origins in a series of field investigations conducted by us, to document the trade and use of medicinal animal species in Brazil, which resulted in the publication of a number of articles focusing on zootherapeutic practices, and their implications for conservation. These led to an invitation, by Dr. Verena Penning, editor of Life Sciences of Springer, to edit a book on the theme.

We readily accepted the challenge, knowing that a book focusing on the different facets of the use of animals in traditional medical systems would fill a significant gap in knowledge, and could become a valuable tool for students, researchers, conservationists, managers, and policy-makers. With the contributions from a distinguished group of international experts, this book aims to introduce the reader to the historical and contemporary uses of animals for medicinal purposes, and to discuss their implications for conservation.

This book owes a debt of gratitude to many people. Thanks are due to the Dr. Verena Penning and Annete Lindqvist (Springer-Verlag), and to all contributors who embraced the "book project" with us. Thanks are also due to all interviewees, who kindly shared their knowledge with the authors of the various chapters which make up this book.

Rômulo Romeu Nóbrega Alves
Ierecê Lucena Rosa

# Foreword

Zootherapy refers to the medicinal use of animals and animal-derived products to treat illnesses and health conditions. Since ancient times animals and the products derived from different organs of their bodies have constituted part of the inventory of medicinal substances used in various cultures; such uses still exist in various types of ethnic folk medicine until the present day. It is important to note that the phenomenon of zootherapy is marked both by a broad geographical distribution and by very deep historical origins.

Zootherapy research as an important field of research in the wider field of ethno-pharmacology was established in the 1990s and evolved during the 2000s. Since then, dozens of ethnozoological studies dealing with zootherapeutic substances have been written and published, many of them during the past decade. The vast majority of these describe the use of medicinal substances of animal origin in various locations and among different communities mainly in South America, Africa, and Asia

The current volume and the 20 chapters collected here reflect the most important issues in the field, stressing not just the historical background nor the traditional or ethnic practice, but also the need to preserve the knowledge and practice on one hand and to conserve the wildlife and its biodiversity on the other. These chapters were written by scholars and scientists devoted to the above-mentioned ideas and using traditional as well as the most modern research methods.

From systematical point of view, this book contains a vast amount of knowledge and information regarding the traditional medical use of hundreds of species in the many systematic zoological divisions including marine invertebrates, fish, reptiles, birds, mammals (including carnivorous as well as aquatic, and even primates). From a geographical point of view, this book describes and analyzes localities in South America such as Brazil (from the Atlantic coastal forest up to the Amazon forest), Europe (Portugal), Africa (Nigeria, Zimbabwe, Benin), West Africa, and South Africa.

However, the main issues this book deals with are the worldwide trade in animals for traditional medicine, bioprospecting, and biodiversity conservation, the use of animals in complementary and alternative medicine, as well as ethno-veterinary issues.

As can be seen so far, the main purpose of the present volume is to not only record historical, ethnic, and traditional knowledge about the medicinal use of animals in various countries, but also to help the international organization and the local authorities in each nation to conserve the species that are used for medicinal substances. This can be done in many cases without having the need to stop the use being done with them by the indigenous people.

The many scholars who have contributed to this volume are all working at the forefront of current research. Some are well known and well-established scientists, while others are at the beginning of their scholarly careers, but all have made a significant mark upon this field; together they represent the spearhead of scientific advancement in the sphere of zootherapy.

The editors, Rômulo Romeu Nóbrega Alves and Ierecê Lucena Rosa and the undersigned, are grateful to all the researchers who willingly agreed to write and present their research in this volume. We are all greatly indebted to Springer's scientific editor and the Springer publishing house without whom this book could not have seen the light of day.

Prof. Efraim Lev
Historian of Medicine and Ethnopharmacologist
Department of Eretz Israel Studies
University of Haifa

# Contents

# Contributors

**Rômulo Romeu Nóbrega Alves** Departamento de Biologia, Universidade Estadual da Paraíba, Campina Grande-PB 58109-753, Brazil, e-mail: romulo_nobrega@yahoo.com.br

**Hugues A. Akpona** Laboratory of Applied Ecology, Faculty of Agronomy Science, University of Abomey-Calavi, Benin, Nigeria, e-mail: hadak01avr@yahoo.fr

**Raynner Rilke Duarte Barboza** Programa de Pós-Graduação em Ciências Biológicas (Zoologia), Departamento de Sistemática e Ecologia, Universidade Federal da Paraíba, João Pessoa-PB 58059-970, Brazil, e-mail: raynner@live.com

**Alpina Begossi** Museu de História Natural, Universidade Estadual de Campinas (UNICAMP), Campinas-SP 13081-970, Brazil, e-mail: alpinab@uol.com.br

**Dandara Monalisa Mariz Bezerra** Departamento de Sistemática e Ecologia, Programa de Pós-Graduação em Ciências Biológicas (Zoologia), Universidade Federal da Paraíba, João Pessoa-PB 58059-970, Brazil, e-mail: dand_mm@yahoo.com.br

**Robin Bruyns** School of Animal, Plant and Environmental Sciences, University of the Witwatersrand, Private Bag 3, Wits 2050, Johannesburg, GP, South Africa, e-mail: robinbruyns@gmail.com

**Luis Miguel Pires Ceríaco** Universidade de Évora, Évora, Portugal, e-mail: luisceriaco@netcabo.pt

**Anthony B. Cunningham** School for Environmental Research, Charles Darwin University, Darwin, NT 0909, Australia, e-mail: tonyc05@bigpond.net.au

**José da Silva Mourão** Ethnozoology, Departamento de Biologia, Conservation and Biodiversity Research Group, Universidade Estadual da Paraíba, Campina Grande-PB, 58109-753, Brazil, e-mail: tramataia@gmail.com

**Kleber da Silva Vieira** Laboratório de Ecofisiologia Animal, Departamento de Sistemática e Ecologia, Universidade Federal da Paraíba, João Pessoa-PB 58051-900, Brazil, e-mail: ksvieira04@yahoo.com.br

**Washington Luiz da Silva Vieira** Laboratório de Ecofisiologia Animal, Departamento de Sistemática e Ecologia, Universidade Federal da Paraíba, João Pessoa-PB, 58051-900, Brazil, e-mail: wlsvieira@yahoo.com.br

**Ulysses Paulino de Albuquerque** Applied Ethnobotany Laboratory, Departamento de Biologia, Universidade Federal Rural de Pernambuco, Dois Irmãos-PE 52171-900, Brazil, e-mail: upa677@hotmail.com

**Gabriela Rocha Defavari** Departamento de Sistemática e Ecologia, Programa de Pós-Graduação em Ciências Biológicas (Zoologia), Universidade Federal da Paraíba, João Pessoa-PB, 58059-970, Brazil, e-mail: gabidefavari@terra.com.br

**Chabi A. M. S. Djagoun** Laboratory of Applied Ecology, Faculty of Agronomy Science, University of Abomey-Calavi, Benin, Nigeria, e-mail: sylvestrechabi@gmail.com

**Toby J. Hibbitts** Department of Wildlife and Fisheries Sciences, Texas A & M University, College Station, TX 77843-2258, USA, e-mail: thibbitts@tamu.edu

**Alan C. Kemp** Research Associate, FitzPatrick Institute of African Ornithology, University of Cape Town, Rondebosch, Cape Town 7700, South Africa, e-mail: leadbeateri@gmail.com

**Maria Franco Trindade Medeiros** Applied Ethnobotany Laboratory, Departamento de Biologia, Universidade Federal Rural de Pernambuco, Dois Irmãos-PE 52171-900, Brazil, e-mail: mariaftm@hotmail.com

**Lívia Emanuelle Tavares Mendonça** Departamento de Sistemática e Ecologia, Programa de Pós-Graduação em Ciências Biológicas (Zoologia), Universidade Federal da Paraíba, João Pessoa-PB, 58059-970, Brazil, e-mail: livia.etm@gmail.com

**Guy A. Mensah** Laboratoire d'Ecologie Appliquée, Faculté des Sciences Agronomiques, Université d'Abomey-Calavi, 01 BP, 526, Cotonou, Benin, e-mail: ga_mensah@yahoo.com

**Paulo Fernando Guedes Montenegro** Laboratório de Ecofisiologia Animal, Departamento de Sistemática e Ecologia, Universidade Federal da Paraíba, João Pessoa-PB 58051-900, Brazil, e-mail: paulomontenegrobr@yahoo.com.br

**Ronnie Enderson Mariano Carvalho Cunha Oliveira** Departamento de Biologia, Universidade Estadual da Paraíba, Campina Grande-PB, 58109-753, Brazil, e-mail: ronnie.enderson@gmail.com

**Tacyana Pereira Ribeiro Oliveira** Departamento de Sistemática e Ecologia, Programa de Pós-Graduação em Ciências Biológicas (Zoologia), Universidade

Federal da Paraíba, João Pessoa-PB, 58059-970, Brazil, e-mail: tpro_imat@yahoo.com.br

**Andrea Pieroni** University of Gastronomic Sciences, Via Amedeo di Savoia, 8I-12060 Pollenzo/Bra, Italy, e-mail: a.pieroni@netcologne.de

**Lorena Cristina Lana Pinto** Universidade Federal de Ouro Preto, Minas Gerais, Brazil, e-mail: lorena_clp@yahoo.com.br

**Cassandra Leah Quave** Department of Microbiology and Immunology, University of Arkansas for Medical Sciences, 4301 W. Markham St., Mail Slot 511, Little Rock, AR 72205, USA, e-mail: cassy.quave@gmail.com

**Milena Ramires** CMU, UNICAMP, Campinas, São Paulo-SP, 6023, Brazil, e-mail: milena.ramires@hotmail.com

**Ierecê Lucena Rosa** Departamento de Sistemática e Ecologia, Universidade Federal da Paraíba, João Pessoa-PB 58059-970, Brazil, e-mail: ierecerosa@yahoo.com.br

**Gindomar Gomes Santana** Laboratório de Ecofisiologia Animal, Departamento de Sistemática e Ecologia, Universidade Federal da Paraíba, João Pessoa-PB 58051-900, Brazil, e-mail: gindomar@yahoo.com.br

**Brice Sinsin** Laboratoire d'Ecologie Appliquée, Faculté des Sciences Agronomiques, Université d'Abomey-Calavi, 01 BP, 526, Cotonou, Benin, e-mail: bsinsin@gmail.com

**Wedson Medeiros Silva Souto** Departamento de Sistemática e Ecologia, Programa de Pós-Graduação em Ciências Biológicas (Zoologia), Universidade Federal da Paraíba, João Pessoa-PB 58059-970, Brazil, e-mail: wedson@live.com

**Martin J. Whiting** School of Animal, Plant and Environmental Sciences, University of the Witwatersrand, Private Bag 3, Wits 2050, Johannesburg, GP, South Africa, e-mail: martin.whiting@mq.edu.au

**Vivienne Williams** University of the Witwatersrand, Johannesburg, Republic of South Africa, e-mail: vivienne.williams@wits.ac.za

# Chapter 1
# Introduction: Toward a Plural Approach to the Study of Medicinal Animals

**Rômulo Romeu Nóbrega Alves and Ierecê Lucena Rosa**

**Abstract** This chapter provides a brief overview on the following chapters and highlights the relevance of research focusing on the historical and current uses of animals in folk medicine.

## 1.1 Introduction

Wildlife and humans coexist in an intricate relationship: people value wildlife as a source of income, food and medicine, as a cultural symbol or as charms (Alves et al. 2010b; Bagniewska and MacDonald 2010). From an ethnomedicinal perspective, wildlife represents an immeasurable source of information and raw materials, which support health systems of different human cultures that depend on nature as a source of medicines to treat and cure illnesses.

Naturally occurring substances of plant, animal, and mineral origin have provided a continuing source of medicines since the earliest times (David and Anderson 1969), and the use of traditional medicines has provided a crucial support and health system to millions of people around the globe for past centuries and this continues to the present day (Grundmann 2011).

R. R. N. Alves (✉)
Departamento de Biologia, Universidade Estadual da Paraíba,
Campina Grande-PB 58109-753, Brazil
e-mail: romulo_nobrega@yahoo.com.br

I. L. Rosa
Departamento de Sistemática e Ecologia, Universidade Federal da Paraíba,
João Pessoa 58059-970, Brazil

R. R. N. Alves and I. L. Rosa (eds.), *Animals in Traditional Folk Medicine*,
DOI: 10.1007/978-3-642-29026-8_1, © Springer-Verlag Berlin Heidelberg 2013

Traditional medical systems throughout the world have been relied on to support, promote, retain, and regain human health for millennia (Alves and Rosa 2007a; Li et al. 2008; Robinson and Zhang 2011; WHO 2002), and it has been estimated that between 70 and 95% of citizens in a majority of developing countries, especially those in Asia, Africa, Latin America, and the Middle East, use traditional medicine for the management of health and as primary health care to address their health care needs and concerns (Robinson and Zhang 2011; WHO 2002).

Although plants and plant-derived materials make up the majority of the ingredients used in most traditional medical systems globally, whole animals, animal parts, and animal-derived products also constitute important elements of the materia medica (Alakbarli 2006; Alves and Rosa 2005; Moquin-Tandon 1861; Scarpa 1981; Stephenson 1832; Unnikrishnan 2004).

Since ancient times, animals and products derived from different organs of their bodies have constituted part of the inventory of medicinal substances used in many cultures, and are still used in traditional folk medicine (Lev 2006). Furthermore, ingredients derived from wild plants and animals are increasingly valued as raw materials in the preparation of modern medicines (Adeola 1992; Alves and Rosa 2005).

Another way in which animals are used to ameliorate human health conditions is through pet therapy or animal-assisted therapy. Both domestic and domesticated animals find medicinal usage as co-therapists as they may improve different kinds of debilitating illnesses or mental deficiencies (Silveira 1998). The contact with animals has proven to be an efficient method to stimulate and help individuals with mobility problems, mental disabilities, or behavior problems. Playing with pets such as dogs, rabbits, and cats, children learn how to command respect, how to develop and how to face new changes, and situations (Pentenero 2001).

In this book, the term zootherapy refers to the medicinal use of animals and animal-derived products to treat illnesses and health conditions. The phenomenon of zootherapy is marked both by a broad geographical distribution and very deep historical origins, and as stated by Marques (1994), "all human culture which presents a structured medical system will utilize animals as medicines." In modern societies, zootherapy constitutes an important alternative among many other known therapies practiced worldwide (Alves and Rosa 2005).

The world's animals and plants—including a number of species used in traditional medicine—face threats from source that range habitat loss to global trade (Cleva 2006). There has been increasing demand for traditional medicines (Alves and Rosa 2007a; Robinson and Zhang 2011), and in recent years, the link between traditional medicine and the loss of certain species has become apparent (Alves and Rosa 2007a; Call 2006). This trend bears important implications for the conservation of the many species of flora and fauna on which traditional remedies are based (Alves and Rosa 2007a; Alves et al. 2007; Lee 1999; Still 2003). Unfortunately, where the use of traditional remedies used to be a localized practice, the globalization of commerce combined with the increased popularity of natural approaches to health worldwide has created a level of demand that threatens the survival of many vulnerable species of wildlife (IFAW 2011).

While the use of floristic resources in traditional medicine has been widely researched, there is a paucity of information regarding the utilization of faunistic resources (Alves and Rosa 2006), and as such, the medical value of animal species has not been included in calculations of the economic value of biodiversity by the World Conservation Union (IUCN) (Pearce and Moran 1994). Nonetheless, the importance of nonbotanical remedies (of animal and mineral origin) is emerging (Alves and Rosa 2007a; Groark 1997). Despite the recent surge in publications about zootherapy, the subject is still far from being well covered, and even more distant from being exhausted. The lack of zootherapeutic studies in the world in general has contributed to an underestimation of the importance of zootherapeutic resources (Alves and Alves 2011; Alves and Rosa 2010).

There are many reasons why research on the traditional medicinal uses of animals should be carried out and recorded. Among several approaches to be considered, this introduction briefly discusses those concerning with the ecological, cultural (traditional knowledge), economical, and sanitary aspects of animals in traditional folk medicine, based on Alves and Rosa (2005).

## 1.2 Ecological Approach

The world is facing potentially massive loss of wildlife due to over-hunting (Bennett et al. 2002; Robinson and Bennett 2000, 2002) and overfishing (Boehlert 1996; Jennings and Kaiser 1998). Transformation of ecosystems wrought through economic activities has been putting severe constraints on the availability and accessibility of specific types of plant and animal species used for medicinal purposes (Anyinam 1995). Regrettably, the demand created by traditional medicine is one of the causes of the overexploitation of the wild population of numerous animal species (Alves and Rosa 2005; Still 2003), notably of vertebrates, such as tigers, rhino, bears, and seahorses. The use of animals in popular medicine certainly provokes pressure on natural resources exploited through traditional forms of collection, mainly due to general acceptance of popular medicine (Almeida and Albuquerque 2002; Alves et al. 2010a). Medically speaking, the one major negative consequence of this trend is that there will be essentially less choice for the future development of medicines (Still 2003).

At present, over 40% of all prescription drugs are substances originally extracted from plants, animals, fungi, and microorganisms (Wilson 1995), natural resource users may be the first to observe depletion. Traditional ecological knowledge is of significance from a conservation perspective and an attribute of societies with continuity in resource use practice (Gadgil et al. 1993), therefore the dissociation of traditional knowledge from managerial ecology may result in the adoption of inadequate management options. Holders of traditional knowledge not only have a role as natural resource managers, but can also provide a model for biodiversity policies. However, as traditional peoples are integrated into the global economy, and come under trade, acculturation, and population pressures, they lose their attachment to their own restricted resource catchments. This may lead to a loss of motivation in sustainable uses of a diversity of local resources, along with the pertinent indigenous knowledge (Alves et al. 2010b; Gadgil et al. 1993).

There is a need to shift the focus from how to obtain the greatest amount of zootherapeutical resources to ensure future uses; there is also a need to foster a transdisciplinary approach to integrate the various aspects of the traditional medicinal uses of animals, in such a way that frameworks or methods to amalgamate ecological and social components of that practice can be increasingly tested. In this context, it is important not only to document the traditional uses of animal species, but also to integrate the cultural and biological aspects of such practices into a broader discourse encompassing conservation, cooperative management, and sustainability.

## 1.3 Cultural Approach

Through their long exposure and experience with natural resources, many human communities have developed health care practices. Thousands of years of observation and experimentation have helped in developing different empirical medical systems, as well as knowledge of plants, animals, and minerals. Medicinal animals are important resources linking people to the environment and their use promotes the traditional lore related to them. There is an increased interest in the knowledge that traditional populations possess on the use of animals for medicinal purposes, partly because the empirical basis developed throughout centuries may have, in many cases, scientific corroboration; but above all due to the historical, economical, sociological, anthropological, and environmental aspects of such a practice (Alves and Rosa 2005; Alves and Rosa 2007a; Lev 2006).

For centuries, healers and indigenous people have been collecting medicines from local plants and animals without threatening the population dynamics of the species because of the low level of harvesting. Medicinal folklore over the years has proved to be an invaluable guide in present day to the screening of important modern drugs (e.g., digitoxin, reserpine, tubocurarine, ephedrine, to name a few) that have been discovered by following leads from folk uses (Anyinam 1995). In view of this, it is evident the need to document the traditional knowledge of human communities, mainly because the majority of such communities are rapidly losing their socioeconomic and cultural characteristics.

The importance of protecting traditional knowledge and its cultural environmental resources is crucial, particularly in the context of globalization and increased demand on natural resources worldwide. Traditional knowledge is valuable not only to those directly involved with it, but also to modern medicine and agriculture, among others. Moreover, protection of traditional knowledge can be used to raise the profile of the knowledge and its custodians. This not only has implications for the continuation of traditional practices within communities, but also for the interactions (e.g., economic, ecological) established outside the communities.

## 1.4 Economical Approach

The value of biodiversity to human health has been highlighted in literature (Beattie et al. 2005; Chivian 2002; Grifo and Rosenthal 1997). The most obvious benefit is the large proportion of the pharmaceutical armamentarium that is derived from the natural world. Over 50% of commercially available drugs are based on bioactive compounds extracted (or patterned) from nonhuman species (Grifo et al. 1997). Almost every class of drug includes a model structure derived from nature, exhibiting the classical effects of the specific pharmacological category. A great number of these natural products have come to us from the scientific study of remedies traditionally employed by various cultures (Holmstedt and Bruhn 1983). In addition to plants and microbes, there has been increasing attention paid to animals, both vertebrates and invertebrates, as sources for new medicines. Animals have been methodically tested by pharmaceutical companies as sources of drugs for modern medical science (Kunin and Lawton 1996), and the current percentage of animal sources for producing essential medicines is quite significant (Alves and Rosa 2005).

Underlying the debate over traditional knowledge may be a much bigger issue such as the position of indigenous communities within the wider economy and society of the country in which they reside, and their access to or ownership of land they have traditionally inhabited. In that sense, concerns about the preservation of traditional knowledge, and the continued way of life of those holding such knowledge, may be symptomatic of the underlying problems that face these communities in the face of external pressures (Commission-on-Intellectual-Property-Rights 2002).

Given the inconsistent terminology and regulatory status of traditional medicines, it is difficult to put a precise figure on the economic value of such products. What data are available are generally considered to significantly under represent the actual economic impact of this branch of the health care marketplace (Robinson and Zhang 2011). What is more certain is that the global market for traditional medicine products has expanded significantly over the last decade with growth in demand, production, and sales. Market estimates suggest that the rate of growth in traditional medicine product sales in recent years amounts to somewhere between 5 and 18% per annum (Calixto 2000).

The trade in wildlife body parts and products includes traditional medicine, and it is well known that the annual global trade in animal-based medicinal products accounts for billions of dollars per year (Alves and Rosa 2007b). Nevertheless, in countries such as Brazil, the trade of animals for medicinal purposes has had little impact on the socioeconomic conditions of collectors, who generally are illiterate, underpaid, and perceive their activity as clandestine or semi-clandestine. The monetary value of animals sold for medicinal purposes in the country increases at each level of trade, and the socioeconomic profile of traders varies accordingly (Alves and Rosa 2005).

Additionally, there is a need to ensure that custodians of traditional knowledge receive fair compensation if the traditional knowledge leads to commercial gain, and to prevent appropriation of traditional knowledge by unauthorized parties (Ashwell and Walston 2008).

## 1.5 Sanitary Approach

The idea that just because traditional medicine products come from natural sources they are completely safe is dangerously false (Calixto 2000). Not everything that is natural is safe; traditional medicine products must be used judiciously and as indicated, just like any other medication, and with awareness of potential herb–herb and herb–drug interactions (Robinson and Zhang 2011). The risks are relatively small when traditional medicines are used correctly, but they are still there, and consumer understanding is generally low. For example, a cross-sectional population survey conducted in Australia found that less than half (46.6%) of traditional, herbal medicine users were even aware that there could be potential risks associated with product use (Zhang et al. 2008).

Traditional drugs and traditional medicine in general represent a still poorly explored field of research in terms of therapeutic potential or clinical evaluation. There is a current preoccupation about this, since it is well-established that all sorts of vegetable, animal, and mineral remedies used in a traditional setting are capable of producing serious adverse reactions. Therefore, it is essential that traditional drug therapies be submitted to an appropriate benefit/risk analysis (De Smet 1991). Unfortunately, little research has been done so far to prove the claimed clinical efficacy of animal products for medicinal purposes (Still 2003).

## 1.6 Why this Book?

Animal-based remedies have constituted part of the inventory of medicinal substances used in various cultures since ancient times, and there is growing recognition that people in different parts of the world still use them as primary or complementary medicine. Conserving the resource base on which traditional medical systems rely is critical both from an ecological and a cultural perspective.

Following this recognition, "Animals in traditional folk medicine: implications for conservation" significantly expands the knowledge base related to the global utilization of animals in traditional medical systems, (1) by bringing together the results of research conducted in different parts of the world, (2) by covering a wide range of animal taxa used in folk medicines, and (3) by providing a compilation of current articles discussing key issues on the use of animal products, their trade, and conservation. This book highlights the importance of inter and cross-disciplinary approaches to increase our comprehension of the interfaces between traditional medicines and the conservation of biodiversity and ecosystem processes. We hope that it will serve as a valuable reference for researchers, students, educators, conservationists, wildlife managers, policymakers, as well as for the general public.

Following this Introduction, Chap. 2 provides an historical overview on use of medicinal animals in medical systems used by different peoples, and emphasizes the role played by historical ethnozoology as a discipline focusing on the relationships between animals and culture over time. Chapter 3 reviews some aspects related to the commercialization of medicinal animals, and explores ways through which conservation and management initiatives can improve dialog with resource users and traders. Chapter 4 reviews the global use of animals in traditional veterinary medicine.

Chapter 5 provides an overview of the interfaces between bioprospecting and biodiversity conservation, while Chap. 6 focuses on native fishing communities in the Atlantic Forest (the Caiçaras) and in the Amazon region (the Caboclos) in Brazil, in the context of medicinal uses of fish and food taboos.

Chapters 7–12 summarize knowledge on the global medicinal use of herpetofauna, primates, carnivorous, seahorses, aquatic mammals (with emphasis on orders Cetacea, Sirenia, and Carnivora) and marine invertebrates, respectively, and discuss the implications of their medicinal uses for conservation.

Chapters 13–20 encompass regional/national case studies of countries in Latin America, Africa, and Europe, as follows: Chap. 13 highlights the importance of animal-derived remedies in Brazil, where at least 354 medicinal animal species are used in folk medicine, while Chap. 14 reviews the current ethnobiological literature on European zootherapy. Chapter 15 provides an overview of the use of animals in traditional medicine, magic rites, and superstitions in Portugal. Chapters 16 explores zootherapeutic practices and biodiversity conservation in Nigeria and is followed (Chap. 17) by an analysis of the trade of mammal species sold in the Benin's traditional medicine market. Chapter 18 reviews the use of birds for traditional medicine in sub-Saharan Africa, while Chap. 19 investigates the trade in animals for traditional medicine at the Faraday traditional medicine market in Johannesburg, South Africa. Chapter 20 focuses on the trade and throughput of the bird species *Bucorvus cafer* in the Makokoba informal market in Bulawayo, Zimbabwe.

**Acknowledgments** We are grateful to Dr. Verena Penning (Editor: Life Sciences) and Dr. Annika Koenig (Editor: Animal Sciences) for entrusting us with the edit of this book. We are also grateful to Anette Lindqvist (Springer Science+Business Media) for her constant support. Special thanks are due to contributors to the book, for their enthusiasm, support and quality work. Thanks are also due to CNPq (Conselho Nacional de Desenvolvimento Científico e Tecnológico), for providing us with research productivity scholarships which helped the development of our studies on the medicinal uses of animals in Brazil.

## References

Adeola MO (1992) Importance of wild animals and their parts in the culture, religious festivals, and traditional medicine, of Nigeria. Environ Conserv 19(02):125–134
Alakbarli F (2006) Medical manuscripts of Azerbaijan. Heydar Aliyev Foundation, Baku
Almeida CFCBR, Albuquerque UP (2002) Uso e conservação de plantas e animais medicinais no Estado de Pernambuco (Nordeste do Brasil): Um estudo de caso. Interciencia 27(6):276–285
Alves RRN, Alves HN (2011) The faunal drugstore: Animal-based remedies used in traditional medicines in Latin America. J Ethnobiol Ethnomed 7(9):1–43
Alves RRN, Rosa IL (2005) Why study the use of animal products in traditional medicines? J Ethnobiol Ethnomed 1(5):1–5. doi:10.1186/1746-4269-1
Alves RRN, Rosa IL (2006) From cnidarians to mammals: the use of animals as remedies in fishing communities in NE Brazil. J Ethnopharmacol 107:259–276
Alves RRN, Rosa IL (2007a) Biodiversity, traditional medicine and public health: where do they meet? J Ethnobiol Ethnomed 3(14):9. doi:10.1186/1746-4269-3-14
Alves RRN, Rosa IL (2007b) Zootherapy goes to town: the use of animal-based remedies in urban areas of NE and N Brazil. J Ethnopharmacol 113:541–555. doi:10.1016/j.jep.2007.07.015
Alves RRN, Rosa IL (2010) Trade of animals used in Brazilian traditional medicine: trends and implications for conservation. Hum Ecol 38(5):691–704
Alves RRN, Rosa IL, Santana GG (2007) The role of animal-derived remedies as complementary medicine in Brazil. Bioscience 57(11):949–955
Alves RRN, Souto WMS, Barboza RRD (2010a) Primates in traditional folk medicine: a world overview. Mamm Rev 40(2):155–180. doi:10.1111/j.1365-2907.2010.00158.x
Alves RRN, Souto WMS, Mourão JS (2010b) A Etnozoologia no Brasil: Importância, Status atual e Perspectivas, vol 7. Estudos & Avanços, 1st edn. NUPEEA, Recife
Anyinam C (1995) Ecology and ethnomedicine: exploring links between current environmental crisis and indigenous medical practices. Soc Sci Med 40(3):321–329
Ashwell D, Walston N (2008) An overview of the use and trade of plants and animals in traditional medicine systems in Cambodia, 1st edn. TRAFFIC Southeast Asia, Greater Mekong Programme, Ha Noi
Bagniewska J, MacDonald D (2010) Executive summary of the report Animal welfare, international development, biodiversity conservation—the road to peaceful coexistence Wildlife Conservation Research Unit University of Oxford, Oxford
Beattie AJ, Barthlott W, Elisabetsky E, Farrel R, Kheng CT, Prance I, Rosenthal J, Simpson D, Leakey R, Wolfson M (2005) New products and industries from biodiversity. In: Hassan R, Scholes R, Ash N (eds) Ecosystems and humanwell-being, vol 1, Millennium ecosystem assessment, Island Press Washington pp 273–295
Bennett EL, Milner-Gulland EJ, Bakarr M, Eves HE, Robinson JG, Wilkie DS (2002) Hunting the world's wildlife to extinction. Oryx 36(04):328–329
Boehlert GW (1996) Biodiversity and the sustainability of marine fisheries. Oceanography 9:28–35
Calixto JB (2000) Efficacy, safety, quality control, marketing and regulatory guidelines for herbal medicines (phytotherapeutic agents). Braz J Med Biol Res 33(2):179–189
Call E (2006) Mending the web of life: Chinese medicine and species conservation. International Fund for Animal Welfare, Massachusetts
Chivian E (2002) Biodiversity: its importance to human health center for health and the global environment. Harvard Medical School, Cambridge
Cleva S (2006) U. S. Conservation laws and treaties: support for the web. In: Call E (ed) Mending the web of life: Chinese medicine and species conservation. Signature Book Printing, Inc, Beijing pp 199–214
Commission-on-Intellectual-Property-Rights (2002) Integrating intellectual property rights and development policy. Commission-on-Intellectual-Property-Rights, London

David JP, Anderson LA (1969) Ethnopharmacology and western medicine. J Ethnopharmacol 25:61–72
De Smet PAGM (1991) Is there any danger in using traditional remedies? J Ethnopharmacol 32:43–50
Gadgil M, Berkes F, Folke C (1993) Indigenous knowledge for biodiversity conservation. Ambio 22(2):151–156
Grifo F, Newman D, Fairfield AS, Bhattacharya B, Grupenhof JT (1997) The origins of prescription drugs. In: Grifo F, Rosenthal J (eds) Biodiversity and human health. Island Press, Washington, pp 131–163
Grifo F, Rosenthal J (1997) Biodiversity and human health. Island Press, Washington
Groark KP (1997) To warm the blood, to warm the flesh: the role of the steambath in Highland Maya (Tzeltal-Tzotzil) Ethnomedicine. J Lat Am Lore 20(1):3–96
Grundmann O (2011) Traditional medicines: bridging the gaps for a better future. Altern Med Stud 1(1):e13
Holmstedt B, Bruhn JG (1983) Ethnopharmacology: a challenge. J Ethnopharmacol 8(3):251–256
IFAW (2011) Traditional medicine. http://www.ifaw.org/sites/default/files/Q&A%20Traditional%20medicine.pdf
Jennings S, Kaiser MJ (1998) The effects of fishing on marine ecosystems. Adv Marine Biol 34:201–212
Kunin WE, Lawton JH (1996) Does biodiversity matter? Evaluating the case for conserving species. In: Gaston KJ (ed) Biodiversity: a biology of numbers and differences. Blackwell, Oxford, pp 283–308
Lee SKH (1999) Trade in traditional medicine using endangered species: an international context. In: Paper presented at the 2nd Australian symposium on traditional medicine and wildlife conservation, Melbourne, Australia
Lev E (2006) Healing with animals in the Levant from the 10th to the 18th century. J Ethnobiol Ethnomed 2(11):9. doi:10.1186/1746-4269-2-11
Li S, Han Q, Qiao C, Song J, Cheng CL, Xu H (2008) Chemical markers for the quality control of herbal medicines: an overview. Chin Med 3(1):7
Marques JGW (1994) A fauna medicinal dos índios Kuna de San Blas (Panamá) e a hipótese da universalidade zooterapica. In: Paper presented at the Anais da 46a Reunião Anual da SBPC, Vitória
Moquin-Tandon A (1861) Elements of medical zoology. Baillière, London
Pearce DW, Moran D (1994) The economic value of biodiversity. Earthscan publications, London
Pentenero L (2001) Los animales ayudan a afrontar problemas. La Nación, Bogotá
Robinson JG, Bennett EL (2000) Carrying capacity limits to sustainable hunting in tropical forests. In: Hunting for sustainability in tropical forests. Columbia University Press, New York, pp 13–30
Robinson JG, Bennett EL (2002) Will alleviating poverty solve the bushmeat crisis? Oryx 36(04):332
Robinson MM, Zhang X (2011) The world medicines situation 2011—traditional medicines: global situation issues and challenges. World Health Organization, Geneva
Scarpa A (1981) Pre-scientific medicines: their extent and value. Soc Sci Med 15A:317–326
Silveira N (1998) Gatos, a emoção de lidar. Editora Leo Christiano, São Paulo, Brasil
Stephenson J (1832) Medical zoology, and mineralogy; or Illustrations and descriptions of the animals and minerals employed in medicine, and of the preparations derived from them: including also an account of animal and mineral poisons. John Wilson, London
Still J (2003) Use of animal products in traditional Chinese medicine: environmental impact and health hazards. Complement Therapies Med 11:118–122. doi:10.1016/S0965-2299(03)00055-4
Unnikrishnan E (2004) Materia medica of the local health traditions of Payyannur. Centre for Development Studies, Thiruvananthapuram
WHO (2002) Traditional medicine strategy 2002–2005. WHO Geneva
Wilson EO (1995) Wildlife: legions of the doomed. *Time* Oct 77–79
Zhang AL, Story DF, Lin V, Vitetta L, Xue CC (2008) A population survey on the use of 24 common medicinal herbs in Australia. Pharmacoepidemiol Drug Saf 17(10):1006–1013

# Chapter 2
# From Past to Present: Medicinal Animals in a Historical Perspective

**Rômulo Romeu Nóbrega Alves, Maria Franco Trindade Medeiros, Ulysses Paulino Albuquerque and Ierecê Lucena Rosa**

**Abstract** Interactions with animals constitute one of the oldest forms of connection between mankind and nature. In general, connections that people establish with nature can be driven by different aspects of their lives, either material (e.g., food), or spiritual (e.g., myths). In this sense, the use of animals in medical practices by various human societies can be perceived as a persisting ancient relationship, either through mainstream medicine (e.g., use of animal toxins) or via complementary, and alternative medical systems. This chapter presents an overview of medical systems used by different peoples, and emphasizes the role played by historical ethnozoology as a discipline focusing on the relationships between animals and culture over time.

## 2.1 Introduction

The World Health Organization defines the broad range of characteristics and elements which make traditional medicine as "health practices, approaches, knowledge and beliefs incorporating plant, animal and/or mineral based medicines,

R. R. N. Alves (✉)
Ethnozoology, Conservation and Biodiversity Research Group, Departamento de Biologia, Universidade Estadual da Paraíba, Campina Grande-PB 58109-753, Brazil
e-mail: romulo_nobrega@yahoo.com.br

M. F. T. Medeiros · U. P. Albuquerque
Applied Ethnobotany Laboratory, Departamento de Biologia, Universidade Federal Rural de Pernambuco, Dois Irmãos, Pernambuco 52171-900, Brazil

I. L. Rosa
Departamento de Sistemática e Ecologia, Universidade Federal da Paraíba, João Pessoa-PB 58059-970, Brazil

R. R. N. Alves and I. L. Rosa (eds.), *Animals in Traditional Folk Medicine*,
DOI: 10.1007/978-3-642-29026-8_2, 

spiritual therapies, manual techniques and exercises applied singularly or in combination to maintain well-being, as well as to treat, diagnose or prevent illness" (Gray 1997). This working definition reflects, among other aspects, the fact that human health and biodiversity have been inextricably connected for thousands of years, and that cultures all over the world have developed characteristic ways of interacting with different components of the biota over time. For instance, until around 12,000 years ago humans derived food and raw materials from wild animals and plants (Serpell 1996), and as recently pointed out by Waters et al. (2011), humans engaged in Mastodon hunting since at least 13,800 years ago.

Animals have long been used by humans, reportedly initially for alimentary purposes; research suggests that humans evolved from a vegetarian lifestyle to the one including meat in their diets around 2.5 million years ago, at the dawn of the genus *Homo* (Holzman 2003; Larsen 2003). Nonetheless, just how much of the prehistoric diet included animals is difficult to tell from archeological evidence (Wing 2000).

At the dawn of recorded history it is known that man often ate, or wore on his person, some portion of an animal that was thought to have a healing or protecting influence (MacKinney 1946), and this aspect highlights that the origin of the medicinal use of faunal elements is intertwined with their use as food. In this same direction, Chemas (2010) remarked that treatment of illnesses using animal-based remedies is an extremely old practice, whose most remote ancestor is a carnivore diet, closely followed by the ritual ingestion of deceased persons (e.g., close relatives, warriors) as a means to absorb their virtues (e.g., courage, virility), and subsequently by a true medicinal use indissociable from magic religious elements. Those observations are in line with the view of nature as a provider of many things for humankind, including the tools for the first attempts at therapeutic intervention (Nakanishi 1999a).

We tend, of course, to brand as pure magic and superstition, primitive customs such as devouring the heart or drinking the blood of a defeated enemy, or eating an animal's testicles, liver, spleen, or brain to strengthen the corresponding human organs; or imbibing human and animal excrement, saliva, teeth, ground bones, and the like (MacKinney 1946). This disconnection may be attributed, among other aspects, to the fact that there is more known about prehistoric man's diseases than about earliest random and desperate efforts to use natural resources as "drugs," given that the latter left scarcely an enduring trace. Equally relevant was the gradual placement into separate categories of magical religious practices and medicine, which happened in our millennium (Kremers and Urdang 1951).

On the other hand, the use of the grease of a poisonous animal as a preventive of poison, suggests that our remote ancestors may have stumbled on a valid principle of immunization. Nor is the use of poultices containing dung, grease, urine, etc., or of potions containing internal organs or secretions, to be dismissed as mere nonsense. A modern age of scientific medicine which has suddenly discovered medical values in all sorts of queer substances, and marvelous qualities in liver, kidneys, snake virus, and maggots, should be slow in condemning as pure magic and superstition the organ therapy, and similarly primitive practices of past

(MacKinney 1946). As pointed out by Kramers and Urdang (1951), given that ancient cultures must be considered from the standpoint of their outlook on life afforded by their spiritual, geographic and economic connotations, it would be unfair to accept as proof of an "unscientific" spirit the fact that Babylonian–Assyrian, ancient Egyptian, and the early past of Greek medicine contained a great deal of "magic," as for those peoples, magic was a part of systematized "science".

The antiquity in the use of medicinal animals and its persistency through times are a testimony to the importance of those therapeutic resources to mankind. Understanding the historical context in which those uses took place is central for tracking the knowledge base associated with the vast array of local or traditional medical practices involving animals throughout the centuries, either through diffusion among cultures, or via colonizers, who often brought with them medical practices which were imposed or integrated into local medical systems.

In this thematic review we survey existing sources and research papers on the medicinal use of animals to provide a summary of knowledge and highlight the contributions made by the field of ethnobiology, aiming to increase our understanding of the connection between humans and nature, and particularly, of traditional medical systems worldwide.

## 2.2 Sources and Avenues to Explore Past Relations Between Humans and Animals

The use of medicinal animals may be found in any historical source that contributes to filling in the complex mosaic of the history of human–animal interactions and the implications of this association constitute a valuable object for studies of historic ethnozoology. Evidences for studies focusing on a historical perspective can be gathered from various sources; for instance, rock paintings and archeological inscriptions can attest the antiquity in the use of animals by humans (Martínez 2008; Sutton 1995), though written documents have more precisely recorded information about the interactions of ancient humans groups with their regional fauna and their uses of those animals (Alves and Souto 2010; Martínez 2008). This vision is corroborated by Sneader (2005), who highlights that ancient paintings on the walls of caves to early use of drugs that had a pronounced effect on the mind. However, it is only the written accounts that communicate something of the reasoning that led our ancestors to seek out remedies for treating disease, such as those compiled by inhabitants of Ancient Egypt, Mesopotamia, and particularly Greece.

Valuable information can be obtained through historical texts produced by chroniclers, writers, doctors, naturalists, as well as by other historic subjects, which contain evidence of the ancient relationships between animals and people (Alves and Souto 2010). Equally important are the iconographic documents which may reveal novel aspects the fauna, and in some cases even assist in the identification of species cited in encyclopedic works.

On delimitation of the research's theme and scope (both temporally and spatially), a search for sources can be conducted, after which analytical and critical readings of the material should clarify the main points related to the central questions of the research. Whenever feasible, this exercise of reading and transcribing should be based on first editions of the works, so that the information and spelling can be best analyzed, and then compared with other sources (Medeiros 2010).

In general, focus should be placed on the identification, documentation, characterization and understanding of the role played by animals in past, and current social contexts. Through this exercise of narrative construction it is possible to fill in gaps in knowledge about the processing of a given animal species, as well as to put forth an analysis of the document's author viewpoint (Albuquerque 1997; Nascimento and Carvalho 2006). Clearly, an important development of historical research in ethnobiology is the contextualization of the document. Consideration of aspects such as authorship, the existence of sponsorship, and the sociocultural context of the period in which the text was produced is pivotal to better apprehend the issues involving use of animals in accordance to style of the examined document. Keeping those facets in mind is crucial to come as close as possible to a synchronicity between interpretation and analysis.

This way of accessing the information provided by the author of the analyzed document makes possible to revisit a particular memory layer in a broader way, thus leaving room for reflections on the ideological and thematic bases of the works under examination. Perhaps this constitutes the greatest challenge of such approach, as the data need to be contextualized in light of the period in which they were produced, thus leading to an additional effort on the part of the researcher. Additionally, a central compartment of the methodological and analytical procedures lies in the taxonomic discussion of the findings, as not all documents accessed by the researcher provide clear clues about the taxon under consideration. In that case, the researcher must resort to an ample arsenal of information to solve the problem.

Historical ethnozoology highlights the fact that the interrelationships between people and animals are determined by ecological and cultural conditions, factors that in turn are dynamic and have suffered both qualitative and quantitative changes through time. This way, the formulation of in-depth inferences about the interfaces involving people and animals should consider a historiographical approach, which should also serve as the starting point for the ethnozoological study.

Nonetheless, it should be noted that the historical ethnozoology intends not only to point out animals which have been useful at a given period of time, but also to characterize a sociocultural community, by accessing its shared set of values and knowledge, and its collective imaginary.

## 2.3 Historical Evidence of the Use of Medicinal Animals

Interest in natural sources to provide treatments for pain, palliatives, or curatives for a variety of maladies or recreational use reaches back to the earliest points of history (Spainhour 2005). In that context, the medicinal use of natural products precedes recorded human history probably by thousands of years (Ji et al. 2009). Surveys of the medicaments used in the past show that, whereas compounders of medicines have invariably used vegetable, animal, and mineral substances, animals are less prevalent than herbs, and more prevalent than minerals (MacKinney 1946). Whole animals, animal parts, and animal-derived products (e.g., urine, fat etc.) have been constituted important elements of the materia medica in different regions of the world (Alakbarli 2006; Alves and Rosa 2005; Lev 2003; Unnikrishnan 2004).

Some of the first records on the use of natural products in medicine were written in cuneiform in Mesopotamia on clay tablets and date to approximately 2600 B.C. (Cragg and Newman 2001a, b; Nakanishi 1999a, b). Others of the earliest known written records of healing with medicinal substances were recorded in Egypt. The ancient Egyptians of 3,000–6,000 years ago are credited with developing an elaborate and effective pharmacological collection of numerous curing materials obtained from natural resources (Halberstein 2005). Nunn (2002 p.137) states: "By far the most common form of treatment recommended in the medical papyri was the use of drugs, drawn from a very wide range of animal, mineral, and vegetable substances and administered in a variety of ways."

Our understanding of Egyptian medicine and pharmacology is based on inscriptions on monuments and graves and papyrus rolls, the most important one being Papyrus Ebers. It is thought to have been written around 1550 B.C. (Lidgard 2005) and includes 800 or more prescriptions comprising various herbs, animals, and minerals, a considerable number contain matter derived from birds and beasts both wild and domestic, from insects, reptiles, and fish. From these animals, a variety of materia medica was derived, ranging from the natural products (milk, eggs, etc.) and excrements of living animals, to the bones, blood, flesh, fat, marrow, and hide of carcasses, and in certain cases special substances such as the feathers of birds, the shell of the tortoise, the quills of the porcupine, the slough of snakes, and dirt from under human fingernails. In the case of insects, worms, and small animals, often the entire body was used; of ordinary animals, only small portions were used: the head, hair, horns, brain, eyes, teeth, jawbone, heart, liver, spleen, gall, uterus, vulva, testicles, legs, feet, hoofs, etc. By all odds the most favored substance was the grease, or fat, particularly of the goose and ox. In over a hundred recipes greasy substances were employed; in some cases, however, as a vehicle, rather than as the active ingredient. Second importance was the excrements which were employed in about half as many prescriptions (MacKinney 1946).

Other evidence of zootherapy has been found in archives of several civilizations of ancient Mesopotamia, mainly the Assyrian, and the Babylonian (Lev 2003). These contain descriptions of fish oil, bee wax and honey, mongoose blood, turtle shell, goat's skin, gazelle sinew and even sheep, deer, and bird excrement and animal fat (Powell 1993; Ritter 1965; Thompson 1923).

In ancient China, various animal substances were used for therapeutic purposes (Gordon 1949). Research on medicinal uses of earthworms has a history of nearly 4,000 years (Zhang et al. 1992). *Compendium of Materia Medica* written by Li Schizhen in 1578 A.D. was a comprehensive summary of pharmacological knowledge accumulated in China up to his time. The use of bear fat as a medicine dates to 3494 B.C. (Ma 1986). Bear gallbladder may have entered the Chinese pharmacopeia as many as 3,000 years ago. Prescriptions for bear gall first appeared in writing in the seventh century (Bensky and Gamble 1986). The medicinal use of rhino horn is recorded from China as early as 2600 B.C. (Nowell et al. 1992), spreading in later years to Western Asia and the Roman Empire (Hirth and Rockhill 1911; Schafer 1963). Elasmobranchs are also employed in traditional Chinese medicine and Read (1939) lists shark flesh, skin, and bile among animal materia medica.

In India, since times immemorial, investigations focused on various zootherapies and traditional medicines, documented in the ancient texts of the *Ayurveda* and *Charaka Samhita* (Jharna et al. 1797; Mahawar and Jaroli 2008). Ancient literature of Ayurveda is full of medicines or formulations made of animal products. *Sharaka Samhita*, the first treatise devoted to the concepts and practice of Indian Ayurveda, was written around 900 B.C. and contains references to nearly 380 types of animal substances (Unnikrisnhan 1998). The next landmark in Ayurvedic literature was the *Sushruta Samhita* (−600 B.C.) (Majumdar 1971), which has special emphasis on surgery. It described 395 medicinal plants, 57 drugs of animal origin, and 64 minerals and metals as therapeutic agents (Dev 1999). Substances which are used in medicine and are derived from animals are grouped under the head of Jangama dravya (material derived from animals). There are hundreds of such formulations in Ayurveda, which utilize a variety of animal products. These animal products are much diverse in their habitat of origin and comprise marine, aquatic, terrestrial, or avian species. Regarding terrestrial animals, the products from wild as well as from domestic animals are also used in medicine. Among a wide variety of products derived from animals and used in medicine, commonly utilized substances are honey, milk and its derivatives, bile, fat, bone marrow, blood, flesh, feces, urine, skin, semen, ligaments, bones, shell, horn, and feathers (Rastogi and Kaphle 2008).

The Bible and other Jewish sources, mainly the Mishna (First to third centuries CE) and the Talmud (Fourth to fifth centuries CE), that is, the literature of the Jewish Sages, mention several animals and their medical uses: honey was used to treat bulimia and goat milk to cure coughing (Lev 2003). Snakes, human urine, pearl, mammal's glands, and several other substances were used for different medical conditions (Preuss and Rosner 1993).

Historical documents also reveal that the use of medicinal animals is directly associated to the history of medicine. Taylor and Tweed (1975) drew attention to the fact that the word medicine owes its origin to insects or at least to the honeybee. The first part of the word medicine has the same root as the word "mead" which is an alcoholic beverage made from honeycomb, which was often consumed as an elixir and thought to possess healing or medicinal powers. Another important taxon associated with the history of medicine are the snakes.

From very ancient times, the snake was a constant companion of medicine men and magicians. In mythical stories, snakes were often said to have knowledge of healing or life restoring plants, and there was a widespread belief that wisdom and medical knowledge could be obtained from the snake in various ways. Asklepios, the Greek god of medicine (better known by his Latin name, Aesculapius), like all Greek healing heroes was closely associated with the snake. Even in later times, when he was portrayed as a man, with snake staff as his emblem, he was said to manifest himself in serpent form and the snakes kept at his temple in Epidaurus were therefore approached with awe by their attendants. Asklepios appeared in 293 B.C.in form of a snake at Rome, where the pestilence was ranging. The outbreak of the plague was arrested and a temple built in his honor (Morris and Morris 1965). The snake became a sign of medicine and healing both by virtue of its godly associations, and also through the metaphorical immortality represented by the shedding of its skin. References to snakes—both actual animals and artistic representations thereof—abound in the healing temples, and it was the snake as cult object that was used to symbolically transfer the power of the healing god from its original geographic locus to the site of a new temple—this was true both of the entrance of Asclepius to Athens during the late fifth century plague (Edelstein and Edelstein 1975), as well as his introduction to Rome in 291B.C. (King 4). The *caduceus*—the snake (and later, two symmetrically opposed snaked) curled around a long staff—is perhaps the best known representation of the Greek sacred snake, one that persists as a symbol of the art of medicine to this day (Retief and Cilliers 2010).

Classical medical literature also indicates animals as remedies (Lev 2003); as an example, Voultsiadou (2010) found records of medicinal uses of marine invertebrates in 40 works of 20 classical authors, covering a period of time of eleventh centuries (5th B.C. to 7th A.D.); Riddle (1987) mentioned a number of animal substances (e.g., cattle milk, chicken's eggs, and mammal's horns) cited in the work of Hippocrates, who left a long list ofopotherapic medications, of which very few have been preserved in modern times. Nonetheless, the combination of bull's bile with honey, a classical Hippocratic remedy for intestinal constipation, remains functional to this day (Chemas 2010).

Ancient Greeks used various animal-based remedies, as exemplified by Homer's narrative tale of the centaur Chiron, who strengthened Ulysses by giving him lion's bone marrow to ingest, and by the priests at the temples of Aesculap, who recommended the ingestion of snake meat by lepers.

The works of Dioscorides, a Greek soldier and traveler who wrote in the age of the Roman Emperor Nero, provide our most detailed collection of Greek materia medica. In about A.D. 65, after much direct observation of plants in their native habitats and careful practical experience on the medicinal uses of herbs, as well as those derived from animals and minerals, Dioscorides wrote *De Materia Medica* (as the original work in Greek is known in Latin) in five books "on the preparation, properties, and testing of drugs" (Preface, I), each chapter dealing with a single substance, its description, preparation, and therapeutic properties (Stockwell 1988).

Although the greater portion of his five book treatise concerns vegetable substances, almost half of book II is devoted to materia medica from animals. Starting with aquatic creatures, Dioscorides discussed the medical properties of each of a number of fish, shell fish, insects, domestic animals, birds, and larger animals; this is followed by a description of the uses of milk, cheese, butter, wool, wax, honey, fat, and of animal substances such as marrow, gall, blood, and the excrements. In the case of insects, worms, and certain smaller animals, often the entire body was crushed, dried, burned, or cooked and applied in the form of a powder, salve, poultice, or potion. Noteworthy among the remedies for both external and internal use, were those made from cantharid beetles, grubs, earthworms, millipedes, roaches, and bugs of various sorts. For example, cockroaches ground up in oil were applied for ear ache. Millipedes, taken internally with wine, were recommended for epilepsy and kidney trouble, with honey for throat trouble, and when ground up with certain herbs and applied directly, they were supposed to cure tooth ache (MacKinney 1946).

Various portions of larger animals were used in equally varied fashion; among them the brain of a chicken, the tusk of an elephant, the hoof of an ass or goat, or internal organs such as the liver, gall bladder, bones or marrow, and testicles. More prominent, however, are substances such as grease, and to a lesser degree blood, excrement, sweat, and other such matter from man or beast (MacKinney 1946). About 10% of the substances mentioned in Dioscorides's *Materia Medica*; (Gillispie 1970–1978) were body parts and products of animals (Gunther 1959; Riddle 1985). Such uses on a smaller scale were common in the Byzantine Empire (Francis 1846).

Among the Roman writers on medicine two are outstanding: Celsus and Pliny the Elder. From the standpoint of animal materia medica Pliny the Elder's *Natural History* has been more influential than most of the works considered thus far. In fact it is one of the most comprehensive of all sources of information concerning the materia medica of the ancient world. Pliny wrote in Rome, at about the middle of the first century A.D. Out of eighteen books in his *Natural History* which deal with materia medica, books XXVIII, XXIX, XXX, XXXI and XXXII are devoted to animal substances. Pliny's *Natural History* marks a high point in the history of animal materia medica in that it records more substances and in greater detail than any extant work prior to the invention of printing. For well over a thousand years after Pliny, the materia medica of animals (along with other medical practices) diminished in quantity and quality (MacKinney 1946).

Even in the eleventh, twelfth, and thirteenth centuries, when Arabic and Salernitan writers introduced progressive ideas in many lines of medicine, materia medica continued to be dominated by the superstition and magic that had been handed down from classical times. Abundant evidence of this assertion can be found in the *Speculum Naturale* of Vincent of Beauvais, a worthy example of the great encyclopedists of the thirteenth century. The *Speculum* is strikingly similar to

Pliny's *Natural History,* so far as animal materia medica is concerned. It is a rather loose and unscientific compilation of material from various sources. A separate section is devoted to each animal, with abstracts of information from classical and Arabian sources. For example, in the treatment of the cantharid beetle, one finds a general description, a brief account of the medical properties, three specific uses, and brief citations of material from Dioscorides, Haly (an Arab writer), and Pliny. In similarly stereotyped fashion, Vincent treats the various animals, large and small, wild and domestic, and also the time honored substances derived there from, and methods of application. In general, Vincent, like his predecessors and contemporaries, placed great faith in animal substances of all kinds. In one passage he asserted that the blood of all animals was medicinal (XXI, 65). His English contemporary, Bartholomew of Glanville, in an encyclopedia of similar character, held that "nothing in a beast's body is devoid of medical value" (XVIII, 1) (MacKinney 1946)

Animal-based medical products were common among Amerindians (Almeida 2007, 2010; Alves et al. 2007; Silva et al. 2004), as described by various travelers who reached the New World; these provided novel information on indigenous groups, their illnesses and treatments. The settlement of the Americas by Europeans in the sixteenth and seventeenth centuries made little fundamental change in the use of animals, or any other substances in materia medica. Just as it was in Europe after the Renaissance, in Central and South America after the coming of the Conquistadores, traditional medical practices persisted (MacKinney 1946), nonetheless the European system came to dominate, relegating local systems to a folk category outside mainstream medicine.

Specifically in the case of Brazil, various writings contain information on the discovery of the fauna, the first ones being texts (e.g. letters) sent to European kings, such as the and letter of Pero Vaz de Caminha to the king D. Manuel I of Portugal, written in 1500 (National Archive, Torre do Tombo, Portugal, 2009); the documents produced by chroniclers such as Hans Staden (1510–1576), whose writings (published in Europe in 1557) contained the description of species he became familiar with through the native Tupinambás. Additional sources are the reports prepared by missionaries, such as the work *História dos Animais e Árvores do Maranhão*, by fray Cristóvão de Lisboa (1583–1652), or documents produced by noblemen and settlers. As pointed out by Miranda (2004), the main concern of those who initially recorded the diversity of the Brazilian fauna was to identify, to described possible uses and treats to humans.

Following colonization (second half of the sixteenth century) Jesuitic medicine and physicians trained in European universities—who represented mainstream medicine—came to Brazil, Jorge Valadares being the first physician to arrive in the country (Scliar 2005). The epithet "Jesuitical medicine" reflects the medical practice of that period, during which the priests of the Society of Jesus served as physicians, nurses and pharmacists, both to natives and settlers. In fact, their infirmaries constituted the first hospitals of the inhabitants of the colony (Duniau 2003; Santos Filho 2005).

Amalgamation of knowledge related to European medicine with indigenous medical practices allowed the Jesuits to describe prescriptions and effects related to the medicinal use of native species of animals and plants, which became incorporated into a global pharmacopaeia (Santos Filho 2005). The Society of Jesus therefore played a role in the process of interculturality, given its presence in various parts of the world, which allowed a comparative systematisation of knowledge gathered in the Americas, Africa and Asia. Among the Jesuits of the sixteenth century based in Brazil, José de Anchieta stands out, according to Santos Filho (2005), as "the best and most important figure in Brazil's sixteenth century medicine. Father Anchieta produced various works, some of them dedicated to the description of the Brazilian fauna (Miranda 2004).

Between the years 1637 and 1644, during the Dutch occupation of NE Brazil, the naturalists William Piso and George Marcgrave explored the states of Pernambuco, Paraíba and Rio Grande do Norte's hinterland, where they collected specimens from flora and fauna. The writings of Piso and Marcgrave were compiled in 12 books grouped as *Historia Naturalis Brasiliae*, first edited in 1648 (Piso and Marcgrave 1648; Piso 1957). *Historia Naturalis Brasiliae* constitutes one of the first comprehensive works dedicated to various aspects of Brazil: flora, animals, illnesses, sugarcane mills, and customs of native indians (van den Berg 1993; Marques 1999; Nogueira 2000). Reviews of the works produced by Piso and Marcgrave revealed that from 43 zootherapeutical prescriptions involving different zoological groups, 30 were recorded by Piso, three by Marcgrave and nine by Joannes de Laet (organizer and editor of the 1648 edition of the *Historia naturalis Brasiliae*). It is worth noting that some of the zootherapeutical prescriptions recorded by Piso are still used in popular medicine—a testimony to their endurance and reproduction power (Almeida 2007).

## 2.4 Final Remarks

Animal substances, although not as numerous as those of vegetable origin, have played (and continue to do so) an important role in healing. Records of the use of animals as remedies can be found in historical sources worldwide, and these highlight that an ample spectrum of animal taxa, from sponges to mammals, have been used as a therapeutic alternative throughout history. In fact, many of the animals prescribed in the past are still in use (being recommended either for the same illness as originally described or for distinct conditions) in traditional medical practiced in different localities.

Increased understanding of medical systems in a historical context can potentially bring new insights into the medical significance of fauna in the past, as well as open new therapeutic perspectives in the future—the historic and sustained use of naturally occurring compounds often has a scientific under-pinning, and the persistence of use of animal resources may indicate the presence of biologically active components.

Nonetheless, perhaps the greatest contribution of a historical approach resides in the possibility of drawing parallels with other disciplines and/or approaches: for instance, medicine, as a means to understanding the evolution of medical practices over time and how these practices have been incorporated into academia (e.g., the use of animal toxins), and of those which have been systematized as complementary or alternative medicine; ecology, as a means to increase our comprehension about species' historical use, management and conservation; anthropology/ethnobiology, as a way for shedding new light into an ancient form of connection between humans and nature.

## References

Alakbarli F (2006) Medical manuscripts of Azerbaijan. Heydar Aliyev Foundation, Baku
Albuquerque UP (1997) Etnobotânica: uma aproximação teórica e epistemológica. Revista Brasileira de Farmácia 78(3):60–64
Almeida AV (2007) Zooterapia indígena brasileira do século XVIII nas obras de Guilherme Piso, Georg Marcgrave e Joannes de Laet. Sitientibus Série Ciências Biológicas 7(3):261–272
Almeida AV (2010) A zooterapia adotada pelos médicos Simão Pinheiro Morão (c. 1618–1685) e João Ferreyra da Rosa (c. 1659–1725) em Pernambuco no final do século XVII. In: Costa-Neto EM, Alves RRN (eds) Zooterapia: Os Animais na Medicina Popular Brasileira, vol 2. 1 edn. NUPEEA, Recife, pp 55–74
Alves RRN, Rosa IL (2005) Why study the use of animal products in traditional medicines? J Ethnobiol Ethnomed 1(5):1–5. doi:10.1186/1746-4269-1
Alves RRN, Rosa IL, Santana GG (2007) The role of animal-derived remedies as complementary medicine in Brazil. Bioscience 57(11):949–955
Alves RRN, Souto WMS (2010) Etnozoologia: conceitos, considerações históricas e importância. In: Alves RRN, Souto WMS, Mourão JS (eds) A Etnozoologia no Brasil: Importância, Status atual e Perspectivas, vol 7. 1st edn. NUPEEA, Recife, pp 19–40
Bensky D, Gamble A (1986) Chinese herbal medicine materia medica. Eastland Press, California
van den Berg ME, (1993) Plantas medicinais na Amazônia (contribuição ao seu conhecimento sistemático). 2nd edn. Museu Paraense Emílio Goeldi, Belém
Cragg GM, Newman DJ (2001a) Medicinals for the millennia: the historical record. Ann N Y Acad Sci 953:3–25
Cragg GM, Newman DJ (2001b) Natural product drug discovery in the next millennium. Pharm Biol 39:8–17
Chemas RC (2010) A zooterapia no âmbito da medicina civilizada. I. Organoterapia humana e animal stricto sensu. In: Costa-Neto EM, Alves RRN (eds) Zooterapia: Os Animais na Medicina Popular Brasileira, vol 2, 1st edn. NUPEEA, Recife, PE, Brazil, pp 75–102
Dev S (1999) Ancient-modern concordance in Ayurvedic plants: some examples. Environ Health Perspect 107(10):783–789
Duniau MCM (2003) Plantas medicinais: da magia à ciência. Brasport, Rio de Janeiro
Edelstein JE, Edelstein L (1975) Asclepius: A collection and interpretation of the testimonies. 2 vols. ARNO Press, New York
Francis A (1846) The seven books of Paulus Aeginata. Sydenham Society, London
Gillispie CC (1970–1978) Dictionary of scientific biography. Charles Scribner's Sons, New York
Gordon BL (1949) Medicine throughout antiquity. F.A. Davis Co, Philadelphia
Gray JS (1997) Marine biodiversity: patterns, threats and conservation needs. Biodivers Conserv 6(1):153–175
Gunther RT (1959) The greek herbal of dioscorides. Hafner, New York

Halberstein RA (2005) Medicinal plants: historical and cross-cultural usage patterns. Ann Epidemiol 15(9):686–699
Hirth F, Rockhill WW (1911) Chau Ju-kua: His work on the Chinese and Arab trade in the twelfth and thirteenth centuries, entitled Chu-fan-chi. Imperial Academy of Sciences, St. Petersburg
Holzman D (2003) Meat eating is an old human habit. *New Scientist* 179:19–40
Jharna C, Meyer-Rochow V, Sampat G (1797) Vertebrates used for medicinal purposes by members of the Nyishi and Galo tribes in Arunachal Pradesh (North-East India). J Ethnobiol Ethnomed 7:13
Ji HF, Li XJ, Zhang HY (2009) Natural products and drug discovery. EMBO Rep 10(3):194–200
Kremers E, Urdang G (1951) History of pharmacy. JB Lippincott Co, Philadelphia
Larsen CS (2003) Animal source foods and human health during evolution. J Nutr 133(11): 3893–3897
Lev E (2003) Traditional healing with animals (zootherapy): medieval to present-day Levantine practice. J Ethnopharmacol 85:107–118. doi:10.1016/S0378-8741(02)00377-X
Lidgard H (2005) Quackery or complementary medicine—a historical approach to the present situation. Master thesis, University of Lund, Sweden
Ma Y (1986) Characteristics and economic value of bears. J Nat Res 6:26–32
MacKinney LC (1946) Animal substances in materia medica. J Hist Med Allied Sci 1(1):149–170
Mahawar MM, Jaroli DP (2008) Traditional zootherapeutic studies in India: a review. J Ethnobiol Ethnomed 4(1):17
Majumdar RC (1971) Medicine. In: Bose DM, Sen SN, Subbarayappa BV (eds) A concise history of science in India. Indian National Science Academy, New Delhi, pp 213–27
Marques VRB (1999) Natureza em boiões: medicinas e boticários no Brasil setecentista. Editora da Unicamp, Campinas
Martínez C (2008) Las aves como recurso curativo en el México antiguo y sus posibles evidencias en la arqueozoología. Arqueobios 2:11–18
Medeiros MFT (2010) Historical ethnobotany: an approach through historical documents and their implications nowadays. In: Albuquerque UP, Hanazaki N (eds) Recent developments and case studies in ethnobotany. NUPEEA, Recife, pp 127–142
Miranda EE (2004) O descobrimento da biodiversidade: a ecologia de índios, jesuítas e leigos no sécolo XVI. Edições Loyola, São Paulo
Morris R, Morris D (1965) Men and snakes. McGraw-Hill, New York
Nakanishi K (1999a) An historical perspective of natural products chemistry. In: Ushio S (ed) Comprehensive natural products chemistry, vol 1. vol 23–40. Elsevier, Amsterdam
Nakanishi K (1999b) An historical perspective of natural products chemistry. Comprehensive Natural Products Chemistry: Isoprenoids including carotenoids and steroids. Pergamon Press, Oxford
Nascimento DR, Carvalho DM (2006) II seminário história das doenças. In: Caderno de resumos. Rio de Janeiro:, Rio de Janeiro. Caderno de resumos. Rio de Janeiro, FIOCRUZ, p 52, FIOCRUZ, p 52
Nogueira E (2000) Uma história brasileira da botânica. Sao Paulo. Marco Zero
Nowell K, Chyi WL, Pei CJ (1992) The horns of a dilemma: The market for rhino horn in Taiwan. TRAFFIC,Taipei
Nunn JF (2002) Ancient egyptian medicine. Univ of Oklahoma, Norman
Piso G (1957) História natural e médica da Índia Ocidental. Instituto Nacional do Livro, Rio de Janeiro, Brasil
Piso W, Marcgraf G (1648) Historia Naturalis Brasiliae. Leiden, Amsterdam
Powell AM (1993) Drugs and pharmaceuticals in ancient Mesopotamia. In: Jacobs I, Jacobs W (eds) The healing past: Pharmaceuticals in the Biblical and Rabbinic world. The Healing Past Brill, Leiden, pp 46–67
Preuss J, Rosner F (1993) Biblical and Talmudic medicine. Jason Aronson, Northvale
Rastogi S, Kaphle K (2008) Sustainable traditional medicine: taking the inspirations from ancient veterinary science. Evidence-based Complementary and Alternative Medicine 5:1–6

Read BE (1939) Chinese materia medica: fish drugs. Peking Natural History Bulletin, Peking
Retief FP, Cilliers L (2010) Snake and staff symbolism in healing. Acta Theologica 26(2):189–199
Riddle JM (1985) Dioscorides on pharmacy and medicine. University of Texas Press, Austin
Riddle JM (1987) Folk tradition and folk medicine: recognition of drugs in classical antiquity. In: Scarboroough J (ed) Folklore and Folk Medicine. American Institute of the History of Pharmacy, Madison
Ritter EK (1965) Magical-expert and physician notes on two complementary professions in Babylonian medicine. Assyriological Studies 16:299–323
Santos Filho L (2005) Anchieta na História da Medicina Brasileira. http://www.novomilenio.inf.br/festas/anchie02.htm.
Schafer EH (1963) The golden peaches of Samarkand: a study of T ang exotics. University of California Pres, California
Scliar M (2005) Uma guerra contra a morte Nossa História 2(21):14–16
Serpell J (1996) In the company of animals: A study of human-animal relationships. Cambridge University Press, Cambridge
Silva MLV, Alves ÑGC, Almeida AV (2004) A zooterapia no Recife (Pernambuco): uma articulação entre as práticas e a história. Biotemas 17(1):95–116
Sneader W (2005) Drug discovery: a history. Wiley, New York
Spainhour CB (2005) Natural products. In: Gad SC (ed) Drug discovery handbook. Wiley, Inc, pp 12–63
Stockwell C (1988) Nature's pharmacy: a history of plants and healing. Century, London
Sutton MQ (1995) Archaeological aspects of insect use. J Archaeol Method Theory 2(3):253–298
Taylor R, Tweed JG (1975) Butterflies in my stomach (or insects in the human nutrition). Woodbridge Press Publishing Co, Santa Barbara
Thompson RC (1923) Assyrian medical texts. Oxford University Press, Oxford
Unnikrishnan E (2004) Materia medica of the local health traditions of Payyannur. Centre for Development Studies, Payyannur
Unnikrisnhan PM (1998) Animals in Ayurveda. Amruth, 1:1–15
Voultsiadou E (2010) Therapeutic properties and uses of marine invertebrates in the ancient Greek world and early Byzantium. J Ethnopharmacol 130(2):237–247
Waters MR, Stafford TW Jr, McDonald HG, Gustafson C, Rasmussen M, Cappellini E, Olsen JV, Szklarczyk D, Jensen LJ, Gilbert MTP (2011) Pre-Clovis Mastodon Hunting 13,800 Years Ago at the Manis Site Washington. Science 334(6054):351–353
Wing ES (2000) Animals used for food in the past: As seen by their remains excavated from archeological sites. In: Kiple KF, Ornelas KC (eds) The cambridge world history of food, vol 1. Cambridge University Press, Cambridge, pp 51–58
Zhang FX, Guo B, Wang HY (1992) The spermatocidal effects of earthworm extract and its effective constituents. Soil Biol Biochem 24(12):1247–1251

# Chapter 3
# Medicine from the Wild: An Overview of the Use and Trade of Animal Products in Traditional Medicines

**Rômulo Romeu Nóbrega Alves, Ierecê Lucena Rosa, Ulysses Paulino Albuquerque and Anthony B. Cunningham**

**Abstract** The use of animals, plants, and fungi as a major source of local medicine for people's well-being in rural areas has been documented in different parts of the world, and has also been recorded in urban environments. In urban settings, the trade of wildlife-based medicinal products is concentrated in local and traditional markets, where various species of medicinal plants and animals are commercialized. The commercialization and use of biological resources as remedies encompass cultural, social, and economic aspects and bear important implications for conservation, especially in relation to the most heavily exploited species. This chapter reviews the literature on commercialization of medicinal animals in local markets, focusing on urban zootherapy and on the social actors involved in these practices. In doing so, we hope to highlight the importance of further research on the topic and to explore ways through which conservation and management initiatives can improve dialog with resource users and traders.

R. R. N. Alves (✉)
Ethnozoology, Conservation and Biodiversity Research Group, Departamento de Biologia, Universidade Estadual da Paraíba, Campina Grande, Paraíba 58109-753, Brazil
e-mail: romulo_nobrega@yahoo.com.br

I. L. Rosa
Departamento de Sistemática e Ecologia, Universidade Federal da Paraíba, João Pessoa 58059-970, Brazil

U. P. Albuquerque
Applied Ethnobotany Laboratory, Departamento de Biologia, Universidade Federal Rural de Pernambuco, Dois Irmãos, Pernambuco 52171-900, Brazil

A. B. Cunningham
Forests and Livelihoods programme, Centre for International Forestry Research, 0113 BOCBD, Bogor, Indonesia

R. R. N. Alves and I. L. Rosa (eds.), *Animals in Traditional Folk Medicine*,
DOI: 10.1007/978-3-642-29026-8_3, © Springer-Verlag Berlin Heidelberg 2013

## 3.1 Introduction

Human populations engage in hunting and fisheries to meet diverse nutritional, economic, medicinal, cultural, and recreational needs (Alves and Souto 2010; Bennett and Robinson 2000; Nooren and Claridge 2001), and these extremely close connections of dependence and co-dependence have existed between humans and animals throughout history (Alves and Souto 2010). The use of biological resources (animals, plants, and fungi) as a major source of local medicine for people's well-being in rural areas has been documented in different parts of the world (Alves 2009; Alves and Rosa 2007a, b; Alves et al. 2007; Heinrich et al. 1998; Kültür 2007; Mahawar and Jaroli 2008; Melo et al. 2009; Oliveira et al. 2007; Parveen et al. 2007), and has also been recorded in urban environments (Alves and Alves 2011; Alves and Rosa 2010; Monteiro et al. 2010). As a large amount of the world's population is concentrated in urban areas, there are changes in the relationship between human activities and biodiversity, and consequently in the way we should think about biodiversity conservation policies (Puppim de Oliveira et al. 2011).

Expanding urban populations in most of the world continue to supplement limited public health facilities and more expensive commercially produced medications with popular remedies (Srivastava et al. 1996), and this has led to an increasing demand for wildlife products for medicinal purposes in urban areas.

Historically used as raw materials in traditional medicine systems, especially by traditional health practitioners, medicinal plants have entered the formal and informal markets worldwide (Bussmann et al. 2007; Cunningham 1993; Laird 1999; Loundou 2008; Monteiro et al. 2010; Oliveira et al. 2010), and as pointed by Cunningham (1997), stimulated by high population growth rates, rapid urbanization, rural unemployment, and the value placed on traditional medicines, commercial trade in traditional medicines is now greater than at any time in the past.

In cities, the trade of medicinal products (plants and animals), especially as raw materials, is concentrated in local and traditional markets (e.g. Albuquerque et al. 2007; Alves and Alves 2011; Alves and Rosa 2010; Monteiro et al. 2010; Van der Berg 1984; Vázquez et al. 2006), which in this text encompasses various denominations used in the literature, such as traditional markets, fairs, open markets (Monteiro et al. 2010). Local markets generally have specific sections where plants and animals are sold for medicinal purposes, and these locales serve to unite, maintain, and diffuse empirical knowledge from different regions and of different origins. The ongoing search for natural products, as part of a collective social strategy, increases the importance of these traditional centers (Albuquerque et al. 2007, 2010).

Although medicinal plant markets have drawn the attention of many ethnobotanists (e.g. Albuquerque et al. 2007; Mati and de Boer 2011; Monteiro et al. 2010; Williams et al. 2000, 2005), the trade in animals for medicinal purposes has been largely overlooked in the literature; only recently that form of trade has attracted the interest of researchers (Alves and Pereira Filho 2007; Alves and

Rosa 2007b, 2010; Ashwell and Walston 2008; Oliveira et al. 2010; Van and Tap 2008; Vázquez et al. 2006; Ferreira et al. 2012), whose studies have revealed that the extensive medicinal use of animal parts and products is sustained by a thriving trade in medicinal animals, mainly conducted by herbalists in markets.

Trade in wildlife products employed in traditional medicines is growing throughout the developing world, resulting in numerous species becoming locally or regionally extinct in the wild (Alves and Rosa 2007a; Marshall 1998a; Olsen 1998). Some authors suggest that commercial demand for traditional medicines is correlated with increasing urbanization, poverty, and related social problems (Alves and Rosa 2010; Cunningham 1992). There is also recognition that health care services are precarious in most developing countries, where many have no access to health care (Smith et al. 2001). On the other hand, Vandebroek and Calewaert (2004) considered that urbanization and the consequent access to services and goods reduces demand for resources, and also negatively affects transmission of traditional botanical knowledge.

Given the conservation status of many of the animal species sold for medicinal use (Alves and Rosa 2005; Alves et al. 2008a, b, c, 2010), the ecological, cultural, social, and public health implications associated with their use, as well as more complete inventories of the species used are urgently needed (Ferreira et al. 2009). An increased understanding of the various facets involved in the medicinal uses of animals is central to better assess how they impact wild populations, and to explore ways through which conservation and management initiatives can improve dialog with resource users and traders. In this context, this chapter reviews the literature on the trade of medicinal animals in local markets, focusing on urban zootherapy and on the social actors involved in these practices. In doing so, we hope to highlight the importance of further research on the topic, and to explore ways through which conservation and management initiatives can improve dialog with resource users and traders.

## 3.2 Commercialization Points and Traders

Animals used for medicinal purposes are openly traded, in conjunction with medicinal plants, in booths located in specific sectors of the local markets (Alves and Rosa 2007b, 2010; Vázquez et al. 2006; Whiting et al. 2011). In addition to formal markets (Fig. 3.1), trade is also conducted through itinerant merchants selling both animals and medicinal plants in stalls located on squares or on the street (Figs. 3.1 and 3.2). In China, Thailand, and Vietnam, where traditional and allopathic medical care systems are closely integrated, large hospitals commonly use both plant and animal products for health care (Fig. 3.3a). In China, there is also extensive trade and price monitoring through the Internet (Fig. 3.3c). Occasionally, medicinal animals or their parts can be found for sale in small dry-goods stores, fish markets, and shops specializing in religious articles (Alves and Rosa 2010).

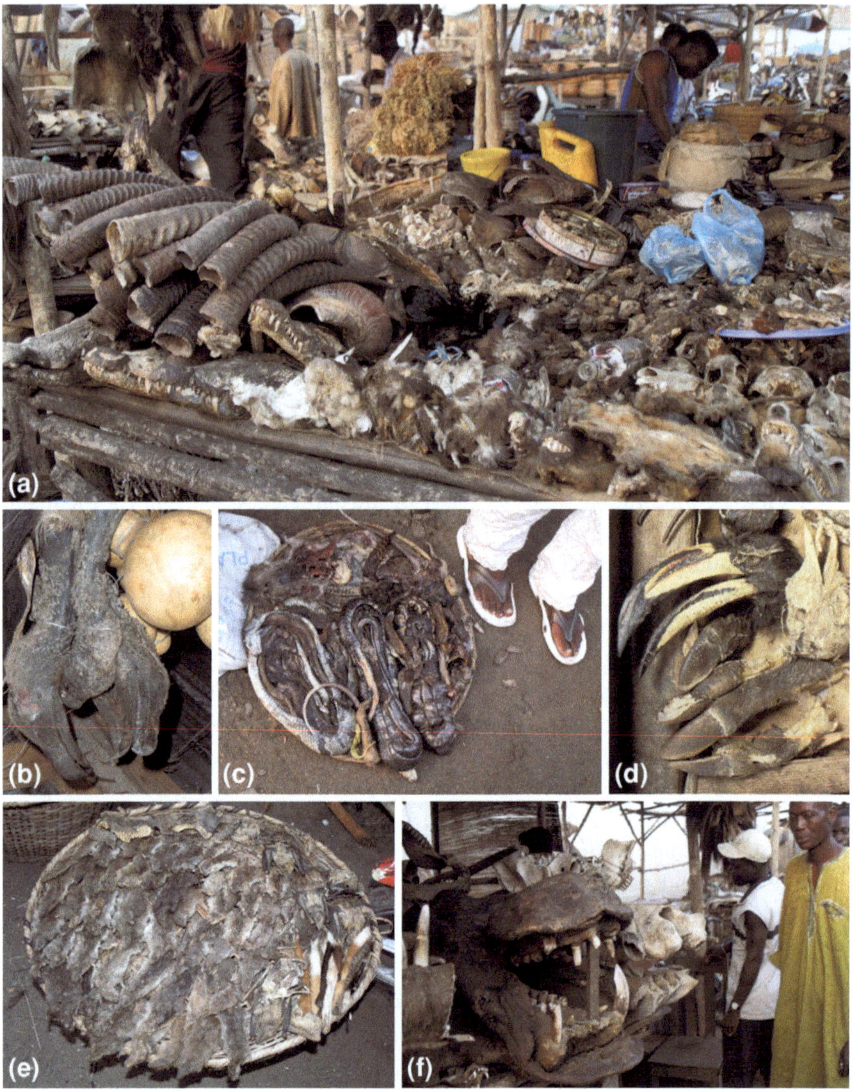

**Fig. 3.1** Informal sector markets characterize the animal medicines trade in Africa. **a** Stalls selling antelope horns, crocodile skulls, and a diversity of carnivore and non-human primate skulls for sale in Benin. Typical animals traded across West and southern Africa are: **b** Vulture heads (Xipamanine market, Mozambique); **c** Pythons (whole, the fat, or the skins) (Benin); **d** Hornbill heads and aardvark claws (Cote d'Ivoire); **e** Small mammals and bats (in this case, *Eidolon helvum*) and **f** Hippopotamus (*H. amphibious*) fat and skin is commonly sold, but in this case, several hippo skulls are displayed in this market in Benin

Besides being the primary source of medicines, medicinal resources are increasingly being harvested for income generation, unemployment, and poverty being among the driving factors for the trade (Alves and Rosa 2005; Alves et al.

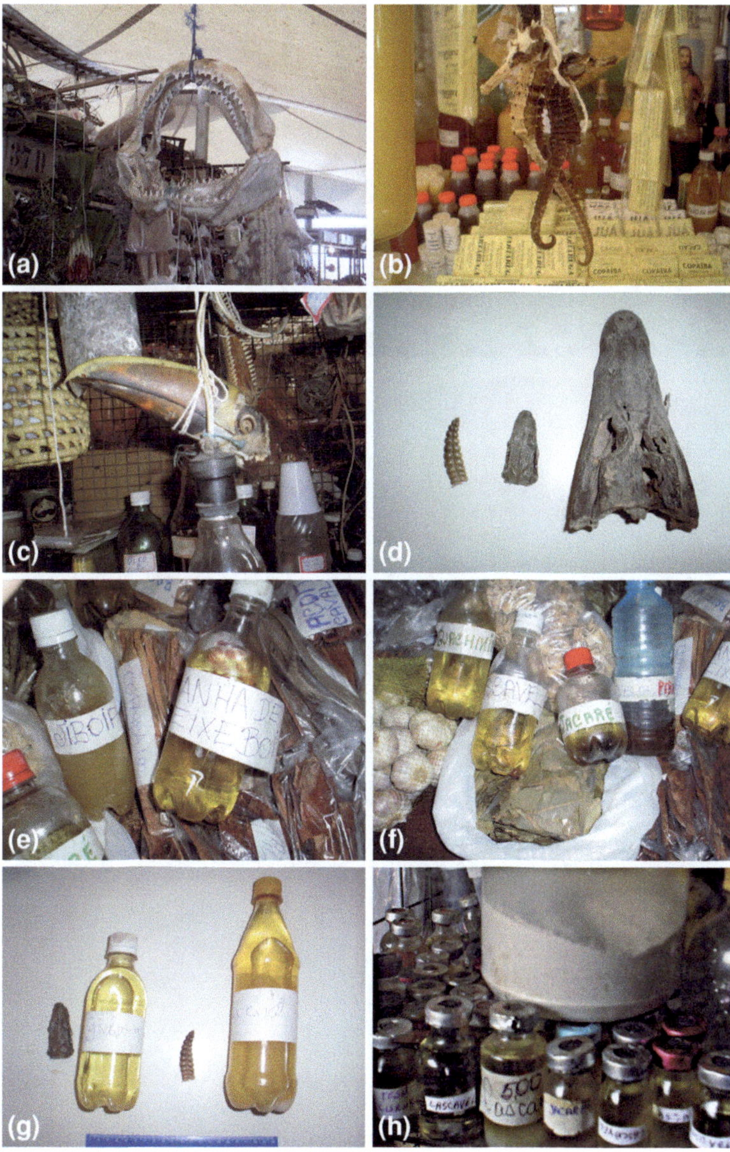


**Fig. 3.2** Medicinal animal products sold in Brazilian cities. **a** Shark jaws and teeth; **b** Dried seahorse and soap produced from honeybee and fat of turtle (*P. expansa)*; **c** Beak of *Ramphastos tucanus* (tucano); **d** Rattle of rattlesnake (*Caudisona durissa*), boas' head (*Boa constrictor*) and cayman's head (*Cayman* sp.); **e** Fat extracted from boa (*Boa constrictor*) and manatee (*Trichecus* sp.); **f** Bottled raccoon fat (*Procyon cancrivorus*), rattlesnake (*Caudisona durissa*), cayman (*Paleosuchus palpebrosus* or *Cayman crocodilus*) and armadillo (*Euphractus sexcintus*); **g** Head and fat of boa (*B. constrictor*) and rattle and fat of rattlesnake (*C. durissa*); **h** Oyster powder (*Crassostrea rhizophorae*) and fats from various species

**Fig. 3.3** With thousands of years of documented use, traditional and modern medical systems are closely integrated in Asia. **a** A hospital in Xi'an, Shaanxi, that integrates both health care systems; **b** The Hehuachi traditional medicine market in Chengdu, Sichuan, China; **c** Interior of the market, which also conducts Internet-based trade; **d** Deer antlers and penises; **e** Ground beetles; **f** Starfish and **g** *Cordyceps sinensis* infected Hepailid moth caterpillars, which sell for US$12,000 per kg

2008b; Loundou 2008; Mander 1998). The high rate of urbanization, resulting in high demand for medicinal plants, combined with the high rate of unemployment, has favored the involvement of commercial gatherers (Cunningham 1993), and as this informal market grows, more and more stakeholders are attracted (Mander 1998). A number of commercial harvesters, mostly unemployed and from poor social strata, are generating income from the informal trade of medicinal plants (Loundou 2008). In Brazil, for instance, traders of zootherapeutic products have a low education level, and this aspect exerts a direct influence on the recruitment of new traders to that activity, which does not require formal education (Alves et al. 2008b). Cultural motivations can also be important determinants for involvement in the trade of medicinal animals. As an example, in NE Brazil a strong family tradition focusing on animal-based remedies persists, as well as the figure of the *doutor de raiz* (roots' doctor)—a popular specialist who not only trades medicinal animal products, but also prescribes and produces various remedies. Although increasingly becoming rare, when present roots' doctors render the local markets interesting places for ethnopharmacological or ethnomedicinal investigations. The impacts of a growing urbanization and the need to generate a cash income, however, seem to be progressively eroding tradition. Cunningham (1992) highlighted that in the past traditional medicines in Africa were predominantly administered by specialized herbalists, a niche now occupied by thousands of non-specialized gatherers operating in the informal economy, who collect plants to supply urban traders or sell directly to subsistence sector consumers (Fig. 3.1). In Brazil, animal-based remedies have become part of a regular business that may involve advertising, the use of books describing, for example, the posology of the remedy, as well as pre-packaged medicines. Nevertheless, such practices are generally looked down upon by traditional traders of folk medicine, who feel proud of their knowledge of the raw materials and remedies (Alves and Rosa 2007b).

## 3.3 Animal Medicinal and Zootherapeutical Products

Recent investigations have revealed that a large number of animal species are traded for medicinal purposes in urban areas worldwide, particularly in African, Asian, and Latin American countries, with many similarities as well as significant differences in the categories of animals selected. Examples of surveys that have documented the diversity of animals sold for traditional medicine are: El-Kamali (2000), who identified 23 animal species, whose products were commercialized in traditional medicine in Central Sudan; Sodeinde and Soewu (1999), who recorded the use of 45 medicinal species in Nigerian markets; Simelane and Kerley (1998), whose results showed that 44 species (eight reptiles, six birds, 30 mammals) were sold in 19 herbalist shops in the Eastern Cape Province of South Africa; Cunningham and Zondi (1991), who examined the trade in animals for traditional medicine in KwaZulu–Nata Province and reviewed the literature reports for South Africa, recording at least 79 species of vertebrate (18 reptiles, 16 birds, 45 mammals), excluding domestic mammals and various marine invertebrates and fishes. More recently, Ngwenya (2001) recorded 132 species of vertebrates

(21 reptiles, 32 birds, 79 mammals) in trade across KwaZulu–Natal Province, of which 50 species were in high demand, especially vultures, chacma baboon, green mamba, southern African python, Nile crocodile, puff adder, striped weasel, and black mamba. Whiting et al. (2011) identified 147 vertebrate species representing about 9% of the total number of vertebrate species in South Africa and about 63% of the total number of documented species (excluding domestic animals) traded in all South African traditional medicine markets. Surveys carried out in 15 Brazilian cities revealed that at least 180 animal species are traded for medicinal purposes (Alves 2010). Ashwell and Waltson (2008) recorded at least 47 animal species being traded for medicinal purposes in Cambodian markets, while Van and Tap (2008) recorded 100 different medicinal products from 68 animal species traded in Ho Chi Minh City, mainly sold as dried products (either the whole animal or parts), soaked in rice wine, or as a gel product which remains after boiling animal remains slowly in water.

The great variety of useful animals encountered in some local markets have been attributed to several factors: Alves and Rosa (2007b) highlighted that the main determinants of the number of animals used for medicinal purposes appear to be the local faunal and cultural diversity, and the size of the regional market. Other authors (e.g. Bolze et al. 1998; Marshall 1998b) have suggested that market expansion induces people to make greater use of wild animals for traditional medicine, and that the practice has spread in developed nations of Asia and the Pacific (e.g. Taiwan, Australia), although there is also evidence that the increasing use of animals for traditional medicine can take place without economic prosperity. As an example, Kritsky (1987) reported that the use of insects for traditional medicine in China increased during the Cultural Revolution. In that same direction, based on an ethnozoological survey of the use of medicinal birds, Joseph (1990) concludes that the use of birds to treat human ailments increased in Madhya Pradesh, Central India, because people could not afford modern treatments.

The indigenous fauna represents most of the species traded in local or even regional markets, and this highlights the role of local biodiversity in furnishing folk medicine, as well as the need for further assessments of the contribution of locally caught species to the medicinal fauna commercialized at the markets. Although there is evidence that the faunal composition, accessibility, and availability directly influence the types of zootherapeutic items used in any given region, studies on market dynamics, socioeconomic and conservationist implications of the use of local resources (either locally or regionally) are virtually nonexistent. Moura and Marques (2008) suggested that the use of medicines derived from animal parts that are not consumed as food and would otherwise be discarded (e.g. teeth, skin, horns), is a means of maximizing the productivity of local ecosystems, while Apaza et al. (2003) remarked that the use of the local fauna generally reduces the cost of acquiring animal products in regions with abundant fauna.

The high number of species involved in medicinal trade reveals that the animals are therapeutic resources culturally important in urban areas. Nevertheless, the lack of zootherapeutic studies has contributed to an underestimation of the importance of zootherapeutic resources. Alves and Rosa (2010) pointed that one of the factors that certainly contribute to the scarcity of information on the subject is the often semi-clandestine or clandestine nature of the trade and use of medicinal animals, which generally results in users and traders being more resistant to provide information—a situation driven by the fact that most medicinal animals are wild-caught and protected by law; some figure in official lists of threatened species.

In contrast to what is observed with medicinal plants, most traders of medicinal animals in Brazil do not directly expose animals or animal parts for public viewing. Whole animals or their parts are generally kept out of sight in shops in closed bags or plastic receptacles that are brought out only when requested by a potential buyer. This is done to avoid the risk of environmental officers seizing specimens, and is related to the fact that traders are aware of the illegality of the trade in wild animals (Alves and Rosa 2007b, 2010).

A similar trend was observed by Whiting et al. (2011), in their study of the medicinal animal at the Faraday market in South Africa, where establishing the impact of traditional medicine on wildlife is notoriously difficult because traders are reluctant to reveal the source of their stock. Miththapala (2006) highlighted that obtaining data on the trade in animals and plants for medicinal purpose and its impacts on biodiversity is difficult because: (1) trade may take place without proper documentation and data are simply lacking in most areas; (2) for threatened species trade is often illegal and therefore there is great secrecy surrounding trade and reliable information is difficult to obtain. This way, as discussed by Padoch (1992), because of the difficulty in obtaining reliable data from market vendors, middlemen, exporters, and local governments, the trade in medicinal plants (and animals) still remains a 'hidden harvest' in many countries. Specifically in the case of the trade in animals for medicinal purposes, the illicit nature of the activity can certainly lead to an underestimation of the number of species commercialized both in rural and urban settings.

Padoch et al. (2008) highlighted that the increasing mobility of human populations globally has enhanced trade of forest goods as people's customs and traditions follow them. As an example, Alves and Rosa (2007) recorded the use of similar resources as medicines in more remote and urban areas of Brazil, and suggested that zootherapeutic practices may function as a social conduit which, in conjunction with other factors, helps to maintain the connections between rural and peasant people living in cities and their own traditional culture and values. Such findings corroborated the studies of Belluck (1996) and O'Connor (1998) who found that within urban centers, members of immigrant and ethnic minority groups typically use a variety of traditional healing resources in conjunction with conventional medicine care. Moreover, as remarked by Cocks (2006), even people who have migrated to (peri) urban areas and have become involved in modern economic sectors still perform certain cultural practices for maintaining a sense of well-being and expressing their identity.

More specifically, it indicates the potential for exchange of materials and information about illnesses and treatments between more remote and urban communities. In recent years, there has been a marked increase in the number of discussions regarding traditional communities in cities. Significant are indigenous populations and other traditional populations living in urban centers (Almada 2011), and many of these try to preserve some of their customs and values, including the use of plants and animals as the basic ingredients of their medical practices. Migration from more remote/rural areas to cities, therefore, has also played a significant role in knowledge transmission of ingredients used in traditional medicine in urban settings, as well as in creating a demand for specific products in local markets.

The prices of the medicinal wildlife items traded showed great variation and reflected various criteria, including species traded, rarity, demand, and size. Larger wild animals, which generally achieved the highest prices, were mainly species especially vulnerable to overhunting, and with limited capacity to recover from population declines (Adeola 1992; Alves et al. 2008a; Ashwell and Walston 2008; Sodeinde and Soewu 1999; Van and Tap 2008; Yinfeng et al. 1997). Modern marketing trends are also reflected in the presentation of some zootherapeutic products, which were either manufactured or pre-packaged. In India, where the Ayurvedic industry is worth an estimated US $1 billion per year, 7500 factories produce thousands of Ayurvedic and Unani formulae (Bode 2006). In China, clinical trials for traditional preparations are now frequent and there are plans to establish standards for these products and a competitive, modern industry in traditional Chinese medicine. Industrial processes and packaging are included for deer products imported from New Zealand and seal and ginseng (*Panax ginseng*) extracts (Fig. 3.4). Seal penises from managed culling of Cape fur seals in Namibia are exported to China.

In Latin America, examples of industrial packaging are the fat extracted from the manatee (*Trichecus* sp.), sold as tablets, and the fat of Amazon River turtle (*Podocnemis expansa* (Schweiger, 1812)—Podocnemididae) sold as manufactured soap in Brazil (Fig. 3.2b) (Alves and Rosa 2007b).

## 3.4 Cultural and Socioeconomics Aspects of Urban Zootherapy

The socioeconomic aspects, the popular culture, and commercial considerations are factors that maintain and drive the market for therapeutic animal products. Recent studies have highlighted the relevant role played by Traditional Medicine in cosmopolitan areas (Botha et al. 2004; Macía et al. 2005; Reiff et al. 2003), where health care needs generally are met by mainstream services, such as hospitals and allopathic pharmacies (Alves and Rosa 2007b). Overall, the urban milieu can contribute to higher levels of chronicity of illness, and low confidence in, or commitment to modern medicine. Quite apart from social or cultural functions, urban folk medicine may offer the most convenient, affordable health care. Folk curers tend to concentrate in ethnic and lower income neighborhoods

**Fig. 3.4** Industrial processes and packing of medicinal animal products is common in China. **a** Red deer products imported from New Zealand and **b** Seal and ginseng (*Panax ginseng*) extracts

and can be seen after work hours. Further, curers and folk remedies are generally cheaper than physicians and prescription drugs. Even where costly, curers may ritualize fees in such a way as to lower patient resistance to payment (Press 1971). In addition, western-type medical facilities have not been able to cope with the rapidly growing urban population.

Animals provide the raw materials for remedies prescribed using the clinical method and are also used in the form of amulets and charms in magic-religious diagnosis (Alves and Rosa 2006). These characteristics, found in various traditional medical systems, can boost the commercialization of animals for treating or illnesses (both physical and spiritual) in urban areas, as observed with plants. Cunningham (1993) highlights that in the stressful environment which is a feature of many urban areas in Africa, it is not surprising that demand has increased for traditional medicinal plant and animal materials which are believed to have symbolic or psychosomatic value. Traditional plant or animal materials which bring luck in finding employment, which guard against jealousy (such as that engendered when one person has a job while his/her peer group are unemployed), or love-charms and aphrodisiacs to keep a wife or girlfriend are popular. A similar trend was observed in Brazil by Alves et al. (2007) and Albuquerque et al. (2007), where different animal and plant species are used in magical-religious practices of Afro-Brazilian cults in the context of rituals that emphasize the holistic nature of traditional medicine and that are designed to address spiritual, physical, and social-psychological problems.

The notable use and commercialization of medicinal animals to alleviate and cure health problems and ailments in the cities reveals the resilience of that therapeutic alternative, in spite of the influence of western medicine. In urban areas, the people brought from their villages to the cities much valuable knowledge of animal-based remedies that is rarely studied.

## 3.5 Sanitary Concerns

The use of medicinal animals is often considered healthy by urban consumers, nevertheless, zoonotic diseases (a still poorly studied topic) have been increasingly mentioned in the literature (van Vliet and Mbazza 2011), as exemplified by Schnurrenberger and Hubbert (1981), who drew attention to the possibility of transmission of serious and widespread zoonotic diseases such as tuberculosis or rabies, an aspect that should be considered whenever animal tissues from unknown sources are handled and used as remedies.

Moreover, the stability and hygiene of medicinal products commercialized in markets is unknown, but probably varies enormously between traders and traditional healers (Mander et al. 2007). In Brazil, Alves and Rosa (2007b) ranked the sanitary conditions of the zootherapeutical products as poor; in the same direction, a study on the commercialization of medicinal plants in São Luís city, NE Brazil (Amaral et al. 2003), revealed that 81.5% of the material analyzed were contaminated by bacteria. In Ouagadougou market, Burkina Faso, traditional healers have been recorded selling animal parts from trophy animals prepared by taxidermists (Fig. 3.5). As this process can use toxic chemicals such as arsenic, this also raises health concerns.

Those observations highlight the need for further assessments of the sanitary conditions of the medicinal products commercialized, as well as for the implementation of measures to address the sanitary aspects of the trade of animals or their parts for medicinal purposes.

Broad categories of sanitary and phytosanitary regulatory measures are recognized for the food trade: (1) information measures which restrict the behavior of suppliers only to the extent that they are required to disclose specified facts about their products; (2) measures that impose prior approval certifying that their products have met some pre-specified safety criteria before they can be released onto the market; and (3) measures that allow suppliers to sell products without any prior official approval but imply that an offense is committed if the products fail to meet certain minimum safety standards (Henson and Heasman 1998). The implementation of equivalent sanitary measures to the trade of animals or their parts for medicinal purposes, however, faces considerable challenges, among them ensuring adequate participation of all stakeholders involved, monitoring of the activity, and combating illegal, unreported, and unregulated trade (Alves and Rosa 2005).

## 3.6 Conservation Implications

Wildlife trade is central to the relationship between sustainable development and biodiversity conservation (Broad 2001). Rapid urbanization and greater demand for traditional medicines result in an increase in harvesting of medicinal plants and

**Fig. 3.5** An old mounted head of an antelope that is being "recycled" in the traditional medicines market of Ougadougou, Burkina Faso, West Africa

animals from rural areas, leading to a depletion of the rural resource base where certain species are vulnerable to over-exploitation, and consequent problems for primary health care (Cunningham 1997).

There is still much that is not known about the wildlife medicinal trade and its impacts, and research will be required on many levels (Marshall 1998b), although assessing the impact of traditional medicine on wildlife is notoriously difficult because this commercialization is often clandestine or semi-cladenstine (Alves and Rosa 2007b, 2010). In this sense, ethnozoological inventories of local markets can provide important information about species that are susceptible to destructive exploitation, species that are exceptionally expensive or hard to find (which may indicate that their natural reserves are becoming exhausted), and species with increased collecting or buying frequencies (which might indicate greater demand) (see for example the chapter on African bird use in this volume). In this context, local markets can provide managers and policy-makers with useful information to assist in the establishment of dialog with users and traders of wildlife, as well as for the implementation of conservation strategies. Ethnozoological studies, however, need to go beyond inventories, and should also test hypotheses about the factors that affect, regulate, and maintain the processes of knowledge transmission, product demand, and particularly of the ecosystems that provide the medicinal products. Identification and characterization of the stakeholders involved in the system of use and trade of wildlife for medicinal products will certainly increase our understanding of such practices.

Wild populations are the principal sources of raw materials for the medicinal animals commercialized in local markets. In addition, many medicinal species are protected by law and/or figure in official lists of threatened species. The case of large, terrestrial animals (e.g. tigers, bears, rhinos, turtles, snakes, tokay geckos, pangolins, monkeys and swiftlets tigers, rhinoceros, bears) threatened by trade for traditional medicine, especially traditional Chinese medicine, is well known. Nevertheless, the list of wild animals believed to have medicinal and curative properties and commercialized is long and the actual number of animal species commercialized in traditional medicine remains unknown (Alves and Alves 2011). There is a growing need for baseline data, for proper quantification of the trade in animal parts for traditional medicine, and for further assessments to evaluate whether this trade makes significant use of species of conservation concern (Whiting et al. 2011).

The reasons for consumption of medicinal animals are complex and integrate economic, cultural, and social reasons that should not be disregarded in efforts to promote the sustainable trade of wildlife. Many people are economically dependent on the traditional commerce of medicinal species. Nevertheless, in countries such as Brazil, the trade of animals for medicinal purposes has had little impact on the socioeconomic conditions of collectors, who generally are illiterate, underpaid, and perceive their activity as clandestine or semi-clandestine. The monetary value of animals sold for medicinal purposes in the country increases at each level of trade, and the socioeconomic profile of traders varies accordingly (Alves and Rosa 2005).

Conservation practices tend to focus on arresting or ameliorating habitat destruction because biodiversity is conserved as a by-product. What has received far less attention from ecologists and conservation biologists is the harvest of animals for use in traditional medicine (Whiting et al. 2011). The commercialization of animals for medicinal purposes is a widespread phenomenon, with significant implications for their conservation and sustainable use. Since this threat is a multi-factorial one, involving the complex socioeconomic conditions in which users and traders and their families live, conservation programs have to be applied within an organizational framework of the social factors involved, and must be based on clearly stated priorities and goals. In this sense, there is an urgent need to (1) increase our understanding of the harvesting and trade of those species; (2) assess the impacts caused by the commercial exploitation; (3) adopt conservation measures as necessary, so that over-collection of such species will not lead to their extirpation and consequently to the loss of source medicinal material.

The cultural context for urban zootherapy comprises a complex network of dynamic interactions between societal elements. The extensive consumptive use of wildlife in the context of a modern-day economy requires the development of conservation strategies that (1) go beyond existing laws that are seldom enforced, and (2) foster dialog among different stakeholders to address issues such as the need for elimination of the illicit aspects of the trade, and the development of some form of collaborative resource management. Furthermore, given the recognition that habitat loss and landscape alterations pose a serious threat to the survival of wildlife, and that species loss may lead to a loss of knowledge regarding their potential value (see Anyinam 1995), the connections between the trade in medicinal animals and ecosystem conservation should be further explored.

## References

Adeola MO (1992) Importance of wild animals and their parts in the culture, religious festivals, and traditional medicine, of Nigeria. Environ Conserv 19(02):125–134

Albuquerque UP, Monteiro JM, Ramos MA, Amorim ELC (2007) Medicinal and magic plants from a public market in Northeastern Brazil. J Ethnopharmacol 110:76–91

Albuquerque UP, Monteiro JM, Ramos MA, Amorim ELC, Alves RRN (2010) A pesquisa etnobiológica em mercados e feiras livres. In: Albuquerque UP, Lucena RFP, Cunha LVFC (eds) Métodos e técnicas na pesquisa etnobiológica e etnoecológica, vol 1. NUPEEA, Recife, pp 207–222

Almada ED (2011) Urban socio-biodiversity: Ethnoecology of cities. Bioremediation, Biodiversity and Bioavailability 5 (Special Issue 1):1–8

Alves RRN (2009) Fauna used in popular medicine in Northeast Brazil. J Ethnobiol Ethnomed 5(1):1–30. doi:10.1186/1746-4269-5-1

Alves RRN (2010) O comércio de recursos zooterápicos. In: Costa-Neto EM, Alves RRN (eds) Zooterapia: os animais na medicina popular brasileira, vol 2. 1st edn. NUPEEA, Recife, pp 159–176

Alves RRN, Alves HN (2011) The faunal drugstore: animal-based remedies used in traditional medicines in Latin America. J Ethnobiol Ethnomed 7(9):1–43

Alves RRN, Lima HN, Tavares MC, Souto WMS, Barboza RRD, Vasconcellos A (2008a) Animal-based remedies as complementary medicines in Santa Cruz do Capibaribe, Brazil. BMC Complement Altern Med 8:44

Alves RRN, Silva CC, Alves HN (2008b) Aspectos sócio-econômicos do comércio de plantas e animais medicinais em área metropolitanas do Norte e Nordeste do Brasil. Revista de biologia e ciências da terra 8(1):181–189

Alves RRN, Vieira WLS, Santana GG (2008c) Reptiles used in traditional folk medicine: conservation implications. Biodivers Conserv 17(8):2037–2049

Alves RRN, Pereira Filho GA (2007) Commercialization and use of snakes in North and Northeastern Brazil: implications for conservation and management. In: Hawksworth DL, Bull AT (eds) Vertebrate conservation and biodiversity, 1st edn. Springer, Amsterdan, pp 143–159

Alves RRN, Rosa IL (2005) Why study the use of animal products in traditional medicines? J Ethnobiol Ethnomed 1(5):1–5

Alves RRN, Rosa IL (2006) From cnidarians to mammals: the use of animals as remedies in fishing communities in NE Brazil. J Ethnopharmacol 107:259–276

Alves RRN, Rosa IL (2007a) Biodiversity, traditional medicine and public health: where do they meet? J Ethnobiol Ethnomed 3(14):9

Alves RRN, Rosa IL (2007b) Zootherapy goes to town: the use of animal-based remedies in urban areas of NE and N Brazil. J Ethnopharmacol 113:541–555

Alves RRN, Rosa IL (2010) Trade of animals used in Brazilian traditional medicine: trends and implications for conservation. Hum Ecol 38(5):691–704

Alves RRN, Rosa IL, Santana GG (2007) The role of animal-derived remedies as complementary medicine in Brazil. Bioscience 57(11):949–955

Alves RRN, Souto WMS (2010) Etnozoologia: conceitos, considerações históricas e importância. In: Alves RRN, Souto WMS, Mourão JS (eds) A etnozoologia no Brasil: importância, status atual e perspectivas, vol 7, 1st edn. NUPEEA, Recife, pp 19–40

Alves RRN, Souto WMS, Barboza RRD (2010) Primates in traditional folk medicine: a world overview. Mammal Rev 40(2):155–180

Amaral FMM, Coutinho DF, Ribeiro MNS, Oliveira MA (2003) Avaliação da qualidade de drogas vegetais comercializadas em São Luís/Maranhão. Revista brasileira de farmacognosia 13:27–30

Anyinam C (1995) Ecology and ethnomedicine: exploring links between current environmental crisis and indigenous medical practices. Soc Sci Med 40(3):321–329

Apaza L, Godoy R, Wilkie D, Byron E, Huanca O, Leonard WL, Peréz E, Reyes-García V, Vadez V (2003) Markets and the use of wild animals for traditional medicine: a case study among the Tsimane'Amerindians of the Bolivian rain forest. J Ethnobiol 23(1):47–64

Ashwell D, Walston N (2008) An overview of the use and trade of plants and animals in traditional medicine systems in Cambodia, 1st edn. TRAFFIC Southeast Asia, Greater mekong programme. Ha Noi, Vietnam

Belluck P (1996) Mingling Two Worlds of Medicine; Some Doctors Work With Folk Healers In Immigrants' Care. New York Times May 9

Bennett EL, Robinson JG (2000) Hunting of wildlife in tropical forests. Environment Department Papers No 76, The World Bank. Washington pp 1–42

Bode M (2006) Taking traditional knowledge to the market: the commoditization of Indian medicine. Anthropol Med 13(3):225–236

Bolze D, Chetkiewicz C, Qiu M, Krakower D (1998) The availability of Tiger-based traditional Chinese Medicine Products and bublic awareness about the threats to the Tiger in New York's Chinese communities: a pilot study. Wildlife Conservation Society Working Paper

Botha J, Witkowski TF, Shackleton CM, Fairbanks HK (2004) Socio-economic differentiation in the trade of wildlife species for traditional medicines in the Lowveld, South Africa: implications for resource management initiatives. Int J Sustain Dev World Ecol 11(3):280–297

Broad S (2001) The nature and extent of legal and illegal trade in wildlife. Hughes Hall, International and Africa Resources Trust. Paper presented at the seminar on wildlife trade regulation and enforcement. Cambridge

Bussmann RW, Sharon D, Vandebroek I, Jones A, Revene Z (2007) Health for sale: the medicinal plant markets in Trujillo and Chiclayo, Northern Peru. J Ethnobiol Ethnomed 3:37

Cocks M (2006) Biocultural diversity: moving beyond the realm of 'indigenous' and 'local' people. Hum Ecol 34(2):185–200

Cunningham AB (1992) Imithi isiZulu: the traditional medicine trade in Natal/KwaZulu. University of Natal, Pietermaritzburg

Cunningham AB (1993) African medicinal plants: setting priorities at the interface between conservation and primary healthcare. People and plants working paper. UNESCO, Paris

Cunningham AB (1997) An Africa-wide overview of medicinal plant harvesting, conservation and health care. In: Bodeker G, Bhat KKS, Burley J, Vantomme P (eds) Medicinal plants for forest conservation and health care. Food and Agriculture Organization of the UN (FAO), Rome, pp 116–129

Cunningham AB, Zondi AS (1991) Use of animal parts for the commercial trade in traditional medicines. Institute of Natural Resources, Pietermaritzburg

El-Kamali HH (2000) Folk medicinal use of some animal products in Central Sudan. J Ethnopharmacol 72:279–282

Ferreira FS, Brito S, Ribeiro S, Saraiva A, Almeida W, Alves RRN (2009) Animal-based folk remedies sold in public markets in Crato and Juazeiro do Norte, Ceara, Brazil. BMC Complement Altern Med 9(1):17

Ferreira FS, Albuquerque UP, Coutinho HDM, Almeida WO, Alves RRN (2012) The trade in medicinal animals in northeastern Brazil. Evidence-based complementary and alternative medicine http://www.hindawi.com/journals/ecam/aip/126938/

Heinrich M, Ankli A, Frei B, Weimann C, Sticher O (1998) Medicinal plants in Mexico: healers' consensus and cultural importance. Soc Sci Med 47(11):1859–1871

Henson S, Heasman M (1998) Food safety regulation and the firm: understanding the compliance process1. Food Policy 23(1):9–23

Joseph ANT (1990) The relevance of traditional bird drugs in relation to modern primary healthcare in Madhya Pradesh. Record Zool Surv India 86:193–210

Kritsky G (1987) Insects in traditional Chinese medicines. In: Proceedings of the Indiana Academy of Science, vol 96, pp 289–291

Kültür Ş (2007) Medicinal plants used in Kırklareli Province (Turkey). J Ethnopharmacol 111:341–364

Laird SA (1999) The botanical medicine industry. In: Kate K, Laird SA (eds) The commercial use of biodiversity: access to genetic resources and benefit-sharing. Earthscan, London, pp 78–116

Loundou PM (2008) Medicinal plant trade and opportunities for sustainable management in the Cape Peninsula, South Africa. University of Stellenbosch, Stellenbosch

Macía MJ, Garcia E, Vidaurre PJ (2005) An ethnobotanical survey of medicinal plants commercialized in the markets of La Paz and El Alto, Bolivia. J Ethnopharmacol 97(2):337–350

Mahawar MM, Jaroli DP (2008) Traditional zootherapeutic studies in India: a review. J Ethnobiol Ethnomed 4(1):17

Mander M (1998) Marketing of indigenous medicinal plants in South Africa: a case study in KwaZulu–Natal. FAO, Rome

Mander M, Ntuli L, Diederichs N, Mavundla K (2007) Economics of the traditional medicine trade in South Africa: health care delivery. In: Harrison S, Bhana R, Ntuli A (eds) South African health review 2007. Health systems trust, Durban, pp 189–196

Marshall NT (1998a) Searching for a cure: conservation of medicinal wildlife resources in East and Southern Africa. Traffic International, Cambridge

Marshall NT (1998b) Searching for a cure: conservation of medicinal wildlife resources in East and Southern Africa. Traffic International, Cambridge

Mati E, de Boer H (2011) Ethnobotany and trade of medicinal plants in the Qaysari market, kurdish autonomous region, Iraq. J Ethnopharmacol 133:490–510

Melo JG, Amorim ELC, Albuquerque UP (2009) Native medicinal plants commercialized in Brazil–priorities for conservation. Environ Monit Assess 156(1):567–580

Miththapala S (2006) An introduction to the trade in wild species used for medicinal purposes. IUCN

Monteiro JM, Araújo EL, Amorim ELC, Albuquerque UP (2010) Local markets and medicinal plant commerce: a review with emphasis on Brazil. Econ Bot 64:352–366

Moura FBP, Marques JGW (2008) Zooterapia popular na Chapada Diamantina: uma medicina incidental? Ciência and Saúde Coletiva 13:2179–2188

Ngwenya MP (2001) Implications for the medicinal animal trade for nature conservation in Kwazulu–Natal. cascades: ezemvelo KZN wildlife Report No. NA/124/04

Nooren H, Claridge G (2001) Wildlife trade in Laos: the end of the game. IUCN—The World Conservation Union, Amsterdam

O'Connor BB (1998) Healing practices. In: Loue S (ed) Handbook of immigrant health. Plenum Press, New York, pp 145–162

Oliveira ES, Torres DF, Brooks SE, Alves RRN (2010) The medicinal animal markets in the metropolitan region of Natal city. Northeastern Brazil. J Ethnopharmacol 130(1): 54–60

Oliveira RLC, Lins Neto EMF, Araújo EL, Albuquerque UP (2007) Conservation priorities and population structure of woody medicinal plants in an area of caatinga vegetation (Pernambuco, Brazil). Environ Monit Assess 132(1–3):189–206

Olsen CS (1998) The trade in medicinal and aromatic plants from central Nepal to northern India. Econ Bot 52(3):279–292

Padoch C (1992) Marketing of non-timber forest products in Western Amazonia: general observations and research priorities. Adv Econ Bot 9:43–50

Padoch C, Brondizio E, Costa S, Pinedo-Vasquez M, Sears RR, Siqueira A (2008) Urban forest and rural cities: multi-sited households, consumption patterns, and forest resources in Amazonia. Ecol Soc 13(2):2

Parveen Upadhyay B, Roy S, Kumar A (2007) Traditional uses of medicinal plants among the rural communities of Churu district in the Thar desert, India. J Ethnopharmacol 113:387–399
Press I (1971) The urban curandero. Am Anthropol 73(3):741–756
Puppim de Oliveira JA, Balaban O, Doll CNH, Moreno-Penaranda R, Gasparatos A, Iossifova D, Suwa A (2011) Cities and biodiversity: perspectives and governance challenges for implementing the convention on biological diversity (CBD) at the city level. Biol Conserv 144(5):1302–1313
Reiff M, O'Connor B, Kronenberg F, Balick M, Lohr P (2003) Ethnomedicine in the urban environment: dominican healers in New York city. Hum Organ 62(1):12–26
Schnurrenberger PR, Hubbert WT (1981) An outline of the zoonoses. Iowa state university press, Ames
Simelane TS, Kerley GIH (1998) Conservation implications of the use of vertebrates by Xhosa traditional healers in South Africa. S Afr J Wildl Res 28(4):121–126
Smith D, Binet L, Bonnevie L, Hakokongas L, Meybaum J (2001) Desequilíbrio fatal: a crise em pesquisa e desenvolvimento de drogas para doenças negligenciadas. Grupo de Trabalho de Drogas para Doenças Negligenciadas, Médicos Sem Fronteiras, Genebra
Sodeinde OA, Soewu DA (1999) Pilot study of the traditional medicine trade in Nigeria. Traffic Bull 18(1):35–40
Srivastava J, Lambert J, Vietmeyer N (1996) Medicinal plants: an expanding role in development. World bank technical paper. N. 320. World bank publications
Van der Berg MA (1984) Ver-o-peso: the ethnobotany of an Amazonian market. In: Prance GT, Kallunki JA (eds) Advances in economic botany. Ethnobotany in the neo tropics, New York botanical garden, vol 1. Botanical garden bronx, New York, pp 140–149
Van NDN, Tap N (2008) An overview of the use of plants and animals in traditional medicine systems in Vietnam, 1 edn. TRAFFIC Southeast Asia, Greater mekong programme, Ha Noi, Vietnam
van Vliet N, Mbazza P (2011) Recognizing the multiple reasons for bushmeat consumption in urban areas: a necessary step toward the sustainable use of wildlife for food in central Africa. Hum Dimens Wildl 16(1):45–54
Vandebroek I, Calewaert JB (2004) Use of medicinal plants and pharmaceuticals by indigenous communities in the Bolivian Andes and Amazon. Bull World Health Organ 82(4):243–250
Vázquez PE, Méndez RM, Guiascón ÓGR, Piñera EJN (2006) Uso medicinal de la fauna silvestre en los altos de chiapas. México. Interciencia 31(7):491–499
Whiting MJ, Williams VL, Hibbitts TJ (2011) Animals traded for traditional medicine at the Faraday market in South Africa: species diversity and conservation implications. J Zool 284:84–96
Williams VL, Balkwill K, Witkowski ETF (2000) Unraveling the commercial market for medicinal plants and plant parts on the Witwatersrand. S Afr Econ Botany 54(3):310–327
Williams VL, Witkowski ETF, Balkwill K (2005) Application of diversity indices to appraise plant availability in the traditional medicinal markets of Johannesburg, South Africa. Biodivers Conserv 14(12):2971–3001
Yinfeng G, Xueying Z, Yan C, Di W, Sung W (1997) Sustainability of wildlife use in Traditional Chinese Medicine. Conserving China's biodiversity: reports of the Biodiversity Working Group (BWG), China council for international cooperation on environment and development

# Chapter 4
# Medicinal Animals in Ethnoveterinary Practices: A World Overview

**Wedson Medeiros Silva Souto, Lorena Cristina Pinto, Lívia Emanuelle Tavares Mendonça, José Silva Mourão, Washington Luiz Silva Vieira, Paulo Fernando Guedes Pereira Montenegro and Rômulo Romeu Nóbrega Alves**

**Abstract** Medicinal animals are used as a source of medicine in virtually all cultures. Such use is not restricted to humans, but also extends to the treatment of livestock diseases. Nevertheless, animal-based medicines in ethnoveterinary practices have been neglected and poorly described. This work assesses the global use of animals in traditional veterinary medicine, as well as the perspective of this field for future research. Our results indicated that at least 98 animal species, of which 95% are wild, are used in worldwide traditional veterinary medicine. A high diversity of animal parts/products is used as remedies in ethnooveterinary practices, fat being the most used product, followed by meat, skin/leather, bones, and honey. Several zootherapeutic products are prescribed to treat similar or identical diseases in animals and humans, confirming a trend of co-evolution between

W. M. S. Souto (✉) · L. E. T. Mendonça
Programa de Pós-Graduação em Ciências Biológicas (Zoologia),
Departamento de Sistemática e Ecologia, Universidade Federal da Paraíba,
João Pessoa-PB 58059-970, Brazil
e-mail: wedson@live.com; wedson.medeiros@gmail.com

L. C. L. Pinto
Programa de Pós-Graduação em Ecologia de Biomas Tropicais,
Departamento de Evolução, Biodiversidade e Meio Ambiente,
Universidade Federal de Ouro Preto, Ouro Preto-MG 35400-000, Brazil

W. M. S. Souto · L. E. T. Mendonça · J. S. Mourão · W. L. S. Vieira · R. R. N. Alves
Ethnozoology, Conservation and Biodiversity Research Group, Departamento de Biologia,
Universidade Estadual da Paraíba, Campina Grande-PB 58109-753, Brazil

W. L. S. Vieira · P. F. G. P. Montenegro
Laboratório de Ecofisiologia Animal, Departamento de Sistemática e Ecologia,
Universidade Federal da Paraíba, Pessoa-PB 58059-970, Brazil
e-mail: paulomontenegrobr@yahoo.com.br

R. R. N. Alves and I. L. Rosa (eds.), *Animals in Traditional Folk Medicine*,
DOI: 10.1007/978-3-642-29026-8_4, © Springer-Verlag Berlin Heidelberg 2013

human Complementary and Alternative Medicine (CAM) and Ethnoveterinary medicine (EVM).

## 4.1 Introduction

Zootherapy is a vital component of traditional medicine (Alves and Rosa 2007a, b; Souto et al. 2009), which is especially evident in rural communities where access to modern health care is very limited. The close interactions of rural populations with their environments over many centuries (usually associated with economic dependence on local natural resources) have fostered the accumulation of a wealth of knowledge about animal-based medicines and medicinal plants that has been passed down through oral traditions (Quave et al. 2010).

Parallel to the traditional use of natural resources for curing human illnesses, people have also treated their animals with the same or similar traditional pharmacological repertoires (see, for instance, Lans et al. 2006; Scarpa 2004; Souto 2009; Souto et al. 2010a). These treatments stem from human beliefs and experiences and, above all, from an insatiable curiosity that has allowed cultures to accumulate rich traditional knowledge bases over many centuries that have been passed from generation to generation through word of mouth and in songs, poems, drawings, paintings, stories, legends, dreams, visions, and initiation ceremonies (Backes 1998; Mathias-Mundy and McCorkle 1989). This rich body of knowledge, skills, methods, practices, and beliefs about animal care has given rise to a field of study known as ethnoveterinary medicine (EVM) (Mathias and McCorkle 2004; McCorkle 1986).

EVM is particularly important in developing countries where conventional remedies for animal health care are inaccessible or simply unaffordable to poor rural farmers (McGaw et al. 2007). According to the World Health Organization (WHO), at least 80% of the people in developing countries depend largely on traditional practices for controlling and treating various diseases affecting both humans and their animals (Alves and Rosa 2005; Shen et al. 2010). High costs and inaccessibility, together with other problems associated with western-style health care systems, have helped maintain traditional treatment practices in these countries and this has stimulated a good deal of interest in this subject (Muhammad et al. 2005). Interestingly, many people in a number of industrialized countries regularly use some form of traditional complementary or alternative medicine (TCAM) for treating humans or livestock/pets, such as Germany (75%), Canada (70%), and England (47%) (Kayne 2010; Thomas et al. 2001).

Studies focusing on EVM have only been undertaken relatively recently, dating principally from the 1980s (Mathias-Mundy and McCorkle 1989; McCorkle 1986), although increasing attention has been paid to ethnoveterinary knowledge and local veterinary (Akerreta et al. 2010; Martin et al. 2001) in recent years—and the numbers of EVM studies have increased dramatically (together with research examining the importance of plants in EVM) (Davis et al. 1995; Dold and Cocks 2001; Lans 2007; Lans et al. 2007a, b; Pieroni et al. 2004, 2006; Viegi et al. 2003). Nevertheless, a considerable proportion of the ethno knowledge and traditional animal health care practices in native and animal stock raising communities remain undocumented despite a growing tendency to integrate them into more widely used primary animal health care delivery systems in rural and peri-urban communities (Wanzala et al. 2005).

According to Orwa (2002), traditional medicine involves the use of plants, animal parts, and minerals, with herbal medicines being most widely used. Although plants are the main source of remedies in traditional pharmacopoeias around the world it is probable that the number of medicinally useful animals is underestimated, as research in this area has often been neglected (Souto et al. 2009). Despite the importance and the implications of uses of medicinal animals and the fact that animal-based medicines are frequently mentioned in ethnobiological studies focused on human complementary and alternative medicine (human CAM), zootherapeutic practices in EVM have still only been poorly covered (Barboza et al. 2007; Confessor et al. 2009; Souto et al. 2010a).

In Brazil, for example, various publications since the 1980s have shown the importance of zootherapy in traditional communities from diverse socio-cultural environmental landscapes, and recent research indicates that more than 300 animal species are used in traditional medicine practices for treating human diseases (Alves 2009; Costa-Neto and Alves 2010), while until 2009 only 17 species had been recorded with traditional veterinary uses (Barboza et al. 2010; Confessor et al. 2009). A large number of species (46) were recently reported being used in traditional veterinary medicine of NE Brazil (Souto et al. 2011), a number very similar to other countries of the world where zootherapeutic pharmacopoeias are relatively well known. More than 1,500 animal species have been recorded to be of some medicinal use in Traditional Chinese Medicine (TCM) (China National Corporation of Traditional and Herbal Medicine 1995) and in India 15–20% of the Ayurvedic medicine is based on animal-derived substances (Unnikrisnhan 1998)—although only scarce information is available in both countries about animal-based remedies employed in traditional veterinary medicine.

Given the importance of animals in traditional medicines in many parts of the world, scholarly investigations of the medicinal uses of animals and their products should not be neglected, and should be considered an important complementary body of knowledge (Alves 2008). Several important questions therefore arise in the context of EVM: (1) What animal species are used in traditional veterinary medicine? (2) Is there any relationship between these EVM uses and traditional human medicine? (3) What are the implications of the use of medicinal animals on livestock health? (4) What are the possible impacts of this harvesting and use on the local fauna? and (5) Is EVM relevant to future studies of the traditional uses of medicinal animals?

Although these subjects can only be fully elucidated by long-term investigations, we provide here a brief literature review that may help assess what animal species are used as remedies in traditional veterinary practices and the numbers of species involved.

## 4.2 Methods

To examine the diversity of animals used in traditional veterinary medicine, all available references or reports of folk remedies based on animal sources were examined. Only taxa that could be identified to species level were included in the database.

Scientific names provided in the publications were updated according to the ITIS/Catalog of Life: 2010 Annual Checklist (Bisby et al. 2010). The conservation status of the wild species follows the International Union for Conservation of Nature (IUCN) Red List (IUCN 2010) and the Convention on International Trade in Endangered Species (CITES) (2008).

This chapter draws on zootherapeutic sources which contained information about ethnoveterinary uses, covering published articles, books and book chapters, theses and dissertations, and including: Abate et al. (2000), Adolph et al. (1996), Alves et al. (2009), Antoine-Moussiaux et al. (2007), Ashwell and Walston (2008), Barboza et al. (2007, 2010), Birdlife International (2010), Bizimana (1994), Confessor et al. (2009), Costa-Neto (1999a, b; 2001), Costa-Neto et al. (2006), Davis et al. (1995), Faure et al. (1944), Ferreira et al. (2009), Froemming (2006), Köhler-Rollefson et al. (2001), Kokwaro (2009), Larrat (1939), Lenko and Papavero (1996), Mares (1954), Marx (1984), Muhammad et al. (2005), Quave et al. (2010), Raziq et al. (2010), Ribeiro et al. (2010), Santos and Costa-Neto (2007), Schinkel (1970), Song and Kim (2010), Souto et al. (2010a, b, 2011), Valdizán and Maldonado (1922) and Xie and Preast (2010).

## 4.3 Results and Discussion

### *4.3.1 Zootherapeutic Species Used in EVM*

Our literature survey revealed that animals are used as sources of folk medicines in EVM in at least 21 countries/regions (Table 4.1). Brazil had the largest number of documented species (52), followed by China (13 species), Sudan (12), the Korean Peninsula (9), Niger (6), Spain (5), Cambodia and Peru (4 each), Albania and West Africa (3 each), Mali, Kenya, and Somalia (2 each), and North and East Africa, Italia, Mauritania, Morocco, Nigeria, Pakistan, Senegal, and Tanzania (1 each).

While the large numbers of species used as therapeutics in EVM in Brazil reflects the ethnobiological research carried out by the Center of Ethnobiology and Ethnoecology at the State University of Paraiba, which has focused on traditional livestock keepers (*sertanejos/vaqueiros*) in the semi-arid region of NE Brazil, the

**Table 4.1** Animals in worldwide ethnoveterinary medicine

| Family/species | Region or country(-ies) | References |
|---|---|---|
| **ANNELIDS** | | |
| **Hirundinidae** | | |
| *Hirundo rustica* (Linnaeus, 1758)[LC] | China, The Korean Peninsula | Song and Kim (2010) |
| **Lumbricidae** | | |
| *Lumbricus rubellus* (Hoffmeister, 1843) | China, The Korean Peninsula | Song and Kim (2010) |
| **ARTHROPODS** | | |
| **Apidae** | | |
| *Apis cerana* (Fabricius, 1793) | China, The Korean Peninsula | Song and Kim (2010), Xie and Preast (2010) |
| *Apis mellifera* (Linnaeus 1758)—Africanised honey bee, "abelha africana"[a] | Brazil | Souto et al. (2011) |
| *Melipona subnitida* (Ducke 1910)—jandaíra bee[a] | Brazil | Souto et al. (2011) |
| *Partamona seridoenses* (Pedro and Camargo 2003)—Cupira bee[a] | Brazil | Souto et al. (2011) |
| *Scaptotrigona sp.*—"abelha canudo" | Brazil | Souto et al. (2011) |
| *Trigona spinipes* (Fabricius, 1793) | Brazil | Costa-Neto (1999b), Costa-Neto et al. (2006) |
| **Polyphagidae** | | |
| *Eupolyphaga sinensis* (Walker, F., 1868)—Wingless cockroach | China | Xie and Preast (2010) |
| **Bothriuridae** | | |
| *Bothriurus asper* Pocock, 1893—black scorpion | Brazil | Souto et al. (2011) |
| **Buthidae** | | |
| *Buthus occitanus* (Amoreux, 1789)—common European scorpion[a] | Spain | Quave et al. (2010) |
| *Rhopalurus rochai* (Borelli, 1910)—"Escorpião amarelo do sertão" | Brazil | Souto et al. (2011) |
| **Gryllidae** | | |
| *Acheta domesticus* (Linnaeus, 1758) | Brazil | Lenko and Papavero (1996) |
| **Pentatomidae** | | |
| *Agonoscelis puberula* Stål, 1853 | North Africa | Abate et al. (2000), Faure (1944) |
| **Termitidae** | | |
| *Nasutitermes corniger* (Motschulsky, 1855)—"cupim preto," black termite | Brazil | Souto et al. (2011) |

(continued)

**Table 4.1** (continued)

| Family/species | Region or country(-ies) | References |
|---|---|---|
| **Vespidae** | | |
| *Polistes canadensis* (Linnaeus, 1758)—"Marimbondo-caboclo" | Brazil | Barboza et al. (2007, 2010) |
| **CRUSTACEA** | | |
| **Astacidae** | | |
| *Astacus sp.*—crayfish[VU] | Peru | Froemming (2006) |
| **Palaemonidae** | | |
| *Cryphiops caementarius* (Molina, 1782)—changallo shrimp | Peru | Valdizán and Maldonado (1922) |
| **ECHINODERMS** | | |
| **Echinasteridae** | | |
| *Echinaster brasiliensis* (Müller and Troschel, 1842)—Starfish, "Estrela-do-mar"[a] | Brazil | Alves et al. (2009) |
| *Echinaster echinophorus* Lamarck, 1816—Starfish, "Estrela-do-mar"[a] | Brazil | Alves et al. (2009) |
| **Luidiidae** | | |
| *Luidia senegalensis* Lamarck, 1916— Starfish, Estrela-das-nove-pontas | Brazil | Alves et al. (2009) |
| **Oreasteridae** | | |
| *Oreaster reticulatus* (Linnaeus, 1758)—cushioned star, "Estrela-do-mar" | Brazil | Alves et al. (2009) |
| **FISHES** | | |
| **Clupeidae** | | |
| *Sardina pilchardus* (Walbaum, 1752)—Sardina europea, sardine | Spain | Quave et al. (2010) |
| **Curimatidae** | | |
| *Prochilodus* sp.[LC] | Brazil | Barboza et al. (2010), Souto et al. (2010a, b) |
| **Electrophoridae** | | |
| *Electrophorus electricus* (Linnaeus, 1766)—"Peixe-elétrico", electric eel [LC] | Brazil | Souto et al. (2011) |
| **Erythrinidae** | | |
| *Hoplias malabaricus* (Bloch, 1794)—Trahira, "traíra"[a] | Brazil | Souto et al. (2011) |

(continued)

**Table 4.1** (continued)

| Family/species | Region or country(-ies) | References |
|---|---|---|
| **Megalopidae** | | |
| *Megalops atlanticus* Valenciennes in Cuvier and Valenciennes, 1847—sábalo, tarpon | Brazil | Costa-Neto (2001) |
| **Syngnathidae** | | |
| *Hippocampus reidi* (Ginsburg, 1933)—Longsnout seahorse, "cavalo-marinho"[a DD/II] | Brazil | Alves et al. (2009) |
| **AMPHIBIANS** | | |
| **Bufonidae** | | |
| *Rhinella jimi* (Stevaux, 2002)—"Cururu toad" or "sapo cururu" [LC] | Brazil | Ferreira et al. (2009) |
| *Rhinella schneideri* (Werner, 1894)—Cururu toad, "sapo cururu"[a LC] | Brazil | Souto et al. (2011) |
| *Bufo bufo gargarizans* Cantor[LC] | China/The Korean Peninsula | Song and Kim (2010) |
| **REPTILES** | | |
| **Alligatoridae** | | |
| *Caiman latirostris* (Daudin, 1801)—Cayman, "jacaré-do-papo-amarelo"[a LC/I] | Brazil | Souto et al. (2011) |
| **Boidae** | | |
| *Boa constrictor* (Linnaeus, 1758)—Boa, "jibóia," "cobra de veado"[a I] | Brazil | Costa-Neto (1999a), Souto et al. (2011) |
| **Chelidae** | | |
| *Phrynops geoffroanus* (Schweigger, 1812)—Geoffroy's side-necked turtle[a] | Brazil | Barboza et al. (2007, 2010), Confessor et al. (2009), Souto et al. (2010a, b, 2011) |
| **Cheloniidae** | | |
| *Caretta caretta* (Linnaeus, 1758)—loggerhead, loggerhead sea turtle[EN] | Brazil | Costa-Neto (2001) |
| **Colubridae** | | |
| *Malpolon monspessulanus* Hermann 1804[LC] | Spain | Quave et al. (2010) |
| *Rhinechis scalaris* (Schinz, 1822)[LC] | Spain | Quave et al. (2010) |
| **Gekkonidae** | | |
| *Gekko gecko* (Linnaeus, 1758)—gecko lizard | China | Xie and Preast (2010) |

(continued)

**Table 4.1** (continued)

| Family/species | Region or country(-ies) | References |
|---|---|---|
| **Iguanidae** | | |
| *Iguana iguana* (Linnaeus, 1758)—common green, "Camaleão"[a] [II] | Brazil | Barboza et al. (2007, 2010), Confessor et al. (2009), Souto et al. (2010a, b, 2011) |
| **Lacertidae** | | |
| *Timon lepidus* Daudin, 1802—ocellated lizard[a] | Spain | Quave et al. (2010) |
| **Teiidae** | | |
| *Tupinambis merianae* (Duméril and Bibron, 1839)—lizard tegu, "tegu", "tejuaçú"[a] [LC/II] | Brazil | Barboza et al. (2010), Confessor et al. (2009), Souto et al. (2010a, b, 2011) |
| **Testudinidae** | | |
| *Chelonoidis carbonaria* (Spix, 1824)—redfooted tortoise, "jabuti" | Brazil | Souto et al. (2011) |
| *Chinemys reevesii*—reeves's terrapin | China | Xie and Preast (2010) |
| **Trionychidae** | | |
| *Trionyx sinensis* Wiegmann, 1835—terrapin, Chinese soft-shelled turtle [VU] | China | Xie and Preast (2010) |
| **Tropiduridae** | | |
| *Tropidurus hispidus* (Spix, 1825)—lizard, "lagartixa," "catenga" | Brazil | Souto et al. (2011) |
| **Viperidae** | | |
| *Caudisona durissa* (Linnaeus, 1758)—South American rattlesnake, "Cascavel"[a] [LC/III] | Brazil | Barboza et al. (2007, 2010), Confessor et al. (2009), Costa-Neto (1999a), Ferreira et al. (2009), Souto et al. (2010a, b, 2011) |
| **BIRDS** | | |
| **Accipitridae** | | |
| *Gypaetus barbatus* (Linnaeus, 1758)—bearded vulture[LC] | Sudan | Adolph et al. (1996) |
| *Gypohierax angolensis* (J. F. Gmelin, 1788)—palm-nut Vulture[LC] | Sudan | Adolph et al. (1996) |
| *Gyps africanus* Salvadori, 1865—white-backed vulture [NT] | Sudan | Adolph et al. (1996) |
| *Gyps rueppellii* (Brehm, 1852)—Rüppell's vulture [NT] | Sudan | Adolph et al. (1996) |

(continued)

**Table 4.1** (continued)

| Family/species | Region or country(-ies) | References |
|---|---|---|
| *Neophron percnopterus* (Linnaeus, 1758)—Egyptian vulture [EN] | Morocco, Sudan | Adolph et al. (1996), Bird Life International (2010) |
| *Torgos tracheliotus* (Forster, 1791)—lappet-faced vulture [VU] | Sudan | Adolph et al. (1996) |
| *Trigonoceps occipitalis* (Burchell, 1824)—white-headed vulture [VU] | Sudan | Adolph et al. (1996) |
| **Cariamidae** | | |
| *Cariama cristata* (Linnaeus, 1766)—"sariema"[LC] | Brazil | Souto et al. (2011) |
| **Cathartidae** | | |
| *Coragyps atratus* (Bechstein, 1793)—black vulture, "urubu," "urubu preto"[LC] | Brazil | Santos and Costa-Neto (2007), Souto et al. (2011) |
| **Ciconiidae** | | |
| *Vultur gryphus* (Linnaeus, 1758)—Andean condor [NT/I] | Peru | Froemming (2006) |
| **Meleagrididae** | | |
| *Meleagris gallopavo* (Linnaeus, 1758)—"peru", domestic turkey[a] [LC] | Brazil | Barboza et al. (2007, Souto et al. (2010a, 2011) |
| **Phasianidae** | | |
| *Gallus gallus* (Linnaeus, 1758)—domestic chicken, "Galinha"[a] [LC] | Albania, Brazil, China,The Korean Peninsula | Barboza et al. (2007, 2010), Confessor et al. (2009), Song and Kim (2010), Souto et al. (2010a, b, 2011), Xie and Preast (2010) |
| **Picidae** | | |
| *Colaptes rupicola* Orbigny, 1840—Andean flicker[a] | Peru | Froemming (2006) |
| **Rheidae** | | |
| *Rhea Americana* (Linnaeus, 1758)—greater rhea, "ema"[a] [NT/II] | Brazil | Souto et al. (2011) |
| **Struthionidae** | | |
| *Struthio camelus* (Linnaeus, 1758)—common ostrich [LC/I] | Sudan | Adolph et al. (1996) |
| **Tinamidae** | | |
| *Nothura maculosa cearensis* Naumburg, 1932—spotted Nothura - "Codorniz"[a] [LC] | Brazil | Barboza et al. (2007), Confessor et al. (2009), Souto et al. (2010a, b, 2011) |

(continued)

**Table 4.1** (continued)

| Family/species | Region or country(-ies) | References |
|---|---|---|
| **MAMMALS** | | |
| **Agoutidae** | | |
| *Agouti paca* (Linnaeus, 1766)—spotted paca, "paca" [LC/III] | Brazil | Souto et al. (2011) |
| **Bovidae** | | |
| *Bos taurus* Linnaeus, 1758—domestic cattle (feral)[a] | Albania, Brazil, Kenya, Mali, Mauritania, Niger, Nigeria, Pakistan, Senegal, Tanzania, East Africa | Antoine-Moussiaux et al. (2007), Barboza et al. (2007, 2010), Bizimana (1994), Confessor et al. (2009), Davis et al. (1995), Kokwaro (2009), Larrat (1939), Muhammad et al. (2005), Quave et al. (2010), Raziq et al. (2010), Souto et al. (2010a, b, 2011) |
| *Capra hircus* Linnaeus, 1758—domestic goat, "cabra"[a] | Brazil, West Africa, Mali, Niger | Bizimana (1994); Köhler-Rollefson et al. (2001), Souto et al. (2011) |
| *Ovis aries* Linnaeus, 1758—"sheep," "carneiro"[a] | Brazil, West Africa, Somalia | Barboza et al. (2007, 2010), Bizimana (1994), Confessor et al. (2009), Köhler-Rollefson et al. (2001), Larrat (1939), Mares (1954), Marx (1984), Schinkel (1970), Souto et al. (2010a, b, 2011) |
| *Tragelaphus spekii* P. L. Sclater, 1863—sitatunga[LC] | Sudan | Adolph et al. (1996) |
| **Camelidae** | | |
| *Camelus dromedarius* Linnaeus, 1758 | Niger, Somalia | Antoine-Moussiaux et al. (2007), Bizimana (1994), Köhler-Rollefson et al. (2001) |
| **Canidae** | | |
| *Canis adustus* Sundevall, 1847—side-striped Jackal[LC] | West Africa | Bizimana (1994) |
| *Canis aureus* Linnaeus, 1758—golden jackal [LC/III] | Niger | Antoine-Moussiaux et al. (2007), Bizimana (1994), Köhler-Rollefson et al. (2001) |
| *Canis familiaris* (Linnaeus, 1758)—dog, "cachorro" | Brazil | Souto et al. (2011) |

(continued)

**Table 4.1** (continued)

| Family/species | Region or country(-ies) | References |
|---|---|---|
| *Cerdocyon thous* (Linnaeus, 1766)—crab-eating fox, "raposa"[a LC/II] | Brazil | Barboza et al. (2007), Confessor et al. (2009), Souto et al. (2010a, b, 2011) |
| *Nyctereutes procyonoides* (Gray 1834)[LC] | China/The Korean Peninsula | Song and Kim (2010) |
| *Vulpes vulpes* (Linnaeus, 1758)[LC/] | China/The Korean Peninsula | Song and Kim (2010) |
| **Caviidae** | | |
| *Cavia aperea* Erxleben 1777—"Preá"[LC] | Brazil | Souto et al. (2011) |
| *Kerodon rupestris* (Wied-Neuwied, 1820)—"Mocó"[a LC] | Brazil | Souto et al. (2011) |
| **Cervidae** | | |
| *Mazama gouazoupira* (G. Fischer, 1814)—gray brocket, "veado catingueiro" | Brazil | Souto et al. (2011) |
| **Dasypodidae** | | |
| *Dasypus novemcinctus* (Linnaeus, 1758)—nine-banded armadillo, 'tatu verdadeiro' [LC] | Brazil | Souto et al. (2011) |
| *Euphractus sexcinctus* (Linnaeus, 1758)—"tatu peba", six-banded armadillo[a LC] | Brazil | Barboza et al. (2010), Barboza et al. (2007), Confessor et al. (2009), Souto et al. (2010a, 2011) |
| **Delphinidae** | | |
| *Delphinus capensis* Gray, 1828—long-beaked common dolphin[DD] | Cambodia | Ashwell and Walston (2008) |
| *Tursiops aduncus* (Ehrenberg, 1833)— Indo-Pacific Bottlenose Dolphin[DD] | Cambodia | Ashwell and Walston (2008) |
| *Tursiops truncatus* (Montagu, 1821)— Bottlenose Dolphin[LC] | Cambodia | Ashwell and Walston (2008) |
| *Sousa chinensis* (Osbeck, 1765)—Chinese white dolphin [NT/I] | Cambodia | Ashwell and Walston (2008) |
| **Erethizontidae** | | |
| *Chaetomys subspinosus* (Olfers, 1818)—"ouriço preto"[VU] | Brazil | Ribeiro et al. (2010) |
| *Sphigurus insidiosus* (Olfers, 1818) | Brazil | Ribeiro et al. (2010) |

(continued)

**Table 4.1** (continued)

| Family/species | Region or country(-ies) | References |
|---|---|---|
| **Felidae** | | |
| *Leopardus tigrinus* (Schreber, 1775)— little tiger cat, "gato-pintado"[a] [VU/I] | Brazil | Souto et al. (2011) |
| *Puma yagouaroundi* (É. Geoffroy Saint-Hilaire, 1803)—jaguarundi, "gato-do-mato vermelho," "gato-domato azul"[a] [LC/II] | Brazil | Souto et al. (2011) |
| **Giraffidae** | | |
| *Giraffa camelopardalis* (Linnaeus, 1758)—giraffe[LC] | Kenya | Bizimana (1994) |
| **Hyaenidae** | | |
| *Crocuta crocuta* (Erxleben, 1777)—spotted hyena[LC] | Niger, Sudan, West Africa | Adolph et al. (1996), Bizimana (1994), Larrat (1939) |
| **Hominidae** | | |
| *Homo sapiens* (Linnaeus, 1758)—human, man[LC] | Niger | Antoine-Moussiaux et al. (2007), Souto et al. (2011) |
| **Hystricidae** | | |
| *Atherurus africanus* Gray, 1842[LC] | Sudan | Adolph et al. (1996) |
| *Hystrix cristata* Linnaeus, 1758—The crested porcupine[LC] | Sudan | Adolph et al. (1996) |
| **Leporidae** | | |
| *Lepus sinensis coreanus* Thomas, 1908[LC] | China, The Korean Peninsula | Song and Kim (2010) |
| **Moschidae** | | |
| *Moschus moschiferus* Linnaeus, 1758[VU/II] | China, The Korean Peninsula | Song and Kim (2010) |
| **Mustelidae** | | |
| *Conepatus semistriatus* (Boddaert, 1785)—Striped hog-nosed skunk, "cangambá," "gambambá," tacaca[a] | Brazil | Souto et al. (2011) |
| **Myrmecophagidae** | | |
| *Tamandua tetradactyla* (Linnaeus, 1758)—collared anteater, 'tamanduá'[LC] | Brazil | Souto et al. (2011) |
| **Procyonidae** | | |
| *Procyon cancrivorus* (G. [Baron] Cuvier, 1798)—crab-eating raccoon, "guaxinim"[LC] | Brazil | Souto et al. (2011) |

(continued)

**Table 4.1** (continued)

| Family/species | Region or country(-ies) | References |
|---|---|---|
| **Suidae** | | |
| *Sus scrofa domestica* Linnaeus, 1758[a] | Brazil, China and The Korean Peninsula, Italia, Albania | Barboza et al. (2007), Confessor et al. (2009), Quave et al. (2010), Song and Kim (2010), Souto et al. (2010a, b, 2011) |

*IUCN Red list status DD* deficient data, *LC* least concern, *NT* near threatened, *EN* Endangered, *VU* vulnerable; CITES appendices—I, II or III

[a] Indicates species that are used both in the treatment of similar or identical diseases in humans or animals

low numbers of medicinal animals registered for countries with consolidated traditions of EVM (e.g., China, many African countries) possibly reflects the limited availability of the literature on zootherapy in EVM, as the use of wildlife in traditional remedies is particularly important in these regions (at least for the treatment of human diseases) (FAO 2008; McKean 2007; Nooren and Claridge 2001; Shepherd 2001; Whiting et al. 2011). This observation is supported by Mathias (2004) who noted that research into traditional EVM has lagged behind its counterpart (modern veterinary medicine) because these practices have often been undertaken in secrecy or the information hidden is in the "gray literature."

Our review indicated that at least 98 animal species (22 invertebrates and 76 vertebrates) are used worldwide in traditional veterinary medicine (Table 4.1). This number is certainly an underestimation of the real situation as the numbers of studies on the theme have been very limited (in both numbers and locations). Mammals were the best represented group used for veterinary purposes (36 species, 36.7%), followed by arthropods and birds (16 species each, 16.3% each), reptiles (15 species, 15.3%), fish (6 species, 6.1%), echinoderms (4 species, 4.1%), amphibians (3 species, 3.1%), and annelids (2 species, 2%). The wide use of vertebrates follows the same trend that was observed in studies of human ethnomedicine (Alves 2009; Alves et al. 2007, 2011; Barboza et al. 2007; Begossi et al. 1999; Branch and Silva 1983; El-Kamali 2000; Mahawar and Jaroli 2008; Souto et al. 2011) and reflects the fact that vertebrates supply larger numbers of parts used as traditional medicines and many of these products can be easily stored (fat, feathers, leather, for example).

Most animal-based remedies were obtained from the wild species (93%), although some domestic species are also used, including domestic cattle—*Bos taurus* Linnaeus, 1758, chickens—*Gallus domesticus* (Linnaeus 1758), dogs—*Canis familiaris* (Linnaeus 1758), goats—*Capra hircus* (Linnaeus 1758), pigs—*Sus scrofa domestica* (Linnaeus 1758), domestic turkeys—*Meleagris gallopavo* (Linnaeus 1758), and sheep—*Ovis aries* (Linnaeus 1758). Elliott et al. (2002) and Aiyadurai (2007) emphasized that wild resources are extremely important to local cultural values and traditions and are often used as foods, in rituals, and as medicines.

A wide diversity of animal parts/products are used as remedies in ethnoveterinary practices: beeswax, blood, bones, bony shells, butter, animal carcasses, eggs, feces, fat/suet, feathers, heads, honey, horns, leather, meat, milk, and its derivatives, musk, nests, 'saburá' (fermented pollen), plastron, skin, skulls, spines, scales, stingers, tails, urine, and viscera. Fat was the most commonly used product, being utilized from 30 of the 98 species mentioned (30.6%). Other important products used as medicines include meat, skin/leather, bones, and honey, which were extracted from 18.4, 16.3, 7.1, and 5%, respectively, of the species recorded. The use of either live or dead animals is determined by the raw materials to be extracted, by popular beliefs, and by the type of illness that is being treated (Alves and Rosa 2006). There is a general tendency for traditional medicines to be administered topically or orally (see Albuquerque et al. 2007; Alves and Rosa 2006; Alves et al. 2007; Alves and Santana 2008; Davis et al. 1995; Ferreira et al. 2009; Kayne 2010; Mahawar and Jaroli 2008; Passalacqua et al. 2006; Pieroni and Price 2005; Quave et al. 2008, 2010; Souto et al. 2011; Viegi et al. 2003), and folk remedies are commonly prepared in the following manners: the whole animal (or its body parts) is dried and then macerated, and the resulting powder is consumed as a tea or ingested together with food; secretions, metabolic products (e.g., urine, honey), and fats are administered as ointments or are ingested (Table 4.1). Some of the procedures registered by previous authors and employed by local experts, however, require considerable skill. In Northeastern Brazil, for example, fat from the crab-eating fox (*Cerdocyon thous*), fat from Geoffroy's side-necked turtle (*Phrynops geoffroanus*), "saburá" (fermented pollen) from Cupira bees (*Partamona seridoensis*), and butter made from cow's milk (*B. taurus*) are used in treating uterine prolapse in livestock, especially in mares and cows (Barboza et al. 2007; Souto 2009), with the exteriorized uterus being washed with the zootherapeutic substance and quickly replaced. A similar procedure for treating uterine prolapse in camels (although involving only washing the uterus with water) was reported by Antoine–Moussiaux et al. (2007) among the Tuaregs in Niger.

Parts such as feather and spines are often extracted without the necessity of killing the animal (Table 4.1), although 93% ($n = 91$) of the recorded species are sources of some medicine that requires the death of the animal. In China, for instance, the internal organs and eyes are removed from dead gecko lizards [*Gekko gecko* (Linnaeus 1758)] and its body is stretched out on bamboo slivers to dry and be transformed into a powder used to make healing teas (Xie and Preast 2010). In Kenya, giraffe skin is toasted and used as smoke treatment for treating *trypanosomiasis* in livestock (Bizimana 1994). The fat and leather of *Puma yagouaroundi* (É. Geoffroy Saint-Hilaire 1803) and *Leopardus tigrinus* (Schreber 1775) are important zootherapeutic remedies in Brazil (Souto 2009).

Zootherapeutic products are used to treat at least 90 animal health problems (or in veterinary practices to prevent animal diseases or medical complications) and these have been classified into 12 broad categories according to the International Classification of Diseases (ICD-10) of the WHO. Several species (41%, $n = 40$) have multiple therapeutic uses. *Apis cerana* (Fabricius 1793), for example, is indicated for treating 23 conditions in Asia; the South American rattlesnake—

*Caudisona durissa* (Linnaeus 1758) and the Crab-eating Fox—*Cerdocyon thous* (Linnaeus 1766) are indicated for treating 13 illnesses in Brazil; and the Gecko lizard and *Apis cerana* are used in China to treat asthma and coughing. One healer interviewed in northeast Brazil mentioned the possibility of using the longsnout seahorse—*Hippocampus reidi* (Ginsburg 1933) and the cushioned star *Oreaster reticulatus* (Linnaeus 1758) to treat asthma in livestock. The local pharmacopeia in Peru includes a great diversity of animals can help increase milk production (used as galactagogues), including the changallo shrimp—*Cryphiops caementarius* (Molina 1782) and the Andean Flicker—*Colaptes rupicola* Orbigny, 1840. Overlapping therapeutic indications of different medicinal animals provide adaptive availability and accessibility to folk medicinal resources (Alves et al. 2007).

Similar to what has been documented in relation to traditional human medicine, religious and magical aspects are of considerable importance in folk veterinary medicine (Scarpa 2000; Souto 2009). There are many examples of magic-religious uses of natural resources in traditional pharmacopoeias (Alves et al. 2007). Although generally representing only a small percentage of the overall use of animals (Viegi et al. 2003), magic uses persist in most countries where traditional medicine is the principal form of treatment of human and/or animal diseases (Alves et al. 2010a; Mathias 1994; Nyamanga et al. 2008). Magic-religious practices cannot be easily separated from other aspects of EVM, although for heuristic purposes, this is sometimes necessary (Mathias-Mundy and McCorkle 1989). This separation seems especially artificial in the domain of ethnopharmacology, where many remedies used for magical or symbolic reasons have in fact been scientifically proven effective (Wanzala et al. 2005). For instance, the Andeans try to maintain their livestock herds away from cold, draughty places so as to keep them out of the haunts of evil winds, and this obviously constitutes a useful management strategy to avoid pneumonias or other respiratory illnesses (McCorkle 1986).

Various traditional medical systems have been influenced by magic-religious practices during their formation. The origin of Chinese folk medicine (for traditional human and veterinary practices) is found in plant lore, religious-magical beliefs, and the rudimentary medical techniques and faith healings of the Neolithic pastoralists and hunter-gatherers who were acutely tuned to survival techniques—and many of these elements can still be observed in cultures throughout China (Schiffeler 1976).

In Latin America, expressions of traditional medicine (CAM/EVM), particularly of zootherapy, represent an interaction of Amerindian, African, and European elements dating to the time of colonization (Almeida 2005). These multicultural interactions resulted in a rich cultural "agglutination" that generated a diverse mixture of ethnomedicine/ethnoveterinary techniques used even today by people belonging to many different social classes (Léo Neto et al. 2009; Souto et al. 2011) and demonstrates how ethnomedicine is deeply rooted in the long history of human culture (Wanzala et al. 2005).

Popular beliefs have implications in the ways that species are used, and protection against the evil eye, or from attacks by other animals are some of the most common of magic-religious applications of medicinal plants and animals (see Alves et al. 2007; Passalacqua et al. 2006). Examples of this can be seen in the use of skulls or horns of *Bos Taurus*, and parts of wild species such as *Cerdocyon thous, Leopardus tigrinus* (Schreber 1775), *Puma yagouaroundi* (É. Geoffroy Saint-Hilaire 1803), and the collared anteater—*Tamandua tetradactyla* (Linnaeus 1758) to protect the animals from attacks by bats in Brazilian EVM. In some situations, magic-religious practices are used to cure *de facto* illnesses, and in some West African countries, herding cultures believe that sacrificing a sheep cures trypanosomiasis in livestock. In Brazil, crickets—*Acheta domesticus* Linnaeus 1758) are collected, put in sacks, and tied in the tails of horses to treat abdominal and urinary tract pains (Lenko and Papavero 1996).

### 4.3.2 Similarities Between Zootherapy in Local Ethnoveterinary Medicine and Human Ethnomedicine

Animal and human medical practices have been closely linked throughout history, with each contributing to the other—and ultimately to the concept of "one medicine" (van Veen 1998). Modern medicine had its origin in the traditional medical practices of ancient China, India, and the Middle East (Schoen and Wynn 1997; Schwabe 1978, 1984); these early medical traditions included mixtures of a wide spectrum of folk knowledge and did not always draw sharp distinctions between animal and human medicine. In early India, physicians treating human beings were also trained in caring for animals. Indian medical treatises such as Charaka Samhita, Susruta Samhita, and Harita Samhita contain chapters or references about caring for both diseased and healthy animals (Somvanshi 2006).

Ancient Europe, Greece, and Rome developed into centers of veterinary and medical knowledge. The veterinary medicine practiced there was derived from (or often performed by) trained veterinary or medical specialists and veterinary empiricists (van Veen 1998). The influence of human medicine in Byzantine veterinary medicine is visible not only in its concepts of pathology and in the descriptions of diseases, but in the therapeutical methods used, and especially in the *materia medica* (Doyen-Higuet 1984).

Although human and veterinary medicines began to diverge in the early nineteenth century, they still largely remain intimately intertwined. The first veterinary schools that reached out to farmers (rather than only to the cavalry) were perforce partly staffed by physicians who had also gained some understanding of animal ailments through their rural practices (van Veen 1998). Even today, traditional medical systems maintain close relationships between human CAM and EVM and ethnoveterinary medicine and ethnomedicine overlap in many cultures, as many healers will treat humans, as well as animals (Mathias-Mundy 1989) and many diseases that affect both animals and humans can be treated with similar remedies (Alves et al. 2010b).

There are many parallels between human and animal ethnomedicines that span not only health care concepts, beliefs, and practitioners, but also nearly all known modes of administration of the materials themselves, the *materia medica*, and ethnomedical techniques and behaviors (with the latter being defined as any act undertaken with a medical intent [emic] and/or with conventionally recognized health implications [etic]) (McCorkle and Martin 1998). This is a common approach and has been well documented with medicinal plants. In Brazil, for example, São Caetano's melon (*Mormodica charantia* L.), the purga potato (*Operculina hamiltonii*), and pumpkin seeds (*Cucurbita pepo* L.) are used as vermifuges in veterinary medicine, in a very similar way that they are prepared in human medicine (Almeida et al. 2006; Faria et al. 2005). About 80% of the plants used in traditional veterinary medicine in Mediterranean Greece are used to treat similar conditions in humans (Pieroni et al. 2006). In Uganda (Africa), 34 of the 38 plants used in treating cattle diseases are also used in human ethnomedicine (Tabuti et al. 2003). Scarpa (2004) reported strong correspondences between the plants used in traditional veterinary and human medicine in the Chaco region of northwestern Argentina, with 60% of the ethnoveterinary therapeutic plants having identical uses in human medicinal therapy.

Animals are employed in folk medicine to treat both human and animal ailments, and at least 31 species recorded in our review are prescribed to treat similar ailments in both humans and animals. In Peru, the Andean flicker (*Colaptes rupicola*) is used as a galactagogue for nursing women and domestic animals (Froemming 2006). The common European scorpion [*Buthus occitanus* (Amoreux 1789)] is traditionally used in Spain as an anticystitis treatment in both mules and women (Quave et al. 2010) and the Ocellated Lizard (*Timon lepidus* Daudin 1802) (either its fat or the live animal) is applied as a healing ointment in livestock and humans (Quave et al. 2010). The fat of *Ovis aries* is used for treating bone fractures and thorn wounds in veterinary medicine in NE Brazil (Alves et al. 2007; Barboza et al. 2007) in a manner very similar to its use in human medicine. Recent research undertaken in the semi-arid region of NE Brazil found a high correlation between the species used as medicines in EVM practices and those employed in human complementary and alternative medicine (Souto et al. 2011). These authors noted that nearly identical descriptions of diseases, symptoms, or treatments were often associated with both humans and animals (especially livestock), and that 24 zootherapeutic species (of the 46 used locally in EVM) were prescribed for treating ailments in livestock based exclusively on similar or identical treatments of diseases affecting humans.

The use of folk remedies to treat diseases or ailments in animals based on similar or identical illness that attack humans has been denominated "human models for animal diseases" by Barboza et al. (2007). The relationships between ethnoveterinary and human ethnomedicine can be easily explained from this perspective, as the main domestic animal stocks (e.g., cattle, sheep, goats, pigs, among others) are mammals (Confessor et al. 2009) and they have many symptoms and health problems that are very similar to humans, and these similarities have been noted by many different cultures (Akerreta et al. 2010; Alves et al. 2010b).

Our results support other studies noted that most traditional societies make no clear division between veterinary and human medicine (Ghirotti 1996).

### 4.3.3 Conservation Aspects

Most of the animals employed in EVM are wild species, and many of them are also of conservation concern. A total of 59% of the species recorded are included in the 2010 IUCN Red List of Threatened Species (version 2010.4), with 2 species being classified as Endangered, 7 as Vulnerable, 5 as Near Threatened, 41 as Least Concern, and 3 as Data Deficient; 17% are also included in CITES (Convention on International Trade in Endangered Species of Wild Fauna and Flora) Appendixes. While the reasons for the inclusion of certain species in the IUCN Redlist or CITES are not necessarily related to medicinal uses, it will still be necessary to include these traditional medical uses in conservation planning.

Although zootherapeutic practices in EVM are less frequent than in human complementary medicine, they constitute yet another form of pressure on wild animal populations. Additionally, although some information about the use of wildlife resources for EVM medicinal purposes is available in published pharmacopoeias and ethnobiological studies, in most cases little is known regarding harvesting techniques or their impacts on fauna conservation. There is, however, a real need to increase our knowledge about the biology and ecology of the species commonly utilized as remedies to better assess the impacts of this use on their wild populations (Alves et al. 2007). Medicinal species that are threatened should receive urgent attention, and efforts to address habitat loss and degradation should emphasize their present and future medicinal uses (Oliveira et al. 2010).

Educational programs (especially on rural communities) can be used to promote the use of alternatives to wild species in EVM or human complementary medicine practices and should encourage caregivers not only to identify, but also to cultivate locally available species with clinically proven medicinal properties that can cure (or ameliorate the symptoms associated with) common ailments such as parasites, diarrhea, skin conditions, and respiratory problems. The idea of sustainability in traditional medicine can be traced through many different cultures and societies (Azaizeh et al. 2008; Kudi 2003; Saad et al. 2005). Zootherapeutic programs, if properly managed, can and should be compatible with environmental conservation programs in which natural resources are used in such a way that both human needs and biodiversity protection are guaranteed (Alves and Rosa 2006).

### 4.3.4 Future Perspectives

To disseminate and preserve traditional ethnoveterinary knowledge and the uses of animal-based medicines, a number of strategies and actions should be adopted; equally important, research and evaluation systems should be implemented.

Scholarly investigations of traditional knowledge of the medicinal uses of animals and their products are important complementary bodies of knowledge (Lev 2003) and ethnobiological/ethnopharmacological research is pivotal to our understanding of the numerous traditional veterinary practices throughout the world.

A growing respect for traditional knowledge has led modern scientists to adopt procedures for assessing the impact of development projects on biological diversity, ecosystem, species, and specific genetic resource monitoring, invasive alien species control, in situ conservation, and the sustainable management of biological diversity, to mention just a few examples (Convenção sobre Diversidade Biológica 2008). The use of animals for medicinal purposes is part of a larger body of traditional knowledge that is becoming increasingly more relevant to discussions of conservation biology, public health policies, the sustainable management of natural resources, biological prospecting, and patent registrations (Alves and Rosa 2005; Cooper 2005). Additional documentation will be beneficial to general health care, ecological control and biodiversity conservation, and may provide leads to animal species with other useful medicinal properties. Further research will also be required to assay the bioactive compounds in animal parts that have already been identified as having medicinal properties. Public policies and legislation will also be needed to address the sustainability of local floras, faunas, and other relevant natural resources to protect them from overexploitation while still promoting the use of proven traditional medicines (Mathias and McCorkle 2004).

Ethnoveterinary medicine can offer inexpensive and readily available alternatives to costly imported drugs, especially in developing countries, and combinations of modern and local remedies and management practices might be more effective for treating some diseases than current techniques, and in-depth studies will be needed to determine how the economic potential of EVM can be best utilized (Mathias 2001). Mathias (2007) noted that two points are particularly important in tapping the potential of EVM: (1) Respect for intellectual property rights, so that local people—the sources of this knowledge—are recognized and appropriately compensated and (2) The conservation of endangered species.

The available literature on zootherapeutic practices in EVM is still incipient and there is a real need for further research in this area focusing on the cultural, socioeconomic, and ecological factors associated with biological resource uses in EVM that can define workable strategies for cultural maintenance and the sustainable use of the local fauna and flora.

## References

Abate T, Van Huis A, Ampofo JKO (2000) Pest management strategies in traditional agriculture: an African perspective. Annu Rev Entomol 45(1):631–659

Adolph D, Blakeway S, Linquist BJ (1996) Operation lifeline Sudan/Southern sector livestock programme, 1st edn. UNICEF OLS/SS Livestock Programme, Nairobi

Aiyadurai A (2007) Hunting in a biodiversity hotspot: a survey on hunting practices by indigenous communities in Arunachal Pradesh, North-east India, 1st edn. The Rufford Small Grants Foundation, UK, Mysore, India
Akerreta S, Calvo MI, Cavero RY (2010) Ethnoveterinary knowledge in Navarra (Iberian Peninsula). J Ethnopharmacol 130(2):369–378. doi:10.1016/j.jep.2010.05.023
Albuquerque UP, Monteiro JM, Ramos MA, Amorim ELC (2007) Medicinal and magic plants from a public market in Northeastern Brazil. J Ethnopharmacol 110:76–91. doi:10.1016/j.jep.2006.09.010
Almeida AV (2005) Prescrições zooterápicas indígenas brasileiras nas obras de Guilherme Piso (1611–1679). In: Alves AGC, Lucena RFP, Albuquerque UP (eds) Atualidades em Etnobiologia e Etnoecologia, 1st edn. Sociedade Brasileira de Etnobiologia e Etnoecologia, Nuppea, Recife, pp 47–60
Almeida KS, Freitas FLC, Pereira TFC (2006) Etnoveterinária: A Fitoterapia na visão do futuro profissional veterinário. Revista Verde 1(1):67–74
Alves RRN (2008) Animal-based remedies as complementary medicine in Brazil. Forsch Komplementmed 15:226–227. doi:10.1159/000144178
Alves RRN (2009) Fauna used in popular medicine in Northeast Brazil. J Ethnobiol Ethnomed 5(1):1–30. doi:10.1186/1746-4269-5-1
Alves RRN, Rosa IL (2005) Why study the use of animal products in traditional medicines? J Ethnobiol Ethnomed 1(5):1–5. doi:10.1186/1746-4269-1
Alves RRN, Rosa IL (2006) From cnidarians to mammals: the use of animals as remedies in fishing communities in NE Brazil. J Ethnopharmacol 107:259–276
Alves RRN, Rosa IL (2007a) Zootherapeutic practices among fishing communities in North and Northeast Brazil: a comparison. J Ethnopharmacol 111:82–103. doi:10.1016/j.jep.2006.10.033
Alves RRN, Rosa IML (2007b) Biodiversity, traditional medicine and public health: where do they meet? J Ethnobiol Ethnomed 3(14):9. doi:10.1186/1746-4269-3-14
Alves RRN, Santana GG (2008) Use and commercialization of podocnemis expansa (Schweiger 1812) (testudines: podocnemididae) for medicinal purposes in two communities in North of Brazil. J Ethnobiol Ethnomed 4(3):6. doi:10.1186/1746-4269-4-3
Alves RRN, Rosa IL, Santana GG (2007) The role of animal-derived remedies as complementary medicine in Brazil. Bioscience 57(11):949–955
Alves RRN, Neto NAL, Brooks SE, Albuquerque UP (2009) Commercialization of animal-derived remedies as complementary medicine in the semi-arid region of Northeastern Brazil. J Ethnopharmacol 124(3):600–608
Alves RRN, Barboza RRD, Souto WMS (2010a) Plants used in animal health care in south and Latin America: an overview. In: Katerere RD, Luseba D (eds) Ethnoveterinary botanical medicine: herbal medicines for animal health, 1st edn. CRC Press, New York, pp 231–256
Alves RRN, Souto WMS, Mourão JS (2010a) A Etnozoologia no Brasil: Importância, Status atual e Perspectivas, vol 7. Estudos & Avanços, 1 edn. NUPEEA, Recife
Alves RRN, Barbosa JAA, Santos SLDX, Souto WMS, Barboza RRD (2011) Animal-based remedies as complementary medicines in the semi-arid region of Northeastern Brazil. Evidence-based complement Altern Med 2011(ID 179876):1–15. doi:10.1093/ecam/nep134
Antoine-Moussiaux N, Faye B, Vias GF (2007) Tuareg ethnoveterinary treatments of camel diseases in Agadez area (Niger). Trop Anim Health Prod 39:83–89. doi:10.1007/s11250-007-4404-1
Ashwell D, Walston N (2008) An overview of the use and trade of plants and animals in traditional medicine systems in Cambodia, 1st edn. TRAFFIC Southeast Asia, Greater Mekong Programme, Ha Noi, Vietnam
Azaizeh H, Saad B, Cooper E, Said O (2008) Traditional Arabic and Islamic medicine, a re-emerging health aid. eCAM:nen039. doi:10.1093/ecam/nen039
Backes M (1998) Indigenous trees in Bukusuland: a selection of useful trees and shrubs for social forestry in Bungoma, 1st edn. Excecutive Printing works Limited, Nairobi
Barboza RRD, Souto WMS, Mourão JS (2007) The use of zootherapeutics in folk veterinary medicine in the district of Cubati, Paraíba State Brazil. J Ethnobiol Ethnomed 3(32):14. doi:10.1186/1746-4269-3-32

Barboza RRD, Alves RRN, Souto WMS, Mourão JS (2010) Etnoveterinária: o conhecimento milenar que cura e trata os animais. In: Costa-Neto EM, Alves RRN (eds) Zooterapia: OS Animais na Medicina Popular Brasileira, vol 2, 1st edn. NUPEEA, Recifel, pp 103–124
Begossi A, Silvano RAM, Amaral BD, Oyakama OT (1999) Uses of fish and game by inhabitants of an extrative reserve (upper Juruá, Acre, Brazil). Environ Dev Sustain 1:73–93
Bird Life International (2010) *Neophron percnopterus*—IUCN Red List of Threatened Species. IUCN. http://www.iucnredlist.org/apps/redlist/details/144347/0. Accessed 14 Dec 2010
Bisby FA, Roskov YR, Orrell TM, Nicolson D, Paglinawan LE, Bailly N, Kirk PM, Bourgoin T, Baillargeon G, Ouvrard D (2010) Catalogue of life 2010. Species 2000 & ITIS—the integrated taxonomic information system. http://www.catalogueoflife.org/. Accessed 12 Aug 2010
Bizimana N (1994) Traditional veterinary practice in Africa, 1st edn. Deutsche Gesellschaft für Technische Zusammenarbeit, Eschborn
Branch L, Silva MF (1983) Folk medicine in Alter do Chão, Pará, Brasil. Acta Amazônica 13:737–797
China National Corporation of Traditional and Herbal Medicine (1995) Materia medica commonly used in China, 1st edn. Science Press, Beijing
CITES (2008) CITES appendix. http://www.cites.org/eng/resources/species.html
Confessor M, Mendonca L, Mourao J, Alves R (2009) Animals to heal animals: ethnoveterinary practices in semi-arid region, Northeastern Brazil. J Ethnobiol Ethnomed 5(1):37
Convenção sobre Diversidade Biológica (2008) Convenção sobre Diversidade Biológica ONU. http://www.onu-brasil.org.br/doc_cdb2.php. Accessed 12 June 2009
Cooper EL (2005) Bioprospecting: a CAM frontier. Evidence-based Complement Altern Med 2(1):1–3
Costa-Neto EM (1999a) Barata é um santo remédio : introdução à zooterapia popular no estado da Bahia, 1st edn. EdUEFS, Feira de Santana
Costa-Neto EM (1999b) Recursos animais utilizados na medicina tradicional dos índios Pankararé que habitam o Nordeste do Estado da Bahia Brasil. Actualidades Biologicas 21(70):69–79
Costa-Neto EM (2001) A cultura pesqueira do Litoral Norte da Bahia: etnoictiologia, desenvolvimento e sustentabilidade, vol 1, 1st edn. EDUFBA/EDUFAL, Maceío
Costa-Neto EM, Alves RRN (2010) Zooterapia: Os Animais na Medicina Popular Brasileira, vol 2. Estudos & Avanços, 1 edn. NUPEEA, Recife
Costa-Neto EM, Ramos-Elorduy J, Pino JM (2006) Los insectos medicinales de Brasil: Primeros resultados. Boletín Sociedad Entomológica Aragonesa 38:395–414
Davis DK, Quraishi K, Sherman D, Sollod A, Stem C (1995) Ethnoveterinary medicine in Afghanistan: an overview of indigenous animal health care among Pashtun Koochi nomads. J Arid Environ 31:483–500
Dold AP, Cocks ML (2001) Traditional veterinary medicine in the Alice district of the Eastern Cape Province, South Africa. S Afr J Sci 97:375–379
Doyen-Higuet A-M (1984) The "hippiatrica" and byzantine veterinary medicine. Dumbarton Oaks Papers 38:111–120
El-Kamali HH (2000) Folk medicinal use of some animal products in Central Sudan. J Ethnopharmacol 72:279–282
Elliott J, Grahn R, Sriskanthan G, Arnold C (2002) Wildlife and poverty study, 1st edn. DFID Rural Livelihoods Department, London
FAO (2008) Wildlife and protected area management. FAO. http://www.fao.org/forestry/wildlife/44520/en/. Accessed 31 May 2011
Faria EB, Sá AVV, Rotondano TEF, Rodrigues OG, Athayde ACR (2005) Difusão da etnoveterinária como alternativa para o controle das verminoses de caprinos e ovinos do Alto Piranhas, PB. In: UFRJ (ed) VIII Congresso Ibero-Americano de Extensão Universitária, Rio de Janeiro, Brasil, Novembro, 2005. UFRJ, pp 610–615
Faure JC (1944) Pentatomid bugs as human food. J Entomol Soc S Afr 8:110–114
Ferreira FS, Brito S, Ribeiro S, Almeida W, Alves RRN (2009) Zootherapeutics utilized by residents of the community Poco Dantas, Crato-CE Brazil. J Ethnobiol Ethnomed 5(1):21

Froemming S (2006) Traditional use of the Andean flicker (Colaptes rupicola) as a galactagogue in the Peruvian Andes. J Ethnobiol Ethnomed 2(23):1–14. doi:10.1186/1746-4269-2-23
Ghirotti M (1996) Recourse to traditional versus modern medicine for cattle and people in Sidama, Ethiopia. In: McCorkle CM, Mathias E, van Veen TWS (eds) Ethnoveterinary research and development, 1st edn. Intermediate Technology Publications, London
IUCN (2010) IUCN Red list of threatened species. http://www.iucnredlist.org
Kayne SB (2010) Traditional medicine: a global perspective, 1st edn. Pharmaceutical Press, London, UK/Chicago
Köhler-Rollefson I, Mundy P, Mathias E (2001) A field manual of camel diseases: traditional and modern health care for the dromedary, 1st edn. ITDG Publishing, London
Kokwaro JO (2009) Medicinal plants of East Africa, 3rd edn. University of Nairobi Press, Nairobi
Kudi CA (2003)Ethno-veterinary, complementary and low cost treatment and management of working animals. In: TAWS workshop held, Silsoe, UK, 24 April 2003, Silsoe Research Institute, p 10
Lans C (2007) Ethnomedicines used in Trinidad and Tobago for reproductive problems. J Ethnobiol Ethnomed 3(13):12. doi:10.1186/1746-4269-3-13
Lans C, Turner N, Brauer G, Lourenco G, Georges K (2006) Ethnoveterinary medicines used for horses in Trinidad and in British Columbia Canada. J Ethnobiol Ethnomed 2(31):20. doi:10.1186/1746-4269-2-31
Lans C, Georges K, Brown G (2007a) Non-experimental validation of ethnoveterinary plants and indigenous knowledge used for backyard pigs and chickens in Trinidad and Tobago. Trop Anim Health Prod 39:375–385. doi:10.1007/s11250-007-9026-0
Lans C, Turner N, Khan T, Brauer G, Boepple W (2007b) Ethnoveterinary medicines used for ruminants in British Columbia Canada. J Ethnobiol Ethnomed 3(11):22. doi:10.1186/1746-4269-3-11
Larrat R (1939) Mèdicine et hygiène Indigènes. Prèvention technique. Bulletin de Service Zootechniques et des Epizooties de l'AOF 2:17–19
Lenko K, Papavero N (1996) Os insetos no folclore, 2nd edn. Editora Plêiade & FAPESP, São Paulo
Léo Neto NA, Brooks SE, Alves RRN (2009) From Eshu to Obatala: animals used in sacrificial rituals at Candomble "terreiros" in Brazil. J Ethnobiol Ethnomed 5(1):1–23
Lev E (2003) Traditional healing with animals (zootherapy): medieval to present-day Levantine practice. J Ethnopharmacol 85:107–118. doi:10.1016/S0378-8741(02)00377-X
Mahawar MM, Jaroli DP (2008) Traditional zootherapeutic studies in India: a review. J Ethnobiol Ethnomed 4(1):17
Mares RG (1954) Animal husbandry, animal industry and animal disease in the Somaliland protectorate, Part II. Br Vet J 110:411–423
Martin M, McCorkle MC, Mathias E (2001) Ethnoveterinary medicine. An annotated bibliography of community animal healthcare, 1 edn. ITDG Publishing, Londres, Inglaterra
Marx W (1984) Traditionelle tierärztliche heilmethoden unter besonderer berücksichtigung der kauterisation in Somalie. Giessener Beiträge zur Entwicklungsforschung, Reihe 1:111–116
Mathias E (1994) Magic, myth and medicine. Econ Bot 48(1):3–7
Mathias E (2001) Introducing ethnoveterinary medicine. http://www.ethnovetweb.com/whatisevm.pdf. Accessed 20 Dec 2007
Mathias E (2004) Ethnoveterinary medicine: harnessing its potential. Vet Bull 74(8):27N–37N
Mathias E (2007) Ethnoveterinary medicine in the era of evidence-based medicine: mumbo-jumbo, or a valuable resource? Vet J 173:241–242. doi:10.1016/j.tvjl.2005.12.005
Mathias E, McCorkle CM (2004) Traditional livestock healers. Rev Sci Tech Off Int Epiz 23(1):277–284
Mathias-Mundy E (1989) Of herbs and healers. Leisa magazine, vol 5. ILEIA
Mathias-Mundy E, McCorkle MC (1989) Ethnoveterinary medicine: an annotated bibliography. Bibliographies in technology and social change, 1 edn. Technology and Social Change Programme, IOWA State University, Ames

McCorkle CM (1986) An introduction to ethnoveterinary research and development. J Ethnobiol 6(1):129–149
McCorkle CM, Martin M (1998) Parallels and potentials in animal and human ethnomedical technique. Agric Hum Values 15:139–144
McGaw LJ, Van der Merwe D, Eloff JN (2007) In vitro anthelmintic, antibacterial and cytotoxic effects of extracts from plants used in South African ethnoveterinary medicine. Vet J 173:366–372. doi:10.1016/j.tvjl.2005.09.004
McKean S (2007) Vultures and traditional medicine. http://www.andersonafrica.co.za/pamphlets/Vultures_and_traditional_medicine.pdf. Accessed 18 June 2011
Muhammad G, Khan MZ, Hussain MH, Iqbal Z, Iqbal M, Athar M (2005) Ethnoveterinary practices of owners of pneumatic-cart pulling camels in Faisalabad city (Pakistan). J Ethnopharmacol 97:241–246. doi:10.1016/j.jep.2004.11.008
Nooren H, Claridge G (2001) Wildlife trade in Laos: the end of the game, 1st edn. Netherlands Committee for IUCN, Amsterdam
Nyamanga PA, Suda C, Aagaard-Hansen J (2008) The socio-cultural context and practical implications of ethnoveterinary medical pluralism in western Kenya. Agric Hum Values 25:513–527. doi:10.1007/s10460-008-9141-1
Oliveira ES, Torres DF, Brooks SE, Alves RRN (2010) The medicinal animal markets in the metropolitan region of Natal city Northeastern Brazil. J Ethnopharmacol 130(1):54–60
Orwa JA (2002) Editorial: herbal medicine in Kenya: evidence safety and efficacy. East Afr Med J 79:341–342
Passalacqua NG, De Fine G, Guarrera PM (2006) Contribution to the veterinary science and of the ethnobotany in Calabria region (Southern Italy). J Ethnobiol Ethnomed 2:1–52. doi:10.1186/1746-4269-2-52
Pieroni A, Price LL (2005) Eating and healing: traditional food as medicine, 1st edn. Food Products Press, Nova Iorque
Pieroni A, Howard P, Volpato G, Santoro RF (2004) Natural remedies and nutraceuticals used in ethnoveterinary practices in Inland Southern Italy. Vet Res Commun 28:55–80
Pieroni A, Giusti ME, Pasquale C, Lenzarini C, Censorii E, Gonzáles-Tejero MR, Sánchez-Rojas CP, Ramiro-Gutiérrez JM, Skoula M, Johnson C, Sarpaki A, Della A, Paraskeva-Hadijchambi D, Hadjichambis A, Hmamouchi M, El-Jorhi S, El-Demerdash M, El-Zayat M, Al-Shahaby O, Houmani Z, Scherazed M (2006) Circum-Mediterranean cultural heritage and medicinal plant uses in traditional animal healthcare: a field survey in eight selected areas within the RUBIA project. J Ethnobiol Ethnomed 2(16):12. doi:10.1186/1746-4269-2-16
Quave CL, Pieroni A, Bennett B (2008) Dermatological remedies in the traditional pharmacopoeia of Vulture-Alto Bradano, Inland Southern Italy. J Ethnobiol Ethnomed 4(1):1–5. doi:10.1186/1746-4269-4-5
Quave CL, Lohani U, Verde A, Fajardo J, Rivera D, Obón C, Valdes A, Pieroni A (2010) A comparative assessment of zootherapeutic remedies from selected areas in Albania, Italy Spain and Nepal. J Ethnobiol 30(1):92–125
Raziq A, de Verdier K, Younas M (2010) Ethnoveterinary treatments by dromedary camel herders in the Suleiman Mountainous Region in Pakistan: an observation and questionnaire study. J Ethnobiol Ethnomed 6(1):1–16. doi:10.1186/1746-4269-6-16
Ribeiro GC, Pereira JPR, Docio L, Alarcon DT, Schiavetti A (2010) Zooterápicos utilizados no sul da Bahia. In: Costa-Neto EM, Alves RRN (eds) Zooterapia: Os Animais na Medicina Popular Brasileira, vol 2, 1st edn. NUPEEA, Recife, pp 221–242
Saad B, Azaizeh H, Said O (2005) Tradition and perspectives of Arab herbal medicine: a review. eCAM 2(4):475–479. doi:10.1093/ecam/neh133
Santos IB, Costa-Neto EM (2007) Estudo Etnoornitológico em uma região do semi-árido do Estado da Bahia. Sitientibus Série Ciências Biológicas 7(3):273–288
Scarpa GF (2000) Plants employed in traditional veterinary medicine by the criollos of the Northwestern Argentine Chaco. Darwiniana 38(3–4):253–265
Scarpa GF (2004) Medicinal plants used by the Criollos of Northwestern Argentine Chaco. J Ethnopharmacol 91(1):115–135

Schiffeler JW (1976) The origin of Chinese folk medicine. Asian Folklore Stud 35(1):17–35

Schinkel HG (1970) Haltung, Zucht und Pflege des Viehs bei den Nomaden Ost-und Nordostafrikas: Ein Beitr. Z. Traditionellen Ökonomie d. Wanderhirten in semiariden Gebieten, 1 edn. Akademie-Verlag, Berlin

Schoen AM, Wynn SG (1997) Complementary and alternative veterinary medicine, 1st edn. Mosby, St. Louis

Schwabe CW (1978) Cattle, priests and progress in medicine, 1st edn. University of Minnesota Press, Minneapolis

Schwabe CW (1984) Veterinary medicine and human health, 3rd edn. Williams and Wilkins, Baltimore

Shen S, Qian J, Ren J (2010) Ethnoveterinary plant remedies used by Nu people in NW Yunnan of China. J Ethnobiol Ethnomed 6(1):1–24. doi:10.1186/1746-4269-6-24

Shepherd CR (2001) Observations on wildlife trade at Golden Rock, Myanmar. TRAFFIC Bull 19:7–10

Somvanshi R (2006) Veterinary medicine and animal keeping in Ancient India. Asian Agri-Hist 10(2):133–146

Song M-J, Kim H (2010) Analysis of ethnoveterinary treatments for cattle (Bos indicus) diseases referred in Sanrimgyeongje including twelve volumes of literature from the 7th to the 18th century. J Ethnopharmacol 133(2):474–479

Souto WMS (2009) Animais de uso etnoveterinário no semi-árido paraibano: implicações para conservação e sustentabilidade. Dissertação (Mestrado), Universidade Federal da Paraíba/ Universidade Estadual da Paraíba, Campina Grande, Paraíba

Souto WMS, Barboza RRD, Mourão JS, Alves RRN (2009) Zootherapy in Brazil: an urgent necessity of interdisciplinary studies. West Indian Med J 58(5):494–495

Souto WMS, Alves RRN, Confessor MVA, Barboza RRD, Mourão JS, Mendonça LET (2010a) A Zooterapia na Etnoveterinária do semi-árido paraibano. In: Alves RRN, Souto WMS, Mourão JS (eds) A Etnozoologia no Brasil: Importância, Status atual e Perspectivas, vol 7, 1st edn. NUPEEA, Recife, pp 423–446

Souto WMS, Mourão JS, Barboza RRD, Rocha MSP, Alves RRN (2012) Animal-based medicines used in ethnoveterinary practices in the semi-arid region of Northeastern Brazil. Anais da Academia Brasileira de Ciências 84:669–678

Souto WMS, Mourão JS, Barboza RRD, Alves RRN (2011) Parallels between zootherapeutic practices in ethnoveterinary and human complementary medicine in NE Brazil. J Ethnopharmacol 134:753–767. doi:10.1016/j.jep.2011.01.041

Tabuti JRS, Dhillion SS, Lye KA (2003) Ethnoveterinary medicines for cattle (Bos indicus) in Bulamogi county, Uganda: plant species and mode of use. J Ethnopharmacol 88:279–286. doi:10.1016/S0378-8741(03)00265-4

Thomas KJ, Nicholl JP, Coleman P (2001) Use and expenditure on complementary medicine in England: a population based survey. Complement Ther Med 9(1):2–11. doi:10.1054/ ctim.2000.0407

Unnikrisnhan PM (1998) Animals in ayurveda. Amruth 1:1–15

Valdizán H, Maldonado A (1922) La medicina popular peruana: contribución al folklore médico del Perú, 1st edn. Imprenta Torres Aguirre, Lima

van Veen TWS (1998) One medicine: The dynamic relationship between animal and human medicine in history and at present. Agric Hum Values 15:115–120

Viegi L, Pieroni A, Guarrera PM, Vangelisti R (2003) A review of plants used in folk veterinary medicine in Italy as basis for a databank. J Ethnopharmacol 89:221–244

Wanzala W, Zessin KH, Kyule NM, Baumann MPO, Mathias E, Hassanali A (2005) Ethnoveterinary medicine: a critical review of its evolution, perception, understanding and the way forward. Livestock Res Rural Dev 17(11):1–31

Whiting MJ, Williams VL, Hibbitts TJ (2011) Animals traded for traditional medicine at the Faraday market in South Africa: species diversity and conservation implications. J Zool 284(2):84–96

Xie H, Preast V (2010) Xie's Chinese veterinary herbology, 1 edn. Wiley-Blackwell, Ames

# Chapter 5
# Animals as a Source of Drugs: Bioprospecting and Biodiversity Conservation

**Rômulo Romeu Nóbrega Alves and Ulysses Paulino Albuquerque**

**Abstract** Plants and animals are undoubtedly the basis of many traditional medicine systems around the world. Although the pharmacological potential of animals used as medicines has been little explored, compared to plants, available studies show that animal natural resources are highly promising in the search for new drugs of medical or pharmaceutical interest. The exploitation of these resources, however, requires a careful strategy that allows the sustainability of the species exploited, since the exploitation of fauna in medicinal bioprospecting can result in overharvesting of target organisms. In view of this reality, economic development associated with animal bioprospecting should be preceded by a broad discussion of the conservation of biodiversity and the sustainable management of natural resources. In this chapter, we review the literature on the potential of animal-based medicines for developing new drugs, and briefly discuss the implications of bioprospecting for the conservation of these bioresources.

## 5.1 Introduction

The identification and evaluation of biological material found in nature to obtain new products is known as bioprospecting (Artuso 2002), a research field that has been defined as the exploration of biodiversity for new biological resources of

R. R. N. Alves (✉)
Ethnozoology, Conservation and Biodiversity Research Group, Departamento de Biologia, Universidade Estadual da Paraíba, Campina Grande, Paraíba 58109-753, Brazil
e-mail: romulo_nobrega@yahoo.com.br

U. P. Albuquerque
Applied Ethnobotany Laboratory, Departamento de Biologia, Universidade Federal Rural de Pernambuco, Dois Irmãos, Pernambuco 52171-900, Brazil

R. R. N. Alves and I. L. Rosa (eds.), *Animals in Traditional Folk Medicine*,
DOI: 10.1007/978-3-642-29026-8_5, © Springer-Verlag Berlin Heidelberg 2013

social and economic value (Beattie et al. 2005). Bioprospecting is carried out by a wide variety of industries, the best known being the pharmaceutical industry, but also by a variety of branches of agriculture, manufacturing, engineering, construction, and many others. Considering that since ancient times humans have had strong dependent or co-dependent interactions with biological resources (Alves and Souto 2010), bioprospecting is not a new activity in human history. Nonetheless, its modern version brings to the scene important aspects of the new economic, social, political, and environmental reality (Ferreira et al. 2010).

In neutral terms, bioprospecting involves the exploration for any biological resource, otherwise termed "biodiversity," for potential commercial use (Pan 2006), and as such it bears ecological implications (Alves and Rosa 2007a; Pan 2006). Obviously, this process involves searching for, identifying, and collecting appropriate biospecimens, and also uses various cutting-edge technologies to process and develop genetic material from these specimens that exhibit characteristics desirable in a commercial product. It is the genetic material, and not the biospecimen itself, that is of interest. Generally then, it would be inefficient, irresponsible, and unnecessary for bioprospectors to collect massive volumes of plants or animals for processing (Pan 2006).

Though the bioprospecting principle may be simple, the interaction between biotechnology use and biodiversity conservation and its sustainability does require a carefully designed strategy to complement other aspects of biodiversity protection and socio-economic development (Sittenfeld et al. 2004). Moreover, because of its nature, bioprospecting is at the intersection of biodiversity conservation and the use of biotechnology and thus has consequences in the areas of legal and regulatory frameworks; technology transfer and business development; intellectual property rights; and facilitation of local, national, and international collaborations (Sittenfeld 1996; Sittenfeld and Lovejoy 1998; Tamayo et al. 1997).

Throughout the ages, humans have relied on nature for their basic needs—for the production of foodstuffs, shelters, clothing, means of transportation, fertilizers, flavors, and fragrances, and last but not least, medicines (Cragg and Newman 2001). In this last case, biodiversity represents an immeasurable source of information and raw materials, which support health systems of different human cultures that depend on nature as a source of medicines to treat and cure illnesses. Plants and animals have been utilized as sources of medications since remote times, and the medicinal use of these resources has been perpetuated; currently, it plays an essential role in health care in traditional medical systems, as well as in the discovery of new drugs (Albuquerque and Hanazaki 2006; Alves and Alves 2011).

Various human cultures have utilized biological resources for therapeutic purposes since ancient times, and this has produced a rich knowledge that has been incorporated into different traditional medical systems around the world (Alves and Rosa 2005, 2007a; Ferreira et al. 2009a; Giday et al. 2007; Heinrich et al. 1998; Lev 2003, 2006; Lev and Amar 2002; Scarpa 2004), and passed on from generation to generation. Thousands of years of observation and experimentation have helped in the development of different empirical medical systems, in which useful elements that have curative potential have been chosen, taxonomies have been determined and different treatments developed to address health issues across all societies (Bhasin 2007).

Human populations represent an enormous reservoir of knowledge of plant and animals drugs (Fitter 1986), as demonstrated in various ethnobotanical and ethnozoological studies (Alves et al. 2007, 2009; Calixto 2005; Confessor et al. 2009; Costa-Neto 2000; Ferreira et al. 2009a; Lev 2003; Parveen et al. 2007; Phillips and Gentry 1993; Uniyal et al. 2006).

During the last 50 years, ethnobiology has emerged as a promising discipline that can play a prominent role in the advancement of many aspects of scientific, sociological, and historical studies, including an increasing number of investigations on medicinal properties and uses of animals and plants, still intact in native cultures in several parts of the world (Padmanabhan and Sujana 2008). Ethnobiological studies in traditional medicine are important not only for the documentation of thousands of medical traditions, contributing to their maintenance, but also for constituting a fundamental step in the discovery of new medications (Alves 2009).

In the specific case of fauna, regrettably, some of the medicinal animals are already included in lists of threatened species (Alves and Rosa 2007a, b; Alves et al. 2007; Ashwell and Walston 2008; Lee 1999; Van and Tap 2008). Unsustainable use of plants and animals in traditional medicine is recognized as one menace to wildlife conservation. The linkage between species endangerment and the demand from traditional medicine has raised much concern among the conservation community worldwide (Alves et al. 2010a, b; Alves and Rosa 2005, 2007a; Ellis 2005; Sodeinde and Soewu 1999; Yinfeng et al. 1997). As pointed out by Marques (1997), considerations of negative impacts on biodiversity should not be limited to the traditional (folk) use of animal or animal-derived products, but must also extend to their exploitation by the pharmaceutical industry (Melo et al. 2009). In Brazil, for example, most plants used industrially for manufacturing phytotherapeutic drugs or herbal medicines are obtained directly from natural stocks through extractivist practices. And if the situation for plants (the resource most studied by scientists under the perspective focused in this chapter) is delicate, the scenario becomes even more problematic when considering animals.

Pan (2006) highlighted that the lack of conservation effort and unsustainable use of biodiversity is a major reservation many have concerning bioprospecting, and drew attention to distinguish pharmaceutical bioprospecting from phytomedicine. In his view, while the general belief that pharmaceutical bioprospecting invariably requires high-volume harvesting of whole organisms for final consumption is not entirely accurate, phytomedicine requires high-volume harvesting, although this does not mean that only a slash and harvest approach must be used (Pan 2006).

The strategic value of natural resources has attracted the attention of so-called bioprospectors, especially in the tropics. The chemical constituents and pharmacological actions of some animal products are already known to some extent, and ethnopharmacological studies focusing on animal remedies could be very important in order to clarify the eventual therapeutic usefulness of this class of biological remedies (Pieroni et al. 2002). Studies on traditional uses of animal-derived resources, however, must be coupled to considerations of conservation biology and public health policies. Sustainable management of natural resources targeted by bioprospecting is of utmost importance (Ferreira et al. 2009b), and in this sense we focused this chapter on the interfaces between the uses of animals as folk remedies in the context of the development of new drugs, and conservation of those bioresources. Our review of the current literature on bioprospecting was guided by the following questions: What is the potential of fauna in medicinal bioprospecting? Which are the principal taxa currently exploited? What are the implications of such uses of the fauna for conservation? What is the role of traditional medical knowledge in animal bioprospecting?

## 5.2 Discovery of New Drugs and the Importance of Knowledge of Traditional Medicine

Traditional medicine has been utilized for a long time for various disorders, and recently Mukherjee et al. (2010) remarked that "now the time has come to use these resources as complementary therapeutics based on scientific validations". A great number of natural products have come to us from the scientific study of remedies traditionally employed by various cultures (Farnsworth and Bingel 1997), and such knowledge is very important in accessing the resources of biodiversity, through providing a short list of potential sources and preventing the need for comprehensive surveys and screening of all available flora and fauna (Alves and Rosa 2006; Soejarto 1996). In recent years, a number of studies have highlighted that knowledge about the utilization of biological resources on the part of traditional communities is an important ally in the investigation of new "active principles," to obtain medications, cosmetics, and other products for the food industry (Demunshi and Chugh 2010; Ferreira et al. 2010; Makhubu 1998; McClatchey 2005). Therefore, the practice of bioprospecting is intimately related to traditional knowledge, particularly from a pharmaceutical perspective.

The traditional use of natural resources may indicate the presence of biologically active components, as historic and maintained use of naturally occurring compounds often has a scientific underpinning (Soejarto 1996). In fact, the World Health Organization recognizes in the biological resources utilized in traditional medicine an important instrument in the development of new pharmaceutical products to fight diseases that distress populations of developing countries (WHO 2002). The number of in vitro studies and clinical trials of traditional or "folk" remedies has been growing rapidly; however, the development of new products based on these studies has unfortunately been almost insignificant. Among the reasons for this, we include: (1) the large number of studies that carry out screenings of many plants that do not show continuity and (2) problems in the experimental design of many studies, especially those based on plant extracts, which overestimate the biological activity found (Gertsch 2009; Houghton et al. 2007).

Overall, however, the results suggest a relatively high degree of efficacy for indigenous pharmacopeia, although some individual remedies seem ineffective or toxic. Nevertheless, since comparatively little attention has been devoted to the *prescription* of traditional remedies, this efficacy has probably been underestimated (Barsh 1997). If appropriately studied, this knowledge may open perspectives for the discovery of new sources of remedies for human well-being, in addition to making it possible to give economic value to animal species otherwise considered harmful and/or useless (Costa-Neto and Alves 2010). The apparently absent relevant activity may also be due to wrong experimental design that does not respect the way medicines are traditionally prepared in a given culture. To discover new sources of drugs from popular information, researchers trained in the techniques and methods of Western science need to abandon an ethnocentric view when exploring knowledge of traditional medicine, investigating the laboratory species indicated as medicines to determine whether compounds of pharmacological value are in fact present.

The evident relation between bioprospecting and traditional knowledge raises an important question with respect to the rights to intellectual properties of the people holding knowledge of traditional medicine. Recently, concern about the ethics of exploiting indigenous knowledge and resources from tropical countries, without sharing the benefits with those who are the traditional custodians of the knowledge and land, has gained attention through the convention on biological diversity (CBD), which requires that such knowledge be protected, respected, and preserved (http://www.biodiv.org/; article 8(j)) and related provisions). Traditional knowledge related to traditional medicine may be protected and conserved through the development of intellectual property rights (IPRs) and/or through benefit sharing (Moran et al. 2001; Zhang and Cheng 2000). Intellectual property rights are rights over intangible information that provides incentives for future innovations (Moran et al. 2001). The compensation of indigenous people can validate their knowledge of the biodiversity they manage and also provide them with an equitable reward for sharing it, thereby compensating biological stewardship and encouraging conservation (see Moran et al. 2001; Svarstad 2000).

Pharmaceutical bioprospecting has been sharply criticized for what has become known as ‘biopiracy’ (Mgbeoji 2006), in which large international pharmaceutical corporations make use of local medicinal knowledge without acknowledging that it is indigenous intellectual property. Thus, profits have accrued solely for the pharmaceutical companies, while indigenous peoples have received little or nothing in return (Finger and Schuler 2004; Laird 2002). The recent history of bioprospecting in Brazil reveals conflicts between bioprospectors and other stakeholders (Beattie et al. 2011). In addition, legislation to protect traditional knowledge from inappropriate use by the private industry has been blamed for stifling systematic and ecological research (Antonelli and Rodriguez 2009).

## 5.3 Animals as a Source of Bioactive Compounds and Drugs

In comparison to research done on plant drugs, modern pharmacology has conducted far fewer studies on animal products and minerals (Paavilainen 2009). Nevertheless, in the previous years, increasing attention has been paid to animals, both vertebrates and invertebrates, as sources of new medicines (Chivian 2002a).

### *5.3.1 Invertebrates*

Invertebrates comprise the majority of animal groups, and their actual and potential medicinal benefits have been highlighted in different studies (Dossey 2010; Haefner 2003; Kellert 1993). Among those that inhabit terrestrial environments, arthropods have drawn attention due to their potential as a source of medicinally relevant substances, mainly the class Insecta (Dossey 2010). Oldfield (1989) states that about 4% of the extracts evaluated in the 1970s from 800 species of terrestrial arthropods (insects included) showed some anticancer activity. Nonetheless, in face of the vast biodiversity that exists in the arthropod world, compared to all other organisms on earth, arthropods should receive a more serious look (Dossey 2010).

Insects are the most diverse group of animals on the planet and include more than a million described species (Chapman 2009). Whole insects and other arthropods, as well as substances extracted from them have been used worldwide as medicinal resources by human cultures (Alves 2009; Alves and Alves 2011; Alves et al. 2007, 2009; Costa-Neto 2005). Their reported properties include immunological, analgesic, anti-bacterial, anti-coagulant, anticancer, diuretic, anesthetic, and anti-rheumatic (Ahn et al. 2006, 2000; Costa-Neto 2005; Dettner 2011; Yamakawa 1998).

Only a small fraction of insect species on earth have been analyzed chemically or explored for the potential presence of medicinally relevant substances (Dossey 2010). In fact, little or nothing is known about the biosynthetic mechanisms or greater ecological significance of these chemicals in all but a handful of species (Morgan 2010). Thus, studying the biosynthesis of various insect chemicals is likely to produce a wealth of useful information with applications in the fields of biochemistry, ecology, and biotechnology (Dossey 2010).

For a variety of reasons, insects and their chemical defense systems present a valuable source of novel chemistry that certainly merits further investigation as source of new medicinal compounds, as well as substances with other applications (Dossey 2010). Insects have proven to be very important as sources of drugs in modern medicine. Chemical screening applied to 14 insect species has confirmed the presence of proteins, terpenoids (triterpenoids and steroids, carotenoids, iridoids, tropolones), sugars, polyols and mucilages, saponins, polyphenolic glycosides, quinones, anthraquinone glycosides, cyanogenic glycosides, and alkaloids (Andary et al. 1996). Kunin and Lawton (1996) have reported that promising anticancer drugs have been isolated from the wings of Asian sulfur butterflies (*Catopsilia crocale*, Pieridae) and from the legs of Taiwanese stag beetles (*Allomyrina dichotomus*, Scarabaeidae)—isoxanthopterin and dichostatin, respectively.

Various studies have reported the presence of antimicrobial substances in insects (Bankova et al. 1999; Bulet and Stocklin 2005; Efem and Iwara 1992; Koru et al. 2007; Kujumgiev et al. 1999; Slocinska et al. 2008), many of which likely function as a defense against microbial attack and infection. It has been shown that in the various steps a microbe must take in order to infect an insect, from the cuticle to the hemocele (body cavity), it encounters a large and varied assortment of antimicrobial substances such as lipids, hydro carbons, diphenols, carbohydrates, melanin, and even the insect's own chitin exoskeleton (Silva 2002). One major example of such substances are insect antimicrobial peptides (e.g. Boman 1995; Bulet and Stocklin 2005; Hultmark et al. 1980; Meylaers et al. 2004; Slocinska et al. 2008). In fact, cecropins were the first antimicrobial peptides discovered (Boman 2000; Boman et al. 1991). Defensins, another common class of insect antimicrobial peptides, are found in a wide variety of organisms from plants to insects to humans, and their biological activity against bacterial and fungal pathogens has been reported (Aerts et al. 2008), while melittin (a major component in the sting venom of honeybees such as the European honeybee, *Apis mellifera;* order Hymenoptera) and cecropins are both effective against Gram-negative bacteria (Wade et al. 1992).

Insect by-products are also very important as potential sources of drugs. A number of laboratory studies, for example, have demonstrated significant antibacterial activity in honey, especially of *Apis* sp. (Molan 1999), and researchers have achieved complete inhibition of important species of infectious bacteria (Willix et al. 1992) using concentrations of honey ranging from 1.8 to 11 % (v/v). The antibacterial activity of honey has also been demonstrated in vivo, with reports of infected wounds dressed with honey becoming sterile in 3–6 days (Branicki 1981; Cavanagh et al. 1970), 7 days (Efem 1993), and 7–10 days (Armon 1980). A recent study found that the honey of *Melipona subnitida* reduces the healing time of wounds, and was effective in eliminating Gram-negative bacteria such as *Pseudomonas* sp., *Klebsiella* sp., *Escherichia coli* and *Acinetobacter* sp. (Sampaio Alves et al. 2008). Park et al. (2000) found anticancer and anti-HIV activities in ethanolic extracts of propolis from *Apis mellifera* collected in different parts of Brazil.

As modern drug discovery efforts move forward, natural products will continue to play a vital role in supplying medicine with chemical compounds that would otherwise be impossible for the human chemist to fathom (Dossey 2010). With modern technologies to analyze and assay ever-smaller amounts of material, it is important that previously neglected taxa and natural matrices are capitalized upon. Clearly, among these are insects and other arthropods, which possess one of the richest and most unexplored reservoirs of potentially useful substances.

From toxins used to defend against attack by predators and other offending opponents to peptides which help to ward off infection by various microbes and other parasites, insects and their defense chemicals hold great promise for the future of natural products drug discovery (Dossey 2010).

Although terrestrial organisms exhibit great species diversity, marine organisms have greater phylogenetic diversity. In fact, thousands of marine species have never been studied and the potential of these marine organisms as sources of medicines has created a renaissance of interest in exploring the deep ocean (Seedhouse 2010).

In the marine environment, various groups have been indicated as sources of natural products. Some 500 marine invertebrates, for example, have been found to possess anticancer properties, and numerous heart, bone, urogenital, and other diseases have been identified as amenable to the biochemical properties of diverse invertebrates (Myers 1979). Approximately 2500 new metabolites were reported from a variety of marine organisms during the decade of 1977–1987 (Ireland et al. 1993). Already, more than 15,000 natural products have been discovered, and this number continues to grow. While bioprospecting and deep-ocean exploration are in their infancy, the novel biology of the organisms discovered to date and their potential for revolutionizing the medical realm means that scientific interest will be increasingly focused on realizing the potential that exists in the deep ocean. And, inevitably, as a growing body of scientific reaffirms that deep-sea biodiversity holds major promise for the treatment of human diseases, exploration will surely venture ever deeper in search of untapped resources (Fusetani 2000; Seedhouse 2010).

Over the past few decades significant efforts have been made, by both pharmaceutical companies and academic institutions, to isolate and identify new marine-derived, natural products (Amador et al. 2003; Kijjoa and Sawangwong 2004; Thomas et al. 2010). As a consequence, in recent years, there has been substantial progress toward developing new pharmaceuticals from marine organisms, particularly invertebrates such as sponges, cnidarians, molluscs, bryozoans, tunicates, etc. (Thakur and Müller 2004). They not only produce a great number of marine natural products currently known, but also show the largest chemical diversity of natural products, including alkaloids, peptides, terpenes, polyketides, etc. Interestingly, of the 13 marine natural products (or analogs derived from them) that are currently in clinical trials as new drug candidates, 12 are derived from invertebrates (Thomas et al. 2010).

Among the marine invertebrates studied, Porifera (sponges) remains the most prolific phylum with regard to novel pharmacologically active compounds (Thakur et al. 2005). Since sponges are simple and sessile organisms, during evolution they have developed potent chemical defensive mechanisms to protect themselves from competitors and predators as well as infectious microorganisms (Thomas et al. 2010). Among metazoans, the phylum Porifera contains the taxa that produce the highest diversity of secondary metabolites (Müller et al. 2003), which play a crucial role in the survival of sponges in the marine ecosystem (Thakur and Müller 2004). The chemical diversity of secondary metabolites isolated from sponges include amino acids, nucleosides, macrolides, porphyrins, terpenoids, aliphatic cyclic peroxides, and sterols (Thakur and Müller 2004).

Because of their prevalence, ease of collection, and ability to synthesize a variety of natural product structural classes, sponges have become one of the dominant sources of biologically active marine natural products (Berlinck et al. 1996; Carté 1996; Ireland et al. 1993). These natural products have interesting biomedical potential, pharmaceutical relevance, and diverse biotechnological applications (Bernan et al. 1997; Haygood et al. 1999; Jensen and Fenical 1994; Lee et al. 2001; Proksch et al. 2003; Thomas et al. 2010), in face of their antiviral, antitumor, antimicrobial, and general cytotoxic properties (Wang 2006). The demonstration of biochemical properties of sponges started with the pioneering work of Bergmann on the extraction of novel bioactive nucleosides from the sponge *Tectitethya crypta* (formerly *Cryptotethya crypta*) (Alvarez et al. 2000), and has been reaffirmed by the discovery of novel compounds with pharmacological properties (Ridley et al. 2005).

Another group of invertebrates with pharmacological potential are the cnidarians (Mariottini and Pane 2010): some bioactive substances were discovered in cnidarians, such as the prostaglandin (15R)-PGA2 in the gorgonian *Plaxaura homomalla* (Weinheimer and Spraggins 1969), and the local anesthetic and vasoconstrictive agent palytoxin in the zoanthid *Palythoa toxica* (Moore and Scheuer 1971). Cytolytic and antitumor substances have also been found, such as prostanoid compounds from the Anthozoan *Clavularia viridis*, which were shown to inhibit the growth of HL-60 leukemic cells (Honda et al. 1985); additionally, the incidence and growth of SNC tumors induced by *N*-ethyl-*N*-nitrosourea were shown to be affected by the crude venom of the scyphozoan *Cassiopea xamachana* (Orduña-Novoa et al. 2003), and the growth of Ehrlich ascites tumors grafted in mice was inhibited by crude extracts of tissues from jellyfish and soft corals (Tabrah et al. 1972). Also, equinatoxin extracted from *Actinia equina* showed antitumor activity in cultured cells (Giraldi et al. 1976); and palytoxin was shown to induce ion currents (channels permeable to Na+ and K+ and slightly permeable to Ca2+, choline and tetramethylammonium) in mouse neuroblastoma cells (Rouzaire-Dubois and Dubois 1990).

Molluscs also feature prominently in a broad range of traditional natural medicines (Alves and Dias 2010; Alves et al. 2007; Costa-Neto 2006), although the active ingredients in the taxa involved are typically unknown. Nonetheless, the potential of molluscs as sources of drugs has been recently recognized in a recent review of the theme published by Benkendorff (2010), who highlighted that some marine gastropods and bivalves have been of great interest to natural products chemists, yielding a diversity of chemical classes and several drug leads currently in clinical trials. Indeed, marine molluscs have become the focus of many chemical studies aimed at isolating and identifying novel natural products, as inferred from the over 1,000 research papers published on molluscan secondary metabolites in the past three decades (Avila 2006). Alam and Thomson (1998) compiled a valuable reference book detailing 585 metabolites isolated from marine mollusks, and as pointed by Benkendorff (2010), there is a need for more targeted research based on specific hypotheses related to molluscan bioactivity and the defense systems of shelled molluscs, to maximize the success rates of future investigations.

### *5.3.2 Vertebrates*

Pharmacological research with vertebrates has been scarce, but the few existing works have shown the potential of animals in this group. The investigation of a variety of antimicrobial peptides, such as magainins, defensins, cathelicidins, and protegrins, generated by vertebrates, has recently become very popular. Cathelicidin-type peptides are a broad range of antimicrobial proteins that have been isolated from rabbits, mice, sheep, and humans. Cathelicidins are composed of two different domains, the cathelin and antimicrobial domains. Purportedly, these materials bind to lipopolysaccharide and neutralize its activity (Spainhour 2005).

Most pharmacological studies on vertebrates have been carried out with amphibians and reptiles, animals that produce venoms and other toxic substances that have been very attractive candidates for drug development (Daly 1998, 2003; Ferreira et al. 2009b, 2011; Santos et al. 2012; Sawyer et al. 1961). Venoms and toxins have found a niche in the pharmaceutical market (Gomes et al. 2010), and several isolated toxins with a known mode of action have practical applications as pharmaceutical agents, diagnostic reagents, or preparative tools (Stocker 1989). Chinese medical companies and traditional drugstores have marketed amphibian parts, such as dried oviducts of the frog *Rana chensinensis* and toad (Bufonid) skin (Jensen and Camp 2003).

One excellent example of successful drug development from a component of snake venom (*Bothrops jararaca* (Wied 1824)) is that of the inhibitors of angiotensin-converting enzyme (ACE). This enzyme is responsible for converting an inactive precursor into the locally active hormone angiotensin, which causes blood vessels to constrict and hence raises blood pressure (Bisset 1991). Another good example is the work initially conducted by Daly during the 1960s on the skin secretions of dendrobatid frogs from Ecuador, and of other "poison dart" frog species in Central and South America. This work has led to the identification of a number of alkaloid toxins that bind to multiple receptors in the membranes of nerve and muscle cells (Chivian 2002b).

Venoms from two Viperidae species (*Bothrops jararaca* and *Crotalus durissus terrificus*) act directly on tumor cells, and their antitumor activity may be due to the indirect phenomenon of inflammatory response mediated by IL2, IL8, and TNF-$\alpha$ (Silva et al. 1996). The anticarcinogenic activities of Indian monocellate cobra and Russell's viper venom were studied in carcinoma, sarcoma, and leukemia models. Under in vivo experiments, it was found that the sublethal doses of the Indian Elapidae (monocellate cobra) and Viperidae (Russell's viper) venom caused cytotoxicity in EAC cells; it increased the life span of EAC-bearing mice and reinforced its antioxidant system (Debnath et al. 2007). Similarly, antitumor activity of Hydrophidae (*Lapemis curtus*) venom was also established against EAC in Swiss albino mice in vivo and HeLa and Hep2 tumor cell cultures in vitro (Karthikeyan et al. 2008).

While lysozymes of the turtles *Trionyx sinensis* Wiegmann, 1835, *Amyda cartilagenea* and *Chelonia mydas* demonstrated a potential bactericidal activity (Thammasirirak et al. 2006), Ferreira et al. (2010) recently evaluated the anti-inflammatory activity of topically administered *Tupinambis merianae* fat in animal models (male and female Swiss mice—*Mus musculus*). In this first experimental test of the anti-inflammatory activity of *T. merianae* fat in vivo models, the authors found that it had significant topical anti-inflammatory activity and reduced inflammation edema in mouse ears caused by croton oil (single and multiple applications), phenol, and arachidonic acid.

Similarly, amphibians have provided compounds capable of offering therapeutic advantage. Many of the compounds contained in granular gland secretions of amphibians have a defensive role (Daly 1998), and their pharmacological effects include cardiotoxic, myotoxic, and neurotoxic activities, and some are vasoconstrictive and hypotensive agents, while others have hallucinogenic effects, all properties that would clearly adversely affect a potential predator (Daly 1998, 2003).

The number and diversity of compounds produced by amphibians in their granular glands is surprisingly high, even within a single species. Dermophin, a novel opioid heptapeptide produced in the skin of the South American leaf frogs, genus *Phyllomedusa*, has a 1000-fold greater effect than morphine at the same dosage level (Clarke 1997).

It is possible that molecules derived from amphibian sources may provide useful alternative or supplementary treatments, primarily as antimicrobial (including antifungal) agents. Additionally, the presence of a wide range of peptides with unknown properties enhances the prospects of using peptide vaccines for viral diseases (Clarke 1997). Other examples of the medical importance of amphibians include: the skin of the poisonous frog, *Epipedobates tricolor*, produces epibatidine, a substance that has ultimately led to the creation of a new class of analgesics (Spainhour 2005); peptides extracted from the scraped secretions of *Phyllomedusa bicolor*, for instance, are used in the treatment of depression, stroke, seizures, and cognitive loss in ailments such as Alzheimer's disease (Amato 1992).

Other vertebrates produce compounds that have been the subjects of substantial research, which offer potential opportunities to identify useful compounds in the areas of cardiovascular function, immune function, and central nervous system function (Spainhour 2005). An example of a compound produced by marine vertebrates is tetrodotoxin, which is water-soluble (Bisset 1991) and found in the liver and reproductive organs of the puffer fish. When diluted, it acts as an extraordinary narcotic and analgesic. Colwell (1997), in evaluating the medical use of this substance, noted that "tetrodotoxin is a very powerful analgesic".

Other examples are provided by: Finkl (1984), who referred to the species *Eptatretus stoutii*, *Dasyatis sabina* and *Taricha* sp. as sources of cardiac stimulants, antitumor agents, and analgesics, respectively; Adeodato (1997), who recorded the use of omega-3 oil, extracted from shark liver, is capable of controlling atherosclerosis and reducing blood pressure, and of alcoxyglycerol to stimulate immune defenses of the organism and wound healing. Among mammal species, Mebs et al. (1996) isolated an anti-hemorrhagic factor (erynacin) from the plasma of the European hedgehog *Erinaceus europaeus,* and But et al. (1991) demonstrated antipyretic activity in aqueous extracts of the horns of the rhinoceros and buffalo.

## 5.4 Bioprospecting and Conservation

The value of biodiversity to human health has been highlighted in a number of publications, one of its most obvious benefits being the large proportion of the pharmaceutical armamentarium that is derived from the natural world (Alves and Rosa 2007a; Chivian 2002b). The immense benefits humans draw from using products and derivatives of microbes, plants, and animals (Fitter 1986), however, has not prevented the continuing exploitation of biodiversity. In fact, species and natural ecosystems are predicted to disappear in the near future at a faster rate than any time in the history of the Earth (Székely and Gaillard 2007).

Ingredients sourced from wild animals are not only widely used in traditional medicines, but are also increasingly valued as raw materials in the preparation of modern medicines. Obviously, as these organisms are obtained from nature, there is an impact on the natural populations, especially for the most exploited species. Therefore, the conservationist implications of such uses should be considered, since the loss of species implies various direct and indirect harmful effects on human health, including the diminution of the possibility of discovering new medicines.

Bioprospecting for pharmaceuticals has been represented typically as an environmentally benign form of resource development that would simultaneously increase our understanding of biodiversity (Quinn et al. 2002) and help fund its conservation, especially in developing countries (Eisner 1990; Rausser and Small 2000; Zilinskas and Lundin 1993). It has been argued that developing nations with high marine biodiversity could benefit financially from granting access for bioprospecting and indirectly by building industry capacity through joint ventures. However, bioprospecting continues to generate considerable debate, as critics dispute the idea that the commodification of nature will contribute to conservation (Simpson 1997), or that natural products have a future in the discovery of new drugs (Firn 2003).

Some animals have been pushed towards extinction because of their alleged curative powers. This has been especially true of those creatures also threatened by habitat loss: musk deer, rhinos, saiga antelope, and tigers, to name a few. When rural communities were harvesting medicinal animals for their own uses, overexploitation was seldom a problem; however, the growing market demands and the promise of quick returns have encouraged hunters to concentrate on species with higher economic value. The current worldwide demand for medicinal animal parts is unprecedented, and the implications for many wildlife populations in Asia and Africa at least are not good (McNeely 2006).

The potential environmental impacts of animal bioprospecting are poorly understood, and it would be ironic if the process of bioprospecting, expected to facilitate conservation of biodiversity, actually reduced it. In terrestrial systems, bioprospecting collections have caused the decline of wild plant populations (Dhillion and Amundsen 2000), such as the *Pilocarpus* spp., which are officially listed as threatened in Brazil because of overexploitation of the leaves to produce the drug pilocarpine (Pinheiro 1997). Although less is known about the marine environment, conservation concerns have been raised about the effects of these bioprospecting collections on marine organisms (Benkendorff 2002; Chivian et al. 2003; Garson 1997).

The threats to medicinal species are often very similar to the threats faced by other species, and many medicinal species are also used for other purposes. These multiple uses enhance the direct impacts on natural populations of animal species, and operate concomitant with various indirect impacts on animal populations, especially habitat degradation and climatic alterations, which have caused the decline or extinction of species in the past years. Thus, conservation measures should consider the medicinal use of animal species in conjunction with other causal factors.

We are living a paradoxical situation when we speak of fauna bioprospection for medicinal purposes. Even if recognizing the role that animals have in many cultures as a source of medicines, we do not have scientific information about the impact of the use of these resources on such species' populations. Additionally, what effects can discovering a promising molecule have on species? Some may argue that this may generate benefits by stimulating conservation measures and/or the sustainable management of natural populations. However, if we use the numerous examples of medicinal plants' industrial exploitation as a reference, such benefits are not always felt.

It is important to consider that human health is dependent on biodiversity and on the natural functioning of healthy ecosystems. Therefore, the use of animals for medicinal purposes is not simply a matter of the pharmaceutical and medical sciences; joint-research programs should be undertaken with experts in the fields of ecology, linguistics, sociology, anthropology, etc. Thus, discussing zootherapy within the multidimensionality of sustainable development turns out to be one of the key elements in achieving the sustenance of medicinal faunistic resources. A fundamental issue is that the potential exists, but the best way to explore this potential must be urgently discussed. Among the possible avenues to address these issues, we highlight:

1. Scientific investigation of the medicinal potential of animals cannot repeat the mistakes already detected in relation to medicinal plant research, which are often marked by an alarming number of studies that do not result in expressive discoveries, mainly due to misdirected investigations and in some cases have negatively impacted wild populations of target species.

2. The study of wild animals as the source of medicine should consider the species' characteristics (abundance, endemism, biological characteristics), and should be coupled with an evaluation of the resource's current and future sustainability (including a scenario in which a new drug may arise).
3. In the case of animals used in traditional medicinal systems, a clear understanding of the dynamics that govern the use of these resources is of surmount importance–especially due to the fact that such uses are seldom exclusively based on animals, and usually also include plants. Are users of medicinal animals aware of plant drugs that can replace animal-based remedies? If so, why do they invest in the use of a resource that demands greater effort to collect and may bring about legal punishment, when there are resources with apparently the same functions and that are easier to obtain? These and other questions need to be addressed in the context of conservation.

## 5.5 Final Considerations

Nature has provided many things for humankind over the years, including the tools for the first attempts at therapeutic intervention (Nakanishi 1999a, b). Although people most commonly think of plants first when talking about natural products, animals can also provide sources of material that could provide the basis of a new therapeutic agent. As remarked by Spainhour (2005), animals, whether highly developed or poorly developed, whether they live on land, sea, or in the air, can be excellent sources of natural products.

Historically, while the use of plants as medicines has been extensively recognized, studied, and reviewed, studies on the use of fauna as a source of drugs have only emerged in the past decades, and they have demonstrated the enormous potential of fauna as a source of natural products and drugs. This potential is perhaps even greater than that of plants, if we consider that the number of animal species is several times greater. For example, Trowell (2003) points out that there are at least 16 times as many insect species as there are plant species, yet plant chemistry has been studied 7,000 times as much as insect chemistry when comparing the amount of research per species. In addition, various other groups of terrestrial animals, particularly reptiles and amphibians, have been of great interest in pharmacological studies. It should also be highlighted that marine animals represent an exceptional source of bioactive natural products, many of which exhibit structural features not found in terrestrial natural products (Faulkner 1998; Ireland et al. 1993). Although marine compounds are underrepresented in current pharmacopeia, it is anticipated that the aquatic environment will become an invaluable source of novel compounds in the future. The marine ecosystem represents 95 % of the biosphere, and all animal phyla, except one, are represented in aquatic environments.

Natural products have been the basis for the development of traditional medicines and for the discoveries of many modern drugs. Although the potential of faunistic biodiversity is well known, the exploitation of these resources requires a careful strategy that allows the sustainability of the exploited species. One of the main conservation concerns about the exploitation of fauna for the search of bioactive compounds is the possible overharvesting of target organisms.

Some taxa with known pharmacological potential are especially susceptible to overexploitation; for example, marine species such as cone shells, mollusks that have been overharvested as sources of clinical neuropharmaceuticals (Sukarmi and Sabdono 2011). There are increasing concerns about harvesting reef organisms for the discovery and development of pharmaceuticals, since it has been perceived by many as not sustainable and threatening to conservation (Hunt and Vincent 2006; Sukarmi and Sabdono 2011).

In another direction, some authors (e.g., Reid et al. 1993; Beattie et al. 2005) argue that well-regulated bioprospecting contributes to the joint goals of ecosystem conservation and social and economic development through partnerships and benefit-sharing. Beattie et al. (2005) listed several ways in which bioprospecting can achieve multiple goals, as follows: generating revenues for protected areas, conservation projects, and local communities; building scientific and technological capacity to study and manage biodiversity; enhancing biodiversity science; raising awareness of the commercial and noncommercial importance of biodiversity; creating businesses dependent upon the sustainable management of resources; and, in rare instances, generating large profits for corporations and shareholders.

Regardless of the perspective adopted, as highlighted by Barrett and Lybbert (2000), the need to conserve precious biodiversity is clear, especially as we begin to appreciate the magnitude of the spiritual, social, and economic services it provides. Furthermore, as remarked by Moran et al. (2001), discussions of biodiversity and its local, national, and international medical utility should never be disaggregated from the rich, complex, and diverse cultural and biological matrix from which it evolved.

Specifically, in relation to bioprospecting, there is an urgent need to take into account bioethical considerations in anticipating the potential consequences of these activities and proposing management options for sustainable use of animals as sources of bioactive compounds (Sukarmi and Sabdono 2011). Hopefully, this message will increasingly resonate among all relevant stakeholders–after all, resources for all bioprospecting rely on biodiversity.

## References

Adeodato S (1997) Os santos remédios do mar. Globo Ciência 4:20–25

Aerts AM, François I, Cammue BPA, Thevissen K (2008) The mode of antifungal action of plant, insect and human defensins. Cell Mol Life Sci 65(13):2069–2079

Ahn MY, Hahn BS, Lee PJ, Wu SJ, Kim YS (2006) Purification and characterization of anticoagulant protein from the tabanus, tabanus bivittatus. Arch Pharm Res 29(5):418–423

Ahn MY, Ryu KS, Lee YW, Kim YS (2000) Cytotoxicity and L-amino acid oxidase activity of crude insect drugs. Arch Pharm Res 23(5):477–481

Alam M, Thomson RH (1998) Handbook of natural products from marine invertebrates. Part 1 phylum mollusca. Harwood Academic Publishers, Australia

Albuquerque UP, Hanazaki N (2006) As pesquisas etnodirigidas na descoberta de novas drogas de interesse médico e farmacêutico: fragilidade e perspectivas. Revista Brasileira de Farmacognosia 16:678–689

Alvarez B, Crisp MD, Driver F, Hooper JNA, Soest RWMV (2000) Phylogenetic relationships of the family Axinellidae (Porifera: Demospongiae) using morphological and molecular data. Zool Scripta 29:169–198

Alves RRN (2009) Fauna used in popular medicine in Northeast Brazil. J Ethnobiol Ethnomed 5(1):1–30. doi:10.1186/1746-4269-5-1

Alves RRN, Alves HN (2011) The faunal drugstore: animal-based remedies used in traditional medicines in Latin America. J Ethnobiol Ethnomed 7(9):1–43

Alves RRN, Barboza RRD, Souto WMS (2010a) A global overview of canids used in traditional medicines. Biodivers Conserv 19(6):1513–1522

Alves RRN, Dias TLP (2010) Usos de invertebrados na medicina popular no Brasil e suas implicações para conservação. Trop Conserv Sci 3(2):159–174

Alves RRN, Rosa IL (2005) Why study the use of animal products in traditional medicines? J Ethnobiol Ethnomed 1(5):1–5. doi:10.1186/1746-4269-1

Alves RRN, Rosa IL (2006) From cnidarians to mammals: the use of animals as remedies in fishing communities in NE Brazil. J Ethnopharmacol 107:259–276

Alves RRN, Rosa IL (2007a) Biodiversity, traditional medicine and public health: where do they meet? J Ethnobiol Ethnomed 3(14):9. doi:10.1186/1746-4269-3-14

Alves RRN, Rosa IL (2007b) Zootherapy goes to town: the use of animal-based remedies in urban areas of NE and N Brazil. J Ethnopharmacol 113:541–555. doi:10.1016/j.jep.2007.07.015

Alves RRN, Rosa IL, Santana GG (2007) The role of animal-derived remedies as complementary medicine in Brazil. Bioscience 57(11):949–955

Alves RRN, Silva CC, Barboza RRD, Souto WMS (2009) Zootherapy as an alternative therapeutic in South America. J Altern Med Res 1(1):21–47

Alves RRN, Souto WMS (2010) Etnozoologia: conceitos, considerações históricas e importância. In: Alves RRN, Souto WMS, Mourão JS (eds) A etnozoologia no Brasil: Importância, Status atual e Perspectivas, 1st edn. vol 7. NUPEEA, Recife, pp 19–40

Alves RRN, Souto WMS, Barboza RRD (2010b) Primates in traditional folk medicine: a world overview. Mamm Rev 40(2):155–180. doi:10.1111/j.1365-2907.2010.00158.x

Amador ML, Jimeno J, Paz-Ares L, Cortes-Funes H, Hidalgo M (2003) Progress in the development and acquisition of anticancer agents from marine sources. Ann Oncol 14(11):1607–1615

Amato I (1992) From'hunter magic', a pharmacopeia? Science 258(5086):1306

Andary C, Motte-Florac E, Ramos-Elorduy J (1996) Privat a chemical screening: updated methodology applied to medicinal insects. In: The 3rd European colloquium on ethnopharmacology and 1st international conference of anthropology and history of health and disease

Antonelli A, Rodriguez V (2009) Brazil should facilitate research permits. Conservation Biol 23(5):1068–1074

Armon PJ (1980) The use of honey in the treatment of infected wounds. Trop Doct 10(2):91

Artuso A (2002) Bioprospecting, benefit sharing, and biotechnological capacity building. World Dev 30(8):1355–1368

Ashwell D, Walston N (2008) An overview of the use and trade of plants and animals in traditional medicine systems in Cambodia, 1st edn. TRAFFIC Southeast Asia, Greater Mekong Programme, Ha Noi

Avila C (2006) Molluscan natural products as biological models: chemical ecology, histology and laboratory culture. In: Cimino G, Gavagnin M (eds) Molluscs: progress in molecular and subcellular biology subseries marine molecular biochemistry. Springer, Berlin, pp 1–23

Bankova V, Christov R, Popov S, Marcucci MC, Tsvetkova I, Kujumgiev A (1999) Antibacterial activity of essential oils from Brazilian propolis. Fitoterapia 70(2):190–193
Barrett CB, Lybbert TJ (2000) Is bioprospecting a viable strategy for conserving tropical ecosystems? Ecol Econ 34(3):293–300
Barsh R (1997) The epistemology of traditional healing systems. Hum Organ 56(1):28–37
Beattie AJ, Barthlott W, Elisabetsky E, Farrel R, Kheng CT, Prance I, Rosenthal J, Simpson D, Leakey R, Wolfson M (2005) New products and industries from biodiversity. In: Hassan R, Scholes R, Ash N (eds) Ecosystems and humanwell-being, vol 1. Millennium Ecosystem Assessment, Island Press, Washington, pp 273–295
Beattie AJ, Hay M, Magnusson B, Nys R, Smeathers J, Vincent JFV (2011) Ecology and bioprospecting. Austral Ecol 36(3):341–356
Benkendorff K (2002) Potential conservation benefits and problems associated with bioprospecting in the marine environment. In: Lumley D, Dickman C (eds) A zoological revolution: using native fauna to assist in its own survival. Royal Zoological Society of New South Wales and Australian Museum, Mosman, pp 90–100
Benkendorff K (2010) Molluscan biological and chemical diversity: secondary metabolites and medicinal resources produced by marine molluscs. Biol Rev 85:757–775
Berlinck RGS, Ogawa CA, Almeida AMP, Sanchez MAA, Malpezzi ELA, Costa LV, Hajdu E, De Freitas JC (1996) Chemical and pharmacological characterization of halitoxin from Amphimedon viridis (Porifera) from the southeastern Brazilian coast. Comparative Biochemistry and Physiology Part C: Comparative Pharmacology and Toxicology 115(2):155–163
Bernan VS, Greenstein M, Maiese WM (1997) Marine microorganisms as a source of new natural products. Adv Appl Microbiol 43:57–90
Bhasin V (2007) Medical anthropology: a review. Stud Ethno-Med 1(1):1–20
Bisset NG (1991) One man's poison, another man's medicine? J Ethnopharmacol 32(1–3):71–81
Boman HG (1995) Peptide antibiotics and their role in innate immunity. Annu Rev Immunol 13(1):61–92
Boman HG (2000) Innate immunity and the normal microflora. Immunol Rev 173(1):5–16
Boman HG, Ingrid F, Gudmundsson GH, Jong-Youn LEE, Lidholm DA (1991) Cell-free immunity in Cecropia. Eur J Biochem 201(1):23–31
Branicki FJ (1981) Surgery in western Kenya. Ann R Coll Surg Engl 63(5):348–352
Bulet P, Stocklin R (2005) Insect antimicrobial peptides: structures, properties and gene regulation. Protein Pept Lett 12(1):3–11
But PP, Tam YK, Lung LC (1991) Ethnopharmacology of rhinoceros horn. II: antipyretic effects of prescriptions containing rhinoceros horn or water buffalo horn. J Ethnopharmacol 33(1–2):45–50
Calixto JB (2005) Twenty-five years of research on medicinal plants in Latin America: a personal view. J Ethnopharmacol 100(1–2):131–134
Carté BK (1996) Biomedical potential of marine natural products. Bioscience 46(4):271–286
Cavanagh D, Beazley J, Ostapowicz F (1970) Radical operation for carcinoma of the vulva. BJOG: An Intern J Obstet Gynaecol 77(11):1037–1040
Chapman AD (2009) Numbers of living species in Australia and the world. Australian Biological Resources Study, Canberra
Chivian E (2002a) Biodiversity: its importance to human health. Harvard Medical School, Boston
Chivian E (2002b) Biodiversity: its importance to human health. Center for Health and the Global Environment. Harvard Medical School, Cambridge
Chivian E, Roberts CM, Bernstein AS (2003) The threat to cone snails. Science 302(5644):391
Clarke BT (1997) The natural history of amphibian skin secretions, their normal functioning and potential medical applications. Biol Rev 72(3):365–379
Colwell RR (1997) Microbial biodiversity and biotechnology. In: Reaka-Kudla ML, Wilson DE, Wilson EO (eds) Biodiversity II: understanding and protecting our biological resources. National Academy Press, Washington, pp 77–98

Confessor M, Mendonca L, Mourao J, Alves R (2009) Animals to heal animals: ethnoveterinary practices in semi-arid region, Northeastern Brazil. J Ethnobiol Ethnomed 5(1):37
Costa-Neto EM (2000) Zootherapy based medicinal traditions in Brazil. Honey Bee 11(2):2–4
Costa-Neto EM (2005) Entomotherapy, or the medicinal use of insects. J Ethnobiol 25(1):93–114
Costa-Neto EM (2006) Os moluscos na zooterapia: medicina tradicional e importância clínico-farmacológica. Biotemas 19(3):71–78
Costa-Neto EM, Alves RRN (2010) Zooterapia: os animais na medicina popular Brasileira, Estudos & Avanços, 1st edn. vol 2. NUPEEA, Recife
Cragg GM, Newman DJ (2001) Medicinals for the millennia: the historical record. Ann N Y Acad Sci 953:3–25
Daly JW (1998) Thirty years of discovering arthropod alkaloids in amphibian skin†. J Nat Prod 61(1):162–172
Daly JW (2003) Ernest Guenther award in chemistry of natural products. Amphibian skin: a remarkable source of biologically active arthropod alkaloids. J Med Chem 46(4):445–452
Debnath A, Chatterjee U, Das M, Vedasiromoni JR, Gomes A (2007) Venom of Indian monocellate cobra and Russell's viper show anticancer activity in experimental models. J Ethnopharmacol 111(3):681–684
Demunshi Y, Chugh A (2010) Role of traditional knowledge in marine bioprospecting. Biodivers Conserv 19:3015–3033
Dettner K (2011) Potential pharmaceuticals from insects and their co-occurring microorganisms. In: Vilcinskas A (ed) Insect biotechnology. Springer, Berlin, pp 95–119
Dhillion SS, Amundsen C (2000) Bioprospecting and the maintenance of biodiversity. In: Svarstad H, Dhillion SS (eds) Responding to bioprospecting: from biodiversity in the south to medicines in the north. Spartacus Forlag A/S, Oslo, pp 103–131
Dossey AT (2010) Insects and their chemical weaponry: new potential for drug discovery. Nat Prod Rep 27:1737–1757
Efem SE (1993) Recent advances in the management of Fournier's gangrene: preliminary observations. Surgery 113(2):200–204
Efem SEE, Iwara CI (1992) The antimicrobial spectrum of honey and its clinical significance. Infection 20(4):227–229
Eisner T (1990) Prospecting for nature's chemical riches. Chemoecology 1(1):38–40
Ellis R (2005) Tiger bone & rhino horn: the destruction of wildlife for traditional Chinese medicine. Shearwater Books, Washington
Farnsworth NR, Bingel AS (1997) Problems and prospects of discovering new drugs from higher plants by pharmacological screening. In: Wagner H, Wolff P (eds) New natural products and plant drugs with pharmacological, biological or therapeutic activity. Springer, Berlin, pp 1–22
Faulkner DJ (1998) Marine natural products. Nat Prod Rep 15(2):113–158
Ferreira FS, Brito S, Ribeiro S, Almeida W, Alves RRN (2009a) Zootherapeutics utilized by residents of the community Poco Dantas, Crato-CE Brazil. J Ethnobiol Ethnomed 5(1):21
Ferreira FS, Brito SV, Costa JGM, Alves RRN, Coutinho HDM, Almeida WdO (2009b) Is the body fat of the lizard Tupinambis merianae effective against bacterial infections? J Ethnopharmacol 126(2):233–237. doi:10.1016/j.jep.2009.08.038
Ferreira FS, Brito SV, Fernandes-Ferreira H, Alves RRN (2010) Prospecção biológica, recursos zooterápicos e sustentabilidade. In: Costa-Neto EM, Alves RRN (eds) Zooterapia: Os Animais na Medicina Popular Brasileira, 1st edn, vol 2. NUPEEA, Recife, pp 141–158
Ferreira FS, Silva NLG, Matias EFF, Brito SV, Oliveira FG, Costa JGM, Coutinho HDM, Almeida WO, Alves RRN (2011) Potentiation of aminoglycoside antibiotic activity using the body fat from the snake Boa constrictor. Revista brasileira de farmacognosia 21(3):503–509
Finger JM, Schuler P (2004) Poor people's knowledge: promoting intellectual property in developing countries. A World Bank Publication, Washington
Finkl CW (1984) Os medicamentos do mar. Enciclopédia dos mares 1:74–75
Firn RD (2003) Bioprospecting–why is it so unrewarding? Biodivers Conserv 12(2):207–216
Fitter RSR (1986) Wildlife for man: how and why we should conserve our species. Collins, London

Fusetani N (2000) Drugs from the Sea. Karger, New York
Garson M (1997) Biodiversity and bioprospecting. Australas J Nat Resour Law and Policy 4(2):227–239
Gertsch J (2009) How scientific is the science in ethnopharmacology? Historical perspectives and epistemological problems. J Ethnopharmacol 122(2):177–183
Giday M, Teklehaymanot T, Animut A, Mekonnen Y (2007) Medicinal plants of the Shinasha, Agew-awi and Amhara peoples in northwest Ethiopia. J Ethnopharmacol 110:516–525. doi:10.1016/j.jep.2006.10.011
Giraldi T, Ferlan I, Romeo D (1976) Antitumour activity of equinatoxin. Chem Biol Interact 13(3–4):199–203
Gomes A, Bhattacharjee P, Mishra R, Biswas AK, Dasgupta SC, Giri B, Debnath A, Gupta SD, Das T (2010) Anticancer potential of animal venoms and toxins. Indian J Exp Biol 48(2):93–103
Haefner B (2003) Drugs from the deep: marine natural products as drug candidates. Drug Discovery Today 8(12):536–544
Haygood MG, Schmidt EW, Davidson SK, Faulkner DJ (1999) Microbial symbionts of marine invertebrates: opportunities for microbial biotechnology. J Mol Microbiol Biotechnol 1(1):33–43
Heinrich M, Ankli A, Frei B, Weimann C, Sticher O (1998) Medicinal plants in Mexico: healers' consensus and cultural importance. Soc Sci Med 47(11):1859–1871
Honda A, Yamamoto Y, Mori Y, Yamada Y, Kikuchi H (1985) Antileukemic effect of coral-prostanoids clavulones from the stolonifer on human myeloid leukemia (HL-60) cells. Biochem Biophys Res Commun 130(2):515–523
Houghton PJ, Howes MJ, Lee CC, Steventon G (2007) Uses and abuses of in vitro tests in ethnopharmacology: visualizing an elephant. J Ethnopharmacol 110(3):391–400
Hultmark D, Steiner H, Rasmuson T, Boman HG (1980) Insect immunity. Purification and properties of three inducible bactericidal proteins from hemolymph of immunized pupae of Hyalophora cecropia. Eur J Biochem 106(1):7–16
Hunt B, Vincent ACJ (2006) Scale and sustainability of marine bioprospecting for pharmaceuticals. AMBIO: A Jo Human Environ 35(2):57–64
Ireland CM, Copp BR, Foster MP, McDonald LA, Radisky DC, Swersey JC (1993) Biomedical potential of marine natural products. Mar Biotechnol 1:1–43
Jensen JB, Camp CD (2003) Human exploitation of amphibians: direct and indirect impacts. In: Semlitsch RD (ed) Amphibian conservation. Smithsonian Books, Washington, pp 199–213
Jensen PR, Fenical W (1994) Strategies for the discovery of secondary metabolites from marine bacteria: ecological perspectives. Annu Rev Microbiol 48(1):559–584
Karthikeyan R, Karthigayan S, Sri Balasubashini M, Somasundaram ST, Balasubramanian T (2008) Inhibition of Hep2 and HeLa cell proliferation in vitro and EAC tumor growth in vivo by Lapemis curtus (Shaw 1802) venom. Toxicon 51(1):157–161
Kellert SR (1993) Values and perceptions of invertebrates. Conserv Biol 7(4):845–855
Kijjoa A, Sawangwong P (2004) Drugs and cosmetics from the sea. Marine Drugs 2:73–82
Koru O, Toksoy F, Acikel CH, Tunca YM, Baysallar M, Uskudar Guclu A, Akca E, Ozkok Tuylu AI, Sorkun K, Tanyuksel M (2007) In vitro antimicrobial activity of propolis samples from different geographical origins against certain oral pathogens. Anaerobe 13(3–4):140–145
Kujumgiev A, Tsvetkova I, Serkedjieva Y, Bankova V, Christov R, Popov S (1999) Antibacterial, antifungal and antiviral activity of propolis of different geographic origin. J Ethnopharmacol 64(3):235–240
Kunin WE, Lawton JH (1996) Does biodiversity matter? Evaluating the case for conserving species. In: Gaston KJ (ed) Biodiversity. A biology of numbers and difference. Blackwell Science, Oxford, pp 367–387
Laird SA (2002) Biodiversity and traditional knowledge: equitable partnerships in practice. Earthscan/James & James, London

Lee SKH (1999) Trade in traditional medicine using endangered species: an international context. Paper presented at the 2nd Australian symposium on traditional medicine and wildlife conservation, Melbourne
Lee YK, Lee JH, Lee HK (2001) Microbial symbiosis in marine sponges. J Microbiol 39:254–264
Lev E (2003) Traditional healing with animals (zootherapy): medieval to present-day Levantine practice. J Ethnopharmacol 85:107–118. doi:10.1016/S0378-8741(02)00377-X
Lev E (2006) Healing with animals in the Levant from the 10th to the 18th cent. J Ethnobiol Ethnomed 2(11):9. doi:10.1186/1746-4269-2-11
Lev E, Amar Z (2002) Ethnopharmacological survey of traditional drugs sold in the Kingdom of Jordan. J Ethnopharmacol 82:131–145
Makhubu L (1998) Bioprospecting in an African context. Science 282(5386):41–42
Mariottini GL, Pane L (2010) Mediterranean jellyfish venoms: a review on scyphomedusae. Marine Drugs 8(4):1122–1152
Marques JGW (1997) Fauna medicinal: Recurso do ambiente ou ameaça à biodiversidade? Mutum 1(1):1–4
McClatchey W (2005) Medicinal bioprospecting and ethnobotany research. Ethnobot Res Appl 3:189–190
McNeely JA (2006) Risks to people of losing medicinal species. In: Miththapala S (ed) Conserving medicinal species: securing a healthy future. IUCN: Ecosystems and Livelihoods Group, Asia, pp 17–31
Mebs D, Omori-Satoh T, Yamakawa Y, Nagaoka Y (1996) Erinacin, an antihaemorrhagic factor from the European hedgehog, Erinaceus europaeus. Toxicon 34(11–12):1313–1316
Melo JG, Amorim ELC, Albuquerque UP (2009) Native medicinal plants commercialized in Brazil—priorities for conservation. Environ Monit Assess 156(1):567–580
Meylaers K, Clynen E, Daloze D, DeLoof A, Schoofs L (2004) Identification of 1-lysophosphatidylethanolamine (C16: 1) as an antimicrobial compound in the housefly, Musca domestica. Insect Biochem Mol Biol 34(1):43–49
Mgbeoji I (2006) Global biopiracy: patents, plants and indigenous knowledge. Cornell University Press, Ithaca
Molan PC (1999) The role of honey in the management of wounds. J Wound Care 8(8):423–426
Moore RE, Scheuer PJ (1971) Palytoxin: a new marine toxin from a coelenterate. Science 172(3982):495–498
Moran K, King SR, Carlson TJ (2001) Biodiversity prospecting: lessons and prospects. Annu Rev Anthropol 30:505–526
Morgan D (2010) Biosynthesis in insects. Royal Society of Chemistry, The
Mukherjee PK, Ponnusankar S, Venkatesh M (2010) Ethno medicine in complementary therapeutics. In: Chattopadhyay D (ed) Ethnomedicine: a source of complementary therapeutics, Research Signpost, pp 29–52
Müller WEG, Brümmer F, Batel R, Müller IM, Schröder HC (2003) Molecular biodiversity. Case study: Porifera (sponges). Naturwissenschaften 90(3):103–120
Myers N (1979) The sinking ark: a new look at the problem of disappearing species. Pergamon Press, New York
Nakanishi K (1999a) An historical perspective of natural products chemistry. In: Sankawa U (ed) Comprehensive natural products chemistry: isoprenoids including carotenoids and steroids. Pergamon, Oxford
Nakanishi K (1999b) An historical perspective of natural products chemistry. In: Ushio S (ed) Comprehensive natural products chemistry. Elsevier Science B.V, Amsterdam, pp 23–40
Oldfield ML (1989) The value of conserving genetic resources. Sinauer Associates, Washington
Orduña-Novoa K, Segura-Puertas L, Sánchez-Rodríguez J, Meléndez A, Nava-Ruíz C, Rembao D, Santamaría A, Galván-Arzate S (2003) Possible antitumoral effect of the crude venom of Cassiopea xamachana (Cnidaria: Scyphozoa) on tumors of the central nervous system induced by N-Ethyl-N-Nitrosourea (ENU) in rats. Proc West Pharmacol Soc 46:85–87
Paavilainen HM (2009) Medieval pharmacotherapy, continuity and change: case studies from Ibn Sina and some of his late Medieval commentators. Brill Academic Publishers, Boston

Padmanabhan P, Sujana KA (2008) Animal products in traditional medicine from Attappady hills of Western Ghats. Indian J Tradit Knowl 7(2):326–329

Pan PG (2006) Bioprospecting: issues and policy considerations. Legislative Reference Bureau, Germplasm resources. Honolulu

Park YK, Inegaki M, Alencar SM, Wang HK, Bastow K, Cosentino M, Lee KH (2000) Determinação das atividades citotóxica e anti-HIV dos extratos etanólicos de própolis coletadas em diferentes regiões do Brasil. Mensagem Doce 56:2–5

Parveen, Upadhyay B, Roy S, Kumar A (2007) Traditional uses of medicinal plants among the rural communities of Churu district in the Thar Desert, India. J Ethnopharmacol 113:387–399. doi:10.1016/j.jep.2007.06.010

Phillips O, Gentry AH (1993) The useful plants of Tambopata, Peru: I. Statistical hypotheses tests with a new quantitative technique. Econ Bot 47(1):15–32

Pieroni A, Giusti ME, Grazzini A (2002) Animal remedies in the folk medicinal practices of the Lucca and Pistoia Provinces, Central Italy. In: Fleurentin J, Pelt JM, Mazars G (eds) Des sources du savoir aux médicaments du futur/from the sources of knowledge to the medicines of the future, 1st edn. IRD Editions, Paris, pp 371–375

Pinheiro CU (1997) Jaborandi (Pilocarpus sp., rutaceae): a wild species. Econ Bot 51(1):49–58

Proksch P, Edrada-Ebel RA, Ebel R (2003) Drugs from the sea-opportunities and obstacles. Marine Drugs 1:5–17

Quinn RJ, Leone PA, Guymer G, Hooper JNA (2002) Australian biodiversity via its plants and marine organisms. A high-throughput screening approach to drug discovery. Pure Appl Chem 74(4):519–526

Rausser GC, Small AA (2000) Valuing research leads: bioprospecting and the conservation of genetic resources. J Political Econ 108(1):173–206

Reid WV, Laird SA, Meyer CA, Gamez R, Gollin MA, Sittenfeld A, Janzen DH, Gollin MA, Juma C (1993) Biodiversity prospecting: using genetic resources for sustainable development. World Resources Institute, Washington

Ridley CP, Faulkner JD, Haygood MG (2005) Investigation of oscillatoria spongeliae-dominated bacterial communities in four dictyoceratid sponges. Appl Environ Microbiol 71(11):7366–7375

Rouzaire-Dubois B, Dubois JM (1990) Characterization of palytoxin-induced channels in mouse neuroblastoma cells. Toxicon 28(10):1147–1158

Sampaio Alves DF, Cabra Júnior FC, Cabral PPAC, Oliveira Junior RM, Rego ACM, Medeiros AC (2008) Efeitos da aplicação tópica do mel de *Melipona subnitida* em feridas infectadas de ratos. Rev Col Bras Cir 35(3):188–193

Santos IJM, Coutinho HDM, Matias EFF, Costa JGM, Alves RRN, Almeida WO (2012) Antimicrobial activity of natural products from the skins of the semiarid living lizards Ameiva ameiva (Linnaeus, 1758) and Tropidurus hispidus (Spix, 1825). J Arid Environ 76:138–141

Sawyer WH, Munsick RA, Van Dyke HB (1961) Pharmacological characteristics of the active principles in neurohypophysial extracts from several species of fishes. Endocrinology 68(2):215–225

Scarpa GF (2004) Medicinal plants used by the Criollos of Northwestern Argentine Chaco. J Ethnopharmacol 91(1):115–135

Seedhouse E (2010) Ocean outpost: the future of humans living underwater. Springer, Berlin

Silva CCA (2002) Aspectos do sistema imunológico dos insetos. Revista Biotecnologia Ciência & Desenvolvimento 24:68–72

Silva RJ, Fecchio D, Barraviera B (1996) Antitumor effect of snake venoms. J Venom Animals and Toxins 2:79–90

Simpson RD (1997) Biodiversity prospecting: shopping the wilds is not the key to conservation. Resources 126:12–15

Sittenfeld A (1996) Special article on issues and strategies for bioprospecting. Genet Eng Biotechnol 4:1–12

Sittenfeld A, Cabrera JM, Mora M (2004) Bioprospecting and biotechnology: some policy issue. Gene Conserve 3(12):198–211

Sittenfeld A, Lovejoy A (1998) Biodiversity prospecting frameworks: The INBio experience in Costa Rica. In: Guruswamy LD, McNeely JA (eds) Protection of global biodiversity, covering strategies. Duke University Press, Durham and London, pp 223–244
Slocinska M, Marciniak P, Rosinski G (2008) Insects antiviral and anticancer peptides: new leads for the future? Protein Pept Lett 15(6):578–585
Sodeinde OA, Soewu DA (1999) Pilot study of the traditional medicine trade in Nigeria. Traffic Bulletin 18(1):35–40
Soejarto DD (1996) Biodiversity prospecting and benefit-sharing: perspectives from the field. J Ethnopharmacol 51(1–3):1–15
Spainhour CB (2005) Drug discovery handbook. Wiley-Interscience, New Jersey
Stocker K (1989) Snake venom proteins affecting hemostasis and fibrinolysis. Sorbate food preservatives. CRC Press, Basel
Sukarmi S, Sabdono A (2011) Ethical perspectives of sustainable use of reef's invertebrates as a source of marine natural products. J Coast Dev 11(3):97–103
Svarstad H (2000) Local Interests and foreign interventions: shaman pharmaceuticals in tanzania. In: Dhillion SS (ed) Svarstad H. Responding to bioprospecting: from biodiversity in the south to medicines in the north, Spartacus Forlag As Oslo, pp 145–153
Székely T, Gaillard A (2007) Conserving biodiversity using patent law. Nat Biotechnol 25(10):1087–1088
Tabrah FL, Kashiwagi M, Norton TR (1972) Antitumor activity in mice of four coelenterate extracts. Int J Clin Pharmacol Ther Toxicol 5(4):420–422
Thakur AN, Thakur NL, Indap MM, Pandit RA, Datar VV, Müller WEG (2005) Antiangiogenic, antimicrobial, and cytotoxic potential of sponge-associated bacteria. Mar Biotechnol 7(3):245–252
Thakur NL, Müller WEG (2004) Biotechnological potential of marine sponges. Curr Sci 86(11):1506–1512
Thammasirirak S, Phonkham P, Preecharram S, Khanchanuan R, Phonyothee P, Daduang S, Srisomsap C, Araki T, Svasti J (2006) Purification, characterization and comparison of reptile lysozymes. Comparative Biochem Physiol Part C 143:209–217
Thomas TRA, Kavlekar DP, LokaBharathi PA (2010) Marine Drugs from sponge-microbe association—a review. Marine Drugs 8(4):1417
Tamayo G, Nader W, Sittenfeld A (1997) Biodiversity for the bioindustries. In: Ford-Lloy BV, New bury HJ, Callow JA (eds) Biotechnology and plant genetic resources: conservation and use. CAB International, Wallingford, pp 255–279
Trowell S (2003) Drugs from bugs: the promise of pharmaceutical entomology. Futurist 37(1):17–19
Uniyal SK, Singh KN, Jamwal P, Lal B (2006) Traditional use of medicinal plants among the tribal communities of Chhota Bhangal, Western Himalaya. J Ethnobiol Ethnomed 2(14):1–8
Van NDN, Tap N (2008) An overview of the use of plants and animals in traditional medicine systems in Viet Nam. 1st edn. TRAFFIC Southeast Asia, Greater Mekong Programme, Ha Noi
Wade D, Andreu D, Mitchell SA, Silveira AMV, Boman A, Boman HG, Merrifield RB (1992) Antibacterial peptides designed as analogs or hybrids of cecropins and melittin. Int J Pept Protein Res 40(5):429–436
Wang G (2006) Diversity and biotechnological potential of the sponge-associated microbial consortia. J Ind Microbiol Biotechnol 33(7):545–551
Weinheimer AJ, Spraggins RL (1969) The occurrence of two new prostaglandin derivatives (15-epi-PGA2 and its acetate, methyl ester) in the gorgonian Plexaura homomalla chemistry of coelenterates XV. Tetrahedron Lett 10:5185–5188
WHO (2002) Traditional medicine strategy 2002–2005, Geneva
Willix DJ, Molan PC, Harfoot CG (1992) A comparison of the sensitivity of wound infecting species of bacteria to the antibacterial activity of manuka honey and other honey. J Appl Microbiol 73(5):388–394

Yamakawa M (1998) Insect antibacterial proteins: regulatory mechanisms of their synthesis and a possibility as new antibiotics. J Seric Sci Jpn 67(3):163–182

Yinfeng G, Xueying Z, Yan C, Di W, Sung W (1997) Sustainability of wildlife use in traditional chinese medicine. Conserving China'Biodiversity: reports of the biodiversity working group (BWG), China Council for International Cooperation on Environment and Development

Zhang D, Cheng Z (2000) Medicine is a humane art: the basic principles of professional ethics in chinese medicine. Hastings Cent Rep 30(4):S8–S12

Zilinskas RA, Lundin CG (1993) Marine biotechnology and developing countries. The International Bank for Reconstruction and Development, Washington

# Chapter 6
# Fish Folk Medicine of Caiçara (Atlantic Coastal Forest) and Caboclo (Amazon Forest) Communities

**Alpina Begossi and Milena Ramires**

**Abstract** Medicines and foods are not always exclusive categories in ethnopharmacology. What is eaten, the diets of the animals being consumed, which foods are healthy, and which foods should not be eaten when one is sick are some examples of the ways in which native populations interact with foods, medicines, and natural resources. This study examined native fishing communities in the Atlantic Forest (the Caiçaras) and in the Amazon region (the Caboclos) regarding their medicinal uses of fish and their food taboos. An analysis of those categories and recommendations for conservation practices are presented, taking into consideration the cultural habits of the studied communities.

## 6.1 Introduction

### *6.1.1 The Importance of Marine Drugs and of the Interactions Between Fields of Study*

The term *medicinal plants* is widely used and constitutes an important and integral part of our popular (folk) vocabulary. The terms *medicinal animals* or *medicinal fish*, on the other hand, generally provoke surprise, since the use of medicinal animals is largely unknown. Nonetheless, local or native populations can provide

A. Begossi (✉) · M. Ramires
Fisheries & Food Institute (FIFO) and ECOMAR, UNISANTA,
Rua Oswaldo Cruz 277, Santos, São Paulo, Brazil
e-mail: alpinab@uol.com.br

A. Begossi
CMU, UNICAMP (CP 6023), Campinas, São Paulo, Brazil

R. R. N. Alves and I. L. Rosa (eds.), *Animals in Traditional Folk Medicine*,
DOI: 10.1007/978-3-642-29026-8_6, 

important leads to the discovery of curative substances through their intimate knowledge of the local fauna.

As an example, some years ago one of the authors (AB) received an email message from a scientist who had to read two of her publications on zootherapy or ethnomedicine (Begossi 1992; Begossi and Braga 1992) and wanted to know if there had been further research on the possible active principles of the animals mentioned. The scientist was informed that, as far as known, no such research had been conducted in Brazil. Some years later that same scientist published an article (Sankaran and Mouly 2007) on bio-active compounds from marine sources—including Lyprinol (from mussels) and Isolutrol (from shark liver)—in which he emphasized the importance of translational knowledge networking and ethno-pharmacology or zootherapy.

Another example of how much still remains to be done is exemplified by a recent article (Ferreira et al. 2010) in which the fat from a congeneric species of *Tupinambis teguixin*—one of the most frequently quoted medicinal animal species on Búzios Island, state of São Paulo (see Begossi 1992)—was found to have anti-inflammatory activity.

These cases highlight how much remains to be discovered in the field of medicinal animals, and serves as a reminder that there are currently too few researchers to effectively study Brazil's mega-biodiversity, or the extent of local folk medicinal knowledge. Collaboration is certainly needed and could bring mutual returns, as the exceptional biodiversity in Brazil provides numerous opportunities to discover new medicines. Several obstacles, however, prevent easy access to the potential benefits of the country's natural resources and potential medicines, many of which linked to overly restrictive federal regulations (Decree 5.459, 2005). The high biodiversity in Brazil provides numerous opportunities for future medicines. However, Brazilians still face obstacles to access their natural resources and future medicines, many of which related to legislation (Decree 5.459, 2005). Considered as xenophobic by many, this Decree has lead to relativly high bureaucracy and obstacles to conduct research on natural resources in Brazil (http://revistapesquisa.fapesp.br/?art=4319&bd=1&pg=1&lg=), and on a personal note, has discouraged us from collecting data on ethnomedicine and ethnobotany in Brazil. In this sense, it is important to stress that the data presented in this chapter come from either published sources or from collections made prior to the year 2000.

We consider any obstacles imposed on data collection as moving in the opposite direction of discoveries made every year with aquatic fauna—especially their medicinal uses. Examples can be found in articles published in popular magazines and official sources:

(1) Newsweek (2005): "*... Scientists... now understand that forests and oceans contain a great deal of useful chemicals";*

(2) Fapesp (07/07/2005): "*... among the molecules analyzed in the mucus that coats the fish skin, proteases and peptides stood out... classes of peptides found in fish skin that can be used to formulate drugs against human diseases*";

(3) Fapesp (5/23/2006): "*... a dip in the Blue Amazon... some compounds from marine organisms have shown good performance against leishmaniasis and tuberculosis*"
(4) Newsweek (11/29/2010): "*Why scientists are heading underwater to search for a new generation of cures*"; "*marine bioprospecting could expand the number of available drugs at least tenfold*"

The importance of local knowledge in association with scientific research has been extensively discussed, but insufficiently developed. In this context, Alves and Rosa (2005) stressed the importance of studying ethnobiology and ethnomedicine in light of the global search for animal- and plant-based medicines, while Alves and Alves (2011) pointed out that the importance of non-botanical remedies (i.e., animal and mineral-based) is growing in spite of frequent neglect.

Several publications focusing on zootherapy in Brazil have mentioned the use of fish in disease treatments. Some studies, including Costa-Neto and Marques (2000), Seixas and Begossi (2001), and Nomura (1984) listed fish and other medicinal animals, while others discussed the relationships between food taboos and medicinal animals (Begossi 1992; Begossi and Braga 1992). Begossi et al. (2006) and Hanazaki et al. (2006) analyzed the relationships between medicines and foods (both animal and plant) in traditional Caiçara and Caboclo communities, and examined the fine line between what is "food" and what is "medicine". Alves and Alves (2011) discussed the importance of traditional medicine in Latin America and listed 110 species of fish (among 13 other taxonomic categories, mostly comprising vertebrates) mentioned by folk practitioners, while Alves and Souto (2011) provided an overview of the ethnozoological knowledge in Brazil, citing various articles related to zootherapy.

This chapter provides an overview of the use of fish species in Brazilian folk medicine, examines some interfaces between animal-based remedies and foods, and presents recommendations for conservation practices.

### 6.1.2 Toxic or Medicinal?

The distinctions between poisons and medicines are often only of degree, and are based on quantitative gradients that differentiate between very small quantities of bioactive compounds. The distinction between what is food and what is a remedy (or poison) is therefore not clear, and there is a great need for experiments that can accurately determine how to use animals (or their subproducts) to prevent illnesses, alleviate symptoms, or treat diseases. Hashimoto (1979) defined toxin as a molecule with a functional group that shows physiological activity.

Despite recognition that there are several toxic marine invertebrates and vertebrates, very little is known about the active ingredients of many poisonous organisms composing the marine fauna. The high zootoxicological and chemical variability of marine toxins was noted by Russell (1965), and can be exemplified by 5-HT (Hydroxytryptamine), which has been identified in a large number of cnidarians and is known as a potent source of pain to humans; despite its unknown molecular origin and bioactive principles, it seems to be an effective vasoconstrictor.

Warnings about poisonous animals at beach areas are relatively common around the world, and Cunningham and Goetz (1996) reported these warnings in the United States and Australia. Many venomous marine animals are known to inhabit the waters of the Pacific Ocean, especially along the coast of Australia (Fig. 6.1a). In Brazil, Haddad (2000) compiled a guide to the medical diagnosis of injuries caused by marine invertebrates and vertebrates as well as freshwater vertebrates, cited 144 accidents treated in the emergency room of the Ubatuba Municipal Hospital in the state of São Paulo, and presented the results of interviews with 63 fishermen from the local Z-10 Colony. Fish accounted for 41 accidents, involving mostly catfish (26) (Ariidae species). Other injuries were caused by stingrays, scorpion fish (*Scorpaena* spp.), toadfish (Batrachoididae), surgeon fish (*Acanthurus chirurgus*), morays, and jaguar fish (*Holocentrus ascension*), among others.

It is interesting to note that many animals involved in poisoning accidents on Buzios Island (located about 25 km off the coast of Ubatuba) are also used in folk medicine, such as stingrays and jaguar fish (jaguariçá) (Begossi 1992). Likewise, fishermen from Siribinha (state of Bahia) cited stingrays, catfish, and toadfish, among others, as medicinal (Costa-Neto and Marques 2000, Costa-Neto 1999), and stingrays and other fish species were mentioned as medicinal animals by the communities at Ilha Grande, state of Rio de Janeiro (Seixas and Begossi 2001). Similar observations concerning freshwater fish were reported by Haddad (2000) and about medicinal fish from Middle Tocantins River by Begossi and Braga (1992) (Tables 6.1, 6.2, 6.3, 6.4, and 6.5). This is not surprising, as poisons and drugs can be very similar, with only their quantities (or dilutions) separating beneficial treatments from intoxication—many trials and bioassays are needed to determine the optimum dose at which a poison becomes a medicinal drug.

Russell (1965) estimated that there are about 200 species of venomous marine fish, most of which are nonmigratory and slow moving; many of them inhabit reef environments. In Brazil, initial studies on venomous marine animals and on poisonous marine fish were conducted by various authors, including Carvalho (1947); Costa (1959); Fonseca (1952); Gonsalves (1905), Hector Fróes (1933), Mello-Leitão (1948), Rocha and Fraga (1962), Silva (1911). More recently, Haddad (2000) published an atlas of dangerous aquatic found in Brazil, which in conjunction with other published sources, such as Goetz (1996) and Russell (1965) provided the basis for Table 6.2. Table 6.2 does not include fish that are toxic when ingested, as is the case with ciguatoxin and tetraodoxin (Begossi 1992; Russell 1965).

There are obvious connections between animals and poisonous drugs, as there are between our diets and health. The food chain is the immediate source of nutrients as well as toxins such as ciguatoxin, which is found in about 300 different

**Fig. 6.1** Fish and folk medicine of Caiçaras and Caboclos. **a** Sign on the famous Bondi Beach in Sydney about the danger of the presence of Bluebottle jellyfish or Portuguese Man of War (*Photo A.* Begossi) (*Physalia physalis or P. utriculus)*; **b** Caiçaras: fishermen in Picinguaba, SP, coast off the Atlantic forest; **c** Caboclos: family from Papaconha island, Middle Tocantins; **d** One species of peixe-porco (filefish), *Stephanolepis hispidus* (Monacanthidae) common on the coast of Brazil. This fish is widely cited by caiçara fishermen for the treatment of asthma and bronchitis. The fish's skin is used in the treatment. This fish was purchased at the Fishermen's Colony of Post 6, Copacabana, Rio de Janeiro

species of marine organisms, most of which are carnivorous or feed on benthic algae (Russell 1965) and tetraodoxin (which is present in *Sphoeroides* blowfish species, among others). Among the carnivorous species, a relationship between the amount of a fish in the diet and toxicity exists (Russell 1965), and this relationship is enhanced, since many toxins accumulate through trophic levels.

### *6.1.3 Food or Medicine?*

Foods and medicines are often complementary in traditional communities that depend on local natural resources for survival, such as the Caiçaras who live in the Atlantic Rain Forest and the Caboclos of the Amazon region. Certain animals are approved for consumption by people suffering from certain illnesses, while in other cases restricted diets are recommended—with certain animals

**Table 6.1** Study sites in southeastern Atlantic Forest and in the Brazilian Amazon

| State | Region | Community |
|---|---|---|
| *Southeast of the Atlantic Forest* | | |
| São Paulo | Ilhabela | Búzios Island |
| | | Vitória Island |
| | Ubatuba | Puruba |
| | | Picinguaba |
| Rio de Janeiro | Sepetiba Bay | Itacuruçá Island |
| | | Jaguanum Island |
| | Ilha Grande Bay | Gipóia Island |
| *Rivers of the Brazilian Amazon* | | |
| Goiás | Araguaia River | Riverines from Barra do Garças to São Felix do Araguaia |
| Acre | Juruá River | Riverines of the Juruá River and effluents of Bagé, Breu and S. João. |
| Goiás, Tocantins and Maranhão | Tocantins River | Riverines from Imperatriz to Estreito |

being prohibited. One example of this type of prohibition is seen in the diets of menstruating women or puerperal women.

Numerous authors have investigated the food and medicinal potentials of animal resources. Etkin and John (1998), for example, noted that foods are chemically very complex and that some of their component substances can have effects on human health—and diet is increasingly perceived as having an important role in treating diseases. Pieroni and Quave (2006) used plants to highlight different food/medicinal gradients, and observed that: (a) many plants have a multifunctional uses as foods and medicines; (b) many plants are eaten because they are considered healthy, and (c) some plants are directly used to treat diseases. Hanazaki et al. (2006) likewise cited examples of this gradient among the inhabitants of fishing communities in the Atlantic Forest and the tributaries of the Araguaia and Rio Negro rivers.

A similar continuum can be found with regard to fish: some are consumed because they are healthy and their consumption may actually be stimulated in the case of some illness. Begossi et al. (2004a, b, 2006) observed that many herbivorous fish are consumed for this reason in fishing communities in the Atlantic Forest and in the Amazon basin. We will highlight in this chapter fish directly used in treating diseases (and their symptoms), or for healing wounds.

Even within the category of "medicinal fish" it is often difficult to determine whether there is a direct relationship between food consumption and the treatment of a given disease, that is, if the individual consumes a given fish because it is a healthy practice to eat it or because it is an effective treatment of a specific disease; this information is quite important to research focusing on active ingredients derived from fish. We emphasized in the Introduction section the close relationship between poisons and medicines, which are usually differentiated only as quantitative gradients. In the early 1500s, Paracelsus delineated the concept of toxin as a specific compound that elicits a specific toxic response (Auerbach and Gehrs 1980).

Table 6.2 Fish toxicity, physiological activity, and active principles: some examples based on Haddad (2000)

| Species | Poison localization and active principles |
|---|---|
| *Aetobatus narirari* (painted stingray), *Dasyatis sp.* (Longnose stingray), *Rhiptera bonasus* (ticonha stingray) | Spikes, which are associated with poisonous glands. Necrotizing, cardiotoxic, and neurotoxic poisons, consisting of polypeptides (serotonin, phosphodiesterase and 5-nucleotidase) |
| *Squalus* (dogfish catfish) | Spines with venom glands |
| *Arius spixii* (Yellow catfish) *Netuma barba* (white catfish) *Bagre marinus* (catfish flag) | Glands scattered in stings: proteins with effects similar to acetylcholine and prostaglandins |
| *Scorpaena brasiliensis* (red scorpion fish) *Scorpaena plumieri* (black scorpion fish) | Venomous spines. Cardiovascular and neurotoxic effects |
| *Pterois volitans* (lion fish) | Poisonous dorsal spines, with cardiotoxic effects |
| *Thalassophryne* sp. (toad fish) *Batrachoides* sp. (toad fish) *Porichthys porosissimus* (toad fish) | Hollow spines attached to venom glands on the back, near the fins and opercle; necrotizing neurotoxic effect |
| *Acanthurus chirurgus* (surgeon fish) | Sharp blades near the tail |
| *Gymnothorax occelatus* (moray) | Glands present on the palate |
| *Astroscopus* (miraceu) | Poisonous glands in spikes—observed in the collections in the Mediterranean Sea, Spain |
| *Holocentrus ascensionis* (jaguar) | Poisonous glands in spikes coated with epithelium allegedly toxic |
| *Conodon nobilis* (snorer) | Dorsal and anal spines covered with poisonous glands |
| *Oligoplites saliens* (guaivira) | The same |
| *Dactylopterus volitans* (turd) | Sharp cephalic spines covered with poisonous glands |
| *Prionotus punctatus* (cabrinha) | Idem |
| *Collorhyncus callorhyncus* (elephant fish) | Poisonous dorsal spine |
| *Lophius* spp. (Frogfish) | Poisonous spine |

With this in mind, it is worth recalling the discovery of digitoxin, one of 30 cardiac glycosides isolated from the plant *Digitalis purpurea* (Balick and Cox 1996). One of the first systematic compilations of medicinal plants was published by John Gerard in 1597 (The Herball) and it described the effects of *D. purpurea*; in 1775 William Withering published an article pointing out the importance of this species in treating heart diseases—and leaves from *D. purpurea* are still recommended for treating heart problems (Balick and Cox 1996). Cotton (1996) has presented a history of ethnopharmacological contributions from ethnobotanical studies and a wide range of phytochemicals and their applications.

In this chapter, we cite fish used directly in treating diseases to aid in organizing information that might be useful in future research on substances with medicinal potential. Fish have occupied a privileged place of honor on dinner tables

**Table 6.3** Historical accounts in the literature on medicinal fish

| Name of quoted fish (Piso 1658) | Name in L. Freire (1939) | Name in Nomura (1984) | Identification according to Froese and Pauly (2005) |
|---|---|---|---|
| Caraúna: marine fish that live on rocks | *Teuthis caerulens*, or acará-pixuna, among others | **Acará-Una**, synonym of Barbeiro Azul (*Teuthis caerulens*, Teuthidae). Reef species, found from Key West (Florida) to Bahia **Caraúna**, synonym of Garoupinha (*Cephalopholis fulvus*, Serranidae) | Species of *Acanthurus* or *Cephalopholis*, reef fish (Acanthuridae) |
| Pacamó, called rã marinha (marine frog), lampreia | Pacamão, or genus *Pseudopimelodus* | There are several "pacamão" cited, among them, the first, like *Amphichthys cryptocentrus*, Batrachoididae, marine species | The figure in Piso resembles *Amphichthys cryptocentrus*, reef (Batrachoididae) |
| Piraumbú, emits grunting sound, riverine, with affinity to cururuca | Piraúna, pirauaca, piratinga | Pirambú, marine species, *Haemulon album*, Pomadasyidae; synonym of Biquara in Ceará, Corcoroca in the Northest and Sargo in Bahia | Cururuca (*Cynoscion, Micropogonias, Umbrina)*; *Plagioscion squamosissimus?* |
| Taiasicá: lives in sandy beaches, light green, with dark spots | Taiabucú: freshwater fish species, Bocarra. Species: *Xyphorhampus herpectus* | Tajasica, or Amoré-Guaçú, *Chonophorus tajasica*, Gobiidae. Occurs in rivers | Taiassica, imborê: *Gobioides* spp. (Gobiiidae) |

throughout history. Alexis Soyer (1809–1859), a renowned Chef in both France and England, wrote several books on the history and art of food preparation. One of his books, published in 1853, noted that the red mullet (probably *Mullus* sp., family *Mullidae*) was considered a delicacy by the Romans and was served alive to maintain its color. Among the Greeks, a meal of sturgeon (*Acipenser* spp.-*Acipenseridae*) was announced to the sound of trumpets and well-dressed slaves would serve in on tables decorated with flowers (Soyer 1853).

Simoons (1994) noted the religious links of fish with human cultures, citing sacred fish species used by the Egyptians in many rituals—such as *Mormyrus* spp. and *Barbus bynni* (Mormyridae and Cyprinidae, respectively). The author also reported that fish were a symbol of fertility to the Hebrews, and their consumption was recommended to pregnant women. The "dance of the fish" retains this symbolism and is still part of the marriage rites among Bosnian Jews. The association between fish and fertility has also been observed in India (Simoons 1994).

**Table 6.4** Fish cited as medicinal in the Buzios Island, Ubatuba (Praia and Sertão do Puruba, Casa de Farinha and Picinguaba) (Puruba Beach and Hinterland of Puruba, Flourmill, and Picinguaba [Coast of São Paulo] and Gipóia Island (Baía da Ilha Grande), and Jaguanum Itacuruçá Island (Bay Sepetiba) [Coast of Rio de Janeiro].

| Folk name | Species or family | Búzios | Ubatuba | Gipóia | Jaguanum | Itacuruçá | Total | Main use |
|---|---|---|---|---|---|---|---|---|
| Arraia-jamanta (juggernaut-ray) | NC | | 1 | | | | 1 | Asthma |
| Bagre (catfish) | *Ariidae* | | | | 1 | | 1 | Bronchitis |
| Baiacu (pufferfish) | *Sphoeroides* spp. | | | | 12 | | 12 | Bronchitis |
| Cação (small shark) | *Several* spp. | | 1 | | | | 1 | – |
| Corvina (drum, corvine) | *Micropogonias furnieri* *Umbrina coroidesi* | | 2 | | 1 | | 3 | Bronchitis |
| Cavalo marinho (seahorse) | *Hippocampus reidi* | 11 | 6 | 11 | 5 | 1 | 34 | Bronchitis |
| Guaivira (leatherjacket) | *Oligoplites saliens* | | | | 1 | | 1 | Bronchitis |
| Jaguarica (squirrelfish) | *Holoscentrus ascensionis* | 1 | | | | | 1 | Injuries |
| Peixe agulha (needle fish) | *Hyporhampus unifasciatus* | | 1 | | | | 1 | Asthma |
| Peixe Elétrico (electric fish) | NC | | 2 | | 1 | 1 | 4 | Rheumatism |
| Peixe galo (look down) | *Selene setapinnis* | | | | | 1 | 1 | Bronchitis |
| Peixe morcego (bat fish) | *Ogcocephalus verpertilio* | 1 | | | | | 1 | Bronchitis |
| Peixe porco (filefish) | Balistidae, Monacanthidae | | 3 | 3 | 72 | 28 | 106 | Asthma, bronchitis |
| Peixe reis (Brazilian silversides) | *Xenomelaniris brasiliensis* | | | | 2 | | 2 | Bronchitis |
| Pegador[rêmora*]*) (remora) NC | *Remora* spp. | | 1 | | 1 | | 2 | |
| Raia[egg] (ray) | *Raja cyclophora* | 4 | 3 | | | | 7 | Bleeding |
| Sardinha (sardine) | Clupeidae | | | | 1 | | 1 | Alcoholism |
| *Interviews* | | 32 | 155 | 25 | 87 | 42 | 341 | |

*NC* not collected
Identification based on Begossi and Figueiredo (1995)

**Table 6.5** Medicinal fish cited by riverine caboclos (Begossi and Braga 1992; Begossi and Garvello 1990)

| Common (folk) name | Species or family | Tocantins | Araguaia | Jurua | Total | Uses |
|---|---|---|---|---|---|---|
| Acará (cará)-pirosca (oscar) | *Astronotus ocellatus* | 3 | 1 | | 4 | Asthma, tuberculosis |
| Bode (catfish) | Loricariidae | | | 1 | 1 | Burns |
| Branquinha | *Psectrogaster amazonica* | 1 | | | 1 | – |
| Cachorra | *Hydrolicus scomberoides* | 1 | | | 1 | Earache |
| Corvina (corvine, drum) | *Plasgioscion squasissimus* | 11 | 7 | | 18 | Urinary tract |
| Cuiú-cuiú | Doradidae | 1 | 7 | | 8 | Lung diseases |
| Curimatá | *Prochilodus nigricans* | 2 | 1 | 5 | 8 | Rheumatism, wounds |
| Dourada | *Brachyplatystoma filamentosum* | 1 | | | 1 | – |
| Jaú | *Paulicea lutkeni* | 47 | | | 47 | Burns |
| Jundiá | *Aguarunichthys tocantinensis* | | | 2 | 2 | Burns |
| Mariana | *Crenicichla johana* | 1 | | | 1 | "Pano-branco"("white cloth") (white spots on the skin) |
| Pacu-manteiga | *Mylossoma duriventre* | 1 | | | 1 | Venereal diseases |
| Piaba-chata | *Steindachnerina* spp. | | | 5 | 5 | Burns, wounds |
| Piabanha | *Brycon* sp. | 1 | | | 1 | Earache |
| Piau | Anostomidae | 1 | | | 1 | – |
| Pirarara | *Phractocephalus hemiliopterus* | 16 | 14 | 1 | 31 | Lung diseases, wounds |
| Pirarucu | *Arapaima gigas* | | | 1 | 1 | Spots on the body |
| Poraquê (electroc knife fish) | *Electrophorus electricus* | 36 | 17 | | 53 | Rheumatism, wounds |
| Raia (ray) | Potamotrygonidae | 76 | 40 | 4 | 120 | Lung diseases |
| Sardinha (sardine) | Engraulidae | 1 | | | 1 | – |
| Surubim | *Pseudoplatystoma fasciatum* | 6 | | 1 | 7 | Burns, flu |
| Surubim chicote | *Sorubimichthys planiceps* | 26 | | | 26 | Leishmaniasis |
| Traíra | *Hoplias malabaricus* | 9 | 9 | 10 | 28 | Earache |
| Ubarana | *Anodus elongatus* | 1 | | | 1 | – |
| Filhote | *Brachyplatystoma filamentosum* | | 1 | | 1 | Bronchitis |
| Matrinchã | *Brycon* spp. | | 1 | | 1 | Bronchitis |
| *Interviews* | | 189 | 95 | 125 | 406 | |

Fish take on a variety of roles, including as a food source, as deities, as symbols of good luck or taboos, and as medicines—and these categories often overlap. Historical searches for divine species and medicinal substances can also be a source of information for biochemical research.

According to Porter (1997), Hippocrates (460–377 B.C.), considered the "father of Western Medicine", focused on the patient as a whole—which is the basis for the concept of *physis*—on an organic entity, treated as a whole and influenced by the environment. The Sicilian philosopher Empedocles (500–430 B.C.) developed the humoral theory that associated the four elements (earth, air, fire, and water) with the four qualities of dryness, cold, hot, and damp. The four fluids, or humors, were derived from the elements and were associated with melancholy, anger, sanguine, and phlegmatic, and the balance of humors in the body would determine one's state of health (Porter 1997). Queiroz (1984) suggested an association between the categories of foods considered "hot" or "cold" and taboos and traditional medicines in Iguape, São Paulo. Among caiçaras and caboclos, food taboos, especially related to fish, are also considered "hot" food (Begossi 1998).

This relationship between diet and medicine, or between food and disease, appears more frequently in rural communities (such as the Caiçaras of the Atlantic Forest or the Caboclos of the Amazon region) than among urban populations. The perceived relationships between diet and medicine can bring about the following reactions: (a) the prohibition of certain animals as foods because they might exacerbate some detrimental state (such as bleeding or allergies); (b) the consumption of recommended animals to provide useful substances to the body that would facilitate digestion or avoid the assimilation of toxic substances; and (c) the use of animals that are strictly medicinal to treat symptoms of illness, the disease itself, or even injuries. The present study addresses the latter aspect in particular.

Many of the fish to be avoided by the infirm are piscivorous—fish that feed on other fish and are settled at the top of the food chain. One fish species avoided by inhabitants of communities along the Atlantic coast is the bonito (Scombridae), whereas several pimelodids, such as the surubim (*Pseudoplatystoma fasciatum*), are avoided by inhabitants in the Amazon. Among the fish recommended for consumption by patients living in communities in the Atlantic Forest are corvinas and pescadas (drums and weakfish, family Sciaenidae). Among the Caboclo of the Amazon, herbivorous species and fish that feed on invertebrates, such as pacus (*Mylossoma duriventre*, *Myleus* and *Metynnis*), and the loricariid species, are recommended (Begossi et al. 2004a, b, 2006). This type of interaction was described by Piso in 1658 (1957:88) as illustrated by the following transcript: "Among lobsters, crabs and cartilaginous fish, we should prefer the easily digested, as well as the montesinho birds, pheasants, pigeons and doves and others, except the carnivores and those flying high in space, because they are too hot".

> "Among lobsters, crabs and cartilaginous fish, we should prefer the easily digested, as well as the montesinho birds, pheasants, pigeons and doves and others, except the carnivores and those flying high in space, because they are too hot".

Researchers in ethnobiology and ethnopharmacology have observed that it is difficult to clearly separate medicines and foods (Begossi et al. 2006; Pieroni and Quave 2006). The use of plants as foods or medicines has been reported in the Amazon and the Atlantic forests, and of the 83 species used as foods and medicines in the Atlantic Coastal Forest and along the Araguaia and the Negro rivers, 53 were introduced (Hanazaki et al. 2006). This type of data is very important in light of the strong influence of Portuguese colonization on the folk medicine of the Caiçaras of the Atlantic Forest and the Caboclos in the Amazon.

## 6.2 Caiçaras and Riverine Caboclos

Although they are very distant and distinct populations, with the Caiçaras living along the shoreline of the Atlantic Forest and the riverine Caboclos living along the banks of the rivers in the Amazon, they have much in common with regard to their cultures and their adaptations to their associated aquatic environments. According to Begossi (2004a, b), the Caiçaras and Caboclos:

- inhabit high diversity aquatic environments,
- cultivate cassava as an important source of carbohydrates (although land clearing along the coast is currently restricted by environmental agencies),
- engage in subsistence or commercial fishing,
- hunt sporadically (although this activity is currently restricted by environmental agencies),
- collect plants for various purposes,
- have food taboos related to various animals, and
- use plants and animals collected from forests, rivers, and the sea for medicinal purposes (Begossi et al. 2004a, b) (Fig. 6.1b, c).

The Caiçaras are the inhabitants of southeastern coastal Atlantic Forest remnants and are descendants of the Tupinambás Amerindians as well as of Portuguese, among others. A good deal of research has been published about the Caiçaras since the 1950s, including studies by França (1954), Mussolini (1980), and Willems (2003). Begossi et al. (2004a, b) and Diegues (2004) provide detailed reviews of this subject.

Caboclos constitute rural populations in northern Brazil and are descendants of native Indians and Portuguese colonists, often with distinctive influences of African migration. Caboclos can be considered equivalent to the Caiçaras of the Atlantic Forest, both in terms of descent and their interactions with regional natural resources. Both populations cultivate *Manihot esculenta* and a wide diversity of cassava varieties (50–100)—resulting in a complex agricultural management regime (Begossi et al. 2000; Peroni 2004; Peroni and Hanazaki 2002; Peroni et al. 2007)—and both populations interact with high-diversity environments. The riverine Caboclos engage in traditional fishing activities using lines, hooks, and gill nets as well as rowboats or motorized canoes. Fish are an important part of the Caboclo diet, and they consume 38–55 kg per capita per year (Cowx et al. 1998). A great deal of information has been accumulated about the Caboclos of the Amazon region (Moran 1979, Nugent 1993 and Wagley 1974).

The present study examined several riverine and coastal communities in Brazil that depend on aquatic resources as food sources and (in many cases) income. The communities studied are listed, by state, in Table 6.1.

### *6.2.1 Medicinal Fish of the Caiçaras and Caboclos*

In this chapter we will discuss fish cited as having medicinal properties (and not those cited as having only beneficial dietary effects) and used for making medicines to treat diseases or infirmities, or ingested in order to cure some disease (or minimize its symptoms). These fish can be used as ointments, ingested as teas, or mixed with food to treat specific illness. Examples are mentioned in (Tables 6.1, 6.2, 6.3, 6.4, and 6.5).

Piso (1658) reported the use of caraúna, pacamó, taiasicá, and piraumbú fish for medicinal purposes (Table 6.3). We attempted to identify the species Piso cited in the seventeenth century, but these fish were not always recognizable by their common names. This information is illustrative, however, of the antiquity of the use of fish in folk medicine.

Folk knowledge about medicinal animals or fish appears to vary regionally, and not all Caboclo or Caiçara residents use them. There are places, such as Vitória Island, where no fish were cited as having medicinal uses (11 interviews). There are also exceptional cases, such as the high citation rate for peixe-porco (filefish, *Monacanthus* or *Stephanolepis*, whose skin is roasted, powdered, and mixed with water to treat bronchitis and asthma) in essentially all of the fishing communities, especially on the islands of Itacuruçá and Jaguanum (Sepetiba Bay) and in Rio de Janeiro (Table 6.4). These same observations about the medical uses of filefish were made by fishermen in Copacabana in the "Fishing Colony of Posto 6" in Rio de Janeiro in a recent field survey (Fig. 6.1d).

Lung diseases were very frequently cited with regard to the medicinal uses of fish. Recipes usually involve either frying or roasting the fish (sea horses), the skin (peixe-porco/filefish), guaivira (leatherjacket), otoliths (corvina, corvine or drum), "beaks" (peixe-agulha, needlefish), tongues (arraia-jamanta, juggernaut-ray), the liver (baiacu, pufferfish), or their tails (peixe-galo, lookdown) and then mixing the powder with food or water.

In some cases, the "treatments" involved actions such as spitting into the mouth of a bagre (catfish), baiacu (pufferfish), or acará (angelfish) and then releasing the fish back into the water (scientific names shown in Tables 6.4 and 6.5). In the Amazon, fish fat is widely used to treat burns, wounds, and earaches and merits further research. In 2008, a group of researchers at UNIFESP reported positive results in experiments using omega-3 (present in fish fat) to regenerate brain tissue and treat epilepsy (www.gl.globo.com/Noticias10/03/2008).

Rays are widely used in the Araguaia and Tocantins River basins, but were only cited in low numbers by Jurua River fishermen because they did not consider the rays to be true fish—and this species was completely excluded a priori from the questionnaires about medicinal fish.

Many of the fish cited by the Caiçaras in southeastern Brazil (as listed in Table 6.4) were already known in the literature, which indicates that they are probably more widely used than our data alone would indicate. Nomura (1996), for example, observed that the eyes of the bagre (catfish) (Ariidae) are crushed or ground to treat injuries caused by the spur of the same fish. Medicinal uses of the seahorse (*Hippocampus reidi*) date from the Roman era, being indicated for treating asthma and bronchitis. The fat from corvine (*Micropogonias furnieri* or *Umbrina coroides*) is applied to burns, and the use of ray eggs (locally called almofadinha or fop) on the island of Buzios was observed by Lenko (1965); *Raja cyclophora* continued to be used on the island of Buzios in 1986 (see Begossi 1992; Camargo and Begossi 2006). Costa-Neto and Marques (2000) noted that fishermen from Siribinha, Bahia, used catfishes, seahorses, and corvine, among others, for medicinal purposes. Seixas and Begossi (2001) reported that seahorses, corvine, and filefish were used to treat bronchitis on Grande Island, Rio de Janeiro (Aventureiro community).

Our studies in the Tocantins, Araguaia, and Jurua river basins in the Amazon are listed in Tables 6.2 and 6.5. Begossi and Braga (1992) documented the use of fish along the Tocantins River as listed in Table 6.5. Some fish, such as Jaú (gilded catfish, *Paulicea lutkeni*), the Pirarara (*Phractochepalus hemiliopterus*, the redtail catfish), and the Poraquê (*Electrophorus electricus*, electric knifefish), among others, were specifically cited as medicinal by riverine Caboclos (Table 6.5). Nomura (1996) likewise noted some of the same fish cited by the riverine Caboclos as having medicinal uses, such as the acará (a species of Loricariidae) and the Pirarara (valued for its fat). Lo Curto and Piermarocchi (1990) described the medicinal use of Poraquê and the Pirarucu in the Amazon region.

## 6.3 Conclusions

Food taboos or prohibitions, as well as food recommended as a treatment for ill persons (or what is allowed, or not as food) can also be determined, among other factors, by the knowledge or perception of food toxicity (in this case, fish toxicity). In examining fish food taboos on Buzios Island, Begossi (1992) noted three classes of prohibitions:

(a) Toxic fish (such as the pufferfish *Sphoeroides spengleri*) are avoided due to poisoning risks. In this particular case, a neurotoxin (tetraodoxina) is present in the gonads, skin, and liver of species of this family (Tetraodontidae);

(b) Medicinal fish (and other medicinal animals). Many medicinal fish used by the community living on the island of Buzios were subject to food taboos, such as *Tupinambis teguixin*, which was considered by 81 % of fishermen interviewed as having medicinal value. This is also the case of many fish found in the Tocantins river, such as the stingray, jau (*Paulicea lutkeni*), and poraquê (*Electrophorus elecricus*) (Begossi and Braga 1992). Similar situations of food taboos being placed on medicinal fish were recorded by Begossi et al. (1999) in the Juruá River and its tributaries. This information led to the development of the "drugstore hypothesis "in which food taboos contribute to the availability of the medicinal animals in nature. Taboos per se are a form to avoid temptations (Harris 1977, 1985); thus, in this case, they can serve to avoid temptations of consuming medicinal fish. As the "pharmacies" of coastal communities are found in nature itself, keeping medicines available to treat diseases requires consuming them less frequently in normal diets. The "drugstore" of available medicinal animals must be maintained in abundance in order to make harvesting (when necessary) easy and rapid (models of optimal foraging ecology are useful here) to promptly treat the symptoms of the diseases and/or the diseases themselves. It is worth remembering that food taboos in the Amazon are thought to be related to conservation concerns (Reichel-Dolmatoff 1976)—which can be considered a 'precautionary approach' (precautionary principle). The pharmacy of wild animals is a way to ensure "quick and inexpensive" services to patients.

(c) Carnivorous fish: Species at the top of the food chain, such as carnivorous fish, that accumulate toxins or pollutants through the ingestion of species lower in the food chain are avoided. Begossi et al. (2004a, b) presented a review of food taboos (illustrated by the consumption patterns of Atlantic Forest and Amazon communities) and found statistically significant relationships between food taboos and carnivorous fish—as opposed to fish recommended for patients and fish that feed on plants, detritus, and invertebrates (or omnivores).

Lung diseases are frequently cited in the use of medicinal fish by Caiçaras living along the Atlantic coast and among riverine Caboclos of the Amazon basin. The high citations of use of medicinal fish for lung diseases may be an indirect form to assess common illness of the populations studied. Thus, it will be worth to investigate the occurrence of lung diseases among these populations of the Atlantic Forest and of the Amazon.

Various fish parts are used medicinally by the Caiçaras of the Atlantic Forest, including skins, otoliths, and tails. The use of fish fat is very widespread among the Caboclos and deserves more attention in searches for bioactive compounds from fish. Omega 3 fatty acid, which is derived from fish fat, is only one substance among many others that may yet be proven to have medicinal value. The folk knowledge of the Caiçara and Caboclo populations may prove to be of great value in guiding this research—and perhaps one day we will be able to respond to the New Zealand researcher cited in the introduction to this chapter: "Yes, there had been numerous studies regarding the use of those species, and some compounds are already on the way to being used to treat diseases!"

Additionally, as noted in the Introduction section, many poisonous fish species such as rays, catfishes, and leatherjackets, among others have venom glands and are used by traditional communities for medicinal purposes. If drugs are viewed as low doses of poisons, there is much research yet to be performed on the substances found in native fish, and it will be worth investigating if the use of these poisonous fish for medicinal purposes has scientific basis—or if it is just one more case for the "freakonomics" analyses of Levitt and Dubner (2005)—who looked for correlations between many factors that might otherwise never have been correlated.

**Acknowledgments** We are grateful for several earlier fruitful collaborations between us and Jose Geraldo W. Marques; we thank FAPESP for grants provided to many of our projects in Ethnobiology, and CNPq (Conselho Nacional de Desenvolvimento Científico e Tecnológico) for the research productivity scholarship provided to the first author.

## References

Alves RRN, Alves H (2011) The faunal drugstore: animal-based remedies used in traditional medicines in Latin America. J Ethnobiol Ethnomed 7(1):1–9. doi:10.1186/1746-4269-7-9

Alves RRN, Rosa IL (2005) Why study the use of animal products in traditional medicines? J Ethnobiol Ethnomed 1(5):1–5. doi:10.1186/1746-4269-1

Alves RRN, Souto WMS (2011) Ethnozoology in Brazil: current status and perspectives. J Ethnobiol Ethnomed 7(1):1–22. doi:10.1186/1746-4269-7-22

Auerbach SI, Gehrs CW (1980) Environmental toxicology: issues, problems, and challenges. In: Witschi H (ed) The scientific basis of toxicity assessment, 1st edn. Elsevier/North-Holland Biomedical Press, Amsterdam/New York, pp 23–29

Balick MJ, Cox PA (1996) Plants, people and culture, 1st edn. Scientific American Library, New York

Begossi A, e Garavello JC (1990) Notes on ethnoicthyology of fishermen from the Tocantins river. Acta Amazônica 20(único):341–352

Begossi A (1992) Food taboos at Búzios Island (SE Brazil): their significance and relation to folk medicine. J Ethnobiol 12(1):117–139

Begossi A (1998) Food taboos: a scientific reason?. In: Prendergast HDV, Etkin NL, Harris DR, Houghton PJ (eds) Plants for food and medicine [symposium 'plants for food and medicine', joint meeting of the society of economic botany and the international society of ethnopharmacology], 1st edn. The Royal Botanic Gardens, Kew, Richmond, Surrey, UK

Begossi A, Braga FMS (1992) Food taboos and folk medicine among fishermen from the Tocantins River (Brazil). Amazoniana 12:101–118

Begossi A, Figueiredo JL (1995) Ethnoichthyology of southern coastal fishermen: cases from Búzios Island and Sepetiba Bay (Brazil). Bull Mar Sci 56(2):710–717

Begossi A, Silvano RAM, Amaral BD, Oyakama OT (1999) Uses of fish and game by inhabitants of an extrative reserve (Upper Juruá, Acre, Brazil). Environ Dev Sustain 1:73–93

Begossi A, Hanazaki N, Peroni N (2000) Knowledge and use of biodiversity in Brazilian hot spots. Environ Dev Sustain 2(3):177–193

Begossi A, Hanazaki N, Ramos R (2004a) Food chain and the reasons for food taboos in the Amazon and in the Atlantic forest coast. Ecol Appl 14(5):1334–1343

Begossi A, Leme A, Seixas CS, Castro F, Pezzuti J, Hanazaki N, Peroni N, Silvano RAM (2004b) Ecologia de Pescadores da Mata Atlântica e da Amazônia, 1st edn. HUCITEC, São Paulo

Begossi A, Hanazaki N, Ramos R (2006) Healthy fish: medicinal and recommended species in the Amazon and in the Atlantic forest coast (Brazil). In: Pieroni A, Price LL (eds) Eating and healing, traditional food as medicine, 1st edn. The Haworth Press, Inc., New York, pp 237–250

Camargo E, Begossi A (2006) Os Diários de campo da Ilha dos Búzios, 1st edn. HUCITEC, São Paulo

Costa-Neto EM (1999) Barata é um santo remédio: introdução à zooterapia popular no Estado da Bahia, 1st edn. UEFS, Feira de Santana
Costa-Neto EM, Marques JGW (2000) Faunistic resources used as medicines by artisanal fishermen from Siribinha Beach, State of Bahia Brazil. J Ethnobiol 20(1):93–109
Cotton CM (1996) Ethnobotany: principles and applications, 1st edn. Wiley, New York
Cowx IG, O'Grady KT, Batista VS, Inhamuns AJ, Freitas CEC, Freire-Brasil D (1998) Characterization of the fishery in river communities in the low-Solimões/high-Amazon region. Fish Manag Ecol 5(5):419–435. doi:10.1046/j.1365-2400.1998.550419.x
Cunningham P, Goetz P (1996) Venomous and toxic marine life of the world, 1st edn. Lonely Planet Pub, Australia
Diegues AC (2004) Enciclopédia caiçara: o olhar do pesquisador. Hucitec, São Paulo
Etkin N, Johns T (1998) "Pharmafoods" and "nutraceuticals": paradigm shifts in biotherapies. In: Prendergast HDV, Etkin NL, Harris DR, Houghton PJ (eds) Plants for food and medicine [Symposium 'plants for food and medicine', joint meeting of the society of economic botany and the international society of ethnopharmacology], 1st edn. The Royal Botanic Gardens, Kew, Richmond, Surrey, UK, pp 3–16
Farnsworth NR (1988) Screening plants for new medicines. In: Wilson EO (ed) Biodiversity. National Academy Press, Washington
Ferreira FS, Brito SV, Saraiva RA, Araruna MKA, Menezes IRA, Costa JGM, Coutinho HDM, Almeida WO, Alves RRN (2010) Topical anti-inflammatory activity of body fat from the lizard Tupinambis merianae. J Ethnopharmacol 130:514–520
França A (1954) A Ilha de São Sebastião. Universidade de São Paulo, SP (Boletim 178)
Freire L (1939) Grande Dicionário da Lingua Portuguesa. Rio de Janeiro, ed. A Noite
Froese R, Pauly D (2005) Fishbase. http://www.fishbase.org
Haddad V Jr (2000) Atlas de Animais Aquáticos Perogosos do Brasil, 1st edn. Editora Roca, São Paulo
Hanazaki N, Peroni N, Begossi A (2006) Edible and healing plants in the ethnobotany of native inhabitants from the Amazon and the Atlantic Forest (Brazil). In: Pieroni A, Price LL (eds) Eating and healing, traditional food as medicine, 1st edn. The Haworth Press, Inc., New York, pp 251–272
Harris M (1977) Cannibals and kings, 1st edn. Vintage Books, New York
Harris M (1985) Good to eat: riddles of food and culture, 1st edn. Simon and Schuster, New York
Hashimoto Y (1979) Marine toxins and other bioactive marine metabolites, 1st edn. Scientific Press, Tokyo
Levitt SD, Dubner SJ (2005) Freakonomics, 1st edn. HarperCollins Pub, New York
Lenko K (1965) Nosso folclore. Chácaras & Quintais (September):453–456
Lo Curto A, Piermarocchi C (1990) Gli animali che curano secondo la medicina indígena dell'Amazzonia, 1st edn. Studio Sintesi, Milan
McDonald DR (1977) Food taboos: a primitive environmental protection agency (South America). Anthropos 72:734–748
Moran E (1979) A ecologia humana das populações da Amazônia, 1st edn. Vozes, Petrópolis
Mussolini G (1980) Ensaios de antropologia Indígena e Caiçara. Paz e Terra, Rio de Janeiro
Nomura H (1984) Dicionário dos Peixes do Brasil, 1st edn. Editerra Editorial, Brasília
Nomura H (1996) Usos, crendices e lendas sobre peixes. Fundação Vingt-Um Rosado Coleção Mossoroense, Mossoró
Nugent S (1993) Amazonian caboclo society: an essay on invisibility and peasant economy, 1st edn. Berg Pub, Providence
Peroni N (2004) Ecologia e genética de mandioca em roças do sudeste de São Paulo: análises espaciais e temporais. PhD Thesis, Universidade Estadual de Campinas, Campinas, SP, Brazil
Peroni N, Hanazaki (2002) Current and lost diversity of cultivated varieties, especially cassava, under swidden cultivation systems in the Brazilian Atlantic Forest. Agric Ecosyst Environ 92:171–183
Peroni N, Kageyama PY, Begossi A (2007) Molecular differentiation, diversity, and folk classification of "sweet" and "bitter" cassava (Manihot esculenta) in Caiçara and Caboclo management systems (Brazil). Genet Resour Crop Evol 54(6):1333–1349

Pieroni A, Quave CL (2006) Functional foods or food medicines? On the consumption of wild plants among Albanans and Southern Italians in Lucania. In: Pieroni A, Price LL (eds) Eating and healing, traditional food as medicine, 1st edn. The Haworth Press, Inc., New York, pp 101–130

Piso G (1658) História natural e médica da India Ocidental, Departamento de Imprensa Nacional, Rio de Janeiro (reprinted in 1957)

Porter R (1997) Medicine, a history of healing, 1st edn. Marlowe & Company, New York

Queiroz MS (1984) Hot and cold classification in traditional Iguape medicine. Ethnology 23:63–72

Reichel-Dolmatoff G (1976) Cosmology as ecological analysis: a view from the rain forest. Man 11:307–318

Russell FE (1965) Poisonous marine animals, 1st edn. Academic Press, Inc., London

Sankaran JK, Mouly VS (2007) Managing innovation in an emerging sector: the case of marine-based nutraceutical. R & D Manag 37:329–344

Seixas CS, Begossi A (2001) Ethnozoology of fishing communities from Ilha Grande (Atlantic Forest coast, Brazil). J Ethnobiol 21(1):107–135

Silvano RAM et al. (2001) Peixes do Alto Juruá. EDUSP, São Paulo

Simoons FJ (1994) Eat not this flesh. Food avoidances from prehistory to the present, 1st edn. The University of Wisconsin Press, Madison

Soyer A (1853) The pantropheon: a history of food and its preparation in ancient times, 1st edn. Simpkin Marshall, London

Wagley C (1974) Man in the Amazon, 1st edn. Indiana University Press, Bloomington

Willems E (2003) A Ilha dos Búzios. Em colaboração com Gioconda Mussolini HUCITEC/USP, São Paulo

# Chapter 7
# Herpetofauna Used in Traditional Folk Medicine: Conservation Implications

**Rômulo Romeu Nóbrega Alves, Washington Luiz Silva Vieira, Gindomar Gomes Santana, Kleber Silva Vieira and Paulo Fernando Guedes Pereira Montenegro**

**Abstract** This chapter provides an overview of the global use of herpetofauna in traditional folk medicine and the implications for conservation. The results indicate that 331 species (284 reptiles and 47 amphibians) are used in traditional folk medicine around the world. Among the species recorded, 182 reptiles and 42 amphibians are listed in the IUCN Red List. Additionally, 93 reptiles are in some of the appendices of CITES. These numbers demonstrate the importance of understanding such medicinal uses in the context of reptile conservation as well as the need for considering sociocultural factors when establishing management plans directed toward the sustainable use of these reptiles.

## 7.1 Introduction

Amphibians and reptiles (collectively known as herpetofauna) represent one of the most important groups of vertebrates. The herpetofauna and human societies have interacted for millennia, virtually wherever they have been in contact. Thereby, amphibians and reptiles are one of the fundamental ethnozoological entities, and

R. R. N. Alves (✉)
Departamento de Biologia, Universidade Estadual da Paraíba, Avenida das Baraúnas, 351/Campus Universitário, Bodocongó, Campina Grande, PB 58109-753, Brazil
e-mail: romulo_nobrega@yahoo.com.br

G. G. Santana · K. S. Vieira
Programa de Pós-Graduação em Ciências Biológicas (Zoologia), Departamento de Sistemática e Ecologia, Universidade Federal da Paraíba, João Pessoa, PB 58051-900, Brazil

W. L. S. Vieira · G. G. Santana · K. S. Vieira · P. F. G. P. Montenegro
Laboratório de Ecofisiologia Animal, Departamento de Sistemática e Ecologia, Universidade Federal da Paraíba, João Pessoa, PB 58051-900, Brazil

R. R. N. Alves and I. L. Rosa (eds.), *Animals in Traditional Folk Medicine*,
DOI: 10.1007/978-3-642-29026-8_7, 

people interpret their relationship with these animals differently depending on culture, environment, and personal experience (Alves et al. 2010a).

Ethnoherpetology, a subdivision of ethnozoology, examines the relationships between human cultures and reptiles/amphibians (Bertrand 1997; Das 1998; Goodman and Hobbs 1994; Speck 1946). Ethnozoological studies can aid in the evaluation of the impacts human populations have on native animal species and in the development of sustainable management plans, and thus, they are essential to conservation efforts (Alves and Souto 2011).

Evidence from a variety of sources has shown that humans have long exploited the eggs, meat, blood, oil, shells, skin, bones, and other parts of reptiles and amphibians to provide food and materials for making tools, ornaments, and religious objects (Alves 2006; Alves et al. 2006, 2008; Alves and Pereira Filho 2007; Alves and Santana 2008; Fitter 1986; Fitzgerald et al. 2004; Franke and Telecky 2001; Frazier 2005; Mohneke et al. 2009; Tyler et al. 2007; Zhou and Jiang 2004). In addition to these uses, the herpetofauna is used for medicinal and cultural purposes (e.g., as totem or fetish or in particular ceremonies), as well as in day-to-day activities (Alves et al. 2008, 2009; Boll 2004; Mohneke et al. 2011; Myers et al. 1978; Morris and Morris 1965).

People have relied on medicinal products derived from natural sources for millennia, and animals have long been an important part of that repertoire (Adeola 1992; Alves and Rosa 2005; Angeletti et al. 1992; Lev 2003). The pharmacopeias of folk societies as well as of traditional (such as those of the Chinese, Ayurvedic, Unani) and Western medical systems contain thousands of uses for medicines made from leaves, herbs, roots, bark, animals, mineral substances, and other materials found in nature (Gesler 1992). Ingredients derived from wild plants and animals are not only widely used in traditional remedies, but are also increasingly valued as raw materials in the preparation of modern medicines and herbal preparations.

Increased demand and the growth of human populations have led to increased and often unsustainable rates of exploitation of natural resources, and some wild species are already threatened with extinction for this reason (Lee 1999). Discussions concerning the links between traditional medicine and biodiversity are therefore becoming imperative, particularly in view of the fact that folk medicine is the source of primary health care for 80% of the world's population (Alves and Rosa 2005; Alves et al. 2007a).

Despite the intensive use of the herpetofauna for medicinal purposes, there is a general lack of detailed information concerning the magnitude of this harvesting and its impact on the species involved (Alves and Pereira Filho 2007; Alves et al. 2008; Mohneke et al. 2011). Demands on the wild sources of traditional medicinal products are increasing as human populations grow inexorably and poorer countries are forced to decrease spending per capita on Western health systems. On the other hand, Western populations are turning to more traditional and homeopathic products, and their demand for natural remedies is increasing (IUCN 2000). Additionally, some species are in danger of extinction due to a combination of factors independent of the growing global demand for traditional medicines and other natural products.

Reptiles are among the animal species most frequently used in traditional folk medicine (Alves and Alves 2011; Alves et al. 2007a, 2008, 2009; Mahawar and Jaroli 2008; Vázquez et al. 2006; Zhou and Jiang 2004). This also applies to amphibians which are used for medicinal purposes in several countries (Alves et al. 2007b; Boll 2004; Mohneke et al. 2011), although to a lesser extent compared to reptiles.

The goal of this chapter is to provide an overview of the global use of the herpetofauna in traditional medicine, to identify those species used as folk remedies, and to discuss the implications of their harvesting. In this context, we address the following questions: (1) which reptile and amphibian species are used in folk medicine; (2) which medicinal species are endangered, and (3) what are the implications of the use of zootherapeutics for reptile and amphibian conservation? We hope to stimulate further discussions about this use of biodiversity and its implications for wildlife conservation.

## 7.2 Methods

In order to examine the diversity of reptiles used in traditional medicine, all available references or reports of folk remedies based on herpetofauna sources were examined. Only taxa that could be identified to the species level were included in the database. Scientific names provided in the publications were updated according to The Reptile Database (2011) and Amphibian Species of the World 5.5 (Frost 2011). The conservation status of the reptile species follows IUCN (2011) and CITES (2011).

The sources analyzed were: Branch and Silva (1983), Begossi (1992), Begossi and Braga (1992), Donadio and Gallardo (1984), Figueiredo (1994), China National Corporation of Traditional and Herbal Medicine (1995), Marques (1995), Freire (1996), Costa-Neto (1996, 1999a, b, c, 2000), SEMARNAP-PROFEPA (1998), Begossi et al. (1999), Sodeinde and Soewu (1999), Chen et al. (2000, 2006, 2009), El-Kamali (2000), Perry (2000), Seixas and Begossi (2001), Almeida and Albuquerque (2002), CITES (2002), Kakati and Doulo (2002), Apaza et al. (2003), Lev (2003), Fitzgerald et al. (2004), Silva et al. (2004), Almeida et al. (2005), Andrade and Costa-Neto (2005), Costa-Neto and Pacheco (2005), Smart et al. (2005), Ashwell and Walston (2008), Alakbarli (2006), Alves (2006), Alves and Rosa (2006; 2007a, b; 2010), Alves et al. (2007a, b), Ives (2006), Kakati et al. (2006), Mahawar and Jaroli (2006), Vázquez et al. (2006), Alves and Pereira Filho (2007), Barzyk (1999), Dharmananda (2007a; 2007b), El Din (2007), Negi and Palyal (2007), Fretey et al. (2007), Highfield and Bayley (2007), Highfield and Slimani (2007), IFAW (2007), Meiling et al. (2008), Quave et al. (2010), Rowley et al. (2010), Martínez and Barboza (2010), Mohneke et al. (2010), Mohneke et al. (2011), Shao-Ke et al. (2010), Alves and Alves (2011), Indian Traditional Medicinal Knowledgebase (2011) and Lohani (2011).

## 7.3 Results and Discussion

### *7.3.1 Medicinal Herpetofauna*

The medicinal herpetofauna includes a total of 331 species, of which 284 are reptiles and 47 are amphibians. These species belonging to 202 genera and 57 families are used in traditional folk medicine.

Among the reptiles, the groups with the largest numbers of species used were snakes (123 species), followed by lizards (71), chelonians (76), and crocodilians (14) (Table 7.1). In relation to amphibians, the family with the largest number of species was Bufonidae (15 species). Of the total amphibians recorded, 42 species are included in the IUCN Red List (Fig. 7.1), but none are in the appendices of CITES. Among the reptiles recorded, a total of 182 species are listed in IUCN (Fig. 7.2) and 93 are found in some of the appendices of CITES. The order Testudines showed the highest number of species included in the IUCN Red List and in the appendices of CITES, with 22 species in the category vulnerable and 26 species in Appendix II, respectively (Fig. 7.3).

These numbers reveal the importance of the medicinal use of the herpetofauna from a conservationist perspective, even though this form of exploitation is not the main threat to the majority of the species of reptiles and amphibians used in traditional medicine. This is an additional factor that increases pressure on exploited species.

The high taxonomic diversity observed among reptiles used in traditional medicine is not surprising, as numerous workers have pointed out that reptiles are among the animals most frequently used in folk medicine (Alves and Alves 2011; Ashwell and Walston 2008; Kakati et al. 2006; Mahawar and Jaroli 2008). However, considering the relatively small number of published studies on this subject, it is possible that the number of medicinal reptile species used is greater than that recorded here.

Some widespread species are used in different countries, such as *Kinosternon scorpioides* (in Mexico and Brazil), *Varanus niloticus* (in India, Sudan, and China), *Varanus bengalensis* (in India and China), and *Dermochelys coriacea* (in Brazil, Mexico, Benin, Cameroon, and Togo). A given animal often has multiple medicinal uses and can be employed to treat more than one ailment, while different species of reptiles and amphibians can likewise be used to treat the same illnesses. Products derived from *Tupinambis merianae* and *T. teguixin*, for instance, are indicated for treating 8 and 18 conditions, respectively, in Brazil (Alves and Rosa 2007a); in India, products derived from the land monitor (*V. bengalensis)* are used to treat hemorrhoids, rheumatism, body pain, and burns, as well as spider and snake bites (Kakati et al. 2006); in Mexico, rattlesnake pills have been indicated for curing a wide variety of ailments, including: skin blotches, cancer, sores, rashes, pimples, welts, itching, rheumatism, varicose veins, face blotches, acne, blackheads, stress, heart disease, diabetes, hemorrhoids, and sexual impotence (Rubio 1998).

**Table 7.1** Herpetofauna species used in traditional folk medicine and conservation status according to IUCN (2011) and CITES Appendix (2011)

| Family/species |
| --- |
| **AMPHIBIA (47 species)** |
| **Caudata (2 species)** |
| **Salamandridae (2 spp)**: *Laotriton laosensis* (Stuart and Papenfuss, 2002), *Salamandra salamandra* (Linnaeus, 1758)[LC]. |
| **Anura (45 species)** |
| **Hyperoliidae (1 sp):** *Kassina fusca* Schiøtz, 1967[LC] |
| **Arthroleptidae (1 sp):** *Leptopelis bufonides* Schiøtz, 1967[LC] |
| **Pyxicephalidae (1 sp):** *Tomopterna cryptotis* (Boulenger, 1907)[LC] |
| **Bufonidae (15 spp):** *Bufo bufo* (Linnaeus, 1758)[LC], "Bufo" *pentoni* Anderson, 1893[LC], *Amietophrynus regularis* (Reuss, 1833)[LC], *A. maculatus* (Hallowell, 1854)[LC], *A. xeros* (Tandy, Tandy, Keith, and Duff-MacKay, 1976)[LC], *Nanorana liebigii* (Günther, 1860)[LC], *Rhinella schneideri* (Werner, 1894)[LC], *R. marina* (Linnaeus, 1758)[LC], *R. jimi* (Stevaux, 2002)[LC], *R. icterica* (Spix, 1824)[LC], *Incilius bocourti* (Brocchi, 1877)[LC], *I. valliceps* (Wiegmann, 1833)[LC], *I. macrocristatus* (Firschein and Smith, 1957)[VU], *Schismaderma carens* (Smith, 1848)[LC], *Duttaphrynus melanostictus* (Schneider, 1799)[LC]. |
| **Leptodactylidae (3 spp):** *Leptodactylus labyrinthicus* (Spix, 1824)[LC], *L. vastus* Lutz, 1930, *L. troglodytes* Lutz, 1926[LC]. |
| **Dicroglossidae (3 spp):** *Hoplobatrachus tigerinus* (Daudin, 1802)[LC], *Nanorana liebigii* (Günther, 1860)[LC], *N. polunini* (Smith, 1951)[LC]. |
| **Craugastoridae (2 spp):** *Craugastor laticeps* (Duméril, 1853)[NT], *C. glaucus* (Lynch, 1967)[CR] |
| **Ranidae (8 spp):** *Lithobates maculatus* (Brocchi, 1877), *L. berlandieri* (Baird, 1859)[LC], *L. montezumae* (Baird, 1854)[LC], *L. sylvaticus* (LeConte, 1825)[LC], *L. spectabilis* (Hillis and Frost, 1985)[LC], *Pelophylax perezi* (López-Seoane, 1885)[LC], *P.* ridibundus (Pallas, 1771)[LC], *Rana amurensis* Boulenger, 1886[LC]. |
| **Hylidae (8 spp):** *Charadrahyla chaneque* (Duellman, 1961)[EN], *Trachycephalus typhonius* (Linnaeus, 1758), *T. resinifictrix* (Goeldi, 1907)[LC], *Phyllomedusa bicolor* (Boddaert, 1772)[LC], *P. burmeisteri* Boulenger, 1882[LC], *Hyla arborea* (Linnaeus, 1758)[LC], *H. cinerea* (Schneider, 1799) [LC], *H. arenicolor* Cope, 1866 [LC]. |
| **Microhylidae (1 sp):** *Hypopachus barberi* Schmidt, 1939[VU]. |
| **Ceratophryidae (1 sp):** *Telmatobius culeus* (Garman, 1876)[CR]. |
| **Calyptocephalellidae (1 sp):** *Calyptocephalella gayi* (Duméril and Bibron, 1841). |
| **REPTILIA (284 species)** |
| **Testudines (76 species)** |
| **Cheloniidae (6 spp):** *Chelonia mydas* (Linnaeus, 1758)[VU/I], *Eretmochelys imbricata* (Linnaeus, 1766)[EN/I], *Caretta caretta* (Linnaeus, 1758)[VU/I], *Lepidochelys olivacea* (Eschscholtz, 1829)[VU/I], L. kempii Garman, 1880 [CR/I], *Natator depressus* Garman 1880[DD//I], |
| **Dermochelyidae (1 spp):** *Dermochelys coriacea* (Vandelli, 1761)[CR/I]. |
| **Emydidae (3 spp):** *Trachemys scripta* (Schoepff, 1792)[LC/II], *Malaclemys terrapin* Schoepff 1793[LC], *Emys orbicularis* (Linnaeus, 1758)[NT] |
| **Chelidae(4 spp):** *Phrynops geoffroanus* (Schweigger, 1812), *P. tuberosus* Peters 1870, *Mesoclemmys tuberculata* (Luederwaldt, 1926), *Chelus fimbriatus* (Schneider, 1783). |
| **Trionychidae (5 spp):** Lissemys punctata (Lacepède, 1788)[LC/II], *Pelochelys bibroni* (Owen 1853)[VU/II], *Pelodiscus sinensis* (Wiegmann 1835)[VU], *Palea steindachneri* (Siebenrock 1906)[EN/III], *Amyda cartilaginea* (Boddaert, 1770)[VU] |
| **Chelydridae (1 sp):** *Platysternon megacephalum* Gray 1831[EN/II] |

(continued)

**Table 7.1** (continued)

| Family/species |
|---|
| **Testudinidae (15 spp):** *Testudo horsfieldii* Gray 1844$^{\mathrm{VU/II}}$, *T. graeca* Linnaeus 1758$^{\mathrm{VU/II}}$, *T. kleinmanni* Lortet 1883$^{\mathrm{CR/I}}$, *Chelonoidis carbonaria* (Spix, 1824)$^{\mathrm{DD/II}}$, *C. denticulata* (Linnaeus, 1766)$^{\mathrm{VU/II}}$, *Geochelone platynota* (Blyth, 1863)$^{\mathrm{CR}}$, *G. elegans* (Schoepff 1795)$^{\mathrm{LC}}$, *Stigmochelys pardalis* (Bell, 1828)$^{\mathrm{LC}}$, *Kinixys belliana* (Gray 1831)$^{\mathrm{LC}}$, K. spekii Gray, 1863$^{\mathrm{LC,}}$ *Indotestudo elongata* (Blyth 1854)$^{\mathrm{EN}}$, *I. forstenii* (Schlegel & Müller 1844)$^{\mathrm{EN/II}}$, *Manouria impressa* (Günther 1882)$^{\mathrm{VU}}$, *Astrochelys radia*ta (Shaw, 1802)$^{\mathrm{CR}}$, Chersina angulata (Schweigger, 1812)$^{\mathrm{LC}}$. |
| **Kinosternidae (5 spp):** *Kinosternon scorpioides* Linnaeus 1766, *K. integrum* (Le Conte 1854)$^{\mathrm{LC}}$, *Pangshura tentoria* (Gray 1834)$^{\mathrm{LC}}$, P. tecta (Gray 1831)$^{\mathrm{LC}}$, *Staurotypus triporcatus* (Wiegmann 1828)$^{\mathrm{NT}}$. |
| **Podocnemididae (4 spp):** *Podocnemis expansa* (Schweiger,1812)$^{\mathrm{LC/II}}$, *P. unifilis* (Troschel, 1848)$^{\mathrm{VU/II}}$, *P. sextuberculata* Cornalia 1849$^{\mathrm{VU/II}}$, *Peltocephalus dumeriliana* (Schweigger 1812)$^{\mathrm{VU/II}}$. |
| **Geoemydidae:** (33 spp): *Rhinoclemmys punctularia* (Daudin, 1802), *Cuora amboinensis* (Daudin 1802)$^{\mathrm{VU}}$, *C. trifasciata* (Bell 1825)$^{\mathrm{CR}}$, C. aurocapitata Luo & Zong, 1988$^{\mathrm{CR}}$ *Cistoclemmys galbinifrons* (Bourret, 1939)$^{\mathrm{CR/II}}$, *C. flavomarginata* (Gray 1863)$^{\mathrm{EN/II}}$, *Leucocephalon yuwonoi* (Mccord, Iverson & Boeadi 1995)$^{\mathrm{CR/II}}$, *Sacalia bealei* (Gray 1831)$^{\mathrm{EN}}$, *S. quadriocellata* (Siebenrock, 1903)$^{\mathrm{EN}}$, *Mauremys reevesii* (Gray, 1831)$^{\mathrm{EN/III}}$, *M. mutica* (Cantor 1842)$^{\mathrm{EN}}$, *M. leprosa* (Schweigger, 1812), *Ocadia sinensis* (Gray 1834)$^{\mathrm{EN}}$, *Morenia petersi* (Anderson, 1879)$^{\mathrm{VU}}$, *M. ocellata* (DumÉril and Bibron 1835)$^{\mathrm{VU/I}}$, *Pyxidea mouhotii* (Gray 1862)$^{\mathrm{EN/II}}$, *Geoemyda spengleri* (Gmelin, 1789)$^{\mathrm{EN/III}}$, *Malayemys subtrijuga* (DumÉril and Bibron 1835)$^{\mathrm{VU/II}}$, *Cyclemys dentata* (Gray 1831)$^{\mathrm{NT}}$, *Geoclemys hamiltonii* (Gray 1831)$^{\mathrm{VU/I}}$, *Hardella thurjii* (Gray 1831)$^{\mathrm{VU}}$, *Heosemys depressa* (Anderson, 1875)$^{\mathrm{CR/II}}$, *H. grandis* (Gray 1831)$^{\mathrm{VU/II}}$, *H. spinosa* (Gray 1831)$^{\mathrm{EN/II}}$, *H. annandalii* (Boulenger, 1903)$^{\mathrm{EN/II}}$, *Chinemys nigricans* (Gray 1834)$^{\mathrm{EN/III}}$, *Melanochelys trijuga* (Schweigger, 1812)$^{\mathrm{NT}}$, *Notochelys platynota* (Gray 1834)$^{\mathrm{VU/II}}$, *Callagur borneoensis* (Schlegel and Müller, 1844)$^{\mathrm{CR}}$, *Pangshura smithii* (Gray 1863)$^{\mathrm{NT/II}}$, *P. tecta* (Gray 1831)$^{\mathrm{LC/I}}$, *Siebenrockiella crassicollis* (Gray 1831)$^{\mathrm{VU/II}}$, Orlitia borneensis Gray, 1873$^{\mathrm{EN}}$. |
| **Platysternidae (1 sp):** *Platysternon megacephalum* Gray 1831$^{\mathrm{EN/II}}$ |
| **Crocodylia (14 species)** |
| **Alligatoridae (7 spp):** *Caiman latirostris* (Daudin, 1801)$^{\mathrm{LC/II}}$, *C. crocodilus* (Linnaeus, 1758)$^{\mathrm{LC/II}}$, *C. yacare* (Daudin 1802), *Paleosuchus palpebrosus* (Cuvier, 1807)$^{\mathrm{LC/II}}$, *P. trigonatus* (Schneider, 1801)$^{\mathrm{DD/II}}$, *Melanosuchus niger* (Spix, 1825)$^{\mathrm{LC/II}}$, Alligator sinensis Fauvel 1879$^{\mathrm{CR/I}}$ |
| **Crocodylidae (6 spp):** *Crocodylus niloticus* Laurenti 1768$^{\mathrm{LC/I}}$, *C. siamensis* Schneider 1801$^{\mathrm{CR/I}}$, *C. porosus* (Schneider, 1801)$^{\mathrm{LC/I}}$, C. palustris Lesson 1831$^{\mathrm{VU/I}}$, *C. moreletii* (Duméril & Bibron 1851)$^{\mathrm{LC/I}}$, *C. acutus* (Cuvier 1807)$^{\mathrm{VU/II}}$. |
| **Gavialidae (1 sp):** *Gavialis gangeticus* (Gmelin 1789)$^{\mathrm{EN/I}}$ |
| **Squamata—Lizards (71 species)** |
| **Phrynosomatidae (8 spp):** *Sceloporus serrifer* Cope 1866$^{\mathrm{LC}}$, *S. taeniocnemis* Cope 1885$^{\mathrm{LC}}$, *S. acanthinus* Bocourt 1873, *S. spinosus* Wiegmann 1828$^{\mathrm{LC}}$, *S. grammicus* Wiegmann 1828$^{\mathrm{LC}}$, *Phrynosoma cornutum* (Harlan 1825)$^{\mathrm{LC}}$, *P. modestum* Girard 1852$^{\mathrm{LC}}$, *P. orbiculare* (Gmelin, 1789)$^{\mathrm{LC}}$. |
| **Anguidae (3 spp):** *Abronia lythrochila* Smith & Alvarez Del Toro 1963$^{\mathrm{LC}}$, *Mesaspis moreletii* Bocourt 1871$^{\mathrm{LC}}$, Ophisaurus harti Boulenger 1899$^{\mathrm{LC}}$. |
| **Scincidae (3 spp):** Acontias plumbeus Bianconi, 1849$^{\mathrm{LC}}$, *Eutropis carinata* (Schneider 1801)$^{\mathrm{LC}}$,Scincus scincus (Linnaeus 1758). |

(continued)

**Table 7.1** (continued)

| Family/species |
|---|
| **Cordylidae (4 spp):** *Cordylus giganteus* Smith 1844$^{VU/II}$, C. tropidosternum (Cope 1869)$^{LC/II}$, C. vittifer (Reichenow, 1887)$^{LC}$, C. warreni (Boulenger 1908) $^{LC/II}$. |
| **Helodermatidae (1 sp):** Heloderma horridum (Wiegmann 1829)$^{VU/II}$. |
| **Agamidae (9 spp):** *Uromastyx hardwickii* Gray, 1827$^{II}$, *U. dispar* Heyden 1827$^{II}$, *U. aegyptia* (Forskal 1775)$^{II}$, *Agama agama* (Linnaeus 1758), *A. impalearis* Boettger 1874, *Calotis versicolor* Daudim 1802, *Laudakia nupta* (De Filippi 1843), *Trapelus mutabilis* (Merrem 1820), Acanthocercus atricollis (Smith, 1849)$^{LC}$. |
| **Lacertidae (4 spp):** Lacerta agilis Linnaeus 1758$^{LC}$, Timon lepidus (Daudin, 1802)$^{NT}$, Podarcis hispanicus (Steindachner, 1870)$^{LC}$, *Zootoca vivipara* (Von Jacquin 1787)$^{LC}$. |
| **Liolaemidae (2 spp):** *Liolaemus pantherinus* Pellegrin 1909, *L. alticolor* Barbour 1909. |
| **Varanidae (5 spp):** *Varanus niloticus* (Linnaeus 1758)$^{LC}$/$^{II}$. *V. bengalensis* (Daudin 1758)$^{LC/I}$, V. salvator (Laurenti 1768)$^{LC/II}$, *V. griseus* (Daudin 1803)$^{I}$, *V. albigularis* (Daudin 1802)$^{LC}$/$^{II}$. |
| **Teiidae (6 spp):** *Tupinambis merianae* (Duméril & Bibron, 1839)$^{LC/II}$, *T. teguixin* (Linnaeus 1758)$^{II}$, *T. rufescens* (Günther,1871)$^{II}$, *Ameiva ameiva* (Linnaeus, 1758), *Cnemidophorus* gr *ocellifer* (Spix, 1825), *Kentropyx pelviceps* (Cope 1868). |
| **Iguanidae (3 spp):** *Iguana iguana* (Linnaeus, 1758)$^{II}$, *Ctenosaura pectinata* Wiegmann 1834, *C. similis* (Gray 1831)$^{LC}$. |
| **Polychrotidae (3 spp):** *Polychrus acutirostris* Spix 1825, *P. marmoratus* (Linnaeus 1758), *Anolis fuscoauratus* D'Orbigny, 1837 |
| **Tropiduridae (4 spp):** *Tropidurus semitaeniatus* (Spix, 1825)$^{LC}$, *T. torquatus* (Wied, 1820)$^{LC}$, *T. hispidus* (Spix, 1825), *Uranoscodon superciliosus* (Linnaeus 1758). |
| **Gekkonidae (4 spp):** *Hemidactylus mabouia* (Moreau de Jonnes, 1818), *H. frenatus* Schlegel 1836$^{LC}$, *Gekko gecko* (Linnaeus 1758)$^{NT}$, G. chinensis (Gray, 1842)$^{LC}$. |
| **Phyllodactylidae (1 sp):** *Tarentola mauritanica* (Linnaeus 1758)$^{LC}$. |
| **Sphaerodactylidae (1 sp):** *Gonatodes hasemani* Griffin 1917$^{LC}$. |
| **Chamaeleonidae (8 spp):** *Chamaeleo senegalensis* Daudin 1802$^{LC/II}$, *C. chamaeleon* (Linnaeus 1758)$^{II}$, C. dilepis Leach, 1819,$^{LC}$ *Furcifer lateralis* (Gray, 1831), Bradypodion dracomontanum Raw 1976$^{II}$, B. nemorale Raw 1978$^{LR/II}$, B. setaroi Raw 1976$^{EN/II}$, B. thamnobates Raw 1976$^{LC/II}$. |
| **Gerrhosauridae (2 spp):** Gerrhosaurus major Duméril, 1851$^{LC}$, G. flavigularis Wiegmann, 1828$^{LC}$. |
| **Squamata—Snakes (123 species)** |
| **Achrochordidae (1 sp):** Acrochordus granulatus (Schneider, 1799)$^{LC}$. |
| **Boidae (8 spp):** *Boa constrictor* Linnaeus, 1758$^{II}$, *Corallus caninus* (Linnaeus, 1758)$^{DD/II}$, *C. hortolanus* (Linnaeus, 1758)$^{DD/II}$, *Epicrates assisi* Machado, 1945, *E. cenchria* (Linnaeus, 1758)$^{II}$, *Eunectes murinus* (Linnaeus, 1758)$^{II}$, *E. notaeus* (Cope, 1862), *Eryx johnii* (Russell 1801)$^{II}$. |
| **Pythonidae (6 spp):** *Python sebae* (Gmelin 1789)$^{II}$, *P. regius* (Shaw 1802)$^{II}$, *P. molurus* (Linnaeus 1758)$^{LC/I}$, *P. natalensis* Smith 1840$^{LC/II}$, P. bivittatus Kuhl, 1820, *Broghammerus reticulatus* (Schneider 1801)$^{II}$. |

(continued)

**Table 7.1** (continued)

| Family/species |
|---|
| **Viperidae (35 spp):** *Lachesis muta* (Linnaeus, 1766), *Caudisona durissa* (Linnaeus, 1758)$^{DD/III}$, *C. molossus* (Baird & Girard 1853)$^{LC}$, *C. basiliscus* (Cope 1864)$^{LC}$, Crotalus. horridus Linnaeus 1758$^{LC}$, *C. scutulatus* (Kennicott, 1861)$^{LC}$, *C. viridis* Rafinesque 1818$^{LC}$, *Hoserea atrox* (Baird and Girard 1853)$^{LC}$, *Uropsophus lepidus* (Kennicott 1861)$^{LC}$, *U. triseriatus* Wagler 1830$^{LC}$, *Aechmophrys pricei* (Van Denburgh 1895)$^{LC}$, *A. willardi* (Meek 1905)$^{LC}$, *A. polystictus* (Cope 1865)$^{LC}$, *A. transversus* (Taylor 1944)$^{LC}$, *Deinagkistrodon acutus* (Günther 1888), *Atropoides nummifer* (Rüppell 1845)$^{LC}$, *Cerrophidion tzotzilorum* (Campbell 1985)$^{LC}$, *Protobothrops mucrosquamatus* (Cantor 1839)$^{LC}$, *Bothrops atrox* (Linnaeus 1758), *B. asper* (Garman 1883), *B. leucurus* Wagler, 1824, *B. lanceolatus* (Bonnaterre, 1790), *Bitis gabonica* (Duméril, Bibron & Duméril 1854), *B. arietans* (Merrem 1820), *Causus rhombeatus* (Lichtenstein 1823), *Agkistrodon contortrix* (Linnaeus, 1766)$^{LC}$, *Gloydius blomhoffi* (Boie 1826), *G. himalayanus* (Günther 1864), *Daboia russelli* (Shaw 1797), *Macrovipera lebetina* (Linnaeus, 1758), Vipera latastei (Bosca, 1878)$^{VU}$, Viridovipera stejnegeri (Schmidt, 1925), V. vogeli (David, Vidal & Pauwels, 2001), Cryptelytrops albolabris (Gray, 1842)$^{LC}$, C. macrops (Kramer, 1977). |
| **Elapidae (25 spp):** *Micrurus spixii* Wagler 1824, *M. surinamensis* (Cuvier 1817), *M. ibiboboca* (Merrem, 1820), *Bungarus multicinctus* Blyth 1861, *B. fasciatus* (Schneider 1801), B. candidus (Linnaeus, 1758), *Naja atra* Cantor 1842$^{II}$, *N. annulifera* Peters 1854, *N. naja* (Linnaeus 1758)$^{II}$, N. mossambica Peters, 1854$^{LC}$, N. kaouthia Lesson, 1831$^{LC}$, N. siamensis Laurenti, 1768$^{LC}$, Elapognathus coronata (Schlegel, 1837), *Ophiophagus hannah* (Cantor 1836)$^{VU/II}$. *Hydrophis cyanocinctus* Daudin 1803$^{LC}$, H. melanocephala Gray, 1849$^{DD}$, Polyodontognathus caerulescens (Shaw, 1802)$^{LC}$, Pelamis platura (Linnaeus, 1766)$^{LC}$, Dendroaspis polylepis Günther, 1864$^{LC}$, D. angusticeps (Smith, 1849)$^{LC}$, Hemachatus haemachatus (Bonnaterre, 1789)$^{LC}$ Lapemis hardwickii Gray, 1834, Praescutata viperina (Schmidt, 1852)$^{LC}$, *Laticauda laticaudata* (Linnaeus, 1758)$^{LC}$, *L. semifasciata* (Reinwardt, 1837)$^{NT}$. |
| **Dipsadidae (8 spp):** *Oxyrhopus trigeminus* Duméril, Bibron & Duméril, 1854, *O. formosus* (Wied 1820), *O. melanogenys* (Tschudi 1845), *Tachymenis peruviana* Wiegmann, 1835, *Imantodes cenchoa* Linnaeus 1758, *Leptodeira annulata* (Linnaeus, 1758), *Philodryas olfersii* (Lichenstein, 1823), *Thamnodynastes strigatus* (Günther, 1858)$^{LC}$. |
| **Homalopsidae (3 sp):** Enhydris enhydris (Schneider, 1799)$^{LC}$, E. chinensis Gray 1842$^{LC}$, *E. plumbea* Boie 1827$^{LC}$. |
| **Colubridae (28 spp):** *Spilotes pullatus* (Linnaeus, 1758), *Leptophis ahetula* (Linnaeus, 1758), *Chironius carinatus* (Linnaeus, 1758), *C. grandisquamis* Peters 1869, *Mastigodryas bifossatus* (Raddi, 1820), *Lampropeltis triangulum* (Lacépède 1789), *Ptyas dhumnades* (Cantor 1842), *P. mucosus* (Linnaeus, 1758)$^{II}$, P. korros (Schlegel 1837), *Drymobius margaritiferus* (Schlegel 1837), *Dinodon rufozonatum* (Cantor 1842), *Orthriophis taeniurus* (Cope 1861), *O. moellendorffi* (Boettger 1886), *Pituophis lineaticollis* (Cope 1861)$^{LC}$, *Rhinechis scalaris* (Schinz, 1822)$^{LC}$, Dispholidus typus (Smith, 1828)$^{LC}$, Ahaetulla nasuta (Bonnaterre, 1790), A. prasina (Boie, 1827)$^{LC}$, Boiga multomaculata (Boie, 1827), Coelognathus radiatus (Boie, 1827), Dendrelaphis pictus (Gmelin, 1789), *Elaphe quatuorlineata* (Wagler, 1833) E. carinata (Günther 1864), *E. radiata* Boie 1827, E. schrenckii Strauch, 1873, E. bimaculata Schmidt, 1925, Euprepiophis mandarinus (Cantor, 1842), Oocatochus rufodorsatus (Cantor, 1842)$^{LC}$. |
| **Cylindrophiidae (1 sp):** Cylindrophis ruffus (Laurenti, 1768). |
| **Natricidae (2 sp):** Xenochrophis flavipunctatus (Hallowell, 1860), Sinonatrix annularis (Hallowell, 1856). |

(continued)

**Table 7.1** (continued)

| Family/species |
|---|
| **Lamprophiidae (3 sp):** Malpolon monspessulanus (Hermann, 1804)$^{LC}$, Lamprophis aurora (Linnaeus, 1758)$^{LC}$, Pseudaspis cana (Linnaeus, 1758)$^{LC}$, |
| **Psammophiidae (3 sp):** Psammophis phillipsi (Hallowell, 1844)$^{LC}$, Psammophylax rhombeatus (Linnaeus, 1758)$^{LC}$, P. tritaeniatus (Günther, 1868)$^{LC}$. |

Categories of IUCN red list: *CR* critically endangered, *EN* endangered, *VU* vulnerable, *LR* lower risk, *DD* deficient data. CITES Appendix (I, II and III)

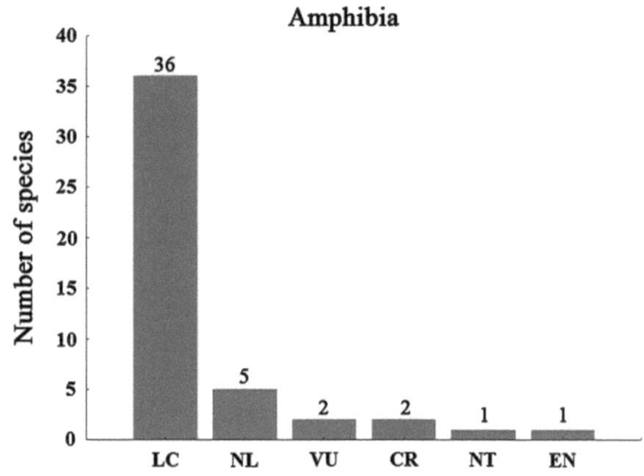


**Fig. 7.1** Status of the conservation of amphibian species used in traditional folk medicine and conservation status according to IUCN (2011). Categories of IUCN Red List: *CR* critically endangered, *EN* endangered, *VU* vulnerable, *LR* lower risk, *DD* deficient data, *NL* not listed

Reptiles are one of the groups most closely associated with the history of medicine. The Greeks and Romans worshipped snakes and the god of medicine is represented holding a snake (Ziemendorff 2008). Historical documents indicated that reptiles have been used in traditional medicines since ancient times (Alakbarli 2006; Almeida 2007; Alves et al. 2007a; MacKinney 1946; Silva et al. 2004). In Brazil, for example, animal species (including reptiles) have been used medicinally by indigenous societies for millennia (Alves et al. 2007a).

In his Compendium of Materia Medica, Li Shizhen, a noted pharmacologist in the Ming Dynasty (1368–1644 A.D.) states that turtle helps "repair internal injury caused by overstrain, strengthens the yin and yang" and "replenishes vital essence, reduces fever, clam the liver and subdues yang, soften and resolve hard masses" (Li et al. 2000). Many ancient Chinese medical books described the therapeutic effects of treating rheumatism, hemiplegia, neuralgia, and muscle poliomyelitis with parts of snakes including gall bladder and liver (Guo et al. 1996).

Similarly, a historical review of the therapeutic uses of animals as described in medieval manuscripts from Alakbarli (2006) revealed a total of 12 species of reptiles with medicinal uses. According to these medieval manuscripts, these reptiles were successfully used to treat ailments that included sexual impotence and leprosy. Among the species mentioned were indigenous species still found in Azerbaijan, such as the Caucasian agama (*Agama caucasica*), the Levantine viper (*Vipera lebetina*), the Mediterranean tortoise (*Testudo graeca*), and the Moorish

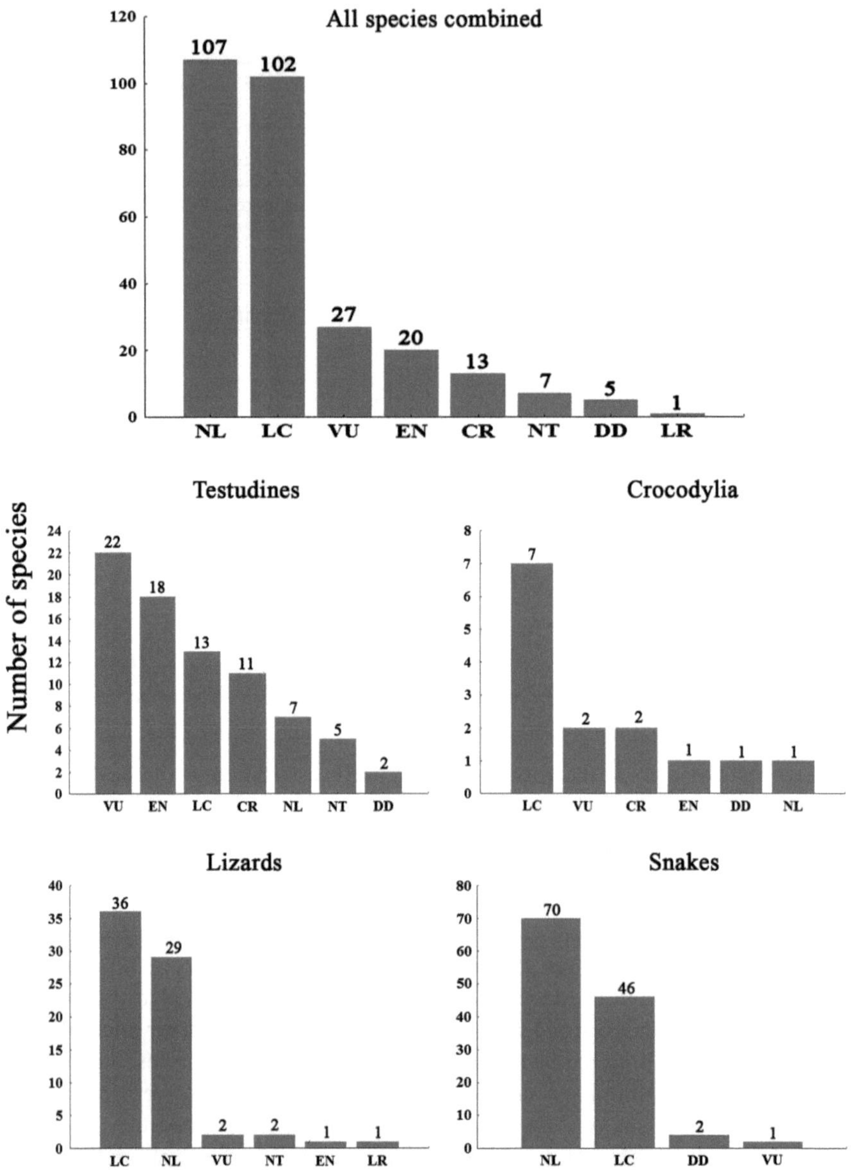


**Fig. 7.2** Status of the conservation of reptile species used in traditional folk medicine and conservation status according to IUCN (2011). Categories of IUCN Red List: *CR* critically endangered, *EN* Endangered, *VU* vulnerable, *LR* lower risk, *DD* deficient data, *NL* not listed

gecko (*Tarentola mauritanica*). Exotic reptiles mentioned included the chameleon (*Chameleo chameleo*), the monitor lizard (*Varanus griseus*), and the crocodile (*Crocodylus niloticus*). The medicines prepared from these reptiles were imported into Azerbaijan from distant countries.

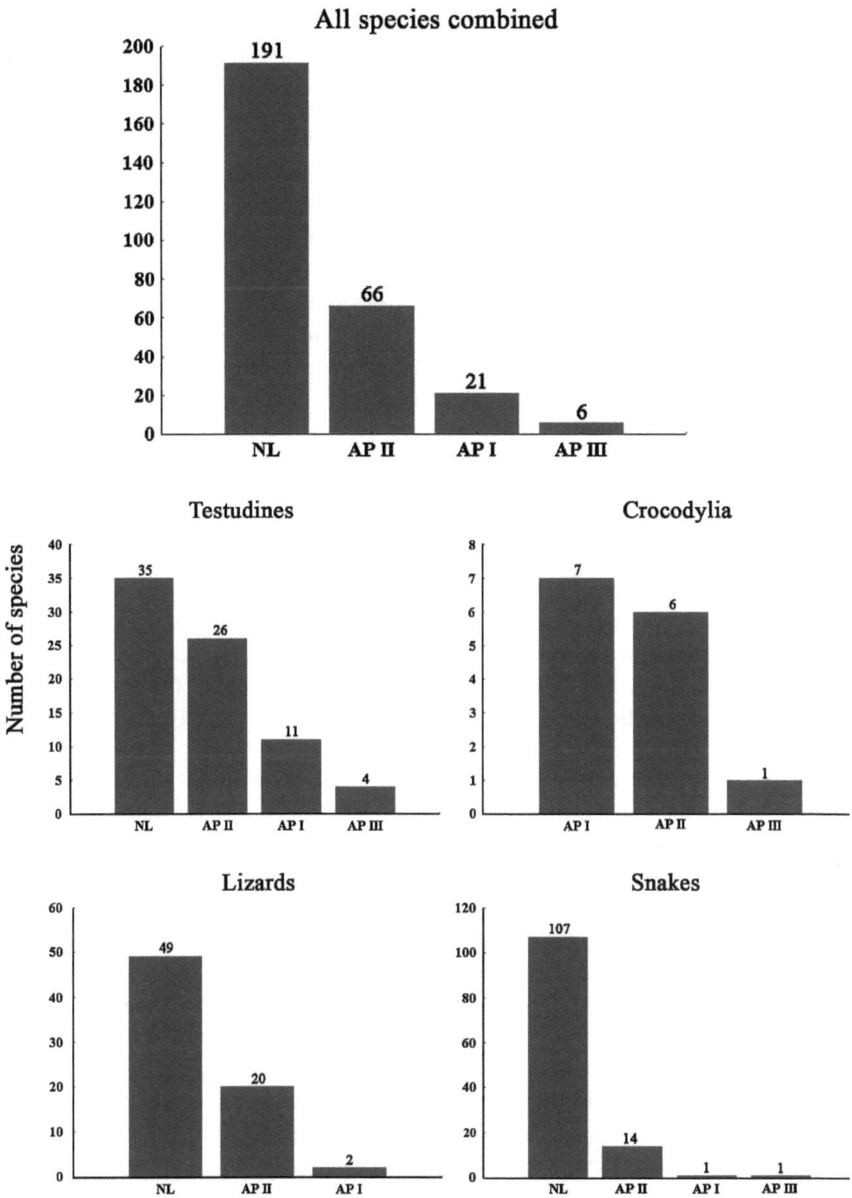


**Fig. 7.3** Reptile species used in traditional folk medicine included in one of the appendices of CITES (2011) *NL* not listed, *AP* I appendix I, *AP* II appendix II, *AP* III appendix III

The medicinal use of amphibians is also very old. Gomes et al. (2007) pointed out that in many ancient cultures, amphibians were believed to possess medicinal properties. For instance, frog potions are used as aphrodisiacs and to prevent impotence and infertility; they are also used as contraceptives and to treat various illnesses (Gomes et al. 2007). Fever could be treated with ground frog heads (Hendricks 1966, p. 44), and whooping cough with a soup made from nine frogs (Hendricks 1980, p. 109). Heart trouble was treated with ground-up toad skins (Hendricks 1980, p. 8). Also, newts are often burned to ashes and then used in medicinal formulas and concoctions since ancient times (Gomes et al. 2007). Alakbarli (2006), highlighted that amphibians were not used in medicine as widely as reptiles, but mentioned that four species of amphibians described in medieval sources have been identified. This group includes frogs, toads, salamanders, and tree-frogs (*Hyla arborea*).

Crocodilians, snakes, lizards, turtles, tortoises, and amphibians serve as important sources of protein for human populations around the world, and the consumption of reptile meat is often intertwined with cultural or medicinal beliefs (Alves et al. 2006, 2008, 2009; Angeletti et al. 1992; Klemens and Thorbjarnarson 1995). Likewise, various medicinal species of amphibians and reptiles are also hunted as food and represent important protein sources for the inhabitants of rural areas (as well as in urban areas where they are often sold), and include *Podocnemis expansa*, *P. unifilis*, *Chelonia mydas*, *Chelonoides denticulata*, *C. carbonaria*, *T. merianae, Paleosuchus palpebrosus, P. trigonatus, Melanosuchus niger, Naja naja, Bungarus fasciatus, Ptyas mucosus, P. korros*, *Bufo pentoni*, *Amietophrynus regularis*, *A. maculatus*, *Kassina fusca*, *Leptodactylus vastus,* and *Leptopelis bufonides,* among others (Alves and Alves 2011; Alves et al. 2006, 2008, 2009; Mohneke et al. 2010, 2011). Other studies have also recorded the use of animal species as foods/medicines (Alves and Rosa 2006, 2007a; Alves et al. 2007a; Pieroni et al. 2002).

Besides their role in healing, natural products often have magical-religious significance, reflecting the different views of health and disease that exist within different cultures. In this context, animal parts are used to prepare clinical remedies as well as to make amulets or charms used in magical/religious diagnoses.

Popular beliefs usually affect the way species are used in zootherapy (Alves and Rosa 2006). Reptiles and amphibians are commonly used in healing through the magical transfer of disease, one of the most engaging subjects in the whole field of folk medicine (Hand 1980). In some situations, it is believed that the mere contact with an animal is sufficient for the disease to pass from the victim to the animal. In Ontario, for example, the mere touching of a live frog to a goiter is sufficient to make the malady pass into the frog; however, final curing depends upon burying the hapless critter head down in the ground until it decays. When this happens the goiter will disappear. Tying or binding a live frog to the affected part will cure a felon, will cause chills to go out of the patient into the frog, will cure asthma, and, in a North Carolina example, spells resembling the hard ague. In the Blue Grass country a live toad is bound to the back to cure rheumatism, the pain passing from the back of the sufferer into the toad (Hand 1980). In Northeast Brazil, jabutis (*Chelonoidis carbonaria*) are usually bred as pets because it is believed that they protect the household from acquiring asthma (Alves et al. 2011).

Another form of spiritual treatment involves the use of amulets containing reptile parts to protect the user from the "evil-eye" or from diseases (Alves et al.

2010b; Alves and Pereira Filho 2007). An example is caiman teeth (*C. latirostris*, *M. niger*, and *P. palpebrosus*) used as protection against snake bites in Brazilian traditional medicine. Alves and Pereira Filho (2007) reported that snakes are sold for medicinal and magic-religious purposes in many Brazilian cities.

Various medicinal amphibians and reptiles are also sold as pets or souvenirs. Large numbers of iguanas (*I. iguana*), for example, are imported to well-established businesses in the United States from El Salvador and other Central American countries and then re-exported for the pet market in Europe and Asia (Gibbons et al. 2000). In many countries in Asia, Africa, and Latin America, amphibians and reptiles are collected for subsistence or local consumption (Altherr et al. 2011; Klemens and Thorbjarnarson 1995; Mohneke et al. 2009). These multiple uses (including medicinal) of the herpetofauna and their impact on animal populations must be properly assessed (Alves et al. 2007a) and taken into consideration when implementing recovery plans for these species, especially those that are highly exploited.

Traditional drugs and medicine in general will require more research and careful evaluations, and it is a well-established fact that many plant, animal, and mineral remedies used in traditional settings are capable of producing serious adverse reactions (Alves and Rosa 2005; De Smet 1991). At least 11 cases of serious extra-gastrointestinal infections by *Salmonella arizona* attributed to the ingestion of a rattlesnake folk remedy have been reported (Fainstein et al. 1982; Fleischman et al. 1989; McIntyre et al. 1982; Riley et al. 1988). In China, people were reported to be reverting to their habit of eating snakes, as the fear of SARS (severe acute respiratory syndrome) was fading (Zhou and Jiang 2005).

A thorough review of the biological risks associated with consumption of reptile products (Magnino et al. 2009) does not include incidents associated with medicinal snake wine, an alcoholic beverage produced by infusing whole snakes in rice wine or grain alcohol (Barzyk 1999). However, human sparganosis, caused by (among others) the ingestion of plerocercoid larvae in raw or insufficiently cooked meat of reptiles (or amphibians), has been reported from Vietnam (Beaver et al. 1984; Magnino et al. 2009). As such, it is essential that traditional drug therapies be submitted to appropriate risk/benefit analyses (De Smet 1991).

Unfortunately, little research has been done so far to prove the claimed clinical efficacy of animal products for medicinal purposes (Still 2003). Moreover, as pointed out by Pieroni et al. (2002), although the chemical constituents and pharmacological actions of certain animal products are known to some extent, more ethnopharmacological studies focusing on animal remedies are needed in order to better define the eventual therapeutic usefulness of this class of biological remedies.

Reptiles and amphibians have been used as sources of drugs for modern medical practices. Reptiles venoms are complex mixtures of bioactive molecules (Chen et al. 2006), and the venom of snakes belonging to the families Viperidae and Elapidae contain analgesic substances that are stronger than morphine and have been used to treat terminal cancer patients (Bisset 1991). These observations are corroborated by Brasil (1937) and Giorgi et al. (1993), who noted that analgesic drugs have been extracted from the venom of *Crotalus durissus*. Batroxobin, extracted from the venom of *Bothrops atrox*, has been found to have significant therapeutic effects on ischemic-reperfused rats in vivo and in clinical trials, and batroxobin, as well as ancrod, is currently being commercially produced. Three other thrombin-like enzyme preparations are also commercially available: reptilase, crotalase, and an enzyme derived from *Agkistrodon contortrix* (Bell and Markland 1997). However, wider clinical use of thrombin-like enzymes has been impeded by immunologic reactions in patients, limited availability of snake venoms, and high production costs as well (Warkentin 1998).

Similarly, studies have demonstrated the enormous potential of amphibians as a source of natural products and drugs (Daly 1998). Amphibian skin has long been known to possess intriguing biological properties, and scientific examination of these secretions has revealed that their components possess a range of medical properties (Shaw 2009). The medicinal activity of various skin components has been confirmed by modern pharmacology as having great potential in ultimately helping to cure various diseases or at least being the basis for derivates and respective anticancer drugs, pain killers, or even agents to prevent cells from being invaded by HIV viruses (Alves et al. 2006; Daly 2003; Garg et al. 2008; Lu et al. 2008). As pointed out by Alves and Rosa (2006), further ethnopharmacological studies are necessary to increase our understanding of the links between traditional uses of faunistic resources, public health policies and sustainable management of natural resources.

It is important to note that behind the perceived efficacy by users, the popularity of animal-based remedies is influenced by cultural aspects, the relations between humans and biodiversity in the form of zootherapeutic practices are conditioned by the social and economic relations between humans themselves (Alves et al. 2008). It has been documented that people sometimes resort to traditional home remedies as a means of resisting urban modern medicine and of asserting their traditional culture (Alves et al. 2007b; Boltanski 1989; Ngokwey 1995). In China, for example, the demand of turtle in traditional medicine is fueled by deeply held cultural beliefs (Rubio 1998).

The medicinal use of the herpetofauna is important to both urban and rural populations, a result in line with Alves and Rosa (2007a, b), who suggested that zootherapeutic practices may function as a social conduit that (in conjunction with other factors) helps rural populations that have migrated to cities to maintain connections with their traditional cultures and values. More specifically, the use of folk remedies indicates an exchange of materials and information on illnesses and treatments between remote rural areas and urban communities.

The commercialization of reptiles and amphibians for medicinal and other purposes has been reported for many parts of the world (Alves et al. 2006; Alves and Rosa 2010; Angeletti et al. 1992; Barzyk 1999; Fitzgerald et al. 2004; Franke and Telecky 2001; Zhou and Jiang 2004). In several Brazilian cities, for example, snakes are widely traded in outdoor markets (that can even have designated stalls for medicinal animals and plants) or in small stores specifically dedicated to this activity (Alves and Pereira Filho 2007; Alves and Rosa 2010). In Mexico, natural

**Fig. 7.4** Geckos (imported from Indonesia) for sale at a market in China. Photo: Anthony B. Cunningham

and traditional remedies derived from reptiles (such as dried rattlesnakes, rattlesnake pills, and rattles) are frequently offered for sale (Fitzgerald et al. 2004). Turtles, snakes, and lizards (especially monitor and gecko lizards—Fig. 7.4) are widely hunted and traded in Vietnam for food and as traditional medicines (Jenkins 1995; Le and Broad 1995; Nash 1997; Stuart 2004). Somaweera and Somaweera (2010) documented the use and trade of snakes in snake wine in four of the most touristic cities in Vietnam. He and Peng (1999) reported that the quantity of snakes consumed in the markets of Guangzhou, Guangdong Province of China, was about $1.4 \times 10^7$ kg each year. From 1990 to 1995, the annual demand for wild snakes from 13 factories producing traditional Chinese medicines (TCM) included 1,656.77 kg of *Zaocys dhumnades*, 234.75 kg of *Deinagkistrodon acutus*, and 20,300 heads and 32.1 kg of *Bungarus multicinctus* (Zheng and Zhang 2000). These examples illustrate the urgent need to increase our knowledge concerning the harvesting and trading of reptiles in traditional medicine and to assess the impacts caused by this commercial exploitation.

## *7.3.2 Implications for Conservation*

Reptile populations are being seriously reduced throughout the world. Factors responsible for these observed declines include the alteration, destruction, or fragmentation of habitat, climate change, disease, impacts from non-indigenous species, ultraviolet radiation, and xenobiotic chemicals (Gibbons et al. 2000). In addition, reptile populations are heavily harvested for human use. The observed population decreases due to human harvesting may be due to the direct physical removal of these animals or due to collection techniques that destroy the habitats used by these reptiles (Alves et al. 2008).

A similar trend is observed for amphibians, which are one of the most threatened groups of animals (Collins and Crump 2009; Stuart 2008; Stuart et al. 2004). Reasons for this are numerous, but besides habitat degradation and loss, disease and rapid enigmatic declines, overexploitation is mentioned as one of the main causes (Alves et al. 2006; Angeletti et al. 1992; Gomes et al. 2007; Halliday 2008). Amphibian species are harvested and used worldwide mainly as a food source, i.e., frog legs are thought to be delicacies in many regions of the world. However, frogs are also collected for leather production and souvenirs, for the pet trade and for cultural reasons including traditional medicine (Gomes et al. 2007; Oza 1990; Warkentin et al. 2009).

The collection of individual animals from the wild for subsistence or commercial and medicinal purposes has been invoked as a factor contributing to the decline of certain species (Alves et al. 2006, 2008; Angeletti et al. 1992; Gibbons et al. 2000), although there has not yet been a comprehensive evaluation of this potential link. The popularity of folk medicine certainly places pressure on these natural resources (Almeida and Albuquerque 2002).

Our results demonstrate that a substantial number of reptile and amphibian species ($n = 331$) are used in traditional medicine throughout the world and that the vast majority of these animals are collected from the wild. Of the medicinal species used, 224 (67.6%) are already included on endangered species lists. Of the species cataloged in this study, 93 (28%) are included in three CITES appendices (see Table 7.1), although the reasons for their inclusion are not necessarily related to medicinal use. These results demonstrate the need to assess the implications of the trade of reptiles used in traditional medicines on their wild populations, and the need for including such uses in discussions about herpetofauna conservation.

The trade of animals for medicinal purposes is a widespread phenomenon, with significant implications for their conservation and sustainable use (Alves and Rosa 2005). The demand for live snakes (and their body parts) for use in traditional medicine appears to have led to significant reductions in their populations in certain parts of the world (Fitzgerald et al. 2004).

Field reports have indicated the southeastern Asian medicinal trade as a growing threat to reptiles, especially turtles and snakes (Klemens and Thorbjarnarson 1995). Asia has a high diversity of turtle species, but its unique fauna is facing a perilous and uncertain future. The main reason for the Asian turtle survival crisis is Chinese demand for turtle products (van Dijk et al. 2000). Directly or indirectly, the medicinal value attributed to chelonians has been one of the main reasons for their trade and overexploitation. In China, turtles are sought as a delicacy because of widespread popular belief, inspired by Traditional Chinese Medicine, that turtle meat or shell possesses especially nutritious or curative properties (Lau and Shi 2000). Such situations demonstrate that cultural aspect should be taken into consideration in the elaboration of conservation plans. A study conducted in China showed that the same (or better) nutritional benefits of turtles can be obtained with cheaper, common, and less-endangered food sources such as domestic animals (Rubio 1998). These authors concluded that given the financial and environmental cost of using turtle products, other options for obtaining the same nutrition should be promoted and that future challenges should involve balancing cultural practices with sustaining biodiversity.

More than one-half of all freshwater tortoise and turtle species from Southeast and Eastern Asia are currently endangered or critically endangered, largely because of overcollection by the food and traditional medicine industries (Garg et al. 2008; Hand 1980; Klemens and Thorbjarnarson 1995; Turtle Conservation Fund 2002). The high demand for crocodile skins, meat, and body parts for traditional medicine have certainly contributed to the observed decline in their populations in Nigeria (Ita 1994), as the demand for live rattlesnakes, skins, and body parts has reduced the populations of these reptiles in Mexico (Fitzgerald et al. 2004).

Many factors affect reptilian and amphibian populations in the world, and the use of these animals for medicinal purposes represents an additional pressure, whose impact varies depending on the species exploited and the cultural factors associated with their exploitation. The medicinal use of the herpetofauna must be considered together with other anthropogenic pressures, such as habitat loss. The depletion of medicinal resources not only poses a challenge for conservation but also represents a serious threat to the health of many human communities, and efforts to stabilize the status of these species are important not only to conservationists but also to millions of people whose health depends on the use of traditional remedies.

Ultimately, the most successful conservation programs are those that identify and deal with the reason a species is endangered and at the same time provide economic benefits to local people (Pough et al. 2004). Therefore, management strategies aimed at herpetofauna conservation need to be established to minimize the impact of the traditional populations that use several species as food and medicine or for other purposes (Alves et al. 2008, 2009).

An alternative proposal can be the creation of reptile breeder cooperatives in rural gatherer communities for raw material supplies and products for medicinal use. These cooperatives could be part of breeding sites of species such as *Podocnemis expansa*, *P. unifilis*, *Chelonia mydas*, *Chelonoidis denticulata*, *C. carbonaria*, *T. merianae*, *Paleosuchus palpebrosus*, *P. trigonatus*, *Boa constrictor*, *Uranoscodon superciliosus*, and *Tupinambis* spp, among others, with the appropriate authorization and regulation by competent governmental bodies, besides the presence of specialists in the area (biologists, veterinarians, and animal husbandry staff) (Alves et al. 2008, 2009). An example of a successful cooperative can be the snake collectors for poison extraction in India (Whitaker 1989).

Sustainable use programs for reptiles have had some success in the world (Pough et al. 2004; Vitt and Caldwell 2009), such as the snake collectors who extract venom in India (Whitaker 1989). Lizards (*Tupinambis*, and certain iguanids) are harvested for local consumption and have experienced sharp population declines in many areas due to overhunting (Vitt and Caldwell 2009).

Breeding programs to raise iguanas for release into the wild have been developed in several countries, including Panama, Costa Rica, Guatemala, Nicaragua, Belize, Honduras, El Salvador, Colombia, and Venezuela (Eilers et al. 2002), and iguana farming has become an attractive economic alternative to cattle breeding and a significant source of food for local populations (Magnino et al. 2009). Managed harvests of crocodilians began about three decades ago to assist the recovery of species and populations that had been devastated by unregulated hunting. The success of managed harvests and captive rearing in Papua New Guinea, Venezuela, and a few other countries stimulated other governments to begin similar programs. These managed species have shown remarkable resilience in many countries, and their populations are no longer endangered. However, with more countries producing skins, supply began to exceed demand and was followed by a declining popularity of crocodilian leather (Vitt and Caldwell 2009).

In addition, the therapeutic indications of wild animals and plants and domestic or cultivated species also overlapped in many cases (Alves et al. 2007a). This aspect opens the possibility of replacing, where suitable, the use of threatened species with other species in traditional medicine formulas. Such substitution of products is of interest from a conservationist perspective, in the context of reducing the pressure on overexploited populations, or legally protected species (Alves and Rosa 2007a). Educational programs are also quite viable alternatives, mainly when focused on rural communities where inhabitants eat reptiles/amphibians and use them in traditional medicine and religious practices.

Projects seeking to train teachers in those communities and old hunters in sites nearby protected areas help to minimize the impacts on the herpetofauna. As pointed out by Pough et al. (2004), education is urgently needed at all levels to maintain viable populations of reptiles. Training in areas of habitat protection, wildlife management, and conservation biology is needed, especially in tropical countries where most species of reptiles are found. The success of conservation and management programs ultimately depends on how well the programs are tailored to the interests and needs of the people on whose land the threatened or endangered animal live.

The use of amphibians and reptiles is an integral part of many cultures (Alves et al. 2006, 2008; Angeletti et al. 1992; Gibbons et al. 2000). The great diversity of interactions between humans and the herpetofauna provide the foundations for the cultural, economic, emotional, intellectual, social, and spiritual motivations that determine how conservation and management activities are designed, conducted, and assessed (Alves et al. 2008, 2009; Frazier 2005).

As described in this chapter, reptiles and amphibians are used globally in traditional folk medicine, which is thereby a form of exploitation that should be taken into consideration. This use of the herpetofauna represents an additional pressure for many species, and for others this has been indicated as an important cause of population decline. Thus, not only should one consider the use of these animals in popular medicine but also their exploitation by the pharmaceutical industry. As pointed out by Shaw (2009), any pharmaceutical scientist who is involved in contemporary natural product research has to get involved in or at the very least become familiar with the global issues of species conservation and/or biodiversity.

Reptiles and amphibians have declined rapidly in both numbers and range in recent decades and their exploitation by humans is noted as an impacting factor for the decline of many species. Hence, an understanding of the cultural, social, and traditional roles of the herpetofauna is essential for establishing management plans directed toward sustainable use. As recorded in this work, medicinal use of the herpetofauna, despite being widely disseminated, has been studied little, limiting the evaluation of the impact of these practices on animal populations. Therefore, studies in ethnoherpetology are essential in conservation strategies and to record associated knowledge of such uses.

## References


Adeola MO (1992) Importance of wild animals and their parts in the culture, religious festivals, and traditional medicine, of Nigeria. Environ Conserv 19(02):125–134

Alakbarli F (2006) Medical manuscripts of Azerbaijan. Heydar Aliyev Foundation, Baku

Almeida AV (2007) Zooterapia indígena brasileira do século XVIII nas obras de Guilherme Piso, Georg Marcgrave e Joannes de Laet. Sitientibus Série Ciências Biológicas 7(3):261–272

Almeida CFCBR, Albuquerque UP (2002) Uso e conservação de plantas e animais medicinais no Estado de Pernambuco (Nordeste do Brasil): Um estudo de caso. Interciencia 27(6):276–285

Almeida AV, Alves AGC, Lucena RFP, Albuquerque UP (2005) Prescrições zooterápicas indígenas brasileiras nas obras de Guilherme Piso (1611–1678). In: Atualidades em Etnobiologia e Etnoecologia. Nupeea, Sociedade Brasileira de Etnobiologia e Etnoecologia, Recife, pp 45–60

Altherr S, Goyenechea A, Schubert DJ (2011) Canapés to extinction: the international trade in frogs' legs and its ecological impact. A report by Pro Wildlife, Defenders of Wildlife and Animal Welfare Institute. Munich (Germany), Washington, D.C. (USA)

Alves RRN (2006) Use of marine turtles in zootherapy in Northeast Brazil. Mar Turt Newslett 112:16–17

Alves RRN, Alves HN (2011) The faunal drugstore: animal-based remedies used in traditional medicines in Latin America. J Ethnobiol Ethnomed 7(9):1–43

Alves RRN, Pereira Filho GA (2007) Commercialization and use of snakes in North and Northeastern Brazil: implications for conservation and management. Biodivers Conserv 16:969–985

Alves RRN, Rosa IL (2005) Why study the use of animal products in traditional medicines? J Ethnobiol Ethnomed 1(5):1–5

Alves RRN, Rosa IL (2006) From cnidarians to mammals: the use of animals as remedies in fishing communities in NE Brazil. J Ethnopharmacol 107:259–276

Alves RRN, Rosa IL (2007a) Zootherapeutic practices among fishing communities in North and Northeast Brazil: A comparison. J Ethnopharmacol 111:82–103

Alves RRN, Rosa IL (2007b) Zootherapy goes to town: the use of animal-based remedies in urban areas of NE and N Brazil. J Ethnopharmacol 113:541–555

Alves RRN, Rosa IL (2010) Trade of animals used in Brazilian traditional medicine: trends and implications for conservation. Human Ecol 38(5):691–704

Alves RRN, Santana GG (2008) Use and commercialization of Podocnemis expansa (Schweiger 1812) (Testudines: Podocnemididae) for medicinal purposes in two communities in North of Brazil. J Ethnobiol Ethnomed 4(3):6
Alves RRN, Souto WMS (2011) Ethnozoology in Brazil: current status and perspectives. J Ethnobiol Ethnomed 7(22):1–19
Alves RRN, Filho GAP, Lima YCC (2006) Snakes used in ethnomedicine in Northeast Brazil. Environ Dev Sustain 9(4):455–464
Alves RRN, Rosa IL, Santana GG (2007a) The Role of animal-derived remedies as complementary medicine in Brazil. Bioscience 57(11):949–955
Alves R, Filho GAP, Lima YCC (2007b) Snakes used in ethnomedicine in Northeast Brazil. Environ Dev Sustain 9(4):455–464
Alves RRN, Vieira WLS, Santana GG (2008) Reptiles used in traditional folk medicine: conservation implications. Biodivers Conserv 17(8):2037–2049
Alves RRN, Léo Neto NA, Santana GG, Vieira WLS, Almeida WO (2009) Reptiles used for medicinal and magic religious purposes in Brazil. Appl Herpetol 6(3):257–274
Alves RRN, Pereira-Filho GA, Vieira KS, Santana GG, Vieira WLS, Almeida WO (2010a) Répteis e as populações humanas no Brasil: uma abordagem etnoherpetológica. In: Alves RRN, Souto WMS, Mourão JS (eds) A Etnozoologia no Brasil: importância, status atual e perspectivas, vol 7, 1st edn. NUPEEA, Recife, pp 121–148
Alves RRN, Pereira-Filho GA, Vieira KS, Santana GG, Vieira WLS, Almeida WO (2010b) Répteis e as populações humanas no Brasil:uma abordagem etnoherpetológica. In: Alves RRN, Souto WMS, Mourão JS (eds) A Etnozoologia no Brasil: importância, status atual e perspectivas futuras. NUPEEA, Recife, pp 121–146
Alves RRN, Barbosa JAA, Santos SLDX, Souto WMS, Barboza RRD (2011) Animal-based remedies as complementary medicines in the semi-arid Region of Northeastern Brazil. eCAM:1–15. doi:10.1093/ecam/nep134
Andrade JN, Costa-Neto EM (2005) Primeiro registro da utilização medicinal de recursos pesqueiros na cidade de São Félix. Estado da Bahia, Brasil Acta Sci Biol Sci 27(2):177–183
Angeletti LR, Agrimi U, French D, Curia C, Mariani-Costantini R (1992) Healing rituals and sacred serpents. Lancet 340(8813):223–225
Apaza L, Godoy R, Wilkie D, Byron E, Huanca T, Leonard WR, Peréz E, Reyes-García V, Vadez V (2003) Markets and the use of wild animals for traditional medicine: a case study among the Tsimané ameridians of the Bolivian rain forest. J Ethnobiol 23(1):47–64
Ashwell D, Walston N (2008) An overview of the use and trade of plants and animals in traditional medicine systems in Cambodia, 1st edn. TRAFFIC Southeast Asia Greater Mekong Programme, Ha Noi
Barzyk JE (1999) Turtles in crisis: the Asian food markets. http://www.tortoisetrust.org/articles
Beaver PC, Jung RC, Cupp EW (1984) Clinical parasitology, vol 9. Lea & Febiger, Philadelphia
Begossi A (1992) Food taboos at Búzios Island (SE Brazil): their significance and relation to folk medicine. J Ethnobiol 12(1):117–139
Begossi A, Braga F (1992) Food taboos and folk medicine among fishermen from the Tocantins River. Amazoniana 12:341–352
Begossi A, Silvano RAM, Amaral BD, Oyakama OT (1999) Uses of fish and game by inhabitants of an extrative reserve (Upper Juruá, Acre, Brazil). Environ Dev Sustain 1:73–93
Bell WR, Markland FS (1997) Defibrinogenating enzymes. Discussion. Drugs 54:18–31
Bertrand H (1997) Contribution à l'étude de l'herpétologie et de l'ethnoherpétologie en Anjou = A study on the herpetology and ethnoherpetology of Anjou province (France). Bulletin de la Société herpétologique de France 82–83:51–62
Bisset NG (1991) One man's poison, another man's medicine? J Ethnopharmacol 32(1–3):71–81
Boll V (2004) The distribution and ethnozoology of frogs (and toad) in North–Eastern Arnhem Land (Australia). Anthropozoologica 39(2):61–72
Boltanski L (1989) As classes sociais eo corpo. Graal, Rio de Janeiro
Branch L, Silva MF (1983) Folk medicine in Alter do Chão, Pará, Brasil. Acta Amazônica 13:737–797

Brasil V (1937) Do emprego da peconha em terapeutica. GB (Brazil), Rio de Janeiro
Chen TH, Lin HC, Chang HC (2000) Current status and utilization of chelonians in Taiwan. In: Van Dijk PP, Stuart BL, Rhodin AGJ (eds) Workshop on conservation and trade of freshwater turtles and tortoises in Asia. Chelonian Research Monographs Chelonian Research Foundation, Lunenburg, pp 45–51
Chen T, Kwok HF, Ivanyi C, Shaw C (2006) Isolation and cloning of exendin precursor cDNAs from single samples of venom from the Mexican beaded lizard (*Heloderma horridum*) and the Gila monster (*Heloderma suspectum*). Toxicon 47(3):288–295
Chen TH, Chang HC, Lue KY (2009) Unregulated trade in turtle shells for Chinese traditional medicine in East and Southeast Asia: the case of Taiwan. Chelonian Conserv Biol 8(1):11–18
China National Corporation of Traditional and Herbal Medicine (1995) Materia medica commonly used in china. Science Press, Beijing
CITES (2002) List of species traded for medicinal purposes. 18th meeting of the animals committee. San José, Costa Rica
CITES (2011) CITES Appendix. http://www.cites.org/eng/resources/species.html. Accessed 7 Aug 2011
Collins JP, Crump ML (2009) Extinction in our times: global amphibian decline. Oxford University Press, USA
Costa-Neto EM (1996) Faunistc Resources used as medicines by an Afro-brazilian community from Chapada Diamantina National Park, State of Bahia-Brazil. Sitientibus 15:211–219
Costa-Neto EM (1999a) Barata é um santo remédio: introdução a zooterapia popular no Estado da Bahia Editora Universitária da UEFS, Feira de Santana
Costa-Neto EM (1999b) Healing with animals in Feira de Santana city, Bahia, Brazil. J Ethnopharmacol 65:225–230
Costa-Neto EM (1999c) Recursos animais utilizados na medicina tradicional dos índios Pankararé que habitam o Nordeste do Estado da Bahia, Brasil. Actualidades Biologicas 21(70):69–79
Costa-Neto EM (2000) Conhecimento e usos tradicionais de recursos faunísticos por uma comunidade Afro-Brasileira. Resultados preliminares. Interciencia 25(9):423–431
Costa-Neto EM, Pacheco JM (2005) Utilização medicinal de insetos no povoado de Pedra Branca, Santa Terezinha, Bahia, Brasil. Biotemas 18(1):113–133
Daly JW (1998) Thirty years of discovering arthropod alkaloids in amphibian skin†. J Nat Prod 61(1):162–172
Daly JW (2003) Ernest Guenther award in chemistry of natural products. Amphibian skin: a remarkable source of biologically active arthropod alkaloids. J Med Chem 46(4):445–452
Das I (1998) The Serpent's tongue: a contribution to the ethnoherpetology of India and adjacent countries. Edn Chimaira, Frankfurt am Main, Germany
De Smet PAGM (1991) Is there any danger in using traditional remedies? J Ethnopharmacol 32:43–50
Dharmananda S (2007a) Endangered species issues affecting turtles and tortoises used in Chinese medicine http://www.itmonline.org/arts/turtles.html
Dharmananda S (2007b) The medicinal use of snakes in China http://www.itmonline.org/arts/snakes.html
Donadio OE, Gallardo JM (1984) Biología y conservación de las especies del género Tupinambis (Squamata, Sauria, Teiidae) en la República Argentina. Revista del Museo Argentino de Ciencias Naturales" Bernardino Rivadavia". Zoologia 13:117–127
Eilers K, Koops W, Udo H, Van Keulen H, Noordhuizen J (2002) Analysis of Iguana iguana farming systems in Nicaragua Costa Rica and Panama. Interciencia 27(11):599–606
El-Din SB (2007) Testudo Kleinmanni in Libya. http://www.tortoisetrust.org/articles/libya.html
El-Kamali HH (2000) Folk medicinal use of some animal products in Central Sudan. J Ethnopharmacol 72:279–282
Fainstein V, Yancey R, Trier P, Bodey GP (1982) Overwhelming infection in a cancer patient caused by Arizona hinshawii: its relation to snake pill ingestion. Am J Infect Control 10(4):147–153

Figueiredo N (1994) Os 'bichos' que curam: os animais e a medicina 'folk' em Belém do Pará. Boletim do Museu Paraense Emílio Göeldi 10(1):75–91
Fitter RSR (1986) Wildlife for man: how and why we should conserve our species. Collins, London
Fitzgerald LA, Painter CW, Reuter A, Hoover C, America TN (2004) Collection, trade, and regulation of reptiles and amphibians of the Chihuahuan desert ecoregion. TRAFFIC North America, Washington
Fleischman S, Haake DA, Lovett MA (1989) Salmonella arizona infections associated with ingestion of rattlesnake capsules. Arch Intern Med 149(3):701–705
Franke J, Telecky TM (2001) Reptiles as pets: an examination of the trade in live reptiles in the United States. Humane Society of the United States, Washington
Frazier J (2005) Traditional and cultural use of marine turtles. In: 3rd Meeting of the signatory states, Bangkok
Freire FC (1996) Répteis utilizados na medicina popular no Estado de Alagoas. Monografia de graduação. Universidade Federal de Alagoas, Maceió
Fretey J, Segniagbeto GH, Soumah M (2007) Presence of sea turtles in traditional pharmacopoeia and beliefs of West Africa. Mar Turt Newsl 116:23–25
Frost DR (2011) Amphibian species of the world: an online reference. Version 5.5. American Museum of Natural History. http://research.amnh.org/vz/herpetology/amphibia/. Accessed 11 Aug 2011
Garg AD, Hippargi RV, Gandhare AN (2008) Toad skin-secretions: potent source of pharmacologically and therapeutically significant compounds. Internet J Pharmacol 5(2):17
Gesler WM (1992) Therapeutic landscapes: medical issues in light of the new cultural geography. Soc Sci Med 34(7):735–746
Gibbons JW, Scott DE, Ryan TJ, Buhlmann KA, Tuberville TD, Metts BS, Greene JL, Mills T, Leiden Y, Poppy S, Winne CT (2000) The global decline of reptiles, Déjà Vu amphibians. BioScience 50(8):653–666
Giorgi R, Bernardi MM, Cury Y (1993) Analgesic effect evoked by low molecular weight substances extracted from Crotalus durissus terrificus venom. Toxicon 31(10):1257–1265
Gomes A, Giri B, Saha A, Mishra R, Dasgupta SC, Debnath A (2007) Bioactive molecules from amphibian skin: Their biological activities with reference to therapeutic potentials for possible drug development. Indian J Exp Biol 45(7):579
Goodman SM, Hobbs J (1994) The distribution and ethnozoology of reptiles of the northern portion of the Egyptian Eastern Desert. J Ethnobiol 14:75–100
Guo Y, Zou X, Chen Y (1996) Tentative survey on sustainable use of medicinal animals. In: Development CICCoEa (ed) Protect the biodiversity of China. China Environmental Science Press, Beijing
Halliday TR (2008) Why amphibians are important. Int Zoo Yearbook 42(1):7–14
Hand WD (1980) Magical medicine: the folkloric component of medicine in the folk belief, custom, and ritual of the peoples of Europe and America: selected essays of Wayland D. Hand. University of California Press, Hand
He H, Peng X (1999) Tentative survey on snake market in Guangzhou. Sichuan J Zool 18(3):139–141
Hendricks GDM (1966) Mice and mus-taches: a sampling of superstitions and popular beliefs in Texas. Texas Folklore Society, Austin
Hendricks GD (1980) Roosters, rhymes and railroad tracks. A second sampling of superstitions and popular beliefs in Texas. Southern Methodist University Press, Dallas
Highfield AC, Bayley JR (2007) Folklore, myth, and exploitation of snakes in Morocco and Tunisia. http://www.tortoisetrust.org/articles/exploit.html
Highfield AC, Slimani T (2007) The Spiny-tailed Lizard at home—Uromastyx acanthinurus in southern Morocco
IFAW—International Fund for Animal Welfare (2007) What's hiding inside your traditional medicine? http://www.ifawct.org/Publications/Program_Publications/Wildlife_Trade/Traditional_Chinese_Medicine/asset_upload_file1_14609.pdf

Indian Traditional Medicinal Knowledgebase (2011) http://www.ncbi.org.in/traditionalmedicine/index.jsp
Ita EO (1994) Aquatic plants and wetland wildlife resources of Nigeria. CIFA occasional paper
IUCN (2000) Approaches to the conservation of species used in traditional medicines. Newslett Species Surviv Comm 33:36–38
IUCN (2011) IUCN Red list of threatened species. IUCN. http://www.iucnredlist.org. Accessed 30 Sept 2011
Ives IE (2006) Conservation of Sulawesi's Two Endemic Chelonians, Leucocephalon yuwonoi and Indotestudo forstenii; an Investigation into In-Situ and Ex-Situ Conservation Concerns. MSc. Thesis, Antioch University, USA
Jenkins M (1995) Tortoises and freshwater turtles: the trade in South East Asia. Traffic International Cambridge, Cambridge
Kakati LN, Doulo V (2002) Indigenous knowledge system of zootherapeutic use by Chakhesang tribe of Nagaland, India. J Hum Ecol 13(6):419–423
Kakati LN, Ao B, Doulo V (2006) Indigenous knowledge of Zootherapeutic use of vertebrate origin by the Ao Tribe of Nagaland. Human Ecol 19(3):163–167
Klemens MW, Thorbjarnarson JB (1995) Reptiles as a food resource. Biodivers Conserv 4(3):281–298
Lau M, Shi H (2000) Conservation and trade of terrestrial and freshwater turtles and tortoises in the People's Republic of China. Chelonian Res Monogr 2:30–38
Le DD, Broad S (1995) Investigations into tortoise and freshwater turtle trade in Vietnam. IUCN Species Survival Commission, IUCN, Gland, Switzerland and Cambridge
Lee SKH (1999) Trade in traditional medicine using endangered species: an international context. Paper presented at the 2nd Australian symposium on traditional medicine and wildlife conservation, Melbourne
Lev E (2003) Traditional healing with animals (zootherapy): medieval to present-day levantine practice. J Ethnopharmacol 85:107–118. doi:10.1016/S0378-8741(02)00377-X
Li G, Tang D, Fang K (2000) An analysis of amino acids in the meat of Cuora trifasciata. Sichuan J Zool 19:165–166
Lohani U (2011) Traditional uses of animals among jirels of Central Nepal. Ethno Med 5(2):115–124
Lu CX, Nan KJ, Lei Y (2008) Agents from amphibians with anticancer properties. Anticancer Drugs 19(10):931–939
MacKinney LC (1946) Animal substances in materia medica. J Hist Med Allied Sci 1(1):149–170
Magnino S, Colin P, Dei-Cas E, Madsen M, McLauchlin J, Nöckler K, Prieto Maradona M, Tsigarida E, Vanopdenbosch E, Van Peteghem C (2009) Biological risks associated with consumption of reptile products. Int J Food Microbiol 134(3):163–175
Mahawar MM, Jaroli DP (2006) Animals and their products utilized as medicines by the inhabitants surrounding the Ranthambhore National Park, India. J Ethnobiol Ethnomed 2(46):1–5
Mahawar MM, Jaroli DP (2008) Traditional zootherapeutic studies in India: a review. J Ethnobiol Ethnomed 4(1):1–17
Marques JGW (1995) Pescando pescadores: etnoecologia abrangente no baixo São Francisco alagoano. NUPAUB-USP, São Paulo
Martínez GJ, Barboza GE (2010) Natural pharmacopoeia used in traditional Toba medicine for the treatment of parasitosis and skin disorders (Central Chaco, Argentina). J Ethnopharmacol 132:86–100
McIntyre KE Jr, Malone JM, Richards E, Axline SG (1982) Mycotic aortic pseudoaneurysm with aortoenteric fistula caused by Arizona hinshawii. Surgery 91(2):173–177
Meiling H, Haitao S, Lirong F, Shiping G, Fong JJ, Parham JF (2008) Scientific refutation of traditional Chinese medicine claims about turtles. Appl Herpetol 5(2):173–187
Mohneke M, Onadeko AB, Rödel MO (2009) Exploitation of frogs—a review with a focus on West Africa. Salamandra 45(4):193–202

Mohneke M, Onadeko AB, Hirschfeld M, Rödel MO (2010) Fried and dried: amphibians in local and regional food markets in West Africa. Traffic Bull 22:117–128
Mohneke M, Onadeko AB, Rödel MO (2011) Medicinal and dietary uses of amphibians in Burkina Faso. Afr J Herpetol 60(1):78–83
Morris R, Morris D (1965) Men and snakes. McGraw-Hill, New York
Myers CW, Daly JW, Malkin B (1978) A dangerously toxic new frog (Phyllobates) used by EmberÃ¡ Indians of western Colombia, with discussion of blowgun fabrication and dart poisoning. Bull Am Museum Nat Hist 161:309–365
Nash SV (1997) Fin, feather, scale and skin: observations on the wildlife trade in Lao PDR and Vietnam. TRAFFIC Southeast Asia, Malaysia
Negi CS, Palyal V (2007) Traditional uses of animal and animal products in medicine and rituals by the shoka tribes of district Pithoragarh, Uttaranchal, India. Ethno-Med 1(1):47–54
Ngokwey N (1995) Home remedies and doctors' remedies in Feira (Brazil). Soc Sci Med 40(8):1141–1153
Oza GM (1990) Ecological effects of the frog's legs trade. Environmentalist 10(1):39–42
Perry A (2000) Global survey of marine medicinals. In: Moreau M-A, Hall HJ, Vincent ACJ (eds) Proceedings of the 1st International workshop on the management and culture of marine species used in traditional medicines, Project Seahorse, Montreal
Pieroni A, Giusti ME, Grazzini A (2002) Animal remedies in the folk medicinal practices of the Lucca and Pistoia Provinces, Central Italy. In: Fleurentin J, Pelt JM, Mazars G (eds) Des sources du savoir aux médicaments du futur/from the sources of knowledge to the medicines of the future, 1st edn. IRD Editions, Paris
Pough FH, Andrews RM, Cadle JE, Crump ML, Savitsky AH, Wells KD (2004) Herpetology, 3rd edn. Prentice Hall, New Jersey
Quave CL, Lohani U, Verde A, Fajardo J, Rivera D, Obón C, Valdes A, Pieroni A (2010) A comparative assessment of zootherapeutic remedies from selected areas in Albania, Italy, Spain and Nepal. J Ethnobiol 30(1):92–125
Riley KB, Antoniskis D, Maris R, Leedom JM (1988) Rattlesnake capsule–associated Salmonella arizona infections. Arch Intern Med 148(5):1207–1210
Rowley J, Brown R, Bain R, Kusrini M, Inger R, Stuart B, Wogan G, Thy N, Chan-Ard T, Trung CT (2010) Impending conservation crisis for Southeast Asian amphibians. Biol Lett 6(3):1–3
Rubio M (1998) Rattlesnake. Portrait of a predator. Smithsonian Institution Press, Washington
Seixas C, Begossi A (2001) Ethnozoology of caiçaras from Aventureiro, Ilha Grande. J Ethnobiol 21(1):107–135
SEMARNAP-PROFEPA (1998) Memorias del curso taller de identificación de productos y subproductos de fauna silvestre de México. Diaproy. CREDES, Mazatlán
Shao-Ke L, Ka-Lok W, Ming L, Paul B, Stephen T, Pang-Chui S (2010) Base: MMDBD (medicinal materials dna barcode database). BMC Genomics 11:1–8
Shaw C (2009) Advancing drug discovery with reptile and amphibian venom peptides: venom-based medicines. Biochem Soc 31:34–37
Silva MLV, Alves ÂGC, Almeida AV (2004) A zooterapia no Recife (Pernambuco): uma articulação entre as práticas e a história. Biotemas 17(1):95–116
Smart R, Whiting MJ, Twine W (2005) Lizards and landscapes: integrating field surveys and interviews to assess the impact of human disturbance on lizard assemblages and selected reptiles in a savanna in South Africa. Biol Conserv 122(1):23–31
Sodeinde OA, Soewu DA (1999) Pilot study of the traditional medicine trade in Nigeria. Traffic Bull 18(1):35–40
Somaweera R, Somaweera N (2010) Serpents in jars: the snake wine industry in Vietnam. J Threatened Taxa 2(11):1251–1260
Speck FG (1946) Ethnoherpetology of the Catawba and Cherokee Indians. Wash Acad Sci, J 36:355–360
Still J (2003) Use of animal products in traditional Chinese medicine: environmental impact and health hazards. Complement Ther Med 11:118–122

Stuart BL (2004) The harvest and trade of reptiles at U Minh Thuong National Park, Southern Vietnam. Traffic Bull 20(1):25–34
Stuart SN (2008) Threatened amphibians of the world. Lynx Edicions, Spain
Stuart SN, Chanson JS, Cox NA, Young BE, Rodrigues ASL, Fischman DL, Waller RW (2004) Status and trends of amphibian declines and extinctions worldwide. Science 306(5702):1783–1786
The Reptile Database (2011) The Reptile database version 2011. Peter Uetz/Jirí Hošek/Jakob Hallerman. http://www.reptile-database.org/. Accessed 27 Aug 2011
Turtle Conservation Fund (2002) A global action plan for conservation of tortoises and freshwater turtles, vol 2007. Strategy and funding prospectus, vol Strategy and funding prospectus 2002–2007. Conservation International and Chelonian Research Foundation, Washington
Tyler MJ, Wassersug R, Smith B (2007) How frogs and humans interact: influences beyond habitat destruction, epidemics and global warming. Appl Herpetol 4(1):1–18
van Dijk PP, Stuart BL, Rhodin AGJ (2000) Asian turtle trade. Chelonian Research Foundation, Lunenburg
Vázquez PE, Méndez RM, Guiascón ÓGR, Piñera EJN (2006) Uso medicinal de la fauna silvestre en los Altos de Chiapas, México. Interciencia 31(7):491–499
Vitt LJ, Caldwell JP (2009) Herpetology: an introductory biology of amphibians and reptiles, 3rd edn. Academic, San Diego
Warkentin TE (1998) Limitations of conventional treatment options for heparin-induced thrombocytopenia. Semin Hematol 35:17–25
Warkentin IG, Bickford D, Sodhi NS, Bradshaw CJA (2009) Eating frogs to extinction. Conserv Biol 23(4):1056–1059
Whitaker Z (1989) Snakeman. India Magazine Books, Bombay
Zheng H, Zhang Q (2000) Medicinal using status and conservation strategy of snakes in China. In: Zhang E, Zheng H (eds) Conservation of endangered wild medicinal fauna and flora resources in China. The Second Military Medical University Press, Shanghai, pp 96–101
Zhou Z, Jiang Z (2004) International trade status and crisis for snake species in China. Conserv Biol 18(5):1386–1394
Zhou Z, Jiang Z (2005) Identifying snake species threatened by economic exploitation and international trade in China. Biodivers Conserv 14(14):3525–3536
Ziemendorff S (2008) Sustancias estimulantes y brebajes afrodisíacos en la tradición de la Amazonía peruana. Culturas populares 7:1–6

# Chapter 8
# Primates in Traditional Folk Medicine: World Overview

**Rômulo Romeu Nóbrega Alves, Wedson Medeiros Silva Souto, Raynner Rilke Duarte Barboza and Dandara Monalisa Mariz Bezerra**

**Abstract** Almost 50% of primates are in danger of becoming extinct, according to the criteria of the IUCN Red List of Threatened Species, one of the reasons being their consumption by humans. The reasons for hunting vary by region. One of these pretexts is the medicinal or magical value of products derived from these animals. This chapter aims at providing an overview of the global use of primates in traditional folk medicines, as well as identifying the species used as remedies associated to folk beliefs. Some important questions relating to their conservation will be addressed. Our results revealed that at least 110 species of primates, belonging to 41 genera and 11 families, are used in traditional folk practices and in magic-religious rituals throughout the world. Of the 110 species of primates recorded in our review, 22 species are classified as Vulnerable, 23 as Endangered, 14 as critically endangered, 7 as Near Threatened, 43 as Least Concern, and 1 classified as data deficient in the most recent IUCN Red List. All species are included in the CITES Appendices I or II also, although the reasons for their inclusion are not necessarily related to medicinal use. The widespread utilization of primates in traditional medicine is evidence of the importance of understanding such uses in the context of primate conservation, as well as the need for considering socio-cultural factors when establishing management plans concerning the sustainable use of these mammals.

R. R. N. Alves (✉)
Departamento de Biologia, Universidade Estadual da Paraíba, Avenida das Baraúnas, Campina Grande-PB, 58109-753, Brazil
e-mail: romulo_nobrega@yahoo.com.br

W. M. S. Souto · R. R. D. Barboza · D. M. M. Bezerra
Programa de Pós-Graduação em Ciências Biológicas, Universidade Federal da Paraíba, Campus 1, Cidade Universitária, João Pessoa-PB, 58059-900, Brazil

R. R. N. Alves and I. L. Rosa (eds.), *Animals in Traditional Folk Medicine*,
DOI: 10.1007/978-3-642-29026-8_8, 

## 8.1 Introduction

Humans and wildlife have interacted for millennia, coexisting in many different ways (Sharma 2002). Because of the biological, phylogenetic, and behavioral overlaps between humans and non-human primates, relationships between the two groups have a special significance (Nadler et al. 2008). As our evolutionary cousins, non-human primates form an integral part of human mythologies, diets, and scientific concepts. So, it is not surprising that in many cultures primates already have significant cultural values. Primates have always had a major ethnozoological role within the symbolic sphere of animistic societies, often identified with human qualities or gods in paintings or other artistic products (Alves et al. 2010; Mittermeier et al. 2007).

In many areas of Asia, South America, and Africa, humans and other primates have shared ecosystems for millennia (Malik and Johnson 1994; Mohnot 1971). In these areas, there are a number of agricultural, religious, and cultural practices that result in a degree of protection for some primate species (Sharma 2002), on the other hand, many of the species are now threatened directly or indirectly by human activities (Bleisch et al. 2008a, b). Primates are among the most persecuted—relentlessly hunted for their meat and fur, shot for stealing crops in fields that were once their home (Li et al. 2007). In addition to being hunted for food, primates are also killed for "medicinal" purposes, and occasionally as bait for large cats and fish, the woolly and spider monkeys being the most targeted species (Mittermeier 1987). Furthermore, in many tourist shops throughout much of the Amazon, hats from monkey skins and necklaces from monkey teeth, skulls, bones, hands, feet, or tails have been observed on sale (Ha 2006). Several primate species are kept as pets, and a small number are rescued and reintroduced into the wild (Bleisch et al. 2008a).

In this sense, the human perception of non-human primates is often one of contradiction, typified by extremes. In some cultures and contexts (e.g. Hindus of Bhutan, India, Nepal) primates are viewed as sacred, in others, such as China or Japan, they are mythical creatures of cunning and deviousness, while for most of the world's subsistence farmers living in close proximity to monkeys and apes, they represent a significant crop pest. In many cultures these two views overlap, resulting in both a love and loathing of the creatures such that they may be worshipped at a temple and killed on the farm next door. In many places, cultural tolerance alone is protecting primate species (e.g. India, Sulawesi). It is within these perceptions and limitations that conservation must work (Srivastava and Mohnot 2001). Human ecologies and societies can influence, and be influenced by, the ecology and behavior of other primates. Crop raiding, fear of personal harm/attack, and habitat alteration by other primates can affect the livelihoods and nutritional statuses of humans. Human alteration of the landscape, human hunting, human religious belief, and even human pet keeping can affect the behavior and ecology of other primates (Pujol 1993). Bi-directional transfer between humans and other primates is well documented and we know that disease introduced by such exchanges can have dramatic effects on humans and nonhuman populations

(PirtaMadhav 1997; Pujol 1993; Southwick and Lindburg 1986). As human pressures on environments continue to increase and primate habitats become smaller and more fragmented, the need for a primatology that considers the impact of human attitudes and behavior on all aspects of primate lives and survival is imperative (Pujol 1993).

The variety of interactions (both past and present) between human cultures and other animals is the subject matter of Ethnozoology, a science that has its roots as deep within the past as humans themselves (Alves and Souto 2010). In case of primates, these interactions are investigated through Ethnoprimatology, a relatively new subdiscipline that bridges cultural anthropology and primatology, exploring the interface between humans and non–human primates (Phillips 1931). As was pointed by Fuentes and Hockings (2010), one of the key goals of ethnoprimatology is to contribute to conservation and management efforts to slow population declines in primates and/or facilitate population recoveries while at the same time recognizing the importance of economic security and the cultural contexts of local human populations that interface with the species being studied (Dunnet et al. 1970; Gron 2008). The importance of a holistic understanding of human–primate interactions is vital to conservation efforts that include environmental education (Pruetz 2002; Takemoto 2002) and conflict resolution/mitigation strategies (Humle 2003a, b). Ethnoprimatology attempts to integrate the interests and concerns of humans with those of non-human primates, and proposes that conservation actions on the behalf of primate species must include measures that construct an ecological framework that includes human dimensions (Gron 2008; Montgomery 1970; Nadler et al. 2008).

Among the animal species most frequently used in traditional folk medicine are the primate order, and their role in folk practices has been recorded in different socio-cultural contexts throughout the world (Alves et al. 2010). Various authors have discussed the conservation concerns of the use of medicinal products from primates, especially because many of them are threatened species (Adeola 1992; Alves and Alves 2011; Alves et al. 2007, 2010; Apaza et al. 2003; Caspary et al. 1998; Duvall and Niagaté 1997; Fruth et al. 2008; Ham 1998; Hanson-Alp et al. 2003; Li et al. 2007; Nisbett et al. 2003; Tashiro et al. 2007). For some of these species, the hunting for medicinal purposes (in association with other uses, such as pets and for food) also impacts the ability of primates to survive in the future (Alves et al. 2010; Bleisch et al. 2008b; Nadler et al. 2008; Tashiro et al. 2007). In this sense, the ethnomedicinal use of primates around the world should be addressed when designing conservation strategies (Alves et al. 2010).

The large part played by primates in traditional folk medicine of many cultures has frequently been overlooked, even though there is widespread anecdotal evidence for their use as medicine. Such use is a serious factor affecting declines in certain primate populations, and there is a need to attempt to improve our limited understanding of the details of the use of primates and the specific impacts of these practices. Such information will be important in developing more realistic and appropriate conservation strategies for affected species. This chapter will aim at providing an overview of the global primate usage in traditional folk medicine, as well as identifying those species used as folk remedies. Some important questions relating to the conservation of these species will be addressed. In this context, we attempt to answer the following questions: (1) How many and which primates species are used in folk medicine around the world? (2) What is the conservation status of the exploited species? and (3) What are the implications of zootherapy for conservation and public health?

## 8.2 Methods

To examine the diversity of primates used in traditional folk medicine, all available references or reports of folk remedies based on primate sources were examined. Only taxa that could be identified to species level were included in the database. Scientific names provided in publications were updated according to the IUCN/SSC Primate Specialist Group (2011). The conservation status of the invertebrate species followed the International Union for Conservation of Nature Red List (IUCN 2011) and the convention on international trade in endangered species of wild fauna and flora (CITES) (CITES 2011).

The sources analyzed were: Banks (1931), Phillips (1931), Westermann (1938), Tang and Li (1957), Tappen (1960), Dunnet et al. (1970), Montgomery (1970), Mohnot (1971), Akowuah et al. (1975), Dunbar (1977), Petter (1977), Kortlandt (1986), Southwick and Lindburg (1986), Albignac (1987), Bennett et al. (1987), Eudey (1987), Mittermeier (1987), Nilsson (1990), Swantz (1990), Wilbert and Simoneau (1990), Ganzhorn and Abraham (1991), Ramakantha (1991), Adeola (1992), Biquand et al. (1992), Ntiamoa-Baidu (1992), Pujol (1993), Thalmann et al. (1993), Figueiredo (1994), King (1994), Malik and Johnson (1994), Mittermeier et al. (1994), Rodríguez-Mahecha et al. (1995), Borang (1996), Indrawan et al. (1996), Duvall and Niagaté (1997), O'Brien and Kinnaird (1997), PirtaMadhav (1997), Seixas (1997), American Zoo and Aquarium Association (1998), Caspary et al. (1998), Gonzalez-Kirchner and Sainz de la Maza (1998), Ham (1998), Horowitz (1998), Morales-Mavil and Villa-Cañedo (1998), Simelane and Kerley (1998), Takeda (1998), Walker (1998), Britt et al. (1999), Duckworth et al. (1999), Garbutt (1999), Knight (1999), Marshall et al. (1999), Mayor and Lehman (1999), Rijksen and Meijaard (1999), Sodeinde and Soewu (1999), Srivastata (1999), Tratado de Cooperacion Amazonica (1999), Wheatley (1999), Geissmann et al. (2000), WCMC (2000), Ahmed (2001), Chiu (2001), Fitch-Snyder (2001), Gray (2001), Srivastava and Mohnot (2001), CITES (2002), Governo do Estado do Paraná (2002), Lizarralde (2002), Pruetz (2002), Sharma (2002), Shepard (2002), Takemoto (2002), Vargas (2002), Winter (2002), Apaza et al. (2003), Cormier (2003), Dang et al. (2003), Hanson-Alp et al. (2003), Humle (2003a, b), Johnson et al. (2003), Kormos et al. (2003), Lammertink et al. (2003), Nadler et al. (2003), Nisbett et al. (2003), Walker and Molur (2003), Nekaris and Jayewardene (2004), Nijman (2004), Shepherd et al. (2004), Solanki and Chutia (2004), Thang Long (2004), Chan et al. (2005), Hilaluddin et al. (2005), Jamir and Lal (2005), Lee and Priston (2005), Nijman (2005), Riley (2005), Silva et al. (2005), Alvarez (2006),

**Table 8.1** Number of primates species used in traditional folk medicine by region

| Region | Diversity | Number of species used in traditional folk medicine | Percentage of species used (%) |
|---|---|---|---|
| Neotropics | 139 species | 25 species | 17,9 |
| Africa and Africa | 158 species | 75 species | 47,46 |
| Madagascar | 93 species | 10 species | 10.7 |
| Total | 390 | 110 | 28.2 |

Dedeke and Abayomi ( 2006), Dedeke et al. (2006), Ha (2006), Loudon et al. (2006), Rodrigues (2006), Tejada et al. (2006), Whittaker (2006), Xuan Canh and Boonratana (2006), Alves and Rosa (2007a, b), Alves et al. (2007), Geissmann (2007), Gonedelé Bi et al. (2007), Hamada et al. (2007), Inogwabini et al. (2007), Jones et al. (2007), Khalid et al. (2007), Li et al. (2007), Mittermeier et al. (2007), Nadler et al. (2007), Negi and Palyal (2007), Tashiro et al. (2007), Wittiger and Sunderland-Groves (2007), Xiang et al. (2007), Ashwell and Walston (2008), Bleisch et al. (2008a, b), Boonratana et al. (2008), Campbell et al. (2008), Fruth et al. (2008), Gron (2008), Hoffmann and Hilton-Taylor (2008), Jones et al. (2008), Kingdon et al. (2008), Kumar et al. (2008), Mahawar and Jaroli (2008), Manh Ha et al. (2008), Meijaard and Nijman (2008), Meijaard et al. (2008), Nadler et al. (2008), Nekaris et al. (2008), Ngoc Thanh et al. (2008a, b), Nijman et al. (2008), Putra et al. (2008), Racero-Casarrubia et al. (2008), Rawson et al. (2008), Soewu (2008), Streicher et al. (2008a, b), Timmins and Boonratana (2008), Van and Tap (2008), Xuan Canh et al. (2008), Ziemendorff (2008), Alves (2009), Arkive.org (2009), Baker et al. (2009), IUCN (2011), Whiting et al. (2011). A database was created containing information about primate species, family names, parts of the animal used, and country where use was recorded.

## 8.3 Results and Discussion

In many parts of the world, primates are killed for consumption by humans (Alves et al. 2010). As relatively large mammals, especially in the neotropics, primates are a common choice for hunters (Alvard et al. 1997; King 1994; Kormos et al. 2003; Lizarralde 2002; Shepard 2002). Reasons for hunting vary by region, and one of these pretexts is the medicinal or magical value of products derived from these animals (Alves et al. 2010).

Our review revealed that at least 110 species of primates belonging to 41 genera and 11 families are used for medicinal purposes or are associated to magical/religious practices. The family with the largest numbers of species recorded were Cercopithecidae (with 54 species), followed by Cebidae (18) and Hylobatidae (8) (Tables 8.1, 8.2 and 8.3). A total of 50 species are used for medicinal purposes, 39 are associated to folk beliefs and magical/religious practices, and 21 for both purposes.

The use of primates in traditional folk medicine practices is widespread, being recorded in 52 countries, mainly in Latin America, Africa, and Asia. This is not surprising given that primate geographic distribution is largely restricted to the tropics (Primate Specialist Group 2011). Most of the medicinal species recorded are from the Old World (68.2%), followed by the New World (22.7%) and Madagascar (about of 9.1%). While human and non-human primates share a number of interconnections, at the most basic level there is a distinction introduced by geography. Primates are prey items for a diverse array of human cultures in zones of sympatry (geographic overlap), and are also captured by people from both zones for various human needs, ranging from ingredients in traditional medicines to being subjected to biomedical research (Marshall et al. 1999).

Sometimes, the whole bodies of primates may be used in traditional medicine, but various parts of these animals, such as fur, legs, fat, oil, eyes, bile, blood, gall bladder, viscera, bone, meat, brain, and skull are used most commonly in preparation of traditional remedies. A similar trend has been observed for other vertebrates in traditional folk medicine, in which extracted zootherapeutical remedies are derived from body parts that are often not used for any purpose (Alves and Alves 2011; Alves et al. 2008; Apaza et al. 2003; El-Kamali 2000; Kakati et al. 2006; Mahawar and Jaroli 2008; Moura and Marques 2008; Sodeinde and Soewu 1999; Vázquez et al. 2006). Thus, many primate zootherapeutical products are the result of hunting these animals for other purposes.

Around the world, zootherapeutical products derived from primates are used to treat a wide variety of health problems. Our review revealed that at least 25.3% (n = 18 of 71) medicinal primates species are reported as used to cure more than one ailment. Among the most versatile species are *Ateles chamek*, *Cebus apella*, *Cercopithecus mona*, *Macaca mulatta*, and *Trachypithecus pileatus*, involved in more than six medicinal indications. Table 8.2 summarizes the species used as remedies and the diseases for which they are prescribed. Different ways of preparing and administering remedies based on primates are used. Hard parts (e.g. bone and fur), are generally sun-dried, grated, and crushed to powder, to be administered as tea or taken during meals. Meat, brain, oil, fat, and blood can be ingested, or used as an ointment. In Indochina, for example, gibbons are sometimes hunted for medicinal purposes (Rijksen and Meijaard 1999) and there have been claims that some will eat the meat of orangutans as an aphrodisiac (Campbell et al. 2008). Adi people (India) occasionally use rhesus monkey (*Macaca mulatta*) meat to treat epidemic diseases like malaria, typhoid, cholera, and pox, etc. (Geissmann 2007), and Marring Nagas grind the loris skull, mix it with water, and take it orally to cure epidemics like cholera (Duckworth et al. 1999).

Some treatment methods recorded are eccentric. In India, it is thought by many that eating monkey (*Macaca assamensis* and *M. mulatta*) brain will treat rheumatism and that drinking monkey blood will cure asthma. For treating rheumatism, a special table with a device is used. The monkey's head (live animal) is introduced through a hole on the top of the table and a chain tied to the bottom part of the table holds its legs. Boiling water is then poured over the head and face of the primate causing extensive scalding and peeling away of the fur and skin. A sharp

**Table 8.2** Primates used as remedies in traditional medicine

| Family/Species | IUCN Red List status | CITES Appendices | Region or country(ies) | Use(s) | Reference(s) |
|---|---|---|---|---|---|
| **Callitrichidae** | | | | | |
| *Saguinus fuscicollis* (Spix, 1823) | LC | II | Bolivia | Impotence | 1, 2 |
| *Saguinus mystax* (Spix, 1823) | LC | II | Bolivia | Impotence | 2 |
| **Cebidae** | | | | | |
| *Alouatta belzebul* (Linnaeus, 1766) | VU | II | Brazil | Whooping cough, sore throat, asthma | 3, 4, 5 |
| *Alouatta fusca* (É. Geoffroy Saint-Hilaire, 1812) | LC | II | Brazil | Used as aphrodisiac; Used to treat any disease | 6, 7 |
| *Alouatta nigerrima* Lönnberg, 1941 | LC | II | Brazil | Whooping cough, inflammation | 4, 8 |
| *Alouatta palliata* (Gray, 1849) | LC | II | Costa Rica | General pains and inflammation | 9 |
| *Alouatta sara* Elliot, 1910 | LC | II | Bolivia | Unspecified | 1 |
| *Alouatta seniculus* (Linnaeus, 1766) | LC | II | Brazil, Colombia | "Whooping cough, inflammation, to accelerate parturition" | 4, 8, 10, 11, 12, 13 |
| *Aotus azarae* (Humboldt, 1811) | LC | II | Bolivia | So that the baby does not dribble | 14 |
| *Aotus griseimembra* Elliot, 1912 | VU | II | Colombia | Unspecified | 15 |
| *Ateles chamek* (Humboldt, 1812) | EN | II | Bolivia | Snake bites, spider bites, fever, cough, cold, shoulder pain, sleeping problems, leishmaniasis | 1, 12, 16 |

(continued)

**Table 8.2** (continued)

| Family/Species | IUCN Red List status | CITES Appendices | Region or country(ies) | Use(s) | Reference(s) |
|---|---|---|---|---|---|
| *Ateles geoffroyi* Kuhl, 1820 | EN | II | Costa Rica, Mexico, Colombia | Rheumatism | 11, 17, 18 |
| *Ateles paniscus* (Linnaeus, 1758) | VU | II | Bolivia | Rheumatism | 19 |
| *Cebus albifrons* (Humboldt, 1812) | LC | II | Venezuela | Used as fortifier | 2, 20 |
| *Cebus apella* (Linnaeus, 1758) | LC | II | Bolivia, Brazil, Peru | Inflammatory processes, insect sting, used for osteomuscular problems, eye infection, and male impotency | 1, 3, 4, 11, 16, 21, 22 |
| *Cebus capucinus* (Linnaeus, 1758) | LC | II | Costa Rica, Colombia | Unspecified | 12, 17 |
| *Lagothrix lagotricha* (Humboldt, 1812) | VU | II | Bolivia, Brazil, Colombia, Ecuador, Peru | Unspecified | 2 |
| **Cercopithecidae** | | | | | |
| *Nasalis larvatus* (Wurmb, 1787) | EN | II | China | Unspecified | 23, 24, 25 |
| *Cercopithecus mitis* Wolf, 1822 | LC | II | Angola, Burundi, Congo DR, Ethiopia, Kenya, Malawi, Mozambique, Rwanda, Somalia, South Africa, Sudan, Swaziland, Tanzania, Uganda, Zambia, Zimbabwe | Unspecified | 26, 27,28, 29 |
| *Cercopithecus mona* (Schreber, 1774) | LC | II | Nigeria | Male impotency, aid in quickening child walking, used by palm wine tappers to climb successfully | 30, 31, 32 |

(continued)

**Table 8.2** (continued)

| Family/Species | IUCN Red List status | CITES Appendices | Region or country(ies) | Use(s) | Reference(s) |
|---|---|---|---|---|---|
| Chlorocebus pygerythrus (Cuvier, 1821) | LC | | South Africa | Unspecified | 29 |
| *Chlorocebus aethiops* (Linnaeus, 1758) | LC | II | Sudan, South Africa | Mental illness | 28, 33 |
| *Erythrocebus patas* (Schreber, 1775) | LC | II | Nigeria, Sudan | Whooping cough, charm preparations, mental illness | 30, 32, 33, 34, 35 |
| *Macaca cyclopis* (Swinhoe, 1863) | LC | II | Taiwan | Unspecified | 36, 37 |
| *Macaca arctoides* (I. Geoffroy Saint-Hilaire, 1831) | VU | II | Vietnam | Unspecified | 38, 39, 40, 41 |
| *Macaca assamensis* (M'#Clelland, 1840) | NT | II | India, Vietnam | Rheumatism, asthma, delivery pains, malaria typhoid, smallpox | 40, 42, 43, 44, 45 |
| *Macaca fascicularis* (Raffles, 1821) | LC | II | Cambodia, Indonesia | Unspecified | 46, 47 |
| *Macaca leonina (Blyth, 1863)* | VU | II | SE Asia | Unspecified | 40, 48 |
| *Macaca mulatta* (Zimmermann, 1780) | LC | II | India, Vietnam | Rheumatism, asthma, malaria, typhoid, cholera, pox, smallpox, stomach disorder, paralysis, adiposity, and parasitic infestation. | 38 39 41 43,49, 50,51,52, 53, 54 |
| *Macaca nemestrina* (Linnaeus, 1766) | VU | II | Lao PDR | Unspecified | 55 |
| *Macaca silenus* (Linnaeus, 1758) | EN | I | India, China | Unspecified | 56, 57 |
| *Mandrillus leucophaeus* (F. Cuvier, 1807) | EN | I | Nigeria | Unspecified | 30 |

(continued)

**Table 8.2** (continued)

| Family/Species | IUCN Red List status | CITES Appendices | Region or country(ies) | Use(s) | Reference(s) |
|---|---|---|---|---|---|
| *Papio ursinus* (Kerr, 1792) | LC | II | South Africa | Treatment of arthritis, protective charm, court case charm, and headache | 27,28, 57, 27 |
| *Presbytis hosei* (Thomas, 1889) | VU | II | Brunei, Indonesia, Malaysia | Unspecified | 58 |
| *Presbytis natunae* (Thomas and Hartert, 1894) | VU | II | Indonesia | Unspecified | 59, 57 |
| *Presbytis frontata* (Müller, 1838) | VU | II | Indonesia | Unspecified | 60, 61 |
| *Presbytis siamensis* (Müller and Schlegel, 1838) | NT | II | Indonesia | Unspecified | 62 |
| *Procolobus verus* (Van Beneden, 1838) | NT | II | Ghana | Diseases in children associated with births | 63 |
| *Procolobus badius* (Kerr, 1792) | EN | II | E. Africa | Bruises and rashes on newly born babies | 27, 28, 63, 64 |
| *Presbytis rubicunda* (Müller, 1838) | VU | II | | Unspecified | 60, 65 |
| *Rhinopithecus avunculus* (Dollman, 1912) | CR | I | Vietnam | Unspecified | 66, 67, 68 |
| *Pygathrix nemaeus* (Linnaeus, 1771) | EN | I | Lao PDR, Vietnam, Thailand | Unspecified | 41, 67, 69, 70 |
| *Pygathrix nigripes* (Milne-Edwards, 1871) | EN | I | Vietnam | Unspecified | 71 |
| *Pygathrix cinerea* Nadler, 1997 | CR | I | Vietnam | Unspecified | 72, 73 |
| *Rhinopithecus bieti* Milne-Edwards, 1897 | EN | II | China | Good for rheumatism and overall health | 74, 75, 76 |

(continued)

**Table 8.2** (continued)

| Family/Species | IUCN Red List status | CITES Appendices | Region or country(ies) | Use(s) | Reference(s) |
|---|---|---|---|---|---|
| *Semnopithecus entellus* (Dufresne, 1797) | LC | I | India | Facilitates delivery | 43, 51, 77, 78 |
| *Trachypithecus delacouri* (Osgood, 1911) | CR | II | Vietnam | Unspecified | 79, 80 |
| *Trachypithecus francoisi* (Pousargues, 1898) | EN | II | China, Vietnam | Unspecified | 41, 67, 81, 82 |
| *Trachypithecus johnii* (Fischer, 1829) | VU | II | India | Unspecified | 40, 51 |
| *Trachypithecus laotum* (Dao, 1970) | VU | II | Lao PDR, Vietnam | Unspecified | 41, 83, 84 |
| *Trachypithecus obscurus* (Reid, 1837) | NT | II | Indonesia | Unspecified | 62 |
| *Trachypithecus phayrei* (Blyth, 1847) | EN | II | China, Vietnam, Thailand, and Lao PDR | Unspecified | 41, 85 |
| *Trachypithecus pileatus* (Blyth, 1843) | VU | I | India | Malaria, typhoid dysentery, and smallpox | 41, 45, 49, 86 |
| *Trachypithecus cristatus* Miller, 1913 | NT | II | Cambodia | Unspecified | 46 |
| *Trachypithecus poliocephalus* (Trouessart, 1911) | CR | II | China, Vietnam | Unspecified | 41, 80 |
| **Galagonidae** | | | | | |
| *Galago moholi* A. Smith, 1836 | LC | II | Unspecified | Charm for love and athletes, fits, stops babies crying | 27 |
| *Otolemur crassicaudatus* (Geoffroy Saint-Hilaire, 1812) | LC | II | South Africa | Unspecified | 29 |

(continued)

**Table 8.2** (continued)

| Family/Species | IUCN Red List status | CITES Appendices | Region or country(ies) | Use(s) | Reference(s) |
|---|---|---|---|---|---|
| **Hominidae** | | | | | |
| *Gorilla gorilla* (Savage and Wyman, 1847) | CR | I | Nigeria | Concoction for ailments, charm preparations, victory in competition, strength and vitality for pregnant women/fetus | 30, 31, 32, 34, 35 |
| *Pan troglodytes* (Blumenbach, 1775) | EN | I | Nigeria, Guinea, Sierra Leone, Liberia, Mali | Male impotency, epilepsy, amulets, concoction for ailments | 31, 32, 34,35, 87,88, 89,90, 91 |
| *Pan paniscus* Schwartz, 1929 | EN | I | Congo DR | Unspecified | 92 |
| *Pongo pygmaeus* (Linnaeus, 1760) | EN | I | Indonesia, Malaysia | Unspecified | 47, 93, 94 |
| **Hylobatidae** | | | | | |
| *Symphalangus syndactylus* Raffles, 1821 | EN | I | Indonesia | Unspecified | 95 |
| *Hylobates hoolock* (Harlan, 1834) | EN | I | India, China | Fever, typhoid, malaria and pox | 40, 44, 45, 96 |
| *Nomascus leucogenys* (Ogilby, 1840) | CR | I | China, Lao PDR, Vietnam, Thailand | Unspecified | 41, 70, 97, 98 |
| *Nomascus hainanus* (Thomas, 1892) | CR | I | China | Unspecified | 80, 99, 100 |
| *Nomascus siki* Delacour, 1951 | EN | I | Lao PDR, Vietnam | Unspecified | 97, 98, 101 |
| **Loridae** | | | | | |
| *Loris lydekkerianus* Cabrera, 1908 | LC | II | Sri Lanka and India | Unspecified | 40, 102, 103 |

(continued)

**Table 8.2** (continued)

| Family/Species | IUCN Red List status | CITES Appendices | Region or country(ies) | Use(s) | Reference(s) |
|---|---|---|---|---|---|
| *Nycticebus bengalensis* (Lacépède, 1800) | VU | I | Bangladesh, Cambodia, China, India, Lao PDR, Myanmar, Thailand, Vietnam | Unspecified | 40, 46, 104, 105, 106, 107 |
| *Nycticebus coucang* (Boddaert, 1785) | VU | I | Cambodia, Malaysia, Indonesia | Asthma | 46, 75, 108, 109 |
| *Nycticebus pygmaeus* Bonhote, 1907 | VU | I | Cambodia, Vietnam, Indonesia | Unspecified | 38, 41, 110 |
| *Perodicticus potto* (Müller, 1766) | LC | II | Ghana | Burns | 63 |

*Legend*: *IUCN Red List Categories—CR* critically endangered; *EN* endangered; *VU* vulnerable; *NT* near threatened; *LC* least concern. *CITES Appendice*—I or II. References: (1) Tejada et al. (2006), (2) Tratado de Cooperacion Amazonica (1999), (3) Alves and Rosa (2007a), (4) Alves et al. (2007), (5) Alves (2009), (6) Governo do Estado do Paraná (2002), (7) Seixas (1997), (8) Figueiredo (1994), (9) Gonzalez-Kirchner and Sainz de la Maza (1998), (10) Rodríguez-Mahecha et al. (1995), (11) Rodrigues (2006), (12) Racero-Casarrubia et al. (2008), (13) Cormier (2003), (14) Vargas (2002), (15) Arkive.org (2009), (16) Apaza et al. (2003), (17) Alvarez (2006), (18) Morales-Mavil and Villa-Cañedo (1998), (19) Winter (2002), (20) Lizarralde (2002), (21) Alves and Rosa (2007b), (22) Ziemendorff (2008), (23) Meijaard et al. (2008), (24) Banks (1931), (25) Westermann (1938), (26) Kingdon et al. (2008), (27) CITES (2002), (28) Simelane and Kerley (1998), (29) Whiting et al. (2011), (30) Adeola (1992), (31) Dedeke et al. (2006), (32) Dedeke and Abayomi (2006), (33) Khalid et al. (2007), (34) Soewu (2008), (35) Sodeinde and Soewu (1999), (36) Chiu (2001), (37) Eudey (1987), (38) Van and Tap (2008), (39) Dang et al. (2003), (40) Walker and Molur (2003), (41) Nadler et al. (2007), (42) Walker (1998), (43) Ahmed (2001), (44) Hilaluddin et al. (2005), (45) Solanki and Chutia (2004), (46) Ashwell and Walston (2008), (47) Nijiman (2005), (48) Boonratana et al. (2008), (49) Borang (1996), (50) Ramakantha (1991), (51) Mahawar and Jaroli (2008), (52) Negi and Palyal (2007), (53) Jamir and Lal (2005), (54) Johnson et al. (2003), (55) Kumar et al. (2008), (56) American Zoo and Aquarium Association (1998), (57) Hoffmann and Hilton-Taylor (2008), (58) Nijman et al. (2008), (59) Lammertink et al. (2003), (60) Putra et al. (2008), (61) Meijaard and Nijman (2008), (62) Nijman (2004), (63) Ntiamoa-Baidu (1992), (64) Swantz (1990), (65) Nijman and Meijaard (2008), (66) Xuan Canh et al. (2008), (67) Nadler et al. (2003), (68) Xuan Canh and Boonratana (2006), (69) Ngoc Thanh et al. (2008b), (70) Timmins et al. (1999), (71) Rawson et al. (2008), (72) Thang Long (2004), (73) Ngoc Thanh et al. (2008a), (74) WCMC (2000), (75) Nilsson (1990), (76) Xiang et al. (2007), (77) Lee and Priston (2005), (78) Sharma (2002), (79) Nadler et al. (2008), (80) Mittermeier et al. (2007), (81) Bleisch et al. (2008b), (82) Li et al. (2007), (83) Timmins and Boonratana (2008), (84) Ha (2006), (85) Bleisch et al. (2008a), (86) Srivastava and Mohnot (2001), (87) Hanson-Alp et al. (2003), (88) Ham (1998), (89) Nisbett et al. (2003), (90) Duvall and Niagaté (1997), (91) Caspary et al. (1998), (92) Fruth et al. (2008), (93) Marshall et al. (1999), (94) Rijksen and Meijaard (1999), (95) Campbell et al. (2008), (96) Geissmann (2007), (97) Duckworth et al. (1999), (98) Geissmann et al. (2000), (99) Tang and Li (1957), (100) Chan et al. (2005), (101) Manh et al. (2008), (102) Nekaris et al. (2008), (103) Nekaris and Jayewardene (2004), (104) Streicher et al. (2008b), (105) Srivastata (1999), (106) Gray (2001), (107) Fitch-Snyder (2001), (108) Bennett et al. (1987), (109) Sherpherd et al. (2004), (110) Streicher et al. (2008a) rap with a small hammer cracks the skull and the 'doctor' pours several tubes of ointment into the skull. The patient, who is seated at the same table, inserts a spoon into the cranium of the monkey and eats the brain, alternately mashing and mixing in medicinal drugs. For asthma, the patient drinks the monkey's blood (Ham 1998). In Vietnam, primates are hunted for food or medicinal purposes and it is a common occurrence to find them in bottles in alcohol, not as zoological specimens, but as tonics for their medicinal value or for sipping as monkey wine (Fruth et al. 2008). Also, in the same country, primates in alcohol are used as 'energy drinks' (Geissmann et al. 2000; Tang and Li 1957). Another curios belief of people in the south of Sri Lanka is that monkey organs (heart and lungs) are being used in the cities for organ transplants in humans due to their close similarity to humans (Horowitz 1998).

Cultural perceptions of primates vary widely over both space and time (Petter 1977). As occurs with other wild animals, primates often have important roles in cultural and religious festivals and ceremonies, many of which seek to promote the good health of human communities (Adeola 1992; Ajayi 1978; Léo Neto et al. 2009). Primates are commonly associated with myths in many different countries and are used in magical/religious rituals (e.g. Hamada et al. 2007; Petter 1977). In a number of regions from Africa to Tibet, primates are revered as guardians of human settlements, as animals that bring good luck, as reincarnations of the spirits of ancestors, and as embodiments of gods (Petter 1977). Table 8.3 summarizes the species associated with folk beliefs or used for magical/religious purposes.

Primates parts are used to prepare clinical remedies as well as to make amulets or charms used in magical/religious diagnoses. One form of spiritual treatment involves the use of amulets containing primate's parts to protect the user against diseases. An example is recorded by Hanson-Alp et al. (2003), in Sierra Leone, where sometimes a piece of dried chimpanzee bone is tied around the waist or wrist of infants in the belief that it makes them stronger as they grow into adulthood. Also in Sierra Leone, chimpanzee central incisors were procured to be worn as amulets around the waist of infants to protect them and give them power over others in their cohort (Chan et al. 2005). In India, the eye of the Hanuman langur (*Semnopithecus entellus*), worn in amulets, is believed to increase courage and strength, and charms prepared from dismembered digits from a monkey's paw, strung together on a thread, are believed to treat paralysis (Ham 1998). The hyoid bone of *Alouatta seniculus* is used by campesinos in the upper Magdalena and Cauca valleys of Colombia as a grinding device, and some believe that it possesses therapeutic properties for curing goiters (Garbutt 1999). In some parts of the Brazilian Amazon, drinking from an *Alouatta* hyoid has been thought to ease labor pains and to treat asthma, and in Surinam, it has been thought to cure stuttering (Albignac 1987; Alves and Rosa 2007a, b, 2010).

In the mythology of different cultures, primates appear in various contexts. In Costa Rica, *Alouatta palliata* is considered as an evil omen that brings bad luck, and, so, when found near a village it is shot (Gonzalez-Kirchner and Sainz de la Maza 1998). In Madagascar, *Daubentonia madagascariensis* (Gmelin 1788) may be killed on sight as crop pests or harbingers of evil (Manh Ha et al. 2008). Due to its evil reputation, the aye-aye (*D. madagascariensis*) is never eaten, but is used in ritualized ceremonies (Nekaris et al. 2008; Nekaris and Jayewardene 2004). In Sri Lanka, valuable material from the eyes is extracted by holding a loris above a fire until the eyes burst. Some people fear their cry as an ill omen (Streicher et al. 2008b). The mournful tone of the cry is thought to bring misfortune, usually resulting in their being stoned to death (Nisbett et al. 2003). In India, skulls of various species of primates are hung outside tribal dwellings in Arunachal Pradesh and other northeastern states to ward off evil spirits or simply as objects of ornamentation.

A broad category of beliefs encountered in Amazonian cultures attribute either positive or negative traits to monkeys that can be transferred to humans (Hamada et al. 2007). Crocker (1971) reported a kind of magical prohibition among the Bororo people that keeps them from eating monkeys, which are considered to epitomize speed and grace. Similarly, according to Lizarralde (2002), the Barí keep spider monkeys as pets and believe that wearing spider monkey teeth necklaces will confer manual dexterity to the owner. Howler monkeys, on the other hand, are considered to be of low intelligence and speed, and they are not kept as pets nor are their teeth valued as charms. A similar perception was recorded in the Matsigenka mythology, in which howler monkeys are considered to be lazy and capuchins are considered thieves, and they believe that these traits can be transferred to humans who eat them (Shepard 2002). This is reflected in another example that bridges couvade and contagion, in which Vilaça (2002) reported a Warí shaman telling two parents that their child was turning into a monkey because they had not followed the correct protocol for preparing and eating capuchins.

Although many primates are killed for magical/religious purposes, folk beliefs in some cases may have a positive influence from a conservationist perspective (Alves et al. 2010). As pointed by Posey (1999), certain native beliefs and practices reflect traditional socioenvironmental concepts that have conservation implications. This is especially the case for primates, many species of which have mythological or symbolic importance and are subject to taboos, restrictions, or dietary avoidance among diverse Amazonian peoples (Ganzhorn and Abraham 1991; Shepard 2002). The reasons for not hunting primates are varied, but are frequently based on religion. Islam, Hinduism, and Buddhism have various restrictions or beliefs that prevent followers from eating or killing primates (Mittermeier 1987). A well-known example is the Hindu-based protection of monkeys in parts of Asia, such as long-tailed macaques (*Macaca fascicularis*) in Bali, Indonesia (Srivastata 1999), and rhesus macaques (*M. mulatta*) and gray langurs (*Semnopithecus* spp.) in India and the neighboring region (Bennett et al. 1987; Fitch-Snyder 2001; Gray 2001). Primates may also be protected by social taboos (Shepherd et al. 2004). Some chimpanzee (*Pan troglodytes*) populations are not hunted because of their physical similarity to humans or folklore regarding an ancestral relationship with humans (Kortlandt 1986; Putra et al. 2008; Silva et al. 2005; Streicher et al. 2008a; Takeda 1998). In the village of Bossou (Republic of Guinea), Manon people consider chimpanzees sacred, as the reincarnation of their ancestors, and believe that their ancestors' souls rest on the sacred hill of Gban (Kortlandt 1986). As the chimpanzee is a totem of the most influential family of Bossou, it is strictly forbidden to hunt or eat the chimpanzee (Yamakoshi 2005). This author proposes that Gban was important for village protection during years of tribal conflict; the current peaceful coexistence between man and chimpanzee may have historical war roots. Chimpanzees regularly crop-raid and are known to attack villagers occasionally. However, due to the local people's strong cultural beliefs, humans and chimpanzees have been able to coexist in such close proximity over many generations (Akowuah et al. 1975; Yamakoshi 2005). Another example was recorded by Shepard (2002), who noted that Matsigenka hunters generally avoided taking woolly and spider monkeys from the peak of the dry season (July–August) through the early rainy season (November–December). This is because fruits are scarce at that time and monkey meat is lean and tough and likely to elicit disparaging comments from their wives. Monkey hunting is instead concentrated in the late rainy season and beginning of the dry season (March–June) when these animals are fat. The Matsigenka also believe that certain monkeys (especially large adult males) and other game animals have vengeful spirits that can "take revenge" on the hunter's family, causing illnesses in young children. Matsigenka women use special fragrant herbs to protect newborn babies from the musk-smell and vengeful spirits of monkeys and other game animals. The examples cited above reinforce the view of Shepard (2002) that traditional socio-environmental concepts could provide ideological frameworks for future conservation measures (Shepard 2002).

The use of primates in traditional folk medicine is influenced by magical beliefs and myths (Alves et al. 2010), corroborating the remark by Alves and Rosa (2006) in which animals provide the raw materials for prescribed remedies using the clinical method, and are also used in the form of amulets and charms in magical/religious diagnosis. As suggested by Alves et al. (2007), in traditional medicine, animals are employed within a magical-prophylactic perspective with the purpose of warding off diseases of 'unnatural' origin—a practice that encompasses perceptions related to the belief that supernatural forces are involved in causing diseases, as well as in their treatment. Links between popular beliefs and zootherapy should be taken into consideration when interpreting results of field surveys, and when designing public health programs for communities where traditional medicine is used. In some cases, integrative approaches encompassing an understanding of traditional cultural views and insights concerning the cause, dissemination, and treatment of a disease might be required to effectively treat it (Alves and Rosa 2006).

In addition to their impacts on cultural patterns and perceptions, spatial overlaps and interactions between humans and non-human primates can create environments where more than space is shared—such as co-mingling of infectious agents. While it is generally assumed that human/non-human primate infectious agent "sharing" has deleterious repercussions, our understandings of the patterns and contexts of these shared environments and their evolutionary implications are actually rather incomplete (Baker et al. 2009; Loudon et al. 2006). Close contacts and range overlaps between humans and non-human primates introduce very real

**Table 8.3** Primates used in magic-religious rituals or practices

| Family/Species | IUCN Red List | CITES Appendices | Region or country(ies) | Use(s) | Reference(s) |
|---|---|---|---|---|---|
| **Callitrichidae** | | | | | |
| *Cebuella pygmaea* (Spix, 1823) | LC | II | Peru | Considered magical and potentially dangerous creatures | 1 |
| *Saguinus midas* (Linnaeus, 1758) | LC | II | Suriname, Brazil | It is considered sacred | 2, 3 |
| **Cebidae** | | | | | |
| *Alouatta seniculus* (Linnaeus, 1766) | LC | II | Peru | Considered to be lazy; this undesirable character trait could be passed on to children who consume their meat. Also known to be infested with botfly larvae, rendering their meat less attractive | 4 |
| *Aotus azarae* (Humboldt, 1811) | LC | II | Bolivia | Implies a potential threat on the spiritual level, used to prevent dribbling in babies | 5 |
| *Aotus azarae* (Humboldt, 1811) | LC | II | Brazil | Folk uses | 3 |
| *Alouatta palliata* (Gray, 1849) | LC | II | Costa Rica | Evil omen that brings bad luck | 6 |
| *Ateles paniscus* (Linnaeus, 1758) | VU | II | Peru, Suriname | Animal spirits considered capable of stealing the souls of children and making them ill. It is considered sacred in Suriname | 1, 2, 3 |
| *Alouatta belzebul* (Linnaeus, 1766) | VU | II | Venezuela | "Folk uses" | 7, 3 |
| Cebus kaapori Queiroz, 1992 | CR | II | Brazil, | Folk uses | 3 |

(continued)

**Table 8.3** (continued)

| Family/Species | IUCN Red List | CITES Appendices | Region or country(ies) | Use(s) | Reference(s) |
|---|---|---|---|---|---|
| *Saimiri sciureus* (Linnaeus, 1758) | LC | | Brazil, | Folk uses | 3 |
| **Cercopithecidae** | | | | | |
| *Cercopithecus mona* (Schreber, 1774) | LC | II | Nigeria, Congo DR, Ghana | Charms or amulets, aid in quickening child walking, used by palm wine tappers to climb successfully, victory in competition, used in funeral rituals, protected by local beliefs and traditions | 8, 9,10, 11 |
| *Cercopithecus neglectus* Schlegel, 1876 | LC | II | Congo DR | Used in funeral rituals | 12 |
| *Cercopithecus petaurista* (Schreber, 1774) | LC | II | Ghana, Ivory Coast | Considered sacred | 13, 14 |
| *Cercopithecus diana* (Linnaeus, 1758) | VU | I | Ghana | Considered sacred | 13 |
| *Chlorocebus tantalus* Ogilby, 1841 | LC | II | Nigeria | Considered sacred | 15 |
| *Chlorocebus sabaeus* Linnaeus, 1766 | LC | II | Ivory Coast | Considered sacred | 14 |
| *Colobus angolensis* Sclater, 1860 | LC | II | Congo DR | To cure supernatural diseases | 12 |
| *Colobus polykomos* (Zimmermann, 1780) | VU | II | Ghana | Purification and strengthening of widow/widower, protected by local beliefs and traditions, spiritual strengthening after birth of tenth child, Considered sacred. Miscarriage. | 11 |

(continued)

**Table 8.3** (continued)

| Family/Species | IUCN Red List | CITES Appendices | Region or country(ies) | Use(s) | Reference(s) |
|---|---|---|---|---|---|
| *Colobus vellerosus* (Geoffroy, 1834) | VU | II | Ghana | It is locally considered sacred | 16, 17 |
| *Erythrocebus patas* (Schreber, 1775) | LC | II | Nigeria, Sudan | Charm preparations | 8, 10, 18, 19 |
| *Macaca fascicularis* (Raffles, 1821) | LC | II | Indonesia | Considered sacred | 20 |
| *Macaca fuscata* (Blyth, 1875) | LC | II | Japan | Has positive religious associations as one of the animal reincarnations of the Buddah; however, in sharp contrast, monkeys also appear as morally dubious or unfavorable animals in Japanese folklore | 21 |
| *Macaca mulatta* (Zimmermann, 1780) | LC | II | India | The palm or finger or skull of *M. mulatta* is hung above the main door to propitiate evil spirits | 22, 23, 24 |
| *Macaca nigra* (Desmarest, 1822) | CR | II | Indonesia | Some incorporation into traditional belief systems | 25 |
| *Macaca pagensis* (Miller, 1903) | CR | II | Indonesia | Part of belief system | 26 |
| *Macaca tonkeana* (Meyer, 1899) | VU | II | Indonesia | It is considered sacred | 27 |
| *Papio hamadryas* (Linnaeus, 1758) | LC | II | Saudi Arabia, Ethiopia, Eritrea, Indonesia | Considered sacred | 26, 28, 29, 30 |
| *Papio anubis* (Lesson, 1827) | LC | II | Nigeria, Kenya, Uganda | Used in cultural ceremonies, viewed as malicious, cunning, and dangerous | 8, 26 |

(continued)

**Table 8.3** (continued)

| Family/Species | IUCN Red List | CITES Appendices | Region or country(ies) | Use(s) | Reference(s) |
|---|---|---|---|---|---|
| *Papio cynocephalus* (Linnaeus, 1766) | LC | II | Kenya, Uganda | Viewed as malicious, cunning, and dangerous | 26 |
| *Papio ursinus* (Kerr, 1792) | LC | II | South Africa | Protective charm, court case charm | 31, 32 |
| *Procolobus badius* (Kerr, 1792) | EN | II | Africa | Spiritual use | 33, 34, 35 |
| *Procolobus verus* (Van Beneden, 1838) | NT | II | Ivory Coast | Considered sacred | 14 |
| *Semnopithecus entellus* (Dufresne, 1797) | LC | I | India | Amulets, sacred primate to Hindus, charms prepared from dismembered digits of a monkey's paw strung together on a thread are believed to treat paralysis, bones subjected to black magic rituals are considered auspicious and bring recognition | 23, 36, 37, 38, 39, 40 |
| *Trachypithecus pileatus* (Blyth, 1843) | VU | I | India | Magical/religious practices | 24, 41, 42 |
| *Theropithecus gelada* (Rüppell, 1835) | LC | II | Not mentioned | To procure their manes for ceremonial headdresses | 29, 43 |
| Pitheciidae | | | | | |
| *Chiropotes satanas* (Hoffmannsegg, 1807) | CR | II | Brazil | Folk uses | 3 |
| *Callicebus torquatus* (Hoffmannsegg, 1807) | LC | | Venezuela, Brazil | Folk uses | 44, 45 |
| Chiropotes chiropotes (Humboldt, 1811) | LC | | Venezuela, Brazil | Folk uses | 44, 45 |
| Hominidae | | | | | |
| *Gorilla gorilla* (Savage and Wyman, 1847) | CR | I | Nigeria, Cameroon | Charms or amulets, to obtain victory in competition. In Cameroon, this species is considered sacred | 8, 9, 10, 18, 46 |

**Table 8.3** (continued)

| Family/Species | IUCN Red List | CITES Appendices | Region or country(ies) | Use(s) | Reference(s) |
|---|---|---|---|---|---|
| *Pan troglodytes* (Blumenbach, 1775) | EN | I | Nigeria, Guinea, Sierra Leone, Liberia, Mali, Guinea, Senegal, Uganda, Tanzania, Cameroon | Amulets, regard the chimpanzee as a sacred totem and a reincarnation of their ancestors, considered sacred, piece of the dried bone of chimpanzees is tied around the waist or wrist of infants in the belief that it makes them stronger as they grow into adulthood, chimpanzee's central incisors were procured to be worn as amulet around the waist of infants to protect them and give them power over others in their cohort, magic rituals | 8, 9, 10,18, 26, 47, 48, 49, 50, 51, 52, 53,54,55, 56, 57, 58, 59 |
| *Pan paniscus* Schwarz, 1929 | EM | I | Congo DR | It is locally considered sacred | 60, 61 |
| *Pongo pygmaeus* (Linnaeus, 1760) | EN | I | Indonesia, Malaysia | Spiritual use. It is considered sacred in Malaysia | 62, 63, 64, 65 |
| *Pongo abelii* Lesson, 1827 | CR | I | Indonesia, Malaysia | Spiritual use | 62, 63, 64 |
| **Loridae** | | | | | |
| *Loris lydekkerianus* Cabrera, 1908 | LC | II | Sri Lanka and India | Religious practices | 66 |
| *Loris tardigradus* Hill, 1942 | EN | II | Sri Lanka, India, Indonesia, Myanmar | Used in religious practices, using its eyes in native medicines and love potions, charms | 66, 67 |
| *Nycticebus bengalensis* (Lacépède, 1800) | VU | I | Lao PDR | They are believed to be a spiritual animal | 68 |
| *Nycticebus pygmaeus* Bonhote, 1907 | VU | I | Lao PDR | They are believed to be a spiritual animal | 68 |

(continued)

**Table 8.3** (continued)

| Family/Species | IUCN Red List | CITES Appendices | Region or country(ies) | Use(s) | Reference(s) |
|---|---|---|---|---|---|
| **Daubentoniidae** | | | | | |
| *Daubentonia madagascariensis* (Gmelin, 1788) | NT | I | Madagascar | *D. madagascariensis* may be killed on sight as crop pests or harbingers of evil. Due to its evil reputation the aye-aye is never eaten, but is used in ritualized ceremonies | 69, 70, 71 |
| **Indriidae** | | | | | |
| *Propithecus verreauxi* A. Grandidier, 1867 | VU | II | Madagascar | Protected or sacred in some areas | 26, 72, 73 |
| *Propithecus tattersalli* Simons, 1988 | EN | II | Madagascar | Used in ritual practices | 5, 73, 74 |
| *Propithecus edwardsi* A. Grandidier, 1871 | EN | II | Madagascar | Species that are considered to embody dead ancestors | 73 |
| *Propithecus perrieri* Lavauden, 1931 | CR | II | Madagascar | It is locally considered sacred | 75 |
| *Avahi laniger* (Gmelin, 1788) | LC | II | Madagascar | Species that are considered to embody dead ancestors. | 73 |
| *Indri indri* (Gmelin, 1788) | EN | I | Madagascar | Protected or sacred in some areas | 73, 76, 77 |
| **Lemuridae** | | | | | |
| *Lemur catta* Linnaeus, 1758 | NT | I | Madagascar | Protected or sacred in some areas | 73,78 |
| *Eulemur rufus* (Audebert, 1799) | DD | I | Madagascar | These are ancestors in animal form | 79 |
| *Eulemur rubriventer* (I. Geoffroy, 1850) | VU | I | Madagascar | These are ancestors in animal form | 79 |
| **Hylobatidae** | | | | | |
| *Hylobates moloch* (Audebert, 1797) | EN | I | Indonesia | It is considered sacred | 16 |

(continued)

**Table 8.3** (continued)

| Family/Species | IUCN Red List | CITES Appendices | Region or country(ies) | Use(s) | Reference(s) |
|---|---|---|---|---|---|
| *Hylobates klossii* (Miller, 1903) | EN | I | Indonesia | It is considered sacred | 17 |
| *Nomascus concolor* Ma and Wang, 1986 | CR | II | Lao PDR | It is considered sacred | 80 |

*CR* critically endangered; *EN* endangered; *VU* vulnerable; *NT* near threatened; *LC* least concern; *DD* Data Deficient. *IUCN Red List Categories*—*CR* Critically Endangered; *EN* Endangered; *VU* Vulnerable; *NT* Near Threatened; *LC* Least Concern; *DD* Data Deficient. *CITES Appendices*—I or II. *References:* (1) Shepard (2002), (2) Mittermeier (1987), (3) Cormier (2003), (4) Silva et al. (2005), (5) Vargas (2002), (6) Gonzalez-Kirchner and Sainz de la Maza (1998), (7) Lizarralde (2002), (8) Adeola (1992), (9) Dedeke et al. (2006), (10) Dedeke and Abayomi (2006), (11) Ntiamoa-Baidu (1992), (12) Takeda (1998), (13) Akowuah et al. (1975), (14), Gonedelé Bi et al. (2007), (15) Baker et al. (2009), (16) Indrawan et al. (1996), (17) Whittaker (2006), (18) Soewu (2008), (19) Khalid et al. (2007), (20) Wheatley (1999), (21) Knight (1999), (22) Hilaluddin et al. (2005), (23) Ahmed (2001), (24) Borang (1996), (25) O'Brien and Kinnaird (1997), (26) Lee and Priston (2005), (27) Riley (2005), (28) Biquand et al. (1992), Dunbar (1977), (30) Tappen (1960), (31) IUCN (2011), (32) Whiting et al. (2011), (33) Swantz (1990), (34) Simelane and Kerley (1998), (35) Pujol (1993), (36) Mahawar and Jaroli (2008), (37) Malik and Johnson (1994), (38) Mohnot (1971), (39) PirtaMadhav (1997), (40) Southwick and Lindburg (1986), (41) Solanki and Chutia (2004), (42) Srivastava and Mohnot (2001), (43) Gron (2008), (44) Montgomery (1970), (45) Wilbert and Simoneau (1990), (46) Wittiger and Sunderland Groves (2007), (47) Hanson-Alp et al. (2003), (48) Ham (1998), (49) Nisbett et al. (2003), (50) Duvall and Niagaté (1997), (51) Caspary et al. (1998), (52) Kortlandt (1986), (53) Dunnet et al. (1970), (54) Takemoto (2002), (55) Pruetz (2002), (56) Humle (2003a), (57) Humle (2003b), (58) King (1994), (59) Kormos et al. (2003), (60) Inogwabini et al. (2007), (61) Tashiro et al. (2007), (62) Marshall et al. (1999), (63) Rijksen and Springer (1999), (64) Nijman (2005), (65) Horowitz (1998), (66) Nekaris and Jayewardene (2004), (67) Phillips (1931), (68) Hamada et al. (2007), (69) Petter (1977), (70) Garbutt (1999), (71) Albignac (1987), (72) Ganzhorn and Abranham (1991), (73) Jones et al. (2008), (74) Mittermeier et al. (1994), (75) Mayor and Lehman (1999), (76) Britt et al. (1999), (77) Thalmann et al. (1993), (78) Loudon et al. (2006), (79) Jones et al. (2007), (80) Geissmann (2007) and potentially dangerous situations for disease transmission (Baker et al. 2009; Indrawan et al. 1996; Loudon et al. 2006; Whittaker 2006). Humans can be exposed to a number of simian viruses, including simian T cell lymphotropic viruses (STLV), simian retrovirus (SRV), simian foamy virus (SFV), and Herpes B virus in addition to other known infectious agents (Baker et al. 2009; Loudon et al. 2006; Wheatley 1999). It is also important to point out that non-human primates are likewise potentially at risk from human pathogens as well, including measles, influenzas, and other respiratory diseases such as para-influenza and tuberculosis (Jones-Engel et al. 2001). Additionally, both humans and non-human primates can act as reservoirs for shared pathogens or potential pathogens including SIV/HIV and Ebola (Rouquet et al. 2005) that move across species boundaries.

The situations described above clearly indicate that health considerations are another important facet of non-human/human primate interfaces (Jones et al. 2008). Perhaps the most notorious case is the long-suspected simian origin of human HIV (Fuentes and Wolfe 2002). For it was recently confirmed that the virus originated in the common chimpanzee (*Pan troglodytes troglodytes*) (Turner and Clifton 2009). A number of recent studies have noted connections between the growth of human populations and areas under cultivation, and bushmeat consumption with increased likelihoods of pathogen exchange with wild primates (e.g. Britt et al. 1999; Jones et al. 2007; Mayor and Lehman 1999; Mittermeier et al. 1994; Thalmann et al. 1993).

Nahallage et al. (2008) pointed out that monkey meat can be fatal if not properly prepared. Lamabaddusuriya et al. (1992) reported an outbreak of salmonella following the consumption of monkey meat in the Southern Province—Sri Lanka (the species eaten was not reported). These authors noted that the meat was probably contaminated because the monkey was already dead when the people found and consumed it. As pointed by Alves et al. (2010), several species of primates have become preferred targets for hunters because their bones, organs, and other body parts are used in traditional remedies. While numerous animal species are hunted (poached) for their meat rather than for medicinal uses (CITES 2002), there is often an overlap between the two, and disease transmission can occur in both situations. In this context, it is appropriate to note that several species of monkey have been identified that harbor infectious diseases with potentially grave effects in humans (Pujol 2006; Still 2003). This situation can be illustrated by a recent case from a Safari Park in Great Britain, where a colony of otherwise clinically healthy macaque monkeys had to be destroyed once it was discovered that they harbored simian herpes B virus—better known as the Ebola virus (Swift 2000). While this agent is not harmful to monkeys, it causes 80% mortality in affected humans (Still 2003).

The effectiveness of most of the medicines from wild animals/wild animal parts has not been scientifically studied and proven, and their potency in many cases may be questionable. As pointed out by Pieroni et al. (2002), the chemical constituents and pharmacological actions of some animal products are already known to some extent, and ethno-pharmacological studies focused on animal remedies could be very important in order to clarify the eventual therapeutic usefulness of this class of biological remedies. It is essential, however, that traditional drug therapies be submitted to an appropriate benefit–risk analysis (De Smet 1991).

## 8.4 Conservation Implications

Our results demonstrate that a substantial number of primate species are used in traditional folk practices throughout the world and that these animals are collected from the wild, evidencing the importance of understanding such uses in the context of primate conservation. Medicinal primates are widely used not only by rural communities, but also by people living in urban environments. The trade of medicinal animals in urban areas represents an important factor for their over-exploitation of many mammal species, and primates are no exception. Mittermeier et al. (2007) highlight that traditional medicines and emerging markets encourage the commercial hunting of primates, threatening them on a scale they have never faced before.

Of the 110 species of primates recorded in our review, 22 are classified as Vulnerable, 23 as Endangered, 14 as critically endangered (CR), 7 as Near Threatened, 43 as Least Concern, and 1 classified as data deficient (DD) in the most recent IUCN Red List. All species are also included in the CITES Appendices I or II, although the reasons for their inclusion are not necessarily related to medicinal use. The local commerce of primates is common in various cities of the world, where parts or products derived from primates is commercialized, mainly in Africa and Asia. In some big cities of Java, for instance, Malang, Surabaya and Denpasar, often people travel around selling medicine with parts of slow loris bodies as the basic material. Slow loris skin and bone are believed to prevent danger and to bring peace to families (Knight 1999). In Vietnam, pygmy lorises are severely threatened, and probably the biggest hazard to these animals is the fact that they are in such demand in China that people offer high prices to have lorises smuggled in for medicinal purposes (Fitch-Snyder 2001). Local hunters capture lorises primarily because they are valuable in the Chinese medicine market. A local villager can get approximately three dollars for a pygmy loris (Riley 2005).

In addition to whole primates or their parts, other products derived from these animals are commercialized in some localities. For instance, in Guangxi Province, China, the Wu-yuan wine, a local wine made from the bone of Francois langur (*Trachypithecus francoisi*), supposedly has magical curative effects for fatigue and rheumatism (Ye 1993). White-headed langurs were hunted for bone, meat, and making medicinal wine. The WuYuan wine was said to be good for digestion, rheumatism, and human health. In the past, five wine factories in Guangxi province were famous for their langur wine, but these factories have been banned by the government since the white-headed langur was put on the National Wild Animal Protection List. Although poaching is seriously punished by the government, it still occurs occasionally (Biquand et al. 1992). In China, tragically, gibbon bones are regarded as more valuable than any others in Chinese traditional medicine; a special cream—Houzi Cream—which is made from gibbon bones is generally believed to cure arthritis and to accelerate recovery (Dunbar 1977).

The trade in primates (whether of live animals or animal products used as medicines or as ornaments) can have dramatic impacts on remnant populations of already rare species (Eudey 1999). Primates in Indochina are especially endangered in areas near the Chinese border—where the market for exotic meats, medicines, and aphrodisiacs seems to be insatiable as the country's economic prosperity increases (Gray 2001). The ultimate cause of the disappearance of chimpanzees from Togo is unknown, but it is likely to have been due to the combination of the many factors listed by Oates (1996), including hunting for bush meat and traditional medicines. Starr et al. (2010) quantified the dramatic impacts of domestic trade on two slow loris species from Cambodia (*Nycticebus bengalensis* and *N. pygmaeus*), demonstrating that their importance in folk medicine may be driving them toward extinction. As highlighted by Nijman et al. (2011), commercial trading in primates is a significant impediment to their conservation. Primates are traded both domestically and internationally to biomedical industries, pharmaceutical markets, entertainment businesses, and pet markets, among others. Primate meat is consumed globally, and their body parts are used as ingredients in traditional medicines (or simply sold as curios) in many parts of the world.

Many populations of Asian langurs are now severely threatened, not only from subsistence hunting and habitat loss, but also from hunting to supply body parts and tissues to the insatiable Asian markets for exotic dishes, amulets, remedies, and aphrodisiacs (Bleisch et al. 2008b). Medicinal usages represent just one additional pressure for some species, but the main threat to others (Alves et al. 2010). Fuentes (2007) stressed that selective hunting for body parts used in traditional medicines is impacting certain primate populations, but it is far from clear if these predation rates are high enough (relative to nutrition-based harvesting of those animals) to have pronounced impacts.

It must be emphasized that many circumstances can affect primate populations, and the use of these animals for medicinal purposes is just one form of exploitation that must be considered in planning conservation efforts (Alves et al. 2010). Within this context, the medicinal uses of primates must be considered together with other anthropogenic pressures. When it comes to human exploitation of primates, the animals are often just as valuable dead as they are alive. Their products may be used in traditional medicine by people in some cultures and their bodies stuffed as hunting trophies by others. Worldwide, the demand for primate skins, meat, and body parts are putting additional strain on their already dwindling populations. Unfortunately, in many cases the punitive fine for killing an animal for trade is much less than what a hunter can earn by selling it (Li et al. 2007). Many indigenous populations use primate meat as a major source of protein in their diets, and primates may also be used for traditional medicine, ornamentation, clothing, or for trade. Heavy hunting for these purposes may lead to local extinction (Bleisch et al. 2008b).

It is also important to point out that habitat losses due to clearing tropical forests for agriculture, logging, and fuel wood continues to be the major factor driving population losses of primates (Bleisch et al. 2008b). Numerous researchers have described threats to non-human primate populations in both the Old and New Worlds due to deforestation (Ganzhorn and Abraham 1991; Tashiro et al. 2007), and there seems to be a general consensus (at least among conservationists and primatologists) that the principal threats to primates are habitat losses and hunting (Kaeslin and Williamson 2010; Lamabaddusuriya et al. 1992; Tashiro et al. 2007). According to Mittermeier et al. (2007), habitat destruction has the most significant influence on primate populations. These authors described agricultural development in tropical forests as a major problem, not only because 90% of primate species live in the tropical forests, but because the poor soils there demand shifting agricultural techniques that lead to successive advances on forest areas. In this context, primates conservation cannot be dissociated from natural habitat and biodiversity protection, and primates can be used as a flag-ship species to help conserve other wildlife.

An overall reduction in hunting pressure appears to represent the ideal conservation and wildlife management strategy, although such reductions may not be feasible in many localities for a variety of cultural and socio-economic reasons (Alves et al. 2010). It is important to recognize that medicines extracted from animals and plants are significant, valuable resources for many human populations that do not have access to industrialized drugs and medical care. For many Asian cultures, and indeed for cultures around the world, plants and animals are seen as objects to serve human needs. Therefore, the concept of wildlife conservation and the practice of prohibiting the use of natural resources can be alien to them. If the need for conservation is to be accepted by people who make their livelihoods from wildlife or use wildlife for necessities such as food and medicine, then care should be taken to avoid what may be seen as ideological or culturally imperialistic approaches. It is important to accept and respect differing views on the value of wildlife, while, at the same time, explaining the necessity of conservation measures (Lee 1999).

The therapeutic properties of wild animals and plants and domestic or cultivated species overlap in many cases (Alves et al. 2007). This aspect opens the possibility of replacing the use of threatened species with less threatened or domestic species in traditional medicine prescriptions (Alves et al. 2008; Still 2003), thereby minimizing the threat to the ecosystem. Educational programs can be used to promote the use of alternatives, focusing on rural communities where inhabitants use primates in traditional medicine and religious practices. In some circumstances, folk beliefs, religious doctrines, and species-specific taboos can be important in the conservation of declining or threatened species, therefore, such practices must be stimulated. In addition, measures that do not directly involve modifying the behavior of the local populations should be considered, such as controlling the illegal wildlife trade (Alves et al. 2010). Implementation of sanitary measures in the use and trade of animal or their parts for medicinal purposes is also necessary. It is remarkable that, in most cases, the medicinal products of primates are by-products from animals hunted for other purposes, thus, these multiple uses (including medicinal) of primates and their impact on animal populations must be properly assessed and taken into consideration when implementing recovery plans for these species, especially those that are highly exploited.

The manners in which natural resources are used by human populations, and the cultural norms associated with those uses, are extremely relevant to the implementation of conservation strategies (Alves et al. 2010). The human–/non-human–primate interface is an important dimension for primate conservation efforts, as well as ecosystem conservation in general (Fuentes and Wolfe 2002).The present study demonstrated that primates have important roles in healing practices in many cultures throughout the world, and that the implications of these uses must be taken into account when planning conservation efforts. As such, the medicinal uses of threatened species should be further investigated and amply discussed by all relevant stakeholders, and the socio-cultural aspects of these uses must be respected when establishing management plans aimed at the sustainable use of medicinal animals.

## References

Adeola MO (1992) Importance of wild animals and their parts in the culture, religious festivals, and traditional medicine, of Nigeria. Environ Conserv 19(2):125–134

Ahmed A (2001) Illegal trade and utilization of primates in India. Wildlife and protected areas: non-human primates of India. ENVIS Bulletin 1:177–184

Ajayi SS (1978) The utilization of tropical forest in West Africa. FAO, Rome

Akowuah DK, Rice K, Merz A, Sackey VA (1975) The children of the gods. J Ghana Wildl Soc 1(2):19–22

Albignac R (1987) Status of the aye-aye in Madagascar. Primate Conserv 8:44–45

Alvard MS, Robinson JG, Redford KH, Kaplan H (1997) The sustainability of subsistence hunting in the neotropics. Conserv Biol 11(4):977–982

Alvarez M (2006) Fortalecimiento de la Conservación de la Naturaleza Mediante las Prácticas Tradicionales de Conservación que Desarrollan las Comunidades Indígenas Cabécares de Bajo Chirripó y Nairi Awari (Technical Report). 1st edn. Asociación IXACAVAA de Desarrollo e Información Indígena, San Jose, Costa Rica

Alves RRN (2009) Fauna used in popular medicine in Northeast Brazil. J Ethnobiol Ethnomed 5(1):1–30. doi:10.1186/1746-4269-5-1

Alves RRN, Alves HN (2011) The faunal drugstore: animal-based remedies used in traditional medicines in Latin America. J Ethnobiol Ethnomed 7(9):1–43

Alves RRN, Rosa IL (2006) From cnidarians to mammals: the use of animals as remedies in fishing communities in NE Brazil. J Ethnopharmacol 107:259–276

Alves RRN, Rosa IL (2007a) Zootherapeutic practices among fishing communities in North and Northeast brazil: a comparison. J Ethnopharmacol 111:82–103. doi:10.1016/j.jep.2006.10.033

Alves RRN, Rosa IL (2007b) Zootherapy goes to town: the use of animal-based remedies in urban areas of NE and N Brazil. J Ethnopharmacol 113:541–555. doi:10.1016/j.jep.2007.07.015

Alves RRN, Rosa IL (2010) Trade of animals used in Brazilian traditional medicine: trends and implications for conservation. Human Ecology 38(5):691–704

Alves RRN, Rosa IL, Santana GG (2007) The role of animal-derived remedies as complementary medicine in Brazil. Bioscience 57(11):949–955

Alves RRN, Souto WMS (2010) Etnozoologia: conceitos, considerações históricas e importância. In: Alves RRN, Souto WMS, Mourão JS (eds) A Etnozoologia no Brasil: Importância, Status atual e Perspectivas, vol 7. 1st edn. NUPEEA, Recife, Brazil, pp 19–40

Alves RRN, Souto WMS, Barboza RRD (2010) Primates in traditional folk medicine: a world overview. Mamm Rev 40(2):155–180. doi:10.1111/j.1365-2907.2010.00158.x

Alves RRN, Vieira WLS, Santana GG (2008) Reptiles used in traditional folk medicine: conservation implications. Biodiversity and Conservation 17(8):2037–2049. doi:10.1007/s10531-007-9305-0

American Zoo and Aquarium Association (1998) Lion-tailed Macaque, Species Survival Plan. American Zoo and Aquarium Association. http://www.nagonline.net/Fact%20Sheet%20pdf/AZA%20-%20Lion-Tailed%20Macaque%20Species%20Survival%20Plan.pdf. Accessed December 18, 2010

Apaza L, Godoy R, Wilkie D, Byron E, Huanca T, Leonard WR, Peréz E, Reyes-García V, Vadez V (2003) Markets and the use of wild animals for traditional medicine: a case study among the Tsimané ameridians of the Bolivian rain forest. J Ethnobiol 23(1):47–64

Arkive.org (2009) Grey-legged night monkey (*Aotus griseimembra*). Arkive.org. http://www.arkive.org/grey-legged-night-monkey/aotus-griseimembra/threats-and-conservation.html. Accessed 12 .May 2009

Ashwell D, Walston N (2008) An overview of the use and trade of plants and animals in traditional medicine systems in Cambodia, 1st edn. Greater Mekong Programme, TRAFFIC Southeast Asia

Baker LR, Tanimola AA, Olubode OS, Garshelis DL (2009) Distribution and abundance of sacred monkeys in Igboland Southern Nigeria. Am J Primatol 71(7):574–586

Banks E (1931) A popular account of the mammals of Borneo. J Malay Branch Royal Asiatic Soc 9:1–120

Bennett EL, Caldecott J, Kavanagh M, Sebastian A (1987) Current status of primates in Sarawak. Primate Conserv 8:184–187

Biquand S, Biquand-Guyot V, Boug A, Gautier JP (1992) The distribution of Papio hamadryas in Saudi Arabia: ecological correlates and human influence. Int J Primatol 13(3):223–243

Bleisch B, Manh Ha N, Khat Quyet L, Yongcheng L (2008a) *Trachypithecus francoisi*. IUCN Red List of Threatened. http://www.iucnredlist.org/details/39853/0

Bleisch B, Brockelman W, Timmins RJ, Nadler T, Thun S, Das J, Yongcheng L (2008b) *Trachypithecus phayrei*. IUCN Red List of Threatened. http://www.iucnredlist.org/details/22040/0

Boonratana R, Das J, Yongcheng L, Htun S, Timmins RJ (2008) *Macaca leonina* (In: IUCN Red List of Threatened Species 2011). IUCN. http://www.iucnredlist.org/apps/redlist/details/39792/0

Borang A (1996) Studies on certain ethno zoological aspects of Adi Tribes of Siang District, A.P. India. Arunachal Forest News 14:1–5

Britt A, Axel A, Young R (1999) Brief surveys of two classified forests in Toamasina province, Eastern Madagascar. Lemur News 4:25–27

Campbell C, Andayani N, Cheyne S, Pamungkas J, Manullang B, Usman F, Wedana M, Taylor-Holzer K (2008) Indonesian gibbon conservation and management workshop final report. IUCN/SSC Conservation Breeding Specialist Group, Apple Valley

Caspary HU, Mertens AD, Niagaté B (1998) Possibilites d'une Exploitation Durable des Ressources Fauniques dans la Réserve de Faune du Bafing. Eschborn, Germany

Chan BPL, Fellowes JR, Geissmann T, Zhang J (2005) Status survey and conservation action plan for the hainan gibbon KFBG. SAR, Hong Kong

Chiu C (2001) *Macaca cyclopis*. University of Michigan Museum of Zoology. http://animaldiversity.ummz.umich.edu/site/accounts/information/Macaca_cyclopis.html. Accessed 14 May 2009

CITES (2002) List of species traded for medicinal purposes. Eighteenth meeting of the animals committee. San José, Costa Rica

CITES (2011) CITES Appendix. http://www.cites.org/eng/resources/species.html. Accessed 7 Aug 2011

Cormier L (2003) Kinship with Monkeys: The Guajá Foragers of Eastern Amazonia. Columbia University Press, Columbia

Crocker JC (1971) The dialectics of Bororo social inversions. In: 38th international Congress of Americanists, Stuttgart-Munich, Germany. Kommissionsverlag Klaus Renner, pp 387–391
Dang NX, Sang NV, La TV, Tuong NX, Truong NQ (2003) Evaluation of the Wildlife Trade in Ba Be and Cho Don Districts, PARC Project VIE/95/G31 & 031. 1st edn. Government of Vietnam (FPD), UNOPS/UNDP, and Scott Wilson Asia-Pacific Ltd, Hanoi
De Smet PAGM (1991) Is there any danger in using traditional remedies? J Ethnopharmacol 32:43–50
Dedeke GA, Aboyami F (2006) Ethnozoological trade and practices among the Ijebu people of southwestern Nigeria and the impact on some mammalian speciea. Indilinga 5(2):175–187
Dedeke GA, Soewu DA, Lawal OA, Ola M (2006) Pilot survey of Ethnozoological utilisation of vertebrates in Southwestern Nigeria, Indilinga. Afr J Indigenous Knowl Syst 5(1):87–96
Duckworth JW, Salter RE, Khounboline K (1999) Wildlife in Lao PDR: 1999 status report IUCN. WCS and CPAWM, Cambodia
Dunbar RIM (1977) The Gelada baboon: status and conservation. In: Prince Rainier III HSH, Bourne GH (eds) Primate Conservation. Academic Press, New York, pp 363–383
Dunnet S, Orshoven JV, Albrecht H (1970) Peaceful coexistence between chimpanzee and man in West Africa. Bijdragen Tot De Dierkunde 40:148–153
Duvall CS, Niagaté B (1997) Inventaire preliminaire des mammiferes, oiseaux, et reptiles de la Reserve de Faune du Bafing. Bamako, Mali. Direction Nationale des Ressources Forestieres, Fauniques, et Halieutiques, Bamako, Mali
El-Kamali HH (2000) Folk medicinal use of some animal products in Central Sudan. J Ethnopharmacol 72:279–282
Eudey AA (1987) Action plan for Asian primate conservation: 1987–1991, 1st edn. IUCN/SSC Primate Specialist Group, Gland
Eudey AA (1999) Asian primate conservation—my perspective. In: Dolhinow P, Fuentes A (eds). In: The nonhuman primates Mayfield. Publishing Co, Mountain View, pp 151–158
Figueiredo N (1994) Os 'bichos' que curam: os animais e a medicina 'folk' em Belém do Pará. Boletim do Museu Paraense Emílio Göeldi 10(1):75–91
Fitch-Snyder H (2001) Scientists: Tiny Primate is Endangered. ABC News. http://dailynews.yahoo.com/h/abc/20010509/ts/endangered_pygmyloris010508_1.html. Acessed 21 Dec 2010
Fruth B, Benishay JM, Bila-Isia I, Coxe S, Dupain J, Furuichi T, Hart J, Hart T, Hashimoto C, Hohmann G, Hurley M, Ilambu O, Mulavwa M, Ndunda. M, Omasombo V, Reinartz G, Scherlis J, Steel L, Thompson J (2008) *Pan paniscus*. IUCN Red List of Threatened. http://www.iucnredlist.org/details/15932/0. Acessed 18 July 2011
Fuentes A (2007) Monkey and human interconnections: the wild, the captive, and the in-between. In: Cassidy R, Mullin M (eds) Where the wild things are now: domestication reconsidered. Berg Publishers, Oxford, pp 123–145
Fuentes A, Hockings KJ (2010) The ethnoprimatological approach in primatology. Am J Primatol 72(10):841–847
Fuentes A, Wolfe LD (2002) Primates face to face: conservation implications of human-nonhuman primate interconnections, vol 29. Cambridge University, Cambridge
Ganzhorn JU, Abraham J (1991) Possible role of plantations for lemur conservation in Madagascar: food for folivorous species. Folia Primatol 56:171–176
Garbutt N (1999) Mammals of Madagascar, 1st edn. Pica Press, London
Geissmann T (2007) Status reassessment of the gibbons: results of the Asian primate red list workshop 2006. Gibbon J 3:5–15
Geissmann T, Dang NX, Lormée N, Momberg F (2000) Vietnam primate conservation status review 2000—Part 1: Gibbons. Fauna & Flora International Indochina Programme, Hanoi
Gonedelé Bi S, Bitty A, Gnangbé F, Bené JC, Koné I, Sangaré A, Zinner D (2007) Conservation status of Geoffroy's pied colobus monkey (*Colobus vellerosus*, Geoffroy 1834) has dramatically declined in Côte d'Ivoire. African Primates 1: without page number
Gonzalez-Kirchner JP, Sainz de la Maza M (1998) Primates hunting by Guaymi amerindians in Costa Rica. Hum Evol 13(1):15–19

Governo do Estado do Paraná (2002) Plano de Manejo do Parque Estadual do Rio Guarani. Instituto Ambiental do Paraná, Governo do Estado do Paraná. http://www.uc.pr.gov.br/arquivos/File/Plano_de_Manejo/PE_Rio_Guarani/10_cap_v_analise_uc.pdf. Accessed 12 Dec 2010

Gray DD (2001) Indochina primates face extinction. http://dailynews.yahoo.com/h/ap/20010603/wl/exp_primates_in_peril_1.html. Accessed 14 Dec 2010

Gron KJ (2008) Primate Factsheets: Gelada baboon (*Theropithecus gelada*) Conservation. http://pin.primate.wisc.edu/factsheets/entry/gelada_baboon/cons. Accessed 21 Dec 2010

Ha NM (2006) Some observations on the Hatinh Langur, *Trachypithecus laotum hatinhensis* (Dao, 1970), in north central Vietnam. Primate Conserv 21:149–154

Ham R (1998) Nationwide chimpanzee survey and large mammal survey. Report for the European Communion and Republic of Guinea

Hamada Y, Malaivijitnond S, Kingsada P, Bounnam P (2007) The distribution and present status of primates in the northern region of Lao PDR. Nat Hist 7(2):161–191

Hanson-Alp R, Bakarr MI, Lebbie A, Bangura KI (2003) Sierra Leone. In: Kormos R, Boesch C, Bakarr M, Butynski T (eds) West African Chimpanzees: status survey and conservation action plan. IUCN/SSC Primate Specialist Group, Gland, pp 77–87

Hilaluddin R, Kaul R, Ghose D (2005) Conservation implications of wild animal biomass extractions in Northeast India. Animal Biodiversity Conserv 28(2):169–179

Hoffmann M, Hilton-Taylor C (2008) *Papio ursinus* (In: IUCN Red List of Threatened Species 2011). http://www.iucnredlist.org/apps/redlist/details/16022/0. Accessed 17 July 2011

Horowitz LS (1998) Integrating indigenous resource management with wildlife conservation: a case study of Batang Ai National Park, Sarawak. Malay Hum Ecol 26(3):371–403

Humle T (2003a) Chimpanzees and crop raiding in West Africa. In: Kormos R, Boesch C, Bakarr MI, Butynski TM (eds) West African Chimpanzees. Status survey and conservation action plan. IUCN/SSC Primate Specialist Group, Gland, pp 147–155

Humle T (2003b) Culture and variation in wild chimpanzee behaviour: a study of three communities in West Africa. PhD Thesis, University of Stirling, Stirling

Indrawan M, Supriyadi D, Supriatna J, Andayani N, Raya JN, Indonesia U (1996) Javan gibbon surviving at a mined forest in Gunung Pongkor, Mount Halimun National Park, West Java: considerable tolerance to disturbances. Asian Primates 6:11–13

Inogwabini BI, Matungila B, Mbende L, Abokome M (2007) Great apes in the Lake Tumba landscape, Democratic Republic of Congo: newly described populations. Oryx 41(4):532–538

IUCN (2011) IUCN Red List of Threatened Species. IUCN. http://www.iucnredlist.org. Accessed 30 Sept 2011

Jamir NS, Lal P (2005) EthnoScientific practices among Naga tribes. Indian J Tradit Knowl 4(1):100–104

Jones JPG, Andriamarovololona MM, Hockley N (2008) The importance of taboos and social norms to conservation in Madagascar. Conserv Biol 22(4):976–986

Jones JPG, Andriamarovolona MA, Hockley NJ (2007) Taboos, social norms and conservation in the eastern rainforests of Madagascar. In: 9th Annual BIOECON conference on economics and institutions for biodiversity conservation, Kings College Cambridge, Cambrindge

Jones-Engel L, Engel GA, Schillaci MA, Babo R, Froehlich J (2001) Detection of antibodies to selected human pathogens among wild and pet macaques (Macaca tonkeana) in Sulawesi Indonesia. Am J Primatol 54(3):171–178

Johnson A, Singh S, Dongdala M, Vongsa O (2003) Wildlife hunting and use in the nam Ha national protected area: implications for rural livelihoods and biodiversity conservation, 1st edn. Wildl Conserv Soc, Vientiane

Kaeslin E, Williamson D (2010) Los bosques, las personas y la vida silvestre. Unasylva 236:3–10

Kakati LN, Ao B, Doulo V (2006) Indigenous knowledge of zootherapeutic use of vertebrate origin by the Ao Tribe of Nagaland. Hum Ecol 19(3):163–167

Khalid HS, El-Kamali HH, Elmanan AMA (2007) Trade of sudanese natural medicinals and their role in human and wildlife health care. *Cropwatch Newsletter* (10):1–15

King S (1994) Utilisation of Wildlife in Bakossiland, West Cameroon, with particular reference to primates. Traffic Bulletin 14:63–73

Kingdon J, Gippoliti S, Butynski TM, Lawes MJ, Eeley H, Lehn C, De Jong Y (2008) *Cercopithecus mitis* (In: IUCN Red List of Threatened Species 2011). IUCN. http://www.iucnredlist.org/apps/redlist/details/4221/0. Accessed 11 May 2011

Knight J (1999) Monkeys on the move: the natural symbolism of people-macaque conflict in japan. J Asian Stud 58(3):622–647

Kormos R, Boesch C, Bakarr MI, Butynski T (2003) West African chimpanzees. Status survey and conservation action plan. IUCN/SSC Primate Specialist Group, Gland

Kortlandt A (1986) The use of stone tools by wild-living chimpanzees and earliest hominids. J Hum Evol 15(2):77–132

Kumar A, Singh M, Molur S (2008) *Macaca silenus* (In: IUCN Red List of Threatened Species 2011). IUCN. http://www.iucnredlist.org/apps/redlist/details/12559/0. Accessed 18 Oct 2011

Lamabaddusuriya SP, Perera C, Devasiri IV, Jayantha UK, Chandrasiri N (1992) An outbreak of salmonellosis following consumption of monkey meat. J tropical med hyg 95(4):292–295

Lammertink M, Nijman V, Setiorini U (2003) Population size, red list status and conservation of the Natuna leaf monkey presbytis natunae endemic to the island of Bunguran, Indonesia. Oryx 37(4):472–479

Lee SKH (1999) Trade in traditional medicine using endangered species. In: An international context. Paper presented at the 2nd Australian symposium on traditional medicine and wildlife conservation, Melbourne, Australia

Lee PC, Priston NEC (2005) Human attitudes to primates: perceptions of pests, conflict and consequences for primate conservation. In: Paterson JD, Wallis J (eds) Commensalism and conflict: the human–primate interface. Am Soc Primatol, Norman, pp 1–23

Léo Neto NA, Brooks SE, Alves RRN (2009) From Eshu to Obatala: animals used in sacrificial rituals at Candomble "terreiros" in Brazil. J Ethnobiol Ethnomed 5(1):1–23

Li Y, Huang C, Ding P, Tang Z, Wood C (2007) Dramatic decline of François' langur *Semnopithecus francoisi* in Guangxi Province, China. Oryx 41:38–43

Lizarralde M (2002) Ethnoecology of monkeys among the Bari of Venezuela: perception, use and conservation. In: Fuentes A, Wolfe D (eds) Primates face to face: the conservation implications of human–nonhuman primate interconnections. Cambridge University Press, Cambridge, pp 85–100

Loudon JE, Sauther ML, Fish KD, Hunter-Ishikawa M, Ibrahim YJ (2006) One reserve, three primates. Ecol Environ Anthropol 2:54–74

Mahawar MM, Jaroli DP (2008) Traditional zootherapeutic studies in India: a review. J Ethnobiol Ethnomed 4(1):1–17

Malik I, Johnson RL (1994) Commensal rhesus in India: the need and cost of translocation. Revue d'écologie 49(3):233–243

Manh Ha N, Rawson B, Geissmann T, Timmins RJ (2008) Nomascus siki (In: IUCN Red List of Threatened 2011). IUCN. http://www.iucnredlist.org/details/39896/0. Accessed 18 Dec 2011

Marshall AJ, Wrangham RW, Arcadi AC (1999) Does learning affect the structure of vocalizations in chimpanzees? Anim Behav 58(4):825–830

Mayor M, Lehman SM (1999) Conservation of perrier's sifaka (Propithecus diadema perrieri) in analamera special reserve, Madagascar. *Lemur News* 4:21–23

Meijaard E, Nijman V (2008) *Presbytis frontata* (In: IUCN Red List of Threatened Species 2011). IUCN. http://www.iucnredlist.org/apps/redlist/details/18127/0. Accessed 23 July 2011

Meijaard E, Nijman V, Supriatna J (2008) *Nasalis larvatus* (In: IUCN Red List of Threatened Species 2011). IUCN. http://www.iucnredlist.org/apps/redlist/details/14352/0. Accessed 1 Nov 2011

Mittermeier RA (1987) Effects of hunting on rain forest primates. In: Marsch CW, Mittermeier R (eds) Primate conservation in the tropical rain forest. Alan R. Liss Inc, New York, pp 109–146

Mittermeier RA, Ratsimbazafy J, Rylands AB, Williamson L, Oates JF, Mbora D, Ganzhorn JU, Rodríguez-Luna E, Palacios E, Heymann EW (2007) Primates in peril: the world's 25 most endangered primates, 2006–2008. Primate Conserv 22(1):2–40

Mittermeier RA, Tattersall I, Konstant WR, Meyers DM, Mast RB (1994) Lemurs of Madagascar. Conservation International, Washington

Mohnot SM (1971) Ecology and behaviour of the Hanuman langur, Presbytis entellus (Primates: Cercopithecidae). Invading fields, gardens and orchards around Jodhpur, Western India. Trop Ecol 12(2):237–249

Montgomery E (1970) With the Shiriana in Brazil. Kendall/Hunt Publishing Company, Dubuque

Morales-Mavil JE, Villa-Cañedo JT (1998) Notas sobre el uso de la fauna silvestre en Catemaco, Veracruz, México. Acta Zoologica Mexicana 73:127–143

Moura FBP, Marques JGW (2008) Zooterapia popular na chapada diamantina: uma medicina incidental? Ciência & Saúde Coletiva 13:2179–2188

Nadler T, Momberg F, Xuan Dang N, Lormée N (2003) Vietnam Primate Conservation Status Review 2002. 1st edn. Frankfurt Zoological Society and Fauna & Flora International-Indochina Programme, Hanoi, Vietnam

Nadler T, Thanh VN, Streicher U (2007) Conservation status of Vietnamese primates Vietnamese. J Primatol 1(1):7–26

Nadler T, Xuan Canh L, Ngoc Thanh V, Khac Quyet L (2008) *Trachypithecus delacouri* (In: IUCN Red List of Threatened 2011) IUCN. http://www.iucnredlist.org/details/22043/0. Accessed 9 May 2010

Nahallage CAD, Huffman MA, Kuruppu N, Weerasingha T (2008) Diurnal primates in Sri Lanka and people's perception of them. Primate Conserv 23:81–87

Negi CS, Palyal V (2007) Traditional uses of animal and animal products in medicine and rituals by the shoka Tribes of district pithoragarh, Uttaranchal India. Ethno-Med 1(1):47–54

Nekaris A, Singh M, Kumar Chhangani A (2008) *Loris lydekkerianus* (In: IUCN Red List of Threatened 2011). IUCN. http://www.iucnredlist.org/details/44722/0. Accessed 18 July 2011

Nekaris KAI, Jayewardene J (2004) Survey of the slender loris (Primates, Lorisidae Gray, 1821: Loris tardigradus Linnaeus, 1758 and Loris lydekkerianus Cabrera, 1908) in Sri Lanka. J Zool 262(4):327–338

Ngoc Thanh V, Lippold L, Nadler T, Timmins RJ (2008a) Pygathrix cinerea (In: IUCN Red List of Threatened Species 2011). IUCN. http://www.iucnredlist.org/apps/redlist/details/39827/0. Accessed 12 Aug 2011

Ngoc Thanh V, Lippold L, Timmins RJ, Manh Ha N (2008b) *Pygathrix nemaeus* (In: IUCN Red List of Threatened Species 2011). Gland, Switzerland. http://www.iucnredlist.org/apps/redlist/details/39826/0. Accessed 14 July 2011

Nijman V (2004) Effects of habitat disturbance and hunting on the density and the biomass of the endemic Hose's leaf monkey *Presbytis hosei* (Thomas, 1889)(Mammalia: Primates: Cercopithecidae) in east Borneo. Contributions Zool 73(4):283–291

Nijman V (2005) In full swing: an assessment of trade in orangutans and Gibbons on Java and Bali, Indonesia. 1st edn TRAFFIC Southeast Asia, Petaling Jaya, Selangor

Nijman V, Meijaard E, Hon J (2008) *Presbytis hosei* (In: IUCN Red List of Threatened Species 2011). IUCN. http://www.iucnredlist.org/apps/redlist/details/18128/0. Accessed 2 July 2011

Nijman V, Meijaard E (2008) Presbytis rubicunda. IUCN 2011. IUCN Red List of Threatened Species. Version 2011.1. http://www.iucnredlist.org/apps/redlist/details/18131/0

Nijman V, Nekaris KAI, Donati G, Bruford M, Fa J (2011) Primate conservation: measuring and mitigating trade in primates. Endange Species Res 13:159–161

Nilsson G (1990) The endangered species handbook, 1st edn. The Animal Welfare Institute, Washington

Nisbett RA, Peal AL, Hoyt RA, Carter J (2003) Liberia. In: Kormos R, Boesch C, Bakarr MI, Butynski TM (eds) West African chimpanzees: status survey and conservation action plan. IUCN/SSC Primate Specialist Group, Gland, pp 89–98

Ntiamoa-Baidu Y (1992) Local perceptions and value of wildlife reserves to communities in the vicinity of forest national parks in Western Ghana, 1st edn. Oxford Environment and Development Group, Oxford

Oates JF (1996) African primates: status survey and conservation action plan. IUCN/SSC Primate Specialist Group, Gland

O'Brien TG, Kinnaird M (1997) Behaviour, diet, and movements of the Sulawesi crested black macaque (*Macaca nigra*). Int J Primatol 18:321–351

Petter JJ (1977) The aye-aye. In: Rainier P III, Bourne GH (eds) Primate conservation. Academic Press, New York, pp 37–57

Phillips WWA (1931) The food of the Ceylon slender loris (Loris tardigradus) in captivity. Spolia Zeylanica 16:205–208

Pieroni A, Giusti ME, Grazzini A (2002) Animal remedies in the folk medicinal practices of the Lucca and Pistoia Provinces, Central Italy. In: Fleurentin J, Pelt JM, Mazars G (eds) Des sources du savoir aux médicaments du futur/from the sources of knowledge to the medicines of the future, 1st edn. IRD Editions, Paris, pp 371–375

PirtaMadhav RS (1997) Management of the rhesus monkey *Macaca mulatta* and Hanuman langur *Presbytis entellus* in Himachal Pradesh India. Biol Conserv 79(1):97–106

Posey DA (1999) Cultural and spiritual values of biodiversity. (Supplement 1, Global Biodiversity Assessment). United Nations Environmental Programme and Intermediate Technology Publications, London. United Nations Environment, Programme. Oxford Centre for the Environment, London

Primate Specialist Group (2011) IUCN/SSC Primate Specialist Group. Primate Specialist Group. http://www.primate-sg.org/index.htm. Accessed 18 July 2011

Pruetz JD (2002) Competition between savanna chimpanzees and humans in southeastern Senegal. Am J Phys Anthropol 34:128

Pujol FH (2006) Virus en primates no humanos: Zoonosis, Antroponosis y Biodiversidad. Interciencia 31 (6):396–402

Pujol J (1993) NaturAfrica:The Herbalist Handbook. African Flora, Medicinal Plants. 1st edn. Jean Pujol Natural Healers Foundation Durban, South Africa

Putra Y, Masy'ud B, Ulfah M (2008) Diversity of medicinal animals in betung kerihun national park, West Kalimantan, Indonesia. Media Konservasi 13 (1):8–15

Racero-Casarrubia JA, Vidal CC, Ruiz OD, Bastelleros JC (2008) Percepción y patronos de uso de la fauna silvestre o comunidades indigenas Embera-Katíos en la cuenca del río San Jorge, zona amortiguadora del PNN-Paramillo. Revista de estudios sociales 31:118–131

Ramakantha V (1991) A note on slow loris survival in N.-E. India with particular reference to native beliefs about its curative and culinary properties. Zoo's Print 6 (1–16)

Rawson B, Lippold L, Timmins R, Ngoc Thanh V, Manh Ha N (2008) *Pygathrix nigripes* . (In: IUCN Red List of Threatened Species 2011). IUCN. http://www.iucnredlist.org/apps/redlist/details/39828/0. Accessed 23 June 2011

Rijksen HD, Meijaard E (1999) Our vanishing relative: the status of wild orangutans at the close of the twentieth century. Springer, Dordrech

Riley EP (2005) Ethnoprimatology of Macaca tonkeana: the interface of primate ecology, human ecology, and conservation in Lore Lindu National Park, Sulawesi, Indonesia. PhD Thesis, University of Georgia

Rodrigues E (2006) Plants and animals utilized as medicines in the Jaú National Park (JNP), Brazilian Amazon. phythér Res 20:378–391. doi:10.1002/ptr.1866

Rodríguez-Mahecha JV, Hernández-Camacho JI, Defler TR, Alberico M, Mast RB, Mittermeier RA, Cadena A (1995) Mamíferos colombianos: sus nombres comunes e indígenas, 1st edn. Conserv Intern, Washington

Rouquet P, Froment JM, Bermejo M, Kilbourn A, Karesh W, Reed P, Kumulungui B (2005) Wild animal mortality monitoring and human Ebola outbreaks, Gabon and Republic of Congo, 2001–2003. Emerg Infect Dis 11:283–290

Seixas CS (1997) Estratégias de pesca e utilização de animais por comunidades pesqueiras da Ilha, RJ. MSc. thesis, Universidade Estadual de Campinas, Campinas

Sharma SK (2002) A Study on Ethnozoology of Southern Rajasthan. In: Trivedi PC, Jaipur (eds) Ethnobotany. AAveshkar Publisher, India, pp 239–253

Shepard GH (2002) Primates in matsigenka subsistence and world view. In: Fuentes A, Wolfe L (eds) Primates face to face: conservation implications of human-nonhuman primate interconnections, vol 101–136. Cambridge University Press, Cambridge, pp 101–136

Shepherd CR, Sukumaran J, Wich SA (2004) Open Season: An analysis of the pet trade in Medan, Sumatra 1997-2001. TRAFFIC Southeast Asia, Kuala Lumpur

Silva MNF, Shepard GH Jr, Yu DW (2005) Conservation implications of primate hunting practices among the Matsigenka of Manu National Park. Neotropical Primates 13(2):31–36
Simelane TS, Kerley GIH (1998) The conservation implications of the use of indigenous vertebrates by traditional healers in Xhosa communities South African. J Wildl Res 28:121–126
Sodeinde OA, Soewu DA (1999) Pilot study of the traditional medicine trade in Nigeria. Traffic Bull 18(1):35–40
Soewu DA (2008) Wild animals in ethnozoological practices among the Yorubas of southwestern Nigeria and the implications for biodiversity conservation. Afr J Agric Res 3(6):421–427
Solanki GS, Chutia P (2004) Ethno zoological and socio-cultural aspects of monpas of Arunachal Pradesh. Hum Ecol 15(4):251–254
Southwick CH, Lindburg DG (1986) The primates of India: Status, trends, and conservation. In: Benirschke K (ed) Primates: the road to self-sustaining populations. Springer New York, pp l71–187
Srivastata A (1999) Primates of Northeast India. Megadiversity Press, Bikaner
Srivastava A, Mohnot SM (2001) Distribution, conservation status and priorities for primates in Northeast India. ENVIS Bulletin, Wildl Prot Areas 1:102–108
Starr C, Nekaris KAI, Streicher U, Leung L (2010) Traditional use of slow lorises Nycticebus bengalensis and N. pygmaeus in Cambodia: an impediment to their conservation. Endanger Species Res 12 (1):17–23
Still J (2003) Use of animal products in traditional Chinese medicine: environmental impact and health hazards. Complementary therapies in medicine 11:118–122. doi:10.1016/S0965-2299(03)00055-4
Streicher U, Ngoc Thanh V, Nadler T, Timmins RJ, Nekaris A (2008a) *Nycticebus pygmaeus* (In: IUCN Red List of Threatened 2011). IUCN. http://www.iucnredlist.org/details/39758/0. Accessed 18 July 2011
Streicher U, Singh M, Timmins RJ, Brockelman W (2008b) *Nycticebus bengalensis* (In: IUCN Red List of Threatened 2011). IUCN. http://www.iucnredlist.org/details/39758/0. Accessed 18 July 2011
Swantz L (1990) The medicine man among the Zaramo of Dar es Salaam. Nordic Africa Institute, Uppsala
Swift G (2000) Park culls colony of diseased monkeys. International Express, London
Takeda J (1998) Plants and animals used on birth and death of the Ngandu (Bongando) in Central Zaire. Afr study monogr Suppl issue 25:135–148
Takemoto H (2002) Feeding ecology of chimpanzees in Bossou, Guinea: coping with the seasonal fluctuation of food supply and micrometeorology in the tropical forest. PhD thesis, Kyoto University, Kyoto
Tang ZY, Li ZF (1957) Brief report on a survey of Hainan's vertebrates. Acta Zool Sinica 1: 246–249
Tappen NC (1960) Problems of distribution and adaptation of the African monkeys. Curr Anthropol 1:91–120
Tashiro Y, Idani G, Kimura D, Bongori L (2007) Habitat changes and decreases in the bonobo population in Wamba, Democratic Republic of the Congo. Afr Study Monograp 28(2):99–106
Tejada R, Chao E, Gómez H, Painter REL, Wallace RB (2006) Evaluación sobre el uso de la fauna silvestre en la Tierra Comunitaria de Origen Tacana Bolivia. Ecología en Bolivia 41(2):138–148
Thalmann U, Geissmann T, Simona A, Mutschler T (1993) The indris of Anjanaharibe-Sud, Northeastern Madagascar. Int J Primatol 14(3):357–381
Thang Long H (2004) Distribution and status of the grey-shanked douc langur (*Pygathrix cinerea*). In: Vietnam. In: Nadler T, Streicher U, Thang Long H (eds) Conservation of Primates in Vietnam. 1st edn. Frankfurt Zoological Society, Hanoi, pp 52–57
Timmins RJ, Boonratana R (2008) *Trachypithecus laotum* (In: IUCN Red List of Threatened Species 2011). http://www.iucnredlist.org/apps/redlist/details/22044/0. Accessed 21 July 2011
Timmins RJ, Do Tuoc TVC, Hendrichsen DK (1999) A preliminary assessment of the conservation importance and conservation priorities of the phong Nha-Ke Bang national park, Quang Binh Province, 1st edn. FFI-Indochina Programme, Hanoi

Tratado de Cooperacion Amazonica (1999) Conservacion y uso de la fauna silvestre en areas protegidas de la Amazonia. 1st edn. Secretaria Pro Tempore (Venezuela)/FAO, Caracas

Turner NJ, Clifton H (2009) "It's so different today": Climate change and indigenous lifeways in British Columbia Canada. Global Environ Change 19(2):180–190

Van NDN, Tap N (2008) An overview of the use of plants and animals in traditional medicine systems in Viet Nam. 1st edn. TRAFFIC Southeast Asia, Greater Mekong Programme, Ha Noi

Vargas LMA (2002) Estudio comparativo de la caza y uso de mamíferos en dos comunidades Tsimane' Undergraduate thesis. Universidad Mayor de San Andrés, La Paz

Vázquez PE, Méndez RM, Guiascón ÓGR, Piñera EJN (2006) Uso medicinal de la fauna silvestre en los Altos de Chiapas México. Interciencia 31(7):491–499

Vilaça A (2002) Making kin out of others in Amazonia. J Roy Anthrop Inst 8(2):347–365

Walker S, Molur S (2003) Summary of the status of south Asian primates. Extracted from the CAMP 2003 Report. 1st edn. Zoo Outreach Organisation, CBSG-South Asia, and WILD, Coimbatore

Walker B (1998) Medical exchange in contested lands: pharmaceutical products as cultural capital. In: Annual meeting of the Assn for Asian Studies, Washington

WCMC—World Conservation Monitoring Centre (2000) Species Under Threat. WCMC. http://www.wcmc.org.uk/species/data/species_sheets/. Accessed 26 May 2011

Westermann JH (1938) 3 Jaren Indisch Natuurleven, 11de Jaarverslag (1936–1938). Nederland Indische Vereniging tot Natuurbescherming, Amsterdam

Wheatley BP (1999) The sacred monkeys of Bali, 1st edn. Waveland Press, Long Grove

Whittaker DJ (2006) A conservation action plan for the Mentawai primates. Primate Conserv 20:95–105

Whiting MJ, Williams VL, Hibbitts TJ (2011) Animals traded for traditional medicine at the Faraday market in South Africa: species diversity and conservation implications. J Zool 284:84–96

Wilbert J, Simoneau K (1990) Folk Literature of the Yanomam Indians. UCLA Latin American Center Publications, Los Angeles

Winter KA (2002) Subsistence use of terrestrial and aquatic animal resources in the Tierra comunitaria de origen Itonama of Lowland Bolivia. PhD. thesis, University of Georgia, Athens

Wittiger L, Sunderland-Groves JL (2007) Tool use during display behavior in wild Cross River Gorillas. Am J Primatol 69:1307–1311

Xiang Z, Huo S, Wang L, Cui L, Xiao W, Quan R, Tai Z (2007) Distribution, status and conservation of the black-and-white snub-nosed monkey *Rhinopithecus bieti* in Tibet. Oryx 41(4):525–531

Xuan Canh L, Boonratana R (2006) A Conservation Action Plan for the Tonkin Snub-nosed Monkey in Viet Nam. 1st edn. Institute for ecology and biological resources & primate conservation Inc., Hanoi

Xuan Canh L, Khac Quyet L, Thanh Hai D, Boonratana R (2008) *Rhinopithecus avunculus* (In: IUCN Red List of Threatened Species 2011). IUCN. http://www.iucnredlist.org/apps/redlist/details/19594/0. Accessed 12 March 2011

Yamakoshi G (2005) What is happening on the border between humans and chimpanzees? Wildlife conservation in West African rural landscapes. Paper presented at the 7th Kyoto University international symposium, Kyoto University, Kyoto

Ye ZZ (1993) Biology of langurs (Presbytis). Yunnan Science and Technology Press, Kunming

Ziemendorff S (2008) Sustancias estimulantes y brebajes afrodisíacos en la tradición de la Amazonía peruana. Culturas populares (Revista Electrónica) (7):1–7

# Chapter 9
# A Global Overview of Carnivores Used in Traditional Medicines

**Rômulo Romeu Nóbrega Alves, Lorena Cristina Lana Pinto, Raynner Rilke Duarte Barboza, Wedson Medeiros Silva Souto, Ronnie Enderson Mariano Carvalho Cunha Oliveira and Washington Luiz Silva Vieira**

**Abstract** Carnivores are among the mammal species most frequently used in traditional folk medicine around the world. Therefore, this chapter assesses the global use of carnivores in traditional folk medicine and its implications. Our review indicated that at least 108 species of carnivores are used in traditional medicine worldwide. Of the species in medicinal use, 14 are classified as Near Threatened, 18 species as Vulnerable, 9 species as Endangered, and one as Critically Endangered on the IUCN Red List. For some species, medicinal use represents an additional direct pressure that may have contributed to declines in natural populations. In addition, the use of medicinal animals may have indirect impacts on the conservation of other species through the spread of disease. To minimize both harvest impact and disease spread, guidance on use of medicinal species may be useful. This could include an exploration of the use of alternatives and implementation of quality measures.

---

R. R. N. Alves (✉)
Departamento de Biologia, Universidade Estadual da Paraíba, Av das Baraúnas 351/Campus Universitário, Bodocongó, Campina Grande-PB, 58109-753, Brazil
e-mail: romulo_nobrega@yahoo.com.br

L. C. L. Pinto
Programa de Pós-Graduação em Ecologia de Biomas Tropicais, Departamento de Evolução Biodiversidade e Meio Ambiente, Universidade Federal de Ouro Preto, Ouro Preto-MG, 35400-00, Brazil

R. R. D. Barboza · W. M. S. Souto · W. L. S. Vieira
Programa de Pós-Graduação em Ciências Biológicas (Zoologia), Departamento de Sistemática e Ecologia, Universidade Federal da Paraíba, João Pessoa-PB, 58059-970, Brazil

R. E. M. C. C. Oliveira
Curso de Licenciatura e Bacharelado em Ciências Biológicas, Universidade Estadual da Paraíba, Av das Baraúnas 351/Campus Universitário, Bodocongó, Campina Grande-PB, 58109-753, Brazil

R. R. N. Alves and I. L. Rosa (eds.), *Animals in Traditional Folk Medicine*,
DOI: 10.1007/978-3-642-29026-8_9, © Springer-Verlag Berlin Heidelberg 2013

**Keywords** Carnivores · Traditional medicine · Animal conservation

## 9.1 Introduction

Throughout human history, people have used various materials from nature to cure their illnesses and improve their health. Substances have been derived from flora, fauna, and mineral sources located both in people's immediate surroundings and in more remote areas (Lev and Amar 2002). Nature has been the source of medicinal agents for thousands of years, and an impressive number of modern drugs have been isolated from natural sources, many based on their use in traditional medicine. In many parts of the world, traditional medicine is the preferred form of health care (World Health Organization 2002), and remains the most available and affordable form of therapy in developing countries. Many animal and plant species are used in traditional medicine practices, and therefore, discussions on the links between traditional medicine and biodiversity are imperative (Alves and Rosa 2007a; Anyinam 1995).

Wild animals and their products constitute essential ingredients in the preparation of traditional medicine (Adeola 1992; Alves and Rosa 2005, 2010; Alves et al. 2007, 2010a; Lee 1999; Whiting et al. 2011). In fact, various authors have pointed out that the phenomenon of zootherapy (the use of animal products in healing) is marked both by a broad geographical distribution and by deep historical origins. In modern societies, zootherapy constitutes an important alternative to many other known therapies practiced worldwide (Alves and Rosa 2005). Currently, examples of animal-derived remedies can be found in many urban, semi-urban, and more remote localities throughout the world (Almeida and Albuquerque 2002; Alves 2009; Alves and Alves 2011; Alves and Pereira Filho 2007; Alves et al. 2008a; Apaza et al. 2003; Sodeinde and Soewu 1999; Vázquez et al. 2006). Nevertheless, while the use of floristic resources has been widely researched, the utilization of faunistic resources in traditional medicine is less well studied (Alves and Rosa 2005).

Among the medicinal animals mammals are most often utilized (Alves and Alves 2011; Ashwell and Walston 2008; Sodeinde and Soewu 1999), and for some species of this class, direct exploitation, including medicinal use, has led to the decline of their natural populations (Bodmer et al. 1997; Corlett 2007; Pantel and Anak 2010; Ray et al. 2008). Mammals have been the focus of human attention for a long time and continue to serve as the traditional focus of government spending on wildlife conservation (Redford et al. 2011). Because of hunting, combined with loss of suitable habitat and other causes, many mammalian species are now considered to be endangered in the wild (e.g., Benson and Nagel 2004; Mittermeier et al. 2007) and one of the reasons is their use for traditional medicinal purposes (Alves et al. 2010a). The case of large carnivores threatened by trade for traditional medicine, especially traditional Chinese medicine, is well-known (Lee 1999). Both tigers and most species of bear have suffered population declines associated with the demand for their body parts in traditional medicines

(Brautigam et al. 1994; Servheen 1999). All of the world's eight species of bears, except the giant panda, have suffered population declines as a result of this traditional medicine trade (Knights 1996).

Carnivores have had long and uneasy relationships with humans (Ray et al. 2005). These animals were exploited for subsistence, medicine, commercial profit, and recreation. Individuals and populations have also been persecuted or "controlled" as vermin or pests because of concerns over personal safety, disease transmission, or depredation on livestock and pets. However, during the last few decades, often under the guise of conservation, an increasing number of more non-invasive or less harmful uses of carnivores have become important, and more emphasis has been placed on protecting and restoring previously decimated carnivore populations (Johnson et al. 2001).

The number of carnivores used in medicine is still unknown. In this context, this work assesses the global use of carnivores in traditional folk medicine around the world. We provide here a review of the literature in order to assess how many and which carnivore species are used as medicine globally. We identify those that are endangered and discuss the implications of the use of zootherapeutics for animal conservation and for public health.

## 9.2 Methods

To examine the global use of carnivores used in traditional medicine, all available references or reports about folk remedies that included carnivores' parts were reviewed. Only taxa that could be identified to the species level were included in the database. Scientific names provided in publications were updated according to the ITIS—Catalogue of Life 2011 Annual Checklist (Bisby et al. 2011) and Wilson and Reeder (2005). The conservation status of the Carnivora species follows the IUCN Red List of Threatened Species 2011 (IUCN 2011) and information on international trade regulation was extracted from CITES (2011). The sources analyzed are listed in the legend of Table 9.1. A database was created containing information on carnivores species, family names, parts of the animal used, and country where use was recorded.

## 9.3 Results and Discussion

Human history and evolution have been shaped and influenced by interactions with a diverse array of carnivores (Johnson et al. 2001; Macdonald and Kays 2005). Our cultural and historical heritage reflects a long-term relationship with carnivores, and our attitude toward them often mirrors our interaction with the natural world (Johnson et al. 2001). Some carnivores have been traditionally viewed as adversaries to be avoided or killed (Alves et al. 2009a; Johnson et al. 2001; Kruuk 2002; Sillero-Zubiri et al. 2004). On the other hand, several carnivore species have

**Table 9.1** Carnivores used in the wordwide traditional folk medicine

| Canidae species | IUCN red list | CITES | Country(-ies) | Use(s) | Reference(s) |
|---|---|---|---|---|---|
| *Canis adustus* (Sundevall, 1847) | LC | | Not mentioned | Epilepsy, to ward off evil spirits | 1 |
| *Canis aureus* (Linnaeus, 1758) | LC | III (Only the populations of India) | India, Viet Nam | Asthma, arthritis, paralysis | 2–4 |
| *Canis latrans* (Say, 1823) | LC | | Mexico | To bring good luck | 5–7 |
| *Canis lupus* (Linnaeus, 1758) | LC | I, II | Brazil, Israel, India, Kingdom of Jordan, Mongolia and Sudan | Chicken pox, mumps, smallpox, asthma, varicella, measles, menstrual cramps, warts | 4, 8–21 |
| *Canis mesomelas* Schreber, 1775 | LC | | Angola, Botswana, Djibouti, Eritrea, Ethiopia, Kenya, Lesotho, Mozambique, Namibia, Somalia, South Africa, Sudan, Swaziland, Tanzania, Uganda, Zimbabwe | Asthma | 22–24 |
| *Cerdocyon thous* (Linnaeus, 1766) | LC | II | Brazil | Earache, sore throat, rheumatism, flu, hemorrhoids, disorders after parturition (to accelerate recovery after parturition), bronchitis, cracks in the sole of the feet, back ache, osteoporosis, eczema, arthritis, womb inflammation | 19–21, 25–36 |
| *Chrysocyon brachyurus* (Illiger, 1815) | NT | II | Bolivia; Brazil; | Bronchitis, kidney disease, snake bites and "to bring good luck", epilepsy | 19, 37–40 |
| *Cuon alpinus* (Pallas, 1811) | EN | II | China | Not mentioned | 41, 42 |
| *Lycalopex culpaeus* (Molina, 1782) | LC | II | Argentina, Bolivia | "*susto*" (fright) | 43 |
| *Lycalopex gymnocercus* (G. Fischer, 1814) | LC | II | Argentina, Bolivia | "*susto*" (fright) | 43, 44 |
| *Lycalopex sechurae* (Thomas, 1900) | NT | | Peru | Used in traditional magic-religious rituals, bronchial illness and stomach disorders, amulets, to attract "good spirits" or "positive energies" | 45, 46 |
| *Lycaon pictus* (Temminck, 1820) | EN | | Parts of West, East and Southern Africa, Zimbabwe, Cameroon | It is considered to have medicinal and magical powers for African traditional cultures | 15, 47–49 |
| *Nyctereutes procyonoides* (Gray, 1834) | LC | | Not mentioned | Not mentioned | 15, 50 |
| *Otocyon megalotis* (Desmarest, 1822) | LC | | South Africa | Not mentioned | 51 |
| *Speothos venaticus* (Lund, 1842) | NT | I | Brazil | Not mentioned | 28, 52 |

(continued)

**Table 9.1** (continued)

| Canidae species | IUCN red list | CITES | Country(-ies) | Use(s) | Reference(s) |
|---|---|---|---|---|---|
| *Vulpes bengalensis* (Shaw, 1800) | LC | III | India | Not mentioned | 39, 53 |
| *Vulpes chama* (A. Smith, 1833) | LC | | Not mentioned | Not mentioned | 15 |
| *Vulpes pallida* (Cretzschmar, 1826) | DD | | Sudan | Asthma | 23 |
| *Vulpes vulpes* (Linnaeus, 1758) | LC | III | Viet Nam | Not mentioned | 2 |
| **Felidae species** | | | | | |
| *Acinonyx jubatus* (Schreber, 1775) | VU | I | Egypt, Jordan, Namibia, Netherlands, South Africa, Zimbabwe | Spiritual | 16, 54–57 |
| *Caracal caracal* (Schreber, 1776) | LC | II (Only the population of Asia; all other populations are included in Appendix II) | Czech republic, Namibia South Africa, Zimbabwe | Not mentioned | 54, 58 |
| *Catopuma temminckii* (Vigors and Horsfield, 1827) | NT | I | Not mentioned | Not mentioned | 54, 59 |
| *Felis manul* (Pallas, 1776) | NT | II | Mongolia, Russia | Not mentioned | 54, 60 |
| *Leopardus colocolo* (Molina, 1782) | NT | II | Not mentioned | Self-encorage | 54, 61 |
| *Leopardus guigna* (Molina, 1782) | VU | II | Chile | Not mentioned | 54, 58 |
| *Leopardus jacobitus* (Cornalia, 1865) | EN | I | Brazil | Self encourage, snake bites | 54, 61; 62 |
| *Leopardus pardalis* (Linnaeus, 1758) | LC | I | Bolivia, Brazil, Colombia, Ecuador Germany, Mexico, Netherlands, Nicaragua, Peru, Venezuela, Viet Nam | Bronchitis, earaches, respiratory disorders, rheumatism, snake bites | 54, 61; 63–67 |
| *Leopardus tigrinus* (Schreber, 1775) | VU | I | Brazil | Joint pain, Snake bite, "Ferida de goelas" (Throat problems), Snake bites, "estrepes" (suck a splinter out of skin), intestinal disorders, swellings, to protect the animals, especially chickens, from bats attack | 26, 27, 30, 54, 67, 68 |
| *Leopardus wiedii* (Schinz, 1821) | NT | I | Bolivia, Brazil, Colombia, Ecuador, Mexico, Nicaragua, Peru, Panama | Snake bites | 54, 56, 65, 67, 69 |
| *Leptailurus serval* (Schreber, 1776) | LC | II | Nigeria, Zimbabwe, West Africa in general | Aphrodisiacs - potency for men, fortune rousers, protection against evil influences/ manipulation, snake poison | 15, 54, 56, 57, 70–72 |

(continued)

**Table 9.1** (continued)

| Canidae species | IUCN red list | CITES | Country(-ies) | Use(s) | Reference(s) |
|---|---|---|---|---|---|
| *Lynx canadensis* (Kerr, 1792) | LC | II | Canada, The United States of America (USA) | Not mentioned | 54, 58 |
| *Lynx lynx* (Linnaeus, 1758) | LC | II | Canada, Czech Republic Germany, Norway Sweden, Switzerland, United States of America | Not mentioned | 54, 58 |
| *Lynx rufus* (Schreber, 1777) | LC | II | Canada, Mexico, United States of America | Not mentioned | 54, 56, 58, 73 |
| *Neofelis nebulosa* (Griffith, 1821) | VU | I | Lao PDR, Netherlands Nigeria, Viet Nam | Liver, kidney problems | 54, 55, 58, 59, 71, 74, 75 |
| *Panthera leo* (Linnaeus, 1758) | VU | I | Austria, Botswana, Canada, China, France, Germany, Ghana, Kenya, Mexico, Nigeria Mozambique, Namibia, South Africa, Sudan, Switzerland, United Kingdom, United Republic of Tanzania, United States of America, Zambia, Zimbabwe | Convulsion, rheumatism, to alleviate abdominal pains, to control ants, to relieve chest pain, vanishing powers, weak child | 17, 23, 54–57, 59, 70, 76–79 |
| *Panthera pardus* (Linnaeus, 1758) | NT | I | Australia, Belgium, Botswana, Cambodia, Canada, China, Comoros, Congo, Côte d'Ivoire, Egypt, Ethiopia, France, Ghana, Hong Kong, India, Ireland, Japan, Kenya, Kingdom of Jordan, Lao PDR, Malaysia, Mozambique, Netherlands, Nigeria, Panama, Republic of Korea, Senegal, Singapore, South Africa, Spain, Swaziland, Switzerland, Sudan, Thailand, United Kingdom, United Republic of Tanzania, United States of America, Vietnam, Zimbabwe. | Aphrodisiac, appeasing witches, body pain, convulsion, foot and mouth diseases, general body weakness, kwashiorkor, malaria, paralysis, promoter of strength and virility, protection against evil influences/ manipulation, rheumatic pain, snake poison, spiritual uses, typhoid, weak child | 3, 4, 15–17, 50, 54, 55, 58, 59, 70, 71, 76, 79–91 |
| *Panthera onca* (Linnaeus, 1758) | NT | I | Bolivia, Brazil, Colombia, Czech Republic, Mexico, Peru, South America, United States of America | Asthma, bone pain, cough, fatigue fever, inflammations, leishmaniosis, osteomuscular problems, pain in bones, pneumonia, protect the wearer against snakebites, wounds | 6, 38, 54, 59, 69, 92–97 |

(continued)

**Table 9.1** (continued)

| Canidae species | IUCN red list | CITES | Country(-ies) | Use(s) | Reference(s) |
|---|---|---|---|---|---|
| *Panthera tigris* (Linnaeus, 1758) | EN | I | Australia, Belgium, Cambodia, Canada, China, Colombia, Côte d'Ivoire, Denmark, Democratic Republic Malaysia, Hong Kong, India, Indonesia, Japan, Lao PDR, Mexico, Nepal, Nigeria, Philippines, Republic of Korea, Russia, Singapore, Sweden, Switzerland, Taiwan, Thailand, United Kingdom, United States of America, Vietnam | Aphrodisiac, treat impotence or restorative tisane, arthritis, asthenia in loins and legs, bone pain, curing carbuncles and ulcerative scrofula, dispelling rheumatism, enhance body strength, generally improve health, hemiplegia, joint sprains, killing evil pathogens, pain in joints, relieving pain, rheumatism, stopping convulsions, strengthening tendons and bones, to help the wearer ward off the common cold, warding off harmful air | 2, 4, 15, 50, 54–56, 59, 71, 75, 84–88, 90, 91, 98–113 |
| *Pardofelis marmorata* (Martin, 1837) | VU | I | SE Asia, Vietnam | Not mentioned | 2, 54, 59, 114 |
| *Prionailurus bengalensis* (Kerr, 1792) | LC | II (Only the populations of Bangladesh, India, and Thailand; all other populations are included in Appendix II) | Cambodia, Japan, Malaysia, Hong Kong, Vietnam | Not mentioned | 2, 54, 84, 86 |
| *Prionailurus planiceps* (Vigors and Horsfield, 1827) | EN | I | Brunei, Indonesia, Malaysia, Myanmar, Singapore, Thailand | Not mentioned | 54, 56 |
| *Puma concolor* (Linnaeus, 1771) | LC | I | Bolivia, Brazil, Canada Colombia, Norway, Switzerland, United Kingdom, United States of America | "Evil eye", arthritis, avoid acne, distend, earache, fever, leishmaniosis, making the child talk; pain in bones, rheumatism, scare, stomach ache, wounds | 30, 43, 54, 56, 58, 59, 61, 73, 92, 94 |
| *Puma yagouaroundi* (É. Geoffroy Saint-Hilaire, 1803) | LC | II (Only the populations of Central and North America; all other populations are included in Appendix II) | Brazil | *Hands hot and swollen, heat in the hands,* 'Estrepes', intestinal disorders, swellings, to protect the animals, especially chickens, from bats attack | 27, 30, 54, 68 |
| *Uncia uncia* (Schereber, 1775) | EN | I | China, SE Asia | Used as a substitute for Tiger Bone and *Panthera pardus* in traditional Asian medicines, liver, kidney | 54, 55, 58, 59, 90, 115–118 |
| **Eupleridae species** | | | | | |
| *Cryptoprocta ferox* Bennett, 1833 | VU | II | Madacascar | Not mentioned | 119 |

(continued)

**Table 9.1** (continued)

| Canidae species | IUCN red list | CITES | Country(-ies) | Use(s) | Reference(s) |
|---|---|---|---|---|---|
| **Mustelidae species** | | | | | |
| *Aonyx cinereus* (Illiger, 1815) | VU | II | Cambodia, Lao PDR | Liver, kidney | 55, 80, 120 |
| *Aonyx capensis* (Schinz, 1821) | LC | II | Not mentioned | Not mentioned | 121 |
| *Arctonyx collaris* Cuvier, 1825 | NT | | Cambodia, China | Not mentioned | 84, 86, 122, 123 |
| *Eira barbara* (Linnaeus, 1758) | LC | III | Bolivia, Costa Rica, Ecuador, Mexico, Suriname | Cure pregnant women "for their children to be good hunters" | 58, 65, 124, 125 |
| *Enhydra lutris* (Linnaeus, 1758) | EN | I | Russia, United States of America | Not mentioned | 58 |
| *Ictonyx striatus* (Perry, 1810) | LC | | South Africa | Not mentioned | 126 |
| *Lontra longicaudis* (Olfers, 1818) | DD | I | Brazil | Thrombosis | 9, 19, 94 |
| *Lontra canadensis* (Schreber, 1777) | LC | II | Canada, Denmark, United States of America | Not mentioned | 58 |
| *Lutra sumatrana* (Gray, 1865) | EN | II | Cambodia | Not mentioned | 120, 127 |
| *Lutra lutra* (Linnaeus, 1758) | NT | II | Cambodia, Canada, Russia, United Kingdom, Viet Nam | Kidney, Liver | 2, 50, 56, 120, 127 |
| *Lutrogale perspicillata* (Geoffroy, 1826) | VU | II | Cambodia | For labor pains | 84, 86, 127 |
| *Mustela frenata* Lichtenstein, 1831 | LC | | Costa Rica, Mexico | Not mentioned | 6, 124 |
| Mustela felipei Izor & de la Torre, 1978 | VU | | Colombia | Asthma, infertility, and 'magical arts' | 128 |
| *Mustela nudipes* Desmarest, 1822 | LC | | Indonesia | Not mentioned | 129 |
| *Mustela sibirica* Pallas, 1773 | LC | III (only India) | Canada, China, France, Germany, Hong Kong, India, Ireland, Italy, Republic of Korea, Russia, Spain, Switzerland, United Kingdom, United States of America | Not mentioned | 50, 55, 58 |
| *Martes zibellina* (Linnaeus, 1758) | LC | | | Not mentioned | 50; 55; 58 |
| *Meles meles* (Linnaeus, 1758) | LC | III | Albania, Bolivia, Czech Republic, Finland, Israel, Mongolia, Korean Peninsula, Slovak Republic, ex-Yugoslavia, | Abscess, snake and scorpion bite, spleen | 13, 14, 50, 55, 58, 130–132 |

(continued)

**Table 9.1** (continued)

| Canidae species | IUCN red list | CITES | Country(-ies) | Use(s) | Reference(s) |
|---|---|---|---|---|---|
| *Mellivora capensis* (Schreber, 1776) | LC | III | Zimbabwe | Not mentioned | 15, 55–57, 77 |
| *Martes flavigula* (Boddaert, 1785) | LC | III (India) and II (China) | China, India | Wounds | 3, 4 |
| *Melogale moschata* (Gray, 1831) | LC | | Lao PDR | Not mentioned | 80 |
| *Poecilogale albinucha* (Gray, 1864) | LC | | South Africa | Not mentioned | 133–135 |
| **Procyonidae species** | | | | | |
| *Nasua narica* (Linnaeus, 1766) | LC | III (Honduras) | Costa Rica, Guatemala, Mexico | Male impotence, sexual weakness | 6, 56, 73, 124, 136 |
| *Nasua nasua* (Linnaeus, 1766) | LC | III | Brazil, Bolivia, other parts of South America not specified | Arthrosis, backache, body strengtheners, calls luck, cold, cough, earache, earache, leg pain, money and love (amulet), neck strain, rheumatism, sexual impotence, skin burns, snake bites, stroke, thrombosis, to have children, to help become pregnant, wooping cough, wounded foot, wounds | 9, 11, 19–21, 26, 29, 39, 55, 56, 61, 65, 67, 69, 92, 94–97, 137–141 |
| *Procyon cancrivorus* (Cuvier, 1798) | LC | | Brazil | Amulet used as a protection against snake bite, snake bite, charm, epilepsy, evil-eye, fever, pitiriase, rheumatism, thrombosis, Tinea versicolor, tonsillitis, to undo enchantment of Umbanda rituals, vitiligo | 10, 19–21, 29, 61, 94, 137, 141, 142 |
| *Procyon lotor* (Linnaeus, 1758) | LC | | Costa Rica, Mexico | Cure diseases of abortion, funerals, rheumatism | 6, 124, 136 |
| *Potos flavus* (Schreber, 1774) | LC | III (Honduras) | Bolivia, Colombia, Ecuador, Peru, Suriname, Venezuela | Ant bite, earache, snake bite | 61, 65, 69, 95 |
| **Ursidae species** | | | | | |
| *Ailuropoda melanoleuca* David, 1869 | EN | I | Canada, South Korea | Clear heat and alleviate spasms, remove toxins from the liver, strengthen the liver function, to treat warm-febrile diseases with high fever and convulsions, to treat delirium associated with extensive burns | 55, 58, 104, 143 |

(continued)

**Table 9.1** (continued)

| Canidae species | IUCN red list | CITES | Country(-ies) | Use(s) | Reference(s) |
|---|---|---|---|---|---|
| *Ursus thibetanus* Cuvier, 1823 | VU | I | Australia, Canada, Cambodia, Fiji, Germany, Hong Kong, Hungary, India, Indonesia, Italy, Japan, Kenya, Lao PDR, Mexico, Mozambique, Myanmar, Nepal, New Zealand, Palau, Philippines, Russia, Singapore, South Korea, Switzerland, Taiwan, Thailand, USA, Vietnam | Heart, Liver, to treat warm-febrile diseases with high fever and convulsions, to treat delirium associated with extensive burns, clear heat and alleviate spasms, strengthen the liver function, remove toxins from the liver, malaria, typhoid, T.B and other serious fevers, myriad diseases, detoxification, inflammation, swelling and pain reduction, cure of carbuncle of heat type, pyocutaneous diseases, hemorrhoid, overabundance of liver-fire, convulsion caused by the overabundance of heat, epilepsy, tic, redness of eyes due to liver heat | 3, 4, 15, 55, 58, 84, 86, 91, 104, 108, 131, 143–147 |
| *Ursus arctos* Linnaeus, 1758 | LC | I | Australia, Canada, Cambodia, China, Czech Republic, Denmark, Estonia, Germany, Hawaii, Hong Kong, Italy, Japan, Kenya, Lao PDR, Mexico, Mongolia, Mozambique, New Zealand, Norway, Palau, Romania, Russia, Singapore, Slovenia, South Korea, Sweden, Switzerland, Thailand, Taiwan, United Kingdom, USA, Vietnam | AIDS, to relieve toxin, to stop endogenous wind, to arrest convulsion, to clear away liver-fire, to improve eyesight in fever fighting, inflammation, swelling and pain reduction. It was also used in the cure of carbuncle of heat type, pyocutaneous diseases, hemorrhoid, overabundance of liver-fire, to convulsion caused by the overabundance of heat epilepsy, tic, redness of eyes due to liver heat, to treat warm-febrile diseases with high fever and convulsions, to treat delirium associated with extensive burns, to clear heat and alleviate spasms, to strengthen the liver function, to remove toxins from the liver, abdominal tumors; to alleviate abdominal pain due to obstruction or bad food intake; cramps in neck and back; chest pains; unsuccesful urge to vomit; irritability; insanity; intermittent fever (malaria-like state); diarrhea. Kills parasite insects, stops vomiting and diffuses smallpox. Also used to treat children's nervous breakdowns, convulsions, stomach ache during pregnancy, abdominal pain after childbirth, produce labor pains. Clears blurred eyes, stops pain of hemorrhoids. Restores vitality and rids impurities from body. | 14, 55, 56, 58, 89, 104, 108, 143, 145–148 |

(continued)

**Table 9.1** (continued)

| Canidae species | IUCN red list | CITES | Country(-ies) | Use(s) | Reference(s) |
|---|---|---|---|---|---|
| *Ursus americanus* Pallas, 1780 | LC | II | Australia, Cambodia, Canada, China, Czech Republic, Denmark, Fiji, Germany, Hawaii, Hong Kong, Hungary, Indonesia, Italy, Japan, Kenya, Lao PDR, Mexico, Mozambique, New Zealand, Norway, Palu, Russia, Singapore, South Korea, Switzerland, Taiwan, Thailand, USA, Vietnam | AIDS, to treat warm-febrile diseases with high fever and convulsions, to treat delirium associated with extensive burns, to clear heat and alleviate spasms, to strengthen the liver function, to remove toxins from the liver, used in fever fighting, detoxification, inflammation, swelling and pain reduction, cure of carbuncle of heat type pyocutaneous diseases, hemorrhoid, to overabundance of liver-fire, convulsion caused by the overabundance of heat, epilepsy, tic, redness of eyes due to liver heat | 55, 56, 58, 104, 143, 145, 147, 149, 150 |
| *Ursus maritimus* Phipps, 1774 | VU | II | Autralia, Cambodia, Canada, China, Denmark, Fiji, France, Germany, Greenland, Hong Kong, Hungary, Indonesia, Italy, Japan, Kenya, Lao PDR, Mexico, Mozambique, New Zealand, Norway, Palau, Russia, Singapore, South Korea, Thailand, Taiwan, USA, Vietnam | Used in fever fighting, detoxification, inflammation, swelling and pain reduction, cure of carbuncle of heat type, pyocutaneous diseases, hemorrhoid, overabundance of liver-fire, convulsion caused by the overabundance of heat, epilepsy, tic, redness of eyes due to liver heat, to treat warm-febrile diseases with high fever and convulsions, to treat delirium associated with extensive burns, to clear heat and alleviate spasms, to strengthen the liver function, remove toxins from the liver | 55, 58, 104, 143, 147 |
| *Melursus ursinus* (Shaw, 1791) | VU | II | Australia, Bangladesh, Bhutan, Cambodia, Canada, China, Fiji, Germany, Hong Kong, Hungary, India, Indonesia, Italy, Japan, Kenya, Lao PDR, Mexico, Mozambique, Nepal, New Zealand, Palau, Singapore, South Korea, Taiwan, Thailand, USA, Vietnam | Similar to prescriptions for *Ursus maritimus* products | 4, 55, 56, 58, 89, 104, 112, 143, 147 |
| *Helarctos malayanus* Raffles, 1821 | VU | I | Australia, Cambodia, Canada, China, Fiji, Germany, Hong Kong, Hungary, Indonesia, Italy, Japan, Kenya, Lao PDR, Mexico, Mozambique, New Zealand, Palau, Singapore, South Korea, Taiwan, Thailand, USA, Vietnam | Used in fever fighting, detoxification, inflammation, swelling and pain reduction, cure of carbuncle of heat type, pyocutaneous diseases, hemorrhoid, overabundance of liver-fire, convulsion caused by the overabundance of heat, epilepsy, tic, redness of eyes due to liver heat | 55, 56, 58, 89, 112, 147 |

(continued)

**Table 9.1** (continued)

| Canidae species | IUCN red list | CITES | Country(-ies) | Use(s) | Reference(s) |
|---|---|---|---|---|---|
| *Tremarctos ornatus* (Cuvier, 1825) | VU | I | Andean Region, Australia, Canada, Hong Kong, Japan, New Zealand, Singapore, South Korea, Taiwan, USA, | Bone bruises, for strength and fertility, to treat warm-febrile diseases with high fever and convulsions, to treat delirium associated with extensive burns, to clear heat and alleviate spasms, for strengthen the liver function, remove toxins from the liver. | 89, 104, 143, 147 |
| **Herpestidae species** | | | | | |
| *Herpestes edwardsii* (Geoffroy, 1818) | LC | III | India | Not mentioned | 4 |
| Herpestes javanicus (Geoffroy, 1818) | LC | III | India | Not mentioned | 151 |
| *Herpestes sanguineus* (Rüppell, 1835) | LC | | Nigeria | Not mentioned | 70, 152 |
| **Mephitidae species** | | | | | |
| *Conepatus chinga* (Molina, 1782) | LC | II | Paraguay | Thrombosis, rheumatism, general diseases | 153, 154 |
| *Conepatus semistriatus (Boddaert, 1785)* | LC | II | Brazil | Rheumatism, acne, scabies, blood problems, bronchial diseases, skin problems, asthma, nervous disturbances | 20, 21, 94, 155–157 |
| *Conepatus leuconotus (Lichtenstein, 1832)* | LC | II | Latin America | Blood disorders, acne, stomach ache, *mal aire*, swelling, undescended testicles, rabies, whooping cough, bone pain, acne, dolor muscular | 154, 158 |
| *Mephitis macroura Lichtenstein, 1832* | LC | | Brazil | Stomach ache, mal aire, swelling, undescended testicles, Rabies, whooping cough, pain in bone, asthma | 94, 137, 154, 159 |
| *Mydaus javanensis* (Desmarest, 1820) | LC | | Indonesia (Jawa, Kalimantan, Sumatera); Malaysia (Sabah, Sarawak) | Used to cure fever or rheumatism | 160 |
| *Spilogale putorius (Linnaeus, 1758)* | LC | | Brazil | Stomach ache, mal aire, swelling, undescended testicles, rabies, whooping cough, pain in bone | 154, 159 |
| **Nandiniidae species** | | | | | |
| *Nandinia binotata* (Gray, 1830) | LC | | Guinea | Not mentioned | 161 |

(continued)

**Table 9.1** (continued)

| Canidae species | IUCN red list | CITES | Country(-ies) | Use(s) | Reference(s) |
|---|---|---|---|---|---|
| **Hyaenidae species** | | | | | |
| *Crocuta crocuta* (Erxleben, 1777) | LC | | Nigeria, Somalia, E and W Africa, SE Asia | Pneumonia, skin diseases. Hyaenas can see and pursue the devil (people with mental disorders are confined in a room, together with a hyaena, in order to exorcize the devil from them) | 55, 56, 58, 70, 77, 162-167 |
| *Hyaena hyaena* (Linnaeus, 1758) | NT | | India, Israel, Jordan, Morocco, Nigeria, Somalia, Sudan | Arthritis, Chest pains and diseases, Diarrhea, Gout Inflammation, Pneumonia, Skin diseases, Stomach ache, Throat diseases, Preparations to invoke witches. Hyaenas can see and pursue the devil (people with mental disorders are confined in a room, together with a hyaena, in order to exorcize the devil from them) | 4, 13, 16, 17, 23, 55, 56, 83, 162, 167 |
| *Hyaena brunnea* Thunberg, 1820 | NT | | South Africa | Used occasionally in traditional medicine and rituals (purposes don't mentioned) | 166, 168 |
| *Proteles cristatus* (Sparrman, 1783) | LC | III | South Africa | Not mentioned | 15, 51, 57, 58, 77, 164 |
| **Prionodontidae species** | | | | | |
| *Prionodon pardicolor* Hodgson, 1842 | LC | I | China, Myanmar, Vietnam | Not mentioned | 2, 169 |
| **Viverridae species** | | | | | |
| *Arctictis binturong* (Raffles, 1821) | VU | III | Vietnam | Not mentioned | 2 |
| *Viverra zibetha* Linnaeus, 1758 | NT | III | Australia, China, Myanmar | Not mentioned | 15, 50, 55, 58, 111 |
| Viverra civettina Blyth, 1862 | CR | III | India | Not mentioned | 170 |
| *Viverricula indica* (Geoffroy, 1803) | LC | III | India, Vietnam, China, Myanmar | Alternativa pra *Viverra zibetha*, used in the preparation of ayurvedic medicine against epilepsy | 2, 55, 111, 171–174 |
| *Genetta genetta* (Linnaeus, 1758) | LC | III | Ghana | Protection against sorcery, TMP clinic ornament | 15, 51, 55, 57, 58 |
| *Genetta tigrina* (Schreber, 1776) | LC | | South Africa | Stick-fight charms, used in treatment of eyes diseases | 121 |
| *Civettictis civetta* (Schreber, 1776) | LC | III | Ethiopia, Nigeria | Not mentioned | 70, 175 |

(continued)

**Table 9.1** (continued)

| Canidae species | IUCN red list | CITES | Country(-ies) | Use(s) | Reference(s) |
|---|---|---|---|---|---|
| *Paradoxurus hermaphroditus* (Pallas, 1777) | LC | III | India, Indonesia, China | To treat body pain | 4, 176, 177 |
| Paradoxurus jerdoni Blanford, 1885 | LC | III | India | Not mentioned | 178 |
| *Chrotogale owstoni* Thomas, 1912 | VU | | Indochina region | Not mentioned | 179–181 |
| Diplogale hosei (Thomas, 1892) | VU | | China | Not mentioned | 182 |

*References* 1. Ginsberg and Macdonald (1990), 2. Van and Tap (2008), 3. Negi and Palyal (2007), 4. Mahawar and Jaroli (2008), 5. Monroy-Vilchis et al. (2008), 6. Vázquez et al. (2006), 7. Morales-Mavil and Villa-Cañedo (1998), 8. Costa-Neto and Oliveira (2000), 9. Costa-Neto (1999a), 10. Marques (1995), 11. Moura and Marques (2008), 12. Souto et al. (2001), 13. Lev (2003), 14. Schmidt et al. (2006), 15. CITES (2002a), 16. Lev and Amar (2002), 17. El-Kamali (2000), 18. Mahawar and Jaroli (2006), 19. Alves (2009), 20. Alves and Rosa (2006), 21. Alves and Rosa (2007b), 22. Loveridge and Nel (2004), 23. Khalid et al. (2007), 24. Loveridge and Nel (2008), 25. Alves et al. (2008b), 26. Moura (2002), 27. Alves et al. (2009a), 28. Alves and Rosa (2007a), 29. Alves and Rosa (2007c), 30. Valle (2007), 31. Ferreira et al. (2009a), 32. Alves et al. (2008c), 33. Ferreira et al. (2009b), 34. Confessor et al. (2009), 35. Alves et al. (2011), 36. Alves et al. (2009e), 37. Rodden et al. (2004), 38. Winter (2002), 39. Costa-Neto (1996), 40. Rodden et al. (2008), 41. Durbin et al. (2004), 42. Durbin et al. (2008), 43. Barbarán (2004), 44. Lucherini et al. (2004), 45. Asa and Cossíos (2004), 46. Asa et al. (2008), 47. Woodroffe et al. (2004), 48. Rasmussen (1999), 49. Breuer (2002), 50. Jiangsu New Medical College (2002), 51. Simelane and Kerley (1998), 52. Figueiredo (1994), 53. Johnsingh and Jhala (2004), 54. Alves et al. (2009b), 55. CITES (2001), 56. CITES (2002b), 57. Marshall (1998), 58. CITES (2002c), 59. TRAFFIC and WWF (2002), 60. Ross et al. (2008), 61. Alves et al. (2009f), 62. Wild Cat Conservation (2011), 63. Almeida and Albuquerque (2002), 64. Hanazaki et al. (2009), 65. Tratado de Cooperacion Amazonica (1999), 66. Hanazaki (2001), 67. Costa-Neto (2000), 68. Souto et al. (2011), 69. Tejada et al. (2006), 70. Soewu (2008), 71. Sodeinde and Soewu (1999), 72. Breitenmoser-Wursten et al. (2008), 73. Roa (1997), 74. Sanderson et al. (2008), 75. Nowell (2000), 76. Mzimela (1995), 77. Pujol (1993), 78. Costa-Neto (2005), 79. Ntiamoa-Baidu (1992), 80. Baird (1995), 81. Carpaneto Germi (1989), 82. Bailey (2000), 83. Adeola (1992), 84. Ashwell and Walston (2008), 85. TRAFFIC (1998), 86. Walston (2005), 87. Yinfeng et al. (1997), 88. Venkataraman (2007), 89. CITES (2000), 90. Nowell and Jackson (1996), 91. Solanki and Chutia (2004), 92. Begossi et al. (1999), 93. Apaza et al. (2003), 94. Alves et al. (2007), 95. Vargas (2002), 96. Rodrigues (2006), 97. Silva (2008), 98. Yi-Ming et al. (2000), 99. But (1995), 100. Patil (2003), 101. Long (2001), 102. Ng and Nemora (2007), 103. Laider and Laider (1996), 104. Kang and Phipps (2003), 105. Yiming and Dianmo (1998), 106. Yates (2005), 107. Do Tat Loi (2004), 108. Bensky et al. (1993), 109. Zhang Enqin et al. (1996), 110. Vietnamese Ministry of Agriculture (2003), 111. Australian Department of Environment and Heritage (2003), 112. TRAFFIC (1999), 113. Duckworth et al. (1999), 114. 2010), 115. Schell (1982), 116. Yanfa and Bangjie (1988), 117. Theile (2003), 118. Alderton (1993), 119. Hawkins and Dollar (2008), 120. Olsson (2007), 121. Cunningham and Zondi (1991), 122. Ramakantha (1994), 123. Bangjie (1989), 124. Alvarez (2006), 125. Costa Rica Government (2002), 126. Maliehe (1993), 127. Campbell et al. (2006), 128. Tirira and González-Maya (2009), 129. Duckworth et al. (2006), 130. CITES (2003), 131. Won and Smith (1999), 132. Lachner-Eitzenberger (2006), 133. Stuart et al. (2008), 134. Rowe–Rowe (1990), 135. Rowe–Rowe (1995), 136. Borge and Castillo (1997), 137. Alves (2006), 138. Costa-Neto (1999b), 139. Costa-Neto (1999c), 140. Andrade and Costa-Neto (2006), 141. Costa-Neto (2004), 142. Luck Mojo Curio Co. (1995), 143. Lee (1995), 144. Enqin (1991), 145. Nilsson (1990), 146. Mills et al. (1995), 147. Feng et al. (2009), 148. Guo et al. (1997), 149. Martin and Phipps (1996), 150. Galster (1996), 151. Chakravorty et al. (2011), 152. Hoffmann (2008), 153. Brooks (1991), 154. Alves and Alves (2011), 155. Alves and Rosa (2010), 156. Branch and Silva (1983), 157. Campos (1960), 158. Lenko and Papavero (1996), 159. Ribeiro et al. (2010), 160. Long et al. (2008), 161. Van Rompaey et al. (2008), 162. Watt and Breyer-Brandwijk (1962), 163. Rinpoche and Kunzang (1973), 164. Ray et al. (2005), 165. Honer et al. (2008), 166. Mills and Hofer (1998), 167. Amir (2006), 168. Republic of South Africa Department of Environmental Affairs (2004), 169. Choudhury (2000), 170. Nandini and Mudappa (2010), 171. Lal Mohan (1994), 172. Sreedevi (2001), 173. Balakrishnan and Sreedevi (2007a), 174. Balakrishnan and Sreedevi (2007b), 175. Bekele et al. (2009), 176. Padmanabham and Sujana (2008), 177. Nijman (2005), 178. Nandini et al. (2002), 179. Animal Diversity Web (2007), 180. World Association of Zoos and Aquariums (2007), 181. Carnivore Conservation (2007), 182. Moutou (2004). *IUCN Red List categories DD* Deficient Data, *LC* Least Concern, *NT* Near Threatened, *VU* Vulnerable, *EN* Endangered, *CR* Critically Endangred; *CITES appendices*—I, II or III been a resource for humans, exploited for subsistence or commercial profit, or hunted for sport (Macdonald and Sillero-Zubiri 2004).

The exploitation and trade of carnivores form an integral part of human cultural heritage (Alves et al. 2009b, 2010b; Johnson et al. 2001). Several carnivores are popular for their supposed medicinal properties, or at least they were in the past (Alves and Alves 2011; Kruuk 2002; Mahawar and Jaroli 2008; Sillero-Zubiri et al. 2004). Bear bile, used in traditional medicine, is historically documented (Foley et al. 2011). In China, the first references to the use of tiger (*Panthera tigris*) bone in traditional medicines date back more than 1,500 years. In the herbal compendium, Mingyi Bielu, or Records of Famous Physicians (Tao Hongjing, circa 500 A.D.), tiger bone was said to have the properties of "warding off harmful air, killing evil pathogens, stopping convulsions, and curing carbuncles and ulcerative scrofula" (But 1995). In Brazil, carnivore species were used medicinally in the first half of the eighteenth century (Souza 2008). Lev (2003, 2006) emphasizes that the fox (*Vulpes* sp.) was used for the treatment of ear diseases in the Levant between the tenth and eighteenth century. In Azerbaijan, there are records of medicinal use of animals such as wolves, foxes, and jackals during the Middle Ages. Various parts (meat, skin, fur, etc.) of these animals were used in medieval Azerbaijan medicine (Alakbarli 2006).

Our review indicated that the use of products derived from carnivores for medicinal purposes has been going on throughout human history. Currently, at least 108 species of carnivores from 64 genera are used in traditional medicine worldwide (see Table 9.1). The families with the largest numbers of medicinal species are: Felidae (with 25 species), followed by Mustelidae (21 spp.), and Canidae (19 spp.) (Table 9.1). Considering that there are 271 known species of carnivores in the world today (Macdonald and Kays 2005), 39.8% are, therefore, used in traditional folk medicines. If we add eight species of carnivores pinnipeds used for medicinal purposes (see Chap. 11 in this volume), this number increases to 42.8%.

The medicinal use of carnivores is widespread, being recorded in 96 countries. Such uses are more prominent in countries such as Africa, Asia, and South America. This is not surprising given that the diversity of carnivores is greater in these areas (Sillero-Zubiri et al. 2004). Some species are used in more than one country, although not necessarily for the same disease. For instance, *Panthera onca* in Bolivia, Mexico, and Brazil (Alves et al. 2007; Apaza et al. 2003; Vázquez et al. 2006), *Puma concolor* in Brazil and Bolivia (Barbarán 2004; Begossi et al. 1999), and *Panthera tigris* in India, China, Vietnam, Sumatra, Cambodia, and Nigeria (Long 2001; Mills et al. 1994; Patil 2003; Sodeinde and Soewu 1999). In Bolivia the species *Panthera onca* is used for its fat in the treatment of human illnesses such as fatigue, fever, and bone pain (Tejada et al. 2006; Vargas 2002). The second and third largest Asian forest felids, the leopard (*Panthera pardus*) and clouded leopard (*Neofelis nebulosa*), are both hunted for their skins and body parts in Asian countries (EIA 2004; Oswell 2010; Shepherd 2001), as well as many of the smaller species. This demonstrates that the utilization of some species is widespread, therefore it is important to understand such uses in the context of biodiversity conservation.

**Fig. 9.1** Examples of bear products used in traditional Asian medicine: **a** One of medicinal bears species—The Asiatic Black Bear (*Ursus thibetanus*) on a bear farm in Luang Prabang, Lao PDR; **b** Bear farm advertisement in Luang Prabang, Lao PDR; **c** Whole gall bladder; **d** Bear bile from a farm in Vietnam; **e** Pills purported to be from bears; **f** Bear claws for sale in Myanmar; **g** Bear bile from China, observed from in downtown Phnom Penh; **h** Bear bile product from China, observed for sale in Melaka, Malaysia. *Photo credits*: A. Oswell, TRAFFIC Southeast Asia (Photos **a** and **b**); C. Yeong, TRAFFIC Southeast Asia (Photos **c**, **e** and **h**); M. Silverberg, TRAFFIC Southeast Asia (Photos **d** and **g**) and C. Shepherd, TRAFFIC Southeast Asia (Photo **f**)

Various carnivore parts and products are used in the preparation of folk medicines, including heart, fur, fat, eyes, bile, blood, viscera, bone, and meat. Carnivores are used to treat approximately 130 conditions. Some species were recorded as having magical religious uses, for example, to bring good luck and in magic and religious rituals. The zootherapeutic products of carnivores are used to prepare clinical remedies as well as to make amulets or charms used in magical/religious diagnoses. For instance, in Brazil, body parts of *C. brachyurus* are believed to bring good luck. In Bolivia, cowboys believe that sitting on the pelt of a maned wolf will protect them from bad luck. Wild dogs (*Lycaon pictus*) are valued in some areas, not only because their kills provide a source of meat, but also due to the medicinal and magical powers that their various body parts are thought to have (Alves et al. 2010b). Another form of spiritual treatment involves the use of amulets containing carnivore parts to protect the user from diseases (Alves et al. 2007; Nowell 2000). An example is the use of the skin, teeth, and claws of the tiger, which is made into magical amulets and novelties (Nowell 2000). In South America, skin and teeth of *Panthera onca* were historically used as amulets. Tooth amulets are believed to protect the wearer against snakebites (Alves et al. 2010b). In Ancient Egypt, the Egyptian cat (*Felis silvestris*) was held in high esteem in the Nile Valley and still today has a large number of admirers, and its role in antiquity has a religious significance by covering the pharaohs with their skins giving them supernatural gifts and protections (Bothmer 1953).

The traditional use of natural resources may indicate the presence of biologically active components, since historic and maintained use of naturally occurring compounds often has a scientific underpinning (Oliveira et al. 2010). Nevertheless, little research has been done so far to prove the claimed clinical efficacy of animal products for medicinal purposes (Still 2003) and the indiscriminate consumption of zootherapeutic products can produce serious adverse effects for users, as a consequence of poor storage conditions or adulteration of the medicinal products used. Espinoza (1994) recorded a good example that illustrates this situation involving the medicinal use of carnivores. According to this author, analyses conducted by the National Forensics Laboratory of the US Fish and Wildlife Service have not yet found any detectable levels of bone in samples of tiger-based pharmaceuticals but have revealed high levels of toxic metals, particularly arsenic. This example reinforces the observation of Alves and Rosa (2005) that emphasizes the need for implementation of quality measures in the trade of animals or their parts for medicinal purposes.

The sale of medicinal products derived from carnivores can occur not only on a domestic scale (at retail stores and open markets) when it involves a large number of species, but also on an international scale when the number of species involved is more limited. Domestic commercial sale of carnivores for medicinal purposes has been documented in a number of countries. Parts or products derived from felidae and canidae are sold in various cities, mainly in Africa and Asia. For example, bones claiming to be of tiger (*Panthera tigris)* and leopard (*Panthera pardus),* or clouded leopard (*Neofelis nebulosa)* (bones or fur) were frequently sold in markets in Nigeria (Sodeinde and Soewu 1999). In Cambodia, *Panthera pardus* is among the species sold in local markets. Johnsingh and Jhala (2004) noted the local trade of the skin, tail, teeth, and claws of the Indian fox [*Vulpes bengalensis* (Shaw, 1800)] for medicinal and charm purposes. In Mexico, the coyote (*Canis latrans* Say, 1823) is sold for medicinal purposes in markets of the city of Chiapas (Vázquez et al. 2006). Durbin et al. (2004) pointed out the need for further research regarding the medicinal use (and commercial exploitation) of *Cuon alpinus*. In Asia, bear (Ursidae) parts and derivatives are heavily traded (Fig. 9.1). These include the paws, skin, claws, canine teeth, skulls, and most prized of all, the gall bladder and bile (Foley et al. 2011). These examples reveal that the sale of carnivores for medicinal purposes is a widespread phenomenon, with significant implications for their conservation and sustainable use.

With regard to the international market, the main species of medicinal carnivores involved are *Panthera tigris*, *Helarctos malayanus,* and *Melursus ursinus*. The intensity of commercialization is greater in Asian countries, particularly China. China is not only a consumer, but also the world's largest exporter of tiger bone and other tiger derivatives, exporting over 27 million units of tiger medicines and wine to 26 countries and territories in 1990–1992 (Mills et al. 1994). Most of the tiger parts sourced in Cambodia were being exported, while remedies in the domestic marketplace usually contained the bones of other animals and herbal substitutes. The reason given for this was that prices of medicine containing tiger bone were too high for locals (Heng 1999). Tiger penis is made into a soup, said to give potency, for rich Asian businessmen who pay as much as US$18,000 for a dinner featuring it (Highley and Highley 1994).

The tradition of brewing medicinal wines in China dates from the Han dynasty (Nowell and Ling 2007). Many medicinal wines are formulated to treat rheumatism, and Tiger bone wine is among the most famous. Many consider medicinal wine particularly effective because it is easy to consume on a daily basis for a chronic condition, and the alcohol is said to have a warming effect (Flaws 1994). Tiger bone wine was a popular form of Tiger medicine, and in the past wine manufacturers consumed large quantities of Tiger bone (Fig. 9.2) (Nowell and Ling 2007). A TRAFFIC survey (Wu 2006) documented seventeen instances of Tiger bone wine for sale on Chinese auction websites, for example, with one seller offering a lot of 5,000 bottles. While the illegal Tiger bone medicine market and Tiger skin market appear to be supplied primarily by wild Tigers, illegal trade in Tiger bone wine may well be supplied primarily from captive Tigers in China (Nowell and Ling 2007).

## *9.3.1 Impacts of Traditional Medicine on Carnivores*

Some biological and ecological traits can create a particular vulnerability to extinction for a species (IUCN 2011; McKinney 1997; Prassack 2003). Among the several attributes that predispose species to the risk of extinction, we can highlight the following types: species with narrow geographical ranges and low population densities, small population sizes and declining population sizes; species in a monophyletic group; species at higher trophic levels, which are more vulnerable to the cumulative effects of disturbance to species lower down the food chain; species

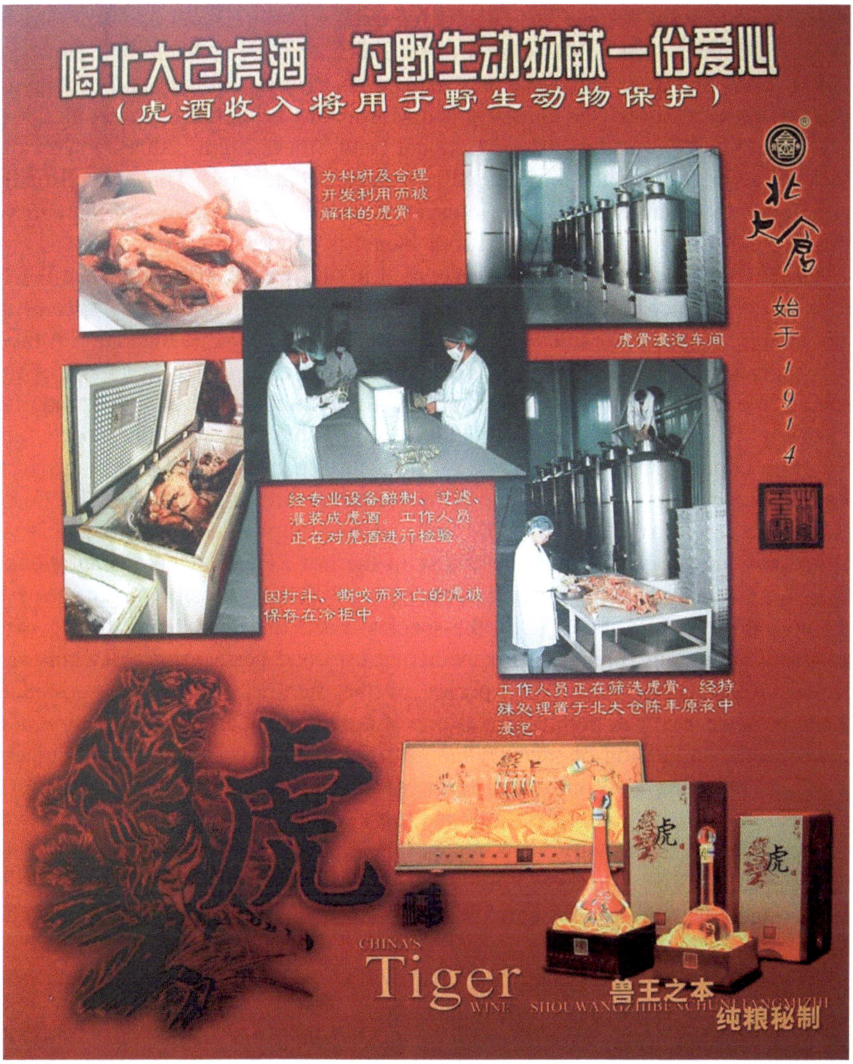


**Fig. 9.2** This poster advertising Bei Da Cang Tiger bone medicinal wine was photographed at the Badaling Safari World outside Beijing in October 2005. The poster claims that the wine is made from captive tigers that have died from fight wounds, and that income from the wine will go toward wildlife conservation. The wine appears to be illegal because Chinese law prohibits trade in and use of tiger parts and products, but Tiger breeders in China are at the forefront of a group of advocates calling for a domestic market in captive-bred tiger bone medicines to be legalized. Photographed at the Badaling Safari World outside Beijing in October 2005. *Photo* Caroline Liou, TRAFFIC. *Source* Nowell and Lingxu (2007)

requiring large home ranges, which are particularly vulnerable to habitat loss and degradation and, in particular, to edge effects; species with large body size, which tend to have low population densities, slower life histories, and larger home ranges; and species that are harvested or hunted by people (Diamond 1984; Gaston 1994; Gittleman et al. 2001; Purvis et al. 2000, 2005; Woodroffe and Ginsberg 1998). Coincidence or not, these are characteristics of the biology and ecology of most living carnivores (see Ginsberg and Macdonald 1990; Gittleman et al. 2001; Macdonald and Loveridge 2010; Macdonald and Sillero-Zubiri 2004; Nowell and Jackson 1996; Servheen et al. 1998; Sillero-Zubire et al. 2004). Thus, it is virtually impossible to separate the various impacts on carnivore populations at the current stage of environmental changes that our planet is undergoing.

Of the 108 species of carnivores recorded in our review, 105 are on the IUCN Red List, of which 14 species are classified as Near Threatened, 18 species as Vulnerable, 9 species as Endangered, and one as Critically Endangered. Seventy-five species are included in the CITES Appendices I, II, or III. For most of the species mentioned in this review, nonetheless, the medicinal usage and trade for medicinal purposes represent so an additional pressure that added by other issues has been contributing to the decline of the natural populations. Despite this fact, there are cases where traditional medicine is a major threat to carnivores. The fossa (*Cryptoprocta ferox* Bennett, 1833), an endemic Eupleridae of Madagascar, is Vulnerable now, because over the course of the last 21 years (three generations), there has been a population reduction exceeding 30% (and possibly much greater), mainly due to habitat loss (given the species' need for intact forest) along with widespread hunting (including for medicinal use) and persecution (Hawkins and Dollar 2008). Particularly for species of the families Felidae and Ursidae, medicinal use has been pointed out as one of the main threats to conservation. For instance, the utilization of tiger (*Panthera tigris*) products in traditional Chinese medicine has certainly contributed to the decline of the South China tiger and is considered by some authorities to be the most serious risk to the tiger's survival, eclipsing habitat loss (Mills et al. 1994; Tilson et al. 1997). Although almost all tiger products, from whiskers to feces, are used in age-old medicines, tiger bone is most valued as treatment for various conditions such as rheumatism (Mainka and Mills 1995).

In Asia, due to a combination of factors, the use of carnivores in traditional medicine takes on a much more dangerous connotation for the survival of exploited species. There exists a powerful underlying belief that wild animal foods and medicine are of higher quality, value, and health benefit (Venkataraman 2007). Consumption of tiger bone has increased in recent years and continues today as both human population and per capita expendable income have increased dramatically in East Asia in concert with a resurgence of interest in traditional cures (Mills et al. 1994).

Demand from retail-level consumers has played an important role in driving the tiger toward extinction (Nowell 2000). In the early 1990s, it became evident that medicinal trade in tiger bone threatened to drive the already endangered tiger *Panthera tigris* to extinction in the wild (Mills et al. 1994). In this period, the local and international trade of tiger bones reached epic proportions, decimating several wild populations. For instance, CITES data for 1990–1992 show that China exported more than 27 million units of tiger products to 26 countries/territories

(Mulliken and Haywood 1994). In South Korea, customs records from before this country acceded to CITES (in 1993) show that 8,951 kg of tiger bone were imported since the 1970s, with around half the bone derived from Indonesia (Loveridge et al. 2010).

Consumer demand for tiger and trade in their parts and derivatives supply luxury markets as well as markets for cultural and medicinal needs. Traditional Chinese medicine (TCM) uses these animal derivatives to prepare medications in two forms—as individually prepared prescriptions and as over-the-counter medicines. Most of these latter medicines are labeled as having been manufactured in China and are sold in markets worldwide. Since 1994, there has been increase in national and international investment in tiger conservation and trade control and promotion of substitutes for tiger bone. However, the progress made has brought new challenges, and some old problems remain to be tackled (Nowell 2000). Traditional Asian medicine uses tiger bone in a number of different formulae. Skin is made into magical amulets and novelties, as are teeth and claws, while tiger penis is an ingredient of allegedly powerful sexual tonics. However, many of these parts and products are fake tiger, made from the parts of other, more common animals. It is not clear what effect the plethora of fakes has on wild tiger populations. Fakes may satisfy some consumer demand which would otherwise have an impact on wild tigers, or they may stimulate increased demand for genuine tiger parts (Nowell 2000).

It is estimated that a century ago there were at least 100,000 tigers in the wild (Jackson and Kemf 1994), whose populations were distributed among nine recognized tiger subspecies: Amur Tiger—*Panthera tigris altaica* Temminck, 1844; South China Tiger—*Panthera tigris amoyensis* (Hilzheimer, 1905); Bali Tiger—*Panthera tigris balica* Schwarz, 1912; Indochinese Tiger—*Panthera tigris corbetti* Mazak, 1968; Malayan Tiger—*Panthera tigris jacksoni* Luo et al. 2004; Javan Tiger—*Panthera tigris sondaica* Temminck, 1844; Sumatran Tiger—*Panthera tigris sumatrae* Pocock, 1929; Bengal Tiger—*Panthera tigris tigris*, and Caspian Tiger—*Panthera tigris virgata* (Illiger, 1815) (IUCN 2011). Actually, the global tiger population is estimated to range from 3,000 to 5,000 in the wild and restricted to the five remaining subspecies: ssp. *altaica,* ssp. *amoyensis*, ssp. *corbetti*, ssp. *jacksoni*, ssp. *sumatrae,* and ssp. *tigris,* which persist in 13 range states: Bangladesh. Bhutan, Cambodia, China, India, Indonesia, the Lao PDR (the Lao People's Democratic Republic), Malaysia, Myanmar, Nepal, Russia, Thailand, and Vietnam (Chundawat et al. 2010; WWF 2009). With tigers so rare, demand has widened to other Asian big cat species (popularly known as leopards), 'including the leopard—*Panthera pardus* (Linnaeus, 1758), snow leopard—*Uncia uncia* (Schreber, 1775), and clouded leopard—*Neofelis nebulosa* (Griffith, 1821) (Nowell and Ling 2007; Shepherd and Magnus 2004).

Throughout the 1990s and in early 2000s, the international black market and local trade of tiger parts and other big cats persisted in range/consumer countries in Asia and in other parts of the world, despite the efforts of most countries in controlling and banning illegal trade. A study performed by TRAFFIC in China, just more than a decade after the ban on trade of products containing wild tiger parts used in traditional medicine, showed that 2.5% ($n = 13$) of 518 shops claimed to stock tiger bone and 5% of the stores visited ($n = 26$) affirmed having products obtained from bones of "leopards" (*N. nebulosa*, *P. pardus* or *U. uncia*) (Nowell and Ling 2007). In Myanmar, between January 2001 and June 2010, observations indicate that a minimum of 91 tigers, 167 leopards, 149 clouded leopards, one Asiatic lion (*Panthera leo persica*), and two snow leopards were recorded in trade (Oswell 2010). Domestic trade is still common in Sumatra (Indonesia), where 86 shops were found to have tiger parts for sale, primarily canines and claws (Shepherd and Magnus 2004), threatening the subspecies *P. tigris sumatrae*, whose population is estimated to be only 342–509. In 2007, nearly 50% of the Ha Noi residents surveyed had at some time used wild animal products (including tiger bone derivatives), of which 45% consumed such products up to three times per year (Venkataraman 2007). In traditional Chinese medicine and other Asian traditional medicine systems, tiger bone is still the most precious part used to treat conditions such as rheumatism and arthritis (TRAFFIC 1997).

Besides felines, particularly big cats, another group of carnivores that deserves mention because of their use and medicinal properties are the Ursidae. In Asia, live bears and their parts and derivatives are openly sold for medicinal purposes (Foley et al. 2011). The trade of bear gall bladders and bear bile is a serious threat to the conservation of Asian bear species (Phillips and Wilson 2002). Although precise population numbers for Asian bears are unknown, it is likely that the massive scale of commercial trade is directly responsible for population declines (Feng et al. 2009; Garshelis 2002).

The traditional medicine market is an important component in the illegal wildlife trade (Alves and Rosa 2010; Foley et al. 2011; Shepherd and Nijman 2008). Although it is a difficult task, combating the illegal trade of medicinal animals is essential for the conservation of wild populations used for medicinal purposes (Alves and Rosa 2010; Alves et al. 2007). Damaging illegal trade in the products of the natural world is often tackled by one of two opposing solutions: a total ban on trade or a controlled trade harvested from the wild (Dutton et al. 2011). In some instances a third option is available: to farm wildlife. Wildlife farming offers, at first glance, an intuitively satisfying solution: a legal trade can in principle be created by farming animals to assuage demand for wild animals, which thus need not be harvested (Alves et al. 2009c; Dutton et al. 2011). However, the success of a policy to ''farm for conservation'' is not certain, and a number of obstacles for its success require consideration (Dutton et al. 2011). In any case, the effectiveness of the CITES international commercial trade ban on Asian big cats, bear species, and several carnivores used as medicines depends upon the prohibition of domestic trade. If domestic trade is legal or even it persists illegally on a large scale, international trade will be difficult to stop.

In addition to the impact of direct loss of species, use and trade in medicinal animals may spread diseases that can affect both people and wildlife (Aguirre 2009; Still 2003). As evidenced in the present review, several organs, bones, and tissues of carnivores are used in traditional medicine, and transmission of diseases can occur through their medicinal use. Carnivores, in particular domestic cats and dogs, are host to several hundred pathogens (Cleaveland et al. 2001; Samuel and Pybus 2001) and parasitic infections (Williams and Barker 2001) that may affect humans and/or livestock. Given the potential for both conservation impacts through harvesting carnivores and of disease transmission in the preparation of medicinal products from carnivores, further action is required to ensure that harvesting is sustainable and to develop guidance on public health aspects.

Besides the fight against illegal trade, other recommendations have been pointed out as ways to minimize the impacts of the medicinal use carnivores. Direct impacts on carnivore populations could be mitigated through educational programs and identification of alternative medicines (see Lee 1999). In fact, the overlap of wild animals and plants and domestic or cultivated species opens the possibility of replacing the use of threatened species with less threatened or domestic species in traditional medicine recipes (Alves et al. 2008a; Still 2003), thereby minimizing the threat to the ecosystem. However, precaution should also be taken when suggesting the replacement of animal products (Alves and Rosa 2007a). The snow leopard (*Uncia uncia*) subject is emblematic. There is demand for snow leopard bones for use as a substitute for tiger bone in traditional Asian medicines. In China, snow leopard bones are much in demand for their medicinal use, and probably represent their primary indigenous value today (Liao and Tan 1988; Schell 1982). Leopard bone has been used for some time as a substitute for tiger bone in the manufacture of drug wine which sells far and wide in China and Southeast Asia. The poaching of snow leopards in China is, therefore, closely connected with this business (Liao and Tan 1988).

It is also noted that there is a need for improvements in laws and their rigorous enforcement. Even with strong legislation for protection of exploited species, many factors such as corruption and ineffective enforcement often allow illegal wildlife trade to occur (Alves and Rosa 2010; Foley et al. 2011). Enforcement efforts to combat the illegal trade of medicinal products are of paramount importance to ensure that trade dynamics do not threaten the survival of exploited species (Foley et al. 2011).

It is important to emphasize that any conservationist measure should consider the cultural aspects of the populations that utilize the resources, which in many cases result from old traditions and passed on through generations. For example, many Vietnamese people still believe that bear bile is a magical traditional medicine, curing many different health problems, ranging from bruises, stomachaches, and digestive ailments to serious illness such as cancer. This belief is the result of the influence of Chinese culture, as China dominated Vietnam from 179 BC until 938 (Nguyen 2006). This situation applies to various species of carnivores. For many cultures around the world, plants and animals are seen as objects to serve human needs. Thus, prohibiting the utilization of natural resources simply does not make sense to many people (Lee 1999). Therefore, the concept of wildlife conservation can be alien to them. If the need for conservation is to be accepted by people who make their livelihood from wildlife or use wildlife for necessities such as food and medicine, care should be taken to avoid what may be seen as ideologically or culturally imperialistic approaches. It is important to accept and respect differing views of the value of wildlife, while, at the same time, explaining the necessity of conservation measures (Lee 1999).

The widespread and growing popularity of traditional medicine threatens the survival of some endangered carnivore species historically used in these traditional formulations (Lee 1999). However, multiple factors have contributed to the significant decline in wild populations of carnivores in the world, and the medicinal use of carnivores must be considered together with other/extra anthropogenic pressures (Alves et al. 2009d, 2010b). As a consequence, carnivore populations are being seriously reduced throughout the world (Nowak 2005). The impacts can be direct or indirect. In the first case, the animals are killed directly for use of their by-products or because they represent a threat to humans and their domestic animals (Davies and Goodall 2005; Johnson et al. 2001; Macdonald and Kays 2005; Nash 1997; Roberton 2007; Shepherd and Magnus 2004). In the second case, habitat destruction is often considered to be the most serious threat to biological diversity worldwide, including carnivores (Michalski and Peres 2005). As predators, carnivore populations need relatively large blocks of habitat and sufficient quantities of suitable wild prey. With the increasing pace of human population growth and development over the last century, both habitat and prey for cats have declined widely. The big cats have been heavily persecuted because they are a danger to humans and livestock, and for their skins, some small cat species have been subject to heavy offtakes for the fur trade. All species have declined in range and number, but the situation is illustrated most acutely by the decline of the big cats over the last century (Nowell 2001).

Most large carnivore species are experiencing global declines, driven almost entirely by human activities and/or conflict with humans (Michalski et al. 2006). It is evident that any measure of conservation needs to consider the human dimension. Use in traditional medicine constitutes one of the different forms of exploitation of wild carnívores by people and has evident conservationist implications. Therefore, educational programs can be used to promote the use of alternatives, focusing on rural communities where inhabitants use carnivores in traditional medicine and religious practices. Additionally, measures that do not directly involve modifying the behavior of the local populations should be considered, such as controlling illegal wildlife trade. Implementation of quality measures in the use and trade of animal or their parts for medicinal purposes is also necessary.

## References

Adeola MO (1992) Importance of wild animals and their parts in the culture, religious festivals, and traditional medicine, of Nigeria. Environ Conserv 19(2):125–134

Aguirre AA (2009) Wild canids as sentinels of ecological health: a conservation medicine perspective. Parasites Vectors 2(Suppl 1):S7

Alakbarli F (2006) Medical manuscripts of Azerbaijan. Heydar Aliyev Foundation, Baku

Alderton D (1993) Wild cats of the world, 1st edn. Facts on File, New York

Almeida CFCBR, Albuquerque UP (2002) Uso e conservação de plantas e animais medicinais no Estado de Pernambuco (Nordeste do Brasil): Um estudo de caso. Interciencia 27(6):276–285

Alvarez M (2006) Fortalecimiento de la Conservación de la Naturaleza Mediante las Prácticas Tradicionales de Conservación que Desarrollan las Comunidades Indígenas Cabécares de Bajo Chirripó y Nairi Awari (Technical report), 1st edn. Asociación IXACAVAA de Desarrollo e Información Indígena, San Jose

Alves RRN (2006) Uso e comércio de animais para fins medicinais e mágico-religiosos no Norte e Nordeste do Brasil. DSc. Thesis, Universidade Federal da Paraíba, João Pessoa, Paraíba

Alves RRN (2009) Fauna used in popular medicine in Northeast Brazil. J Ethnobiol Ethnomed 5(1):1–30. doi:10.1186/1746-4269-5-1

Alves RRN, Alves H (2011) The faunal drugstore: animal-based remedies used in traditional medicines in Latin America. J Ethnobiol Ethnomed 7(1):1–9. doi:10.1186/1746-4269-7-9

Alves RRN, Pereira Filho GA (2007) Commercialization and use of snakes in North and Northeastern Brazil: implications for conservation and management. Biodivers Conserv 16:969–985. doi:10.1007/s10531-006-9036-7

Alves RRN, Rosa IL (2005) Why study the use of animal products in traditional medicines? J Ethnobiol Ethnomed 1(5):1–5. doi:10.1186/1746-4269-1

Alves RRN, Rosa IL (2006) From cnidarians to mammals: the use of animals as remedies in fishing communities in NE Brazil. J Ethnopharmacol 107:259–276

Alves RRN, Rosa IL (2007a) Biodiversity, traditional medicine and public health: where do they meet? J Ethnobiol Ethnomed 3(14):9. doi:10.1186/1746-4269-3-14

Alves RRN, Rosa IL (2007b) Zootherapeutic practices among fishing communities in North and Northeast Brazil: a comparison. J Ethnopharmacol 111:82–103. doi:10.1016/jjep2006.10.033

Alves RRN, Rosa IL (2007c) Zootherapy goes to town: the use of animal-based remedies in urban areas of NE and N Brazil. J Ethnopharmacol 113:541–555. doi:10.1016/j.jep2007.07.015

Alves RRN, Rosa IL (2010) Trade of animals used in Brazilian traditional medicine: trends and implications for conservation. Human Ecol 38(5):691–704

Alves RRN, Rosa IL, Santana GG (2007) The role of animal-derived remedies as complementary medicine in Brazil. Bioscience 57(11):949–955

Alves RRN, Vieira WLS, Santana GG (2008a) Reptiles used in traditional folk medicine: conservation implications. Biodivers Conserv 17(8):2037–2049. doi:10.1007/s10531-007-9305-0

Alves RRN, Lima HN, Tavares MC, Souto WMS, Barboza RRD, Vasconcellos A (2008b) Animal-based remedies as complementary medicines in Santa Cruz do Capibaribe Brazil. BMC Complementary Alternative Med 8:44. doi:10.1186/1472-6882-8-44

Alves RRN, Soares TC, Mourão JS (2008c) Uso de animais medicinais na comunidade de Bom Sucesso, Soledade Paraíba. Sitientibus Série Ciências Biológicas 8(2):142–147

Alves RRN, Mendonça LET, Confessor MVA, Vieira WLS, Lopez LCS (2009a) Hunting strategies used in the semi-arid region of northeastern Brazil. J Ethnobiol Ethnomed 5(12):1–50. doi:10.1186/1746-4269-5-12

Alves RRN, Barboza RRD, Souto WMS (2009b) Endangered felidae used in traditional medicine. In: Columbus A, Kuznetsov L (eds) Endangered species: new research. Nova Science Publishers, Inc., Hauppauge, pp 343–356

Alves RRN, Léo Neto NA, Santana GG, Vieira WLS, Almeida WO (2009c) Reptiles used for medicinal and magic religious purposes in Brazil. Appl Herpetol 6(3):257–274

Alves RRN, Barboza RRD, Souto WMS (2009d) Endangered felidae used in traditional medicines. Int J Med Biol Frontiers 15:357–370

Alves RRN, Oliveira MGG, Barboza RRD, Singh R, Lopez LCS (2009e) Medicinal animals as therapeutic alternative in a semi-arid region of Northeastern Brazil. Forsch Komplementmed/ Res Complement Med 16:305–312. doi:10.1159/000235855

Alves RRN, Silva CC, Barboza RRD, Souto WMS (2009f) Zootherapy as an alternative therapeutic in South America. J Altern Med Res 1(1):21–47

Alves RRN, Souto WMS, Barboza RRD (2010a) Primates in traditional folk medicine: a world overview. Mammal Rev 40(2):155–180. doi:10.1111/j.1365-2907.2010.00158.x

Alves RRN, Barboza RRD, Souto WMS (2010b) A global overview of canids used in traditional medicines. Biodivers Conserv 19(6):1513–1522

Alves RRN, Barbosa JAA, Santos SLDX, Souto WMS, Barboza RRD (2011) Animal-based remedies as complementary medicines in the semi-arid region of Northeastern Brazil. Evidence-Based Complement Altern Med 2011(ID 179876):1–15. doi:10.1093/ecam/nep134

Amir OG (2006) Wildlife trade in Somalia, 1st edn. IUCN/SSC/Antelope Specialist Group/ Northeast African Subgroup, Landau in der Pfalz

Andrade JN, Costa-Neto EM (2006) O comércio de produtos zooterápicos na cidade de Feira de Santana, Bahia, Brasil. Sitientibus 6:37–43

Animal Diversity Web (2007) Chrotogale owstoni. Animal diversity web. http://animaldiversity.ummz.umich.edu/site/accounts/information/Chrotogale_owstoni.html. Accessed 21 Dec 2008

Anyinam C (1995) Ecology and ethnomedicine: exploring links between current environmental crisis and indigenous medical practices. Soc Sci Med 40(3):321–329

Apaza L, Godoy R, Wilkie D, Byron E, Huanca T, Leonard WR, Peréz E, Reyes-García V, Vadez V (2003) Markets and the use of wild animals for traditional medicine: a case study among the Tsimané ameridians of the Bolivian rain forest. J Ethnobiol 23(1):47–64

Asa CS, Cossíos ED (2004) Sechuran fox (*Pseudalopex sechurae*). In: Sillero-Zubire C, Hoffmann M, MacDonald DW (eds) Canids: foxes, wolves, jackals and dogs: status survey and conservation action plan, 1st edn. IUCN Canid Specialist Group, Cambridge

Asa CS, Cossíos ED, Williams R (2008) Pseudalopex sechurae. IUCN. http://www.iucnredlist.org/apps/redlist/details/6925/0. Accessed 18 Sept 2011

Ashwell D, Walston N (2008) An overview of the use and trade of plants and animals in traditional medicine systems in Cambodia, 1st edn. TRAFFIC Southeast Asia Greater Mekong Programme, Ha Noi

Australian Department of Environment and Heritage (2003) Wildlife conservation and complementary medicines. Australian Government. http://www.environment.gov.au/biodiversity/trade-use/publications/traditional-medicine/index.html. Accessed 21 July 2011

Bailey N (2000) Global and historical perspectives on market hunting: implications for the African bushmeat crisis. Sustainable Development and Conservation Biology—University of Maryland and Bushmeat Crisis Task Force Silver Spring, Maryland

Baird IG (1995) Lao PDR: an overview of traditional medicines derived from wild animals and plants. TRAFFIC Southeast Asia, Kuala Lumpur

Balakrishnan M, Sreedevi MB (2007a) Captive breeding of the small Indian Civet *Vivericula indica* (É. Geoffroy Saint-Hillaire, 1803). Small Carniv Conserv 36:5–8

Balakrishnan M, Sreedevi MB (2007b) Husbandry and management of the small Indian Civet *Vivericula indica* (É. Geoffroy Saint-Hillaire, 1803) in Kerala, India. Small Carniv Conserv 36:9–13

Bangjie T (1989) Natural distribution and ecology of mustelids and vivenids in Manipur, north-eastem India. Small Carniv Conserv 1:5–8

Barbarán FR (2004) Usos mágicos, medicinales y rituales de la fauna en la Puna del Noroeste Argentino y Sur de Bolivia. Contribuciones al manejo de vida silvestre en Latinoamérica 1(1):1–26

Begossi A, Silvano RAM, Amaral BD, Oyakama OT (1999) Uses of fish and game by inhabitants of an extrative reserve (Upper Juruá, Acre, Brazil). Environ Dev Sustain 1:73–93

Bekele T, Bekele A, Balakrishnan M (2009) Feeding ecology of the African Civet *Civettictis civetta* in the Menagesha–Suba state forest, Ethiopia. Small Carniv Conserv 39:19–24

Bensky D, Gamble A, Kaptchuk T (1993) Chinese herbal medicine materia medica revised edition, 1st edn. Eastland Press, Seattle

Benson S, Nagel R (2004) Endangered species. 2nd edn. The Gale Group, Inc., a division of Thomson Learning, Inc., Farmington Hills

Bisby FA, Roskov YR, Orrell TM, Nicolson D, Paglinawan LE, Bailly N, Kirk PM, Bourgoin T, Baillargeon G, Ouvrard D (2011) Catalogue of life 2011. Species 2000 & ITIS—the integrated taxonomic information system. http://www.catalogueoflife.org/. Accessed 12 Aug 2010

Bodmer RE, Eisenberg JF, Redford KH (1997) Hunting and the likelihood of extinction of Amazonian mammals. Conserv Biol 11(2):460–466

Borge C, Castillo R (1997) Cultura y Conservación en la Talamanca indígena, 1st edn. UNED, San Jose

Bothmer BV (1953) Ptolemaic reliefs. IV. A Votive tablet. Bull Mus Fine Arts 51(286):80–84

Branch L, Silva MF (1983) Folk medicine in Alter do Chão, Pará, Brasil. Acta Amazônica 13:737–797

Brautigam AJH, Humphreys T, Hutton J (1994) Analysis of the market for tigers, bears and musk deer in the Russian far east. Traffic Bull 15(1):23–30

Breitenmoser-Wursten C, Henschel P, Sogbohossou E (2008) *Leptailurus serval*. Gland, Switzerland. http://www.iucnredlist.org/apps/redlist/details/11638/0. Accessed 1 Sept 2011

Breuer T (2002) Distribution, feeding ecology and conservation of the African wild dog (Lycaon pictus) in Northern Cameroon. 1st edn. North Carolina Zoological Park/WWF/Project Gef-Biodiverte, Yaoundé

Brooks D (1991) Some notes on terrestrial mustelids in the central Paraguayan Chaco. Small Carniv Conserv 4:5–6

But P (1995) Conservation of tigers and bears for sustainable use in Chinese medicine. In: TRAFFIC, WWF, Government of Hong Kong (eds) International symposium on traditional Chinese medicine and wildlife conservation. TRAFFIC, Hong Kong

Campbell IC, Poole C, Giesen W, Valbo-Jorgensen J (2006) Species diversity and ecology of Tonle Sap Great lake Cambodia. Aquatic Sci 68(3):355–373

Campos E (1960) Folclore do Nordeste, 1st edn. Edições O Cruzeiro, Rio de Janeiro

Carnivore Conservation (2007) The Owston's palm Civet conservation program. http://carnivoreconservation.portland.co.uk/pdf/owstons%20program%20summary.pdf. Accessed 3 Aug 2011

Carpaneto GM, Germi FP (1989) The mammals in the zoological culture of the Mbuti pygmies in North-Eastern Zaire. Hystrix 1:1–83

Chakravorty J, Meyer-Rochow VB, Ghosh S (2011) Vertebrates used for medicinal purposes by members of the Nyishi and Galo tribes in Arunachal Pradesh (North-East India). J Ethnobiol Ethnomed 7(1):1–13. doi:10.1186/1746-4269-7-13

Choudhury AU (2000) Some small carnivore records from Nagaland, India. Small Carniv Conserv 23:7–9

Chundawat RS, Habib B, Karanth U, Kawanishi K, Ahmad Khan J, Lynam T, Miquelle D, Nyhus P, Sunarto S, Tilson R, Wang S (2010) *Panthera tigris* (In: IUCN Red List of Threatened Species). IUCN. http://www.iucnredlist.org/apps/redlist/details/15955/0. Accessed August 1, 2011

CITES (2000) Summary of the status of wild populations of species listed on cites appendix I and the difficulty of keeping or breeding specimens of these species in captivity. CITES. http://www.cites.org/common/com/AC/16/E16-Inf-15.pdf. Accessed 9 Sept 2011

CITES (2001) List of animal species used in traditional medicine. CITES. http://www.cites.org/eng/com/AC/17/E17i-05Rev.doc. Accessed 14 Jan 2011

CITES (2002a) List of species traded for medicinal purposes. http://www.cites.org/eng/com/AC/17/E17-13-1.pdf. Accessed 2 May 2008

CITES (2002b) List of species traded for medicinal purposes. http://www.cites.org/eng/com/ac/18/E18-13-1.pdf. Acessed 2 May 2008

CITES (2002c) Trade in products, possibly used as medicinals, of the species listed in document AC18 Doc. 13.1. CITES. http://www.cites.org/common/com/ac/18/E18i-08.doc. Acessed 3 May 2008

CITES (2003) Document Doc. AC. 16.7.4. http://www.cites.org/eng/com/ac/16/16-7-4.pdf. Accessed 12 July 2011

CITES (2011) CITES Appendix. http://www.cites.org/eng/resources/species.html. Acessed 8 Sept 2011

Cleaveland S, Laurenson MK, Taylor LH (2001) Diseases of humans and their domestic mammals: pathogen characteristics, host range and the risk of emergence. Philos Trans R Soc Lond B Biol Sci 356(1411):991–999
Confessor MVA, Mendonça LET, Mourão JS, Alves RRN (2009) Animals to heal animals: ethnoveterinary practices in semi-arid region, Northeastern Brazil. J Ethnobiol Ethnomed 5(1):1–37
Corlett RT (2007) The impact of hunting on the mammalian fauna of tropical Asian forests. Biotropica 39(3):292–303
Costa Rica Government (2002) Estrategia Nacional de Conservación y Uso Sostenible de la Biodiversidad: Estudio Nacional de Biodiversidad (Atualización). Ministerio del Ambiente y Energía de Costa Rica. http://www.inbio.ac.cr/es/biod/minae/Estudio_Pais/estudio/indice.html. Accessed 12 Jan 2011
Costa-Neto EM (1996) Faunistc Resources used as medicines by an Afro-brazilian community from Chapada Diamantina National Park, State of Bahia-Brazil. Sitientibus 15:211–219
Costa-Neto EM (1999a) Barata é um santo remédio : introdução à zooterapia popular no estado da Bahia, 1st edn. EdUEFS, Feira de Santana
Costa-Neto EM (1999b) Healing with animals in Feira de Santana City, Bahia, Brazil. J Ethnopharmacol 65:225–230
Costa-Neto EM (1999c) Traditional use and sale of animals as medicines in Feira de Santana City, Bahia Brazil. Indig Knowl Dev Mon 7(2):6–9
Costa-Neto EM (2000) Conhecimento e usos tradicionais de recursos faunísticos por uma comunidade Afro-Brasileira. Resultados preliminares Interciencia 25(9):423–431
Costa-Neto EM (2004) Implications and applications of folk zootherapy in the state of Bahia, Northeastern Brazil. Sustain Dev 12(3):161–174
Costa-Neto EM (2005) Animal-based medicines: biological prospection and the sustainable use of zootherapeutic resources. Anais da Academia Brasileira de Ciências 77(1):33–43
Costa-Neto EM, Oliveira MVM (2000) Cockroach is good for asthma: zootherapeutic practices in Northeastern Brazil. Human Ecol Rev 7(2):41–51
Cunningham AB, Zondi AS (1991) Use of animal parts for the commercial trade in traditional medicines. Institute of Natural Resources, Univeristy of Natal, Pietermaritzburg
Davies B, Goodall J (2005) Black market: inside the endangered species trade in Asia. Mandala Pub Group, San Rafael
Diamond JM (1984) 'Normal' extinctions of isolated populations. In: Nitecki MH (ed) Extinctions. Chicago University Press, Chicago, pp 191–246
Do Tat Loi (2004) Vietnamese medicinal plants and remedies/substances, 12th edn. Health Publishing House, Ha Noi
Duckworth JW, Salter RE, Khounboline K (1999) Wildlife in Lao PDR: 1999 status report, 1st edn. IUCN, Gland
Duckworth JW, Lee BPY-H, Meijaard E, Meiri S (2006) The Malay Weasel Mustela nudipes: distribution, natural history and a global conservation status review. Small Carniv Conserv 34(35):2–21
Durbin LS, Venkataraman S, Hedges W, Duckworth W (2004) *Cuon alpinus* (Pallas, 1811). In: Sillero-Zubire C, Hoffmann M, MacDonald DW (eds) Canids: foxes, wolves, jackals and dogs: status survey and conservation action plan. 1st edn. IUCN Canid Specialist Group, Cambridge, pp 210–218
Durbin LS, Hedges S, Duckworth JW, Tyson M, Lyenga A, Venkataraman A (2008) *Cuon alpinus*. IUCN. http://www.iucnredlist.org/apps/redlist/details/5953/0. Accessed 18 Sept 2011
Dutton AJ, Hepburn C, Macdonald DW (2011) A stated preference investigation into the Chinese Demand for farmed vs. wild bear bile. PLoS ONE 6(7 e21243):1–10
EIA (2004) The tiger skin trail. Environmental Investigation Agency, London
El-Kamali HH (2000) Folk medicinal use of some animal products in Central Sudan. J Ethnopharmacol 72:279–282
Enqin Z (1991) Rare Chinese materia medica, 1st edn. Science Press, Shanghai

Espinoza EP, Mann MJ, Yates B, Elliot L (1994) Preliminary forensic analysis of the wildlife content of patented Asian medicines. In: Gaski AL, Johnson KA (eds) Prescription for extinction: endangered species and patented oriental medicines in trade. TRAFFIC USA, Washington DC, pp 94–132

Feng Y, Siu K, Wang N, Ng K-M, Tsao S-W, Nagamatsu T, Tong Y (2009) Bear bile: dilemma of traditional medicinal use and animal protection. J Ethnobiol Ethnomed 5(2):1–45. doi:10.1186/1746-4269-5-2

Ferreira FS, Brito S, Ribeiro S, Saraiva A, Almeida W, Alves RRN (2009a) Animal-based folk remedies sold in public markets in Crato and Juazeiro do Norte, Ceara, Brazil. BMC Complement Altern Med 9(1):1–17

Ferreira FS, Brito S, Ribeiro S, Almeida W, Alves RRN (2009b) Zootherapeutics utilized by residents of the community Poco Dantas, Crato-CE, Brazil. J Ethnobiol Ethnomed 5(1): 1–21

Figueiredo N (1994) Os 'bichos' que curam: os animais e a medicina 'folk' em Belém do Pará. Boletim do Museu Paraense Emílio Göeldi 10(1):75–91

Flaws B (1994) Chinese medicinal wines & elixirs. Blue Poppy Press, Boulder

Foley KE, Stengel CJ, Shepherd CR (2011) Pills, powders, vials and flakes: the bear bile trade in Asia. TRAFFIC Southeast Asia, Petaling Jaya

Galster S (1996) Russia's final roar. Criminal threats to the Siberian Tiger and local communities: an inside look at the new fight for survival. 1st edn. The Investigative Network (Global Survival Network), Washington, DC

Garshelis DL (2002) Misconceptions, ironies, and uncertainties regarding trends in bear populations: invited paper. Ursus 13:321–334

Gaston KJ (1994) Rarity, 1st edn. Springer, Berlin

Ginsberg JR, Macdonald DW (1990) Foxes, wolves, jackals, and dogs: an action plan for the conservation of canids, 1st edn. IUCN, Gland

Gittleman JL, Funk SM, Macdonald DW, Wayne RK (2001) Why 'carnivore conservation'? In: Gittleman JL, Funk SM, Macdonald DW, Wayne RK (eds) Carnivore conservation, 1st edn. Cambridge University Press, Cambridge, UK, pp 1–8

Guo Y, Zou X, Chen Y, Wang D, Wang S (1997) Sustainability of wildlife use in traditional Chinese medicine. In: Xie Y (ed) Conserving China's biodiversity. BWG/China Council for International Cooperation on Environment and Development, Beijing, pp 190–220

Hanazaki N (2001) Ecologia de Caiçaras: Uso de recurso e dieta. PhD. Thesis, Universidade Estadual de Campinas, Campinas, SP, Brazil

Hanazaki N, Alves R, Begossi A (2009) Hunting and use of terrestrial fauna used by Caicaras from the Atlantic Forest coast (Brazil). J Ethnobiol Ethnomed 5(1):1–36

Hawkins AFA, Dollar L (2008) *Cryptoprocta ferox* (In: IUCN Red List of threatened species 2011). IUCN/SSC http://www.iucnredlist.org/apps/redlist/details/5760/0/. Accessed 20 Sept 2011

Heng K (1999) Interviews with traditional medicine pharmacists and traders in Cambodia. Cat Action Treas, Los Gatos

Highley K, Highley S (1994) Bear farming and trade in China and Taiwan. Humane Society of the United States, Washington

Hoffmann M (2008) *Herpestes sanguineus* (In: IUCN Red List of Threatened Species 2011). IUCN. http://www.iucnredlist.org/apps/redlist/details/41606/0. Accessed 11 Aug 2011

Honer O, Holekamp KE, Mills G (2008) *Crocuta crocuta* (In: IUCN Red list of threatened species 2011). IUCN. http://www.iucnredlist.org/apps/redlist/details/5674/0. Accessed 23 Aug 2011

IUCN (2011) IUCN Red list of threatened species. http://www.iucnredlist.org. Accessed 30 Sept 2011

Jackson P, Kemf E (1994) Wanted alive: tigers in the wild, 1st edn. WWF International, Gland

Jiangsu New Medical College (2002) The Chinese materia medica dictionary, 2nd edn. Jiangsu New Medical College, Shanghai

Johnsingh A, Jhala Y (2004) Vulpes bengalensis (Shaw 1800). In: Sillero-Zubire C, Hoffmann M, MacDonald DW (eds) Canids: foxes, wolves, jackals and dogs: status survey and conservation action plan. IUCN Canid Specialist Group, Cambridge, UK, pp 219–222

Johnson WE, Eizirik E, Lento GM (2001) The control, exploitation and conservation of carnivores. In: Gittleman JL, Funk S, MacDonald DW, Wayne RK (eds) Carnivore conservation. Cambridge University Press, Cambridge, pp 196–220

Jun L, Defu H, Liangliang Y (2010) Legal status and management of cat species in China. Cat News Special Issue 5:5–6

Kang S, Phipps M (2003) A question of attitude: South Korea's traditional medicine practitioners and wildlife conservation, 1st edn. TRAFFIC East Asia, Hong Kong

Khalid HS, El-Kamali HH, Elmanan AMA (2007) Trade of Sudanese natural medicinals and their role in human and wildlife health care. Cropwatch Newsl 10:1–15

Knights P (1996) From forest to pharmacy. The Global Underground Trade in Bear Parts. The Investigative Network and The Humane Society of the United States, Washington, DC

Kruuk H (2002) Hunter and hunted: relationships between carnivores and people. Cambridge University Press, Cambridge

Lachner-Eitzenberger UM (2006) Medicina traditional Ergebnisse einer ethnomedizinischen Feldstudie bei den Chiquitano in Bolivien. PhD thesis, Ludwig-Maximilians-Universität zu München, Munich, Germany

Laider L, Laider K (1996) China's threatened wildlife, 1st edn. Blandford A. Cassell Imprint, London

Lal Mohan RS (1994) Trade in civetone from the Indian small Civet (*Viverricula indica*) from Malabar, India). Small Carniv Conserv 10:13

Lee TH (1995) Measures and current usage status in Korea of the medicine made with endangered species. In: TRAFFIC (ed) Seminar on international trade in endangered wild fauna and flora, Seoul, Republic of Korea, 1995. TRAFFIC East Asia/Ministry of Environment of South Korea/Ministry of Health and Welfare of South Korea, unpaginated

Lee SKH (1999) Trade in traditional medicine using endangered species: an international context. Paper presented at the 2nd Australian symposium on traditional medicine and wildlife conservation, Melbourne, Australia

Lenko K, Papavero N (1996) Os insetos no folclore, 2nd edn. Editora Plêiade & FAPESP, São Paulo

Lev E (2003) Traditional healing with animals (zootherapy): medieval to present-day Levantine practice. J Ethnopharmacol 85:107–118. doi:10.1016/S0378-8741(02)00377-X

Lev E (2006) Healing with animals in the Levant from the 10th to the 18th cent. J Ethnobiol Ethnomed 2(11):9. doi:10.1186/1746-4269-2-11

Lev E, Amar Z (2002) Ethnopharmacological survey of traditional drugs sold in the Kingdom of Jordan. J Ethnopharmacol 82:131–145

Liao Y, Tan B (1988) A preliminary study of the geographic distribution of snow leopards in China. In: Seattle (ed) Proceedings of the 5th international snow leopard symposium, pp 51–63

Long B (2001) Tiger conservation and survey training for the Central Vietnam tiger corridor, 1st edn. WWF, Ha Noi

Long B, Hon J, Azlan MJ, Duckworth JW (2008) *Mydaus javanensis* (In: IUCN Red list of threatened species 2011). IUCN. http://www.iucnredlist.org/apps/redlist/details/41628/0. Accessed 2 Sept 2011

Loveridge AJ, Nel JAJ (2004) Black-backed jackal (*Canis mesomelas*). In: Sillero-Zubire C, Hoffmann M, MacDonald DW (eds) Canids: foxes, wolves, jackals and dogs: status survey and conservation action plan, 1st edn. IUCN Canid Specialist Group, Cambridge, UK

Loveridge AJ, Nel JAJ (2008) *Canis mesomelas*. IUCN Canid Specialist Group. http://www.iucnredlist.org/details/3755. Accessed 17 May 2009

Loveridge AJ, Wang SW, Laurence Frank G, Seidensticker J (2010) People and wild felids: conservation of cats and management of conflicts. In: Macdonald DW, Loveridge AJ (eds)

The biology and conservation of wild felids, 1st edn. Oxford University Press, Oxford, UK/ New York, pp 161–197
Lucherini M, Pessino M, Farias AA (2004) *Canis mesomelas*. In: Sillero-Zubire C, Hoffmann M, MacDonald DW (eds) Canids: foxes, wolves, jackals and dogs: status survey and conservation action plan, 1st edn. IUCN Canid Specialist Group, Cambridge, UK
Luck Mojo Curio Co. (1995) Raccoon penis bones. http://www.luckymojo.com/raccoonpenis. 452html. Accessed 6 Aug 2011
Macdonald DW, Kays RW (2005) Carnivores of the world: an introduction. In: Nowak RM (ed) Walker's carnivores of the world. The Johns Hopkins University Press, London, pp 1–67
Macdonald DW, Loveridge AJ (2010) The biology and conservation of wild felids, 1st edn. Oxford University Press, Oxford, UK/New York
Macdonald DW, Sillero-Zubiri C (2004) The biology and conservation of wild canids, 1st edn. Oxford University Press, New York
Mahawar MM, Jaroli DP (2006) Animals and their products utilized as medicines by the inhabitants surrounding the Ranthambhore National Park, India. J Ethnobiol Ethnomed 2(46):1–5. doi:10.1186/1746-4269-2-46
Mahawar MM, Jaroli DP (2008) Traditional zootherapeutic studies in India: a review. J Ethnobiol Ethnomed 4(1):1–17
Mainka SA, Mills JA (1995) Wildlife and traditional Chinese medicine: supply and demand for wildlife species. J Zoo Wildlife Med 26(2):193–200
Maliehe TM (1993) Non-wood forest products in South Africa (paper prepared for the Regional Expert Consultation on Non-wood forest products for English-speaking African Countries in Tanzania), 1st edn. Commonwealth Science Council, Arusha
Marques JGW (1995) Pescando Pescadores: Etnoecologia abrangente no baixo São Francisco Alagoano. 1 st edn. NUPAUB/USP, São Paulo
Marshall NT (1998) Searching for a cure, conservation of medicinal wildlife resources in East and Southern Africa, 1st edn. TRAFFIC International, Cambridge, UK
Martin ES, Phipps M (1996) A review of the wild animals trade in Cambodia. Traffic Bull 16(2):45–60
McKinney ML (1997) Extinction vulnerability and selectivity: combining ecological and paleontological views. Annu Rev Ecol Sys 28:495–516. doi:10.1146/annurev.ecolsys.28.1.495
Michalski F, Peres CA (2005) Anthropogenic determinants of primate and carnivore local extinctions in a fragmented forest landscape of southern Amazonia. Biol Conserv 124(3):383–396
Michalski F, Boulhosa RLP, Faria A, Peres CA (2006) Human–wildlife conflicts in a fragmented Amazonian forest landscape: determinants of large felid depredation on livestock. Animal Conserv 9(2):179–188
Mills MGL, Hofer H (1998) Hyaenas: status survey and conservation action plan, 1st edn. IUCN/ SSC Hyaena Specialist Group, Gland
Mills JA, Jackson P, Gray J (1994) Killed for a cure: a review of the worldwide trade in tiger bone. Traffic International, Cambridge
Mills J, Chan S, Ishihara A, Kelso BJ (1995) The bear facts: the East Asian market for bear gallbladder, 1st edn. TRAFFIC International, Cambridge, UK
Mittermeier RA, Valladares-Pádua C, Rylands AB, Eudey AA, Butynski TM, Ganzhorn JU, Kormos R, Aguiar JM, Walker S (2007) Primates in peril: the world's 25 most endangered primates, 2004–2006. Primate Conserv 22:1–40
Monroy-Vilchis O, Cabrera L, Suárez P, Zarco-González MM, Rodríguez-Soto C, Urios V (2008) Uso tradicional de vertebrados silvestres en la Sierra Nanchititla, México. Interciencia 33(4):308–313
Morales-Mavil JE, Villa-Cañedo JT (1998) Notas sobre el uso de la fauna silvestre en Catemaco, Veracruz, México. Acta Zoologica Mexicana 73:127–143
Moura FP (2002) Entre o peixe e o dendê:etnoecologia do povo dos Marimbús (Chapada Diamantina-BA). Doutorado, Universidade Federal de São Carlos, São Carlos

Moura FBP, Marques JGW (2008) Zooterapia popular na Chapada Diamantina: uma Medicina incidental? Ciência Saúde Coletiva 13:2179–2188
Moutou F (2004) The possible role of oriental Civets in the recent SARS epidemic. Small Carniv Conserv 31:10–12
Mulliken T, Haywood M (1994) Recent data on trade in rhino and tiger products, 1988–1992. Traffic Bull 14(3):99–106
Mzimela T (1995) The muthi business. Indic South Afr 12(2):81–85
Nandini R, Mudappa D (2010) Mystery or myth: a review of history and conservation status of the Malabar Civet *Viverra civettina* Blyth, 1862. Small Carniv Conserv 43:47–59
Nandini R, Mudappa D, Van Rompaey H (2002) Distribution and status of the Brown Palm Civet in the Western Ghats, South India. Small Carniv Conserv 27:6–11
Nash SV (1997) Fin, feather, scale and skin: observations on the wildlife trade in Lao PDR and Vietnam. TRAFFIC Southeast Asia, Malaysia
Negi CS, Palyal V (2007) Traditional uses of animal and animal products in medicine and rituals by the Shoka Tribes of District Pithoragarh, Uttaranchal, India. Ethno-Med 1(1):47–54
Ng J, Nemora (2007) Tiger trade revisited in Sumatra, Indonesia, 1st edn. TRAFFIC Southeast Asia, Petaling Jaya
Nguyen NQ (2006) History of Vietnam, 1st edn. Education Publishing House, Ha Noi
Nijman V (2005) In full swing: an assessment of trade in orangutans and gibbons on Java and Bali, Indonesia, 1st edn. TRAFFIC Southeast Asia, Petaling Jaya, Selangor, Malaysia
Nilsson G (1990) The Endangered species handbook, 1st edn. The Animal Welfare Institute, Washington, D.C
Nowak RM (2005) Walker's carnivores of the world. Johns Hopkins University Press, Baltimore
Nowell K (2000) Far from a cure: The tiger trade revisited, 1st edn. TRAFFIC Southeast Asia, Kuala Lumpur
Nowell K (2001) Status and conservation of the Felidae, Handbook of the Mammals of the World. Lynx Edicions, Madrid
Nowell K, Jackson P (1996) Wild cats: status survey and conservation action plan, 1st edn. IUCN, Gland
Nowell K, Ling X (2007) Taming the tiger trade: China's markets for wild and captive tiger products since the 1993 domestic trade ban, 1st edn. TRAFFIC East Asia, Hong Kong
Ntiamoa-Baidu Y (1992) Local perceptions and value of wildlife reserves to communities in the vicinity of forest national parks in Western Ghana. In: Oxford Environment and Development Group (ed) Protected area development in South-west Ghana Report. 1st edn. Oxford Environment and Development Group, Oxford, UK
Oliveira ES, Torres DF, Brooks SE, Alves RRN (2010) The medicinal animal markets in the metropolitan region of Natal City, Northeastern Brazil. J Ethnopharmacol 130(1):54–60
Olsson A (2007) Mitigating threats to Otters in Cambodia: Concept note. Furget-Me-Not. http://www.furgetmenot.org.uk/documents/Cambodia_Otter_fur_concept_note_Nov_2007.pdf. Accessed 18 Aug 2011
Oswell AH (2010) The big cat trade in Myanmar and Thailand, 1st edn. TRAFFIC Southeast Asia, Petaling Jaya, Selangor, Malaysia
Padmanabham P, Sujana KA (2008) Animal products in traditional medicine from Attappady hills of Western Ghats. Indian J Tradit Knowl 7(2):326–329
Pantel S, Anak NA (2010) A preliminary assessment of pangolin trade in Sabah, 1st edn. TRAFFIC Southeast Asia, Petaling Jaya
Patil SH (2003) Ethno-medico-zoological studies on Nandurbar district of Maharashtra. Indian J Tradit Knowl 2(3):297–299
Phillips T, Wilson PJ (2002) The bear bile business: the global trade in bear products from China to Asia and beyond. WSPA Rep, London
Prassack KA (2003) Hunting across a mosaic landscape: The effects of fragmentation, edge effect, and intraguild interactions as an explanation for low population densities of *Lycaon pictus*. Rutgers, The State University of New Jersey. http://crssa.rutgers.edu/courses/lse/Web_Patch/final/Keri/Kero_Final.htm. Accessed 2 Oct 2011

Pujol J (1993) Natur Africa: The herbalist handbook. 1st edn. Jean Pujol Natural Healers Foundation, Durban

Purvis A, Gittleman JL, Cowlishaw G, Mace GM (2000) Predicting extinction risk in declining species. Proc R Soc Lond B Biol Sci 267(1456):1947–1952. doi:10.1098/rspb.2000.1234

Purvis A, Gittleman JL, Brooks T (2005) Phylogeny and conservation, 1st edn. Cambridge University Press, Cambridge, UK

Ramakantha V (1994) Natural distribution and ecology of mustelids and vivenids in Manipur, north-eastem India. Small Carniv Conserv 11:16–19

Rasmussen GSA (1999) Livestock predation by the painted hunting dog *Lycaon pictus* in a cattle ranching region of Zimbabwe: a case study. Biol Conserv 88(1):133–139. doi:10.1016/s0006-3207(98)00006-8

Ray JC, Hunter L, Zigouris j (2005) Setting conservation and research priorities for larger African carnivores WCS working paper No. 24, Wildlife Conservation Society, New York

Ray J, Gaubert P, Hoffmann M (2008) Civettictis civetta (In: IUCN Red list of threatened species 2011). IUCN. http://www.iucnredlist.org/apps/redlist/details/41695/0. Accessed 23 Aug 2011

Redford KH, Ray JC, Boitani L (2011) Mapping and navigating mammalian conservation: from analysis to action. Philos Trans Royal Soc B Biol Sci 366(1578):2712–2721

Republic of South Africa Department of Environmental Affairs (2004) Mammals list of threatened species (Draft). Cape Town, South Africa. http://www.environment.gov.za/Documents/Documents/2004Dec15_12/Revised_NEMBAListing_Mammals.pdf. Accessed 2 Dec 2010

Ribeiro GC, Pereira JPR, Docio L, Alarcon DT, Schiavetti A (2010) Zooterápicos utilizados no sul da Bahia. In: Costa-Neto EM, Alves RRN (eds) Zooterapia: Os Animais na Medicina Popular Brasileira, vol 2., 1NUPEEA, Recife, pp 221–242

Rinpoche VR, Kunzang J (1973) Tibetan medicine, 1st edn. University of California Press, Berkeley

Roa MA (1997) XV Simposio sobre fauna Silvestre, 1st edn. Facultad de Medicina Verein aria y Zootecnia, UNAM, Mexico City

Roberton SI (2007) The status and conservation of small carnivores in Vietnam. PhD thesis, University of East Anglia

Rodden M, Rodrigues F, Bestelmeyer S (2004) Maned wolf, *Chrysocyon brachyurus*. In: Sillero-Zubire C, Hoffmann M, MacDonald DW (eds) Canids: foxes, wolves, jackals and dogs: status survey and conservation action plan, 1st edn. IUCN Canid Specialist Group, Cambridge, UK

Rodden M, Rodrigues F, Bestelmeyer S (2008) *Chrysocyon brachyurus*. IUCN. http://www.iucnredlist.org/details/4819. Accessed 17 May 2009

Rodrigues E (2006) Plants and animals utilized as medicines in the Jaú National Park (JNP), Brazilian Amazon. Phytother Res 20:378–391. doi:10.1002/ptr.1866

Ross S, Murdoch J, Mallon D, Sanderson J, Barashkova A (2008) *Otocolobus manul*. IUCN. http://www.iucnredlist.org/apps/redlist/details/15640/0. Accessed 23 July 2011

Rowe–Rowe DT (1990) Action plan for African otters. In: Foster-Turley P, Macdonald S, Mason C (eds) Otters: an action plan for their conservation. IUCN, Gland, pp 41–51

Rowe–Rowe DT (1995) Distribution and status of African otters. Habitat 11:8–10

Samuel WM, Pybus MJ (2001) Parasitic diseases of wild mammals. Wiley-Blackwell, London

Sanderson J, Khan JA, Grassman L, Mallon DP (2008) *Neofelis nebulosa* (In: IUCN Red list of threatened species). IUCN. http://www.iucnredlist.org/apps/redlist/details/14519/0. Accessed 1 Aug 2011

Schell J (1982) The Fate of the earth, 1st edn. Alfred A. Knopf, New York

Schmidt S, Altanchimeg C, Tungalagtuya K, Narangerel Y, Ganchimeg D, Erdenechimeg B, Dambayuren S, Battogoo D (2006) Depleting natural wealth—perpetuating poverty: rural livelihoods and access to forest resources in Mongolia, 1st edn. FAO, Rome

Servheen C (1999) The trade in bears and bear parts. In: Servheen C, Herrero S, Peyton B (eds) Bears: status survey and conservation action plan, pp 33–38

Servheen C, Herrero S, Peyton B (1998) Bears: status survey and conservation action plan. 1st edn. IUCN, Gland, Switzerland, and Cambridge, UK

Shepherd CR (2001) Observations on wildlife trade at Golden Rock, Myanmar. Traffic Bull 19(1):7–10
Shepherd CR, Magnus N (2004) Nowhere to hide: The trade in Sumatran Tiger. TRAFFIC Southeast Asia, Petaling Jaya, Selangor, Malaysia
Shepherd CR, Nijman V (2008) The trade in bear parts from Myanmar: an illustration of the ineffectiveness of enforcement of international wildlife trade regulations. Biodivers Conserv 17(1):35–42
Sillero-Zubire C, Hoffmann M, MacDonald DW (2004) Canids: foxes, wolves, jackals and dogs: status survey and conservation action plan, 1st edn. IUCN Canid Specialist Group, Cambridge, UK
Sillero-Zubiri C, Hoffmann M, Macdonald DW (2004) Canids, foxes, wolves, jackals, and dogs. World Conservation Union, Gland
Silva AL (2008) Animais medicinais: conhecimento e uso entre as populações ribeirinhas do rio Negro, Amazonas, Brasil. Boletim do Museu Paraense Emílio Göeldi 3(3):343–357
Simelane TS, Kerley GIH (1998) The conservation implications of the use of indigenous vertebrates by traditional healers in Xhosa communities. South Afr J Wildlife Res 28:121–126
Sodeinde OA, Soewu DA (1999) Pilot study of the traditional medicine trade in Nigeria. Traffic Bull 18(1):35–40
Soewu DA (2008) Wild animals in ethnozoological practices among the Yorubas of southwestern Nigeria and the implications for biodiversity conservation. Afr J Agric Res 3(6):421–427
Solanki GS, Chutia P (2004) Ethno zoological and socio-cultural aspects of Monpas of Arunachal Pradesh. Human Ecol 15(4):251–254
Souto FJB, Andrade CTS, Souza AF (2001) Uma abordagem etnoecológica sobre a zooterapia na medicina popular em Andaraí, Chapada Diamantina, Bahia. I Encontro Baiano de Etnobiologia e Etnoecologia, Feira de Santana, Bahia, 1999. UEFS, Feira de Santana, Bahia, pp 1–18
Souto WMS, Mourão JS, Barboza RRD, Alves RRN (2011) Parallels between zootherapeutic practices in ethnoveterinary and human complementary medicine in NE Brazil. J Ethnopharmacol 134:753–767. doi:10.1016/j.jep2011.01.041
Souza RF (2008) Medicina e fauna silvestre em Minas Gerais no século XVIII. Varia hist 24(39):273–291
Sreedevi MB (2001) A study on certain aspects of breeding and behaviour of the Small Indian Civet *Viverricula indica* (Desmarest). Ph.D. thesis, University of Kerala, Thiruvananthapuram, India
Still J (2003) Use of animal products in traditional Chinese medicine: environmental impact and health hazards. Complement Ther Med 11:118–122. doi:10.1016/S0965-2299(03)00055-4
Stuart C, Stuart T, Hoffmann M (2008) *Poecilogale albinucha* (In: IUCN Red list of threatened species 2011). IUCN. http://www.iucnredlist.org/apps/redlist/details/41662/0/. Accessed 20 Sept 2011
Tejada R, Chao E, Gómez H, Painter REL, Wallace RB (2006) Evaluación sobre el uso de la fauna silvestre en la Tierra Comunitaria de Origen Tacana, Bolivia. Ecología en Bolivia 41(2):138–148
Theile S (2003) Fading footsteps: the killing and trade of snow leopards, 1st edn. TRAFFIC International, Cambridge, UK
Tilson R, Traylor Holzer K, Jiang QM (1997) The decline and impending extinction of the South China tiger. Oryx 31(4):243–252
Tirira DG, González-Maya JF (2009) Current state of knowledge of the least-known carnivore in South America: Colombian Weasel Mustela felipei in Colombia and Ecuador. Small Carniv Conserv 41:46–50
TRAFFIC (1997) Tiger Progress? The Response to CITES—Resolution Conf. 9.13. TRAFFIC International. http://www.traffic.org/species-reports/traffic_species_mammals17.pdf. Accessed 22 Sept 2010
TRAFFIC (1998) A survey of wildlife trade in Russia. TRAFFIC. http://www.traffic.org/general-reports/traffic_pub_trade11.pdf. Accessed 18 Sept 2011

TRAFFIC (1999) Seizures and prosecutions. Traffic Bull 18(1):1–7

TRAFFIC, WWF (2002) A CITES Priority: Conservation of tigers and other Asian big cats. http://www.traffic.org/cop12/ABC_CoP12.pdf. Acessed 19 Aug 2010

Tratado de Cooperacion Amazonica (1999) Conservacion y uso de la fauna silvestre en areas protegidas de la Amazonia. 1st edn. Secretaria Pro Tempore (Venezuela)/FAO, Caracas, Venezuela

Valle YBM (2007) Vaqueiros do Sítio do Meio (Lagoa Grande/PE) & Mamíferos Nativos das Caatingas Pernambucanas: Percepções e Interações MSc. Thesis. Universidade Federal de Pernambuco, Recife

Van Rompaey H, Gaubert P, Hoffmann M (2008) *Nandinia binotata* (In: IUCN Red list of Threatened species 2011). IUCN. http://www.iucnredlist.org/apps/redlist/details/41589/0. Accessed 1 Aug 2011

Van NDN, Tap N (2008) An overview of the use of plants and animals in traditional medicine systems in Viet Nam, 1st edn. TRAFFIC Southeast Asia, Greater Mekong Programme, Ha Noi

Vargas LMA (2002) Estudio comparativo de la caza y uso de mamíferos en dos comunidades Tsimane' undergraduate thesis. Universidad Mayor de San Andrés, La Paz

Vázquez PE, Méndez RM, Guiascón ÓGR, Piñera EJN (2006) Uso medicinal de la fauna silvestre en los Altos de Chiapas, México. Interciencia 31(7):491–499

Venkataraman B (2007) A matter of attitude: The consumption of wild animal products in Ha Noi, Viet Nam, 1st edn. TRAFFIC Southeast Asia, Ha Noi

Vietnamese Ministry of Agriculture (2003) Conservation and management of wildlife in Vietnam, 1st edn. Vietnam Government, Hanoi

Walston N (2005) An overview of the use of Cambodia's wild plants and animals in traditional medicine systems. TRAFFIC Southeast Asia—Indochina, 1st edn. TRAFFIC, Petaling Jaya

Watt JM, Breyer-Brandwijk MG (1962) The medicinal and poisonous plants of Southern and Eastern Africa, 2nd edn. Publisher E and Livingstone LTD, Edinburg and London, UK

Whiting MJ, Williams VL, Hibbitts TJ (2011) Animals traded for traditional medicine at the Faraday market in South Africa: species diversity and conservation implications. J Zool 284:84–96

Wild Cat Conservation (2011) Andean Cat *Leopardus jacobita* International Society for endangered cats (ISEC) Canada. http://www.wildcatconservation.org/Andean-Cat.html. Accessed 6 July 2011

Williams ES, Barker IK (2001) Infectious diseases of wild mammals. Wiley-Blackwell, Iowa State University Press, Ames

Wilson DE, Reeder DM (2005) Mammal species of the world: a taxonomic and geographic reference, 3rd edn. Johns Hopkins University Press, Baltimore

Winter KA (2002) Subsistence use of terrestrial and aquatic animal resources in the Tierra comunitaria de origen Itonama of Lowland Bolivia. PhD. thesis, University of Georgia, Athens, GA, USA

Won C, Smith K (1999) History and current status of mammals of the Korean Peninsula. Mammal Rev 29:3–33

Woodroffe R, Ginsberg JR (1998) Edge effects and the extinction of populations inside protected areas. Science 280(5372):2126–2128

Woodroffe R, McNutt JW, Mills MGL (2004) African wild dog *Lycaon pictus* (Temminck, 1820). In: Sillero-Zubire C, Hoffmann M, MacDonald DW (eds) Canids: foxes, wolves, jackals and dogs: status survey and conservation action plan. 1st edn. IUCN Canid Specialist Group, Cambridge, UK

World Association of Zoos and Aquariums (2007) Conservation projetcs. http://www.waza.org/conservation/projects/projects.php?id=57. Accessed 20 July 2011

World Health Organization (2002) Traditional Medicine Strategy 2002–2005, 1st edn. WHO, Geneva

Wu J (2006) World without borders: The internet as hotspot of Illegal wildlife trade. TRAFFIC East Asia Taipei, Taipei

WWF (2009) Tiger Farms: A ticket to extinction. http://www.worldwildlife.org/what/globalmarkets/wildlifetrade/tigerfarms.html. WWF. Accessed October 12, 2011

Yanfa L, Bangjie T (1988) A preliminary study on the geographical distribution of snow leopards in China. In: Freeman H (ed) Proc fifth Int Snow Leopard Symp Seattle. Wildlife Institute of India/International Snow Leopard Trust, Bombay, India, pp 1–11

Yates BC (2005) Distinguishing real vs fake tiger penises. Identif Guides Wildlife Law Enforc 6:1–8

Yiming L, Dianmo L (1998) The dynamics of trade in live wildlife across the Guangxi border between China and Vietnam during 1993–1996 and its control strategies. Biodivers Conserv 7(7):895–914

Yi-Ming L, Zenxiang G, Xinhai L, Sung W, Niemelä J (2000) Illegal wildlife trade in the Himalayan region of China. Biodivers Conserv 9(7):901–918

Yinfeng G, Xeying Z, Chen Y, Di W, Sung W (1997) Sustainability of wildlife in traditional Chinese medicine. In: Xie Y, Biodiversity Working Group, hui Zhjyfzgjhzwy (eds) Conserving China's biodiversity. BWG, Beijing, pp 190–216

Zhang Enqin (1996) The Chinese materia medica, Shanghai, China

# Chapter 10
# Seahorses in Traditional Medicines: A Global Overview

**Ierecê Lucena Rosa, Gabriela Rocha Defavari, Rômulo Romeu Nóbrega Alves and Tacyana Pereira Ribeiro Oliveira**

**Abstract** Seahorses (*Hippocampus* spp.) are under threat worldwide because of the global demand for them and products arising from them, and by destruction of their habitats. A landmark report published in 1996 exposed the massive trade in seahorses, involving at least 32 countries, and established that the market for these fishes was threatening their wild populations. A subsequent set of surveys identified a much larger number of countries (nearly 80) involved in the seahorse trade, thus furthering concern over its sustainability. As a means of ensuring that trade is not detrimental to the survival of wild populations, all species of *Hippocampus* were added to Appendix II of the Convention on International Trade in Endangered Species of Wild Fauna and Flora (CITES), effective May 2004. Given that traditional Chinese medicine (TCM) and its derivatives account for the largest consumption of seahorses, further understanding of traditional medicinal uses of seahorses is central to conservation efforts, both on national and international levels. This chapter summarizes knowledge about the medicinal uses of seahorses, focusing on the species used, illnesses and ailments to which species are prescribed, and implications for conservation.

I. L. Rosa (✉) · G. R. Defavari · T. P. R. Oliveira
Departamento de Sistemática e Ecologia, Universidade Federal da Paraíba,
João Pessoa-PB 58059-970, Brazil
e-mail: ierece@dse.ufpb.br

R. R. N. Alves
Departamento de Biologia, Ethnozoology, Conservation and Biodiversity Research Group,
Universidade Estadual da Paraíba, Campina Grande-PB 58109-753, Brazil

R. R. N. Alves and I. L. Rosa (eds.), *Animals in Traditional Folk Medicine*,
DOI: 10.1007/978-3-642-29026-8_10, 

## 10.1 Introduction

Traditional medicines are used worldwide, especially in developing countries, where 70–95 % of the population count on folk medicines in their primary health needs (Robinson and Zhang 2011). Although much less studied in comparison to medicinal plants (Solavan et al. 2004), animals and their by-products have constituted part of the inventory of medicinal substances used in various cultures since ancient times, and such uses still exist in ethnic folk medicine (Lev 2003).

The endurance of that ancient custom has been recorded in a number of studies (Begossi 1992; Begossi and Braga 1992; But et al. 1991; Costa-Neto and Marques 2000; El-Kamali 2000; Huxtable 1992; Lev 2003; Marques 1997; Seixas and Begossi 2001), and as pointed out by Lev (2003), scholarly investigation and the study of the medicinal uses of animals and their products, as well as of inorganic materials, should not be neglected and should be considered as an important complementary body of knowledge.

Another important reason for furthering knowledge about the traditional uses of animals for medicinal purposes is related to the growing concern over the global demand for animal-based remedies, whose use has contributed to reducing the numbers of many species to critical levels (Alves et al. 2010b; Brautigam et al. 1994; Ellis 2005; Lee 1999; Mainka and Mills 1995; Mills and Servheen 1994; Servheen 1999; Wemmer 1998; Wu et al. 2004).

Seahorses (*Hippocampus* spp.) are increasingly providing a focus and leverage for discussing important conservation issues, among which the trade in wildlife for medicinal purposes. Threatened on a global scale by overfishing, habitat damage and bycatch (Lourie et al. 2004), these mythical and charismatic fishes (Scales 2009) are globally used as remedies (e.g., Baum and Vincent 2005; Lourie et al. 1999, 2004; Moreau et al. 2000; Perry et al. 2010; Rosa 2005; Rosa et al. 2010, 2011; Scales 2009; Vincent 1995a, b, 1996; Vincent et al. 2011a, b), to treat a range of medical conditions.

International concern over the sustainability of the seahorse trade culminated in the inclusion of the entire genus *Hippocampus* in Appendix II of the Convention on International Trade in Endangered Species of Wild Fauna and Flora (CITES), effective May 2004, as a means of ensuring that trade is not detrimental to the survival of wild populations (Lourie et al. 2004). Seahorses stand out as the first marine fish species of commercial importance to be listed under CITES in decades (Vincent et al. 2011a), and may represent the most widely and voluminously traded CITES animal species (Bruckner et al. 2005).

The major exploitation of seahorses is attributed mainly to Traditional Chinese Medicine and other traditional medical systems derived from TCM, such as *jamu* (Indonesia), *hanyak* (Korea), and *kanpo* (Japan) (Vincent 1996). Additionally, seahorses are used in folk medicines practiced in Africa (e.g., McPherson and Vincent 2011), India, Philippines (Vincent 1996), and Latin America (Baum and Vincent 2005). In Brazil, seahorses are among the most traded species for medicinal purposes (Alves and Rosa 2010; Rosa 2005), and are regarded as one of

**Fig. 10.1** Seahorses in dried form, for sale in: **a–b** TCM shop in Singapore; **c–d** Public market in Brazil. Photos: Ierecê Rosa

the most versatile fish species in terms of therapeutic indications (El-Deir et al. 2012). The global demand for seahorses is mirrored by an international trade in which large volumes of primarily dried specimens (Fig. 10.1) are commercialized (Lourie et al. 2004; Vincent 1996).

Available information on the medicinal uses of seahorses was first summarized by Vincent (1996), who conducted an extensive survey of the seahorse trade in Asia, and established that the market for seahorses was very large, economically important, and threatening to wild populations. Subsequent trade surveys conducted in Asia (Giles et al. 2006; Perry et al. 2010), Africa (McPherson and Vincent 2004, 2011), Latin America (Baum and Vincent 2005, 2011a, b; Baum et al. 2011; Rosa 2005; Rosa et al. 2010, 2011) and Australia (Martin-Smith and Vincent 2006) expanded, among other aspects, knowledge about the medicinal uses attributed to seahorses. In addition to the information gathered in aforementioned trade surveys, Scales (2009) presented an overview of the use of seahorses as medicines (mainly focusing on historical therapeutic uses and TCM), in her book chapter entitled "A Seahorse Cure."

The use of seahorses for medicinal purposes has also been recorded, with varying degrees of detailing, in publications focusing on ethnobiology, fisheries management, aquaculture, taxonomy, genetics, pharmacology, and wildlife conservation (see Table 10.2). Collectively, those publications attest to the importance of further systematizing knowledge on the various aspects related to the use of seahorses in traditional medicine, to assist conservation efforts.

This chapter summarizes and updates knowledge on the global use of seahorses in traditional medical systems. First, we contextualize the medicinal uses of seahorses in a historical perspective. Then, we summarize knowledge on the medicinal uses (species used, illnesses and health conditions to which seahorses are prescribed, ways of administration). Finally, we discuss possible avenues for the conservation of seahorses in the context of traditional medicine.

### 10.1.1 Background

Most modern *Materia Medica* indicate the *Omissions from the [Classic of the] Materia Medica* (Ben Cao Shi Yi), elaborated by *Chen Zang-Qi* in 720 A.D., as the text in which *Hai ma* (seahorse) first appeared in TCM (e.g., Bensky et al. 2004). Nonetheless, as remark by May and Tomoda (2002), the *Nan Zhou Yi Wu Zhi (Record of strange things from the southern regions)*, written by Wan Qian sometime between 270 and 310 A.D., appears to be the earliest clearly datable record of the use of seahorses in medicines, with the following indication of use: "when a woman has difficult childbirth it is necessary to cut her open so the baby can be born, [instead] grasp *Hai Ma* in the hand and this will make the birth easy just as it is for a sheep."

The first comprehensive discussion on seahorses' [medicinal uses] is found in translations of the *Compendium of Materia Medica* (*Ben Cao Gang Mu*), written by Li Shi Zhen in the sixteenth century (May and Tomoda 2002), and regarded as the best known example of Chinese *Materia Medica* literature (Bensky et al. 2004). In that work, the medicinal use of seahorses is described as: "Supplements kidney. Strengthens yang [male sexual function]. Removes lumps and masses in the lower abdomen of women. Treats furuncles and toxic swellings, dystocia [difficult labour], and pain due to malfunctions in blood and vital energy" (Vincent 1996) or "Warms the water organ, fortifies the way of yang [male sexual function], reduces mobile and fixed abdominal masses, and treats toxic swollen sores" (Bensky et al. 2004).

The therapeutic uses of seahorses have also been described in the treatment of a wide range of diseases for hundreds of years, in many other cultures around the world. As pointed out by Eastman (1915), the alleged medicinal properties of *Hippocampus*, which are gravely set forth by Matthioli, Belon, Rondelet, Gesner, and others, are traceable to a number of late Greek and Roman writers, the more important of whom are Menander, Strabo, Philostratus, Dioscorides, Aelian, and Pliny.

Alakbarli (2006) remarked that ancient Greeks recommended the use of seahorse ashes to cure bald spots. Indeed, the use of seahorses in the treatment of baldness was reported in the first century A.D., described by the Greek physician Dioscorides, as follows: "*being burnt & the ashes thereof taken either in Axungia* (goose grease)*, or liquid pitch, or unguentum Amaracinum* (marjoram ointment)*, & anointed on, doth fill up the Alopecia with haire*" (Gunther 1959).

The Roman author Claudius Aelianus (also known as Aelian) reported the use of seahorses for treating rabies (Jacobs 1832), while the Roman naturalist Pliny the Elder, in his *Natural History*, reported the use of seahorses as aphrodisiac, to cure alopecia, neutralize the poison of the sea-hare, remove lichens and leprous spots, for urinary incontinence, pains in the sides, chills, and fevers (Bostock and Riley 1855).

Adverse effects of the use of seahorse-based remedies were mentioned by Aelian, who reported that the ingestion of seahorse stomach in wine could lead, among other symptoms, to a dry and violent cough, swelling of the abdomen, hot flushes in the head, nostrils discharges with a fishy odor, inflamed eyes and, if the victim did not vomit, it could lead to death (Jacobs 1832; Scales 2009).

Seahorses were also reportedly used by European women in the eighteenth century, as published in Gentleman's Magazine (Urban 1753), "to increase their milk"; in northern Europe, Roman "drug boxes" contained, among a range of other ingredients (such as mistletoe, mushrooms, calcinated shells), perhaps ground-up seahorses (Scarborough 1996). While Pitrè (1896) recorded the use of seahorses in Italy for treating renal colic and as amulets against recurrent fever, in Portugal, in the 16th and 17th centuries, there are records of seahorses used for treating hair loss (Rasteiro 2003), and as amulets to prevent melancholy (Vasconcelos 1985).

In Brazil, the first record we found of the medicinal use of seahorses dates back to the nineteenth century, as follows: seahorse "when dried, is worn next to the skin, and is powerful in driving off the headache as well as the devils" (Ewbank 1856). Nonetheless, given that the Brazilian traditional medicine, particularly zootherapy, comprises an interaction between African and European elements, which, coupled with the indigenous component, take part of the history of traditional medicine in the country since the beginning of colonization (Rocha 1960), perhaps the medicinal use of seahorses in Brazil is an older practice.

## 10.2 Data Compilation

The relevant information and data were gathered through an extensive literature review. Our main sources were: published articles, books and book chapters, as well as reports, available at international online databases such as Web of Science, Scopus, Google Scholar, or journals' web sites. In addition, we consulted the seahorse online database HippocampusInfo web site (http://hippocampusinfo.org).

The resulting database encompasses information on seahorse species used as medicines, countries where their medicinal use was recorded, conditions to which remedies were prescribed, and the way medicines were taken. Scientific names provided in publications were updated according to Lourie et al. (2004).

## 10.3 Seahorses in Traditional Medicines

Overall, 20 species of *Hippocampus* were cited in the literature as having medicinal use (Table 10.2). Records of medicinal use of "seahorse" or "seahorses," without indication of the species used, were also found (cited as *Hippocampus* sp. on Table 10.2). Records of medicinal uses of seahorses were found for at minimum 33 countries worldwide (Australia, Belize, Brazil, Cambodia, Canada, China, Costa Rica, Guatemala, Honduras, Hong Kong SAR, India, Indonesia, Italy, Japan, Korea, Laos, Malaysia, Mexico, New Zealand, Nicaragua, Panama, Peru, Philippines, Portugal, Singapore, Sri Lanka, South Africa, Taiwan, Tanzania, Thailand, United States, Venezuela, Vietnam).

Seahorses were related to 284 therapeutic properties and uses, described in Table 10.1. From the 18 health conditions categorized by the International Statistical Classification of Diseases and Related Health Problems (ICD-10), *Mental and behavior diseases* were the most quoted, followed by *Genitourinary system diseases*, reflecting the widespread use of seahorses for treating sexual dysfunctions (e.g., impotence), and diseases of the sexual organs and urinary tract, respectively (see Vincent 1996). Interestingly, May and Tomoda (2002) drew attention to the fact that, although easily regarded as sexual stimulants, seahorses had a long tradition of use in difficult childbirth, and argued that a probable major factor in popularizing seahorses as medicines and highlighting them as sex tonics was the widespread sale of manufactured pills containing seahorses throughout Asia since the 1970s.

In addition to the diseases which could be categorized, we found records of several other TCM-specific terms, which could not be classified according to ICD-10, as exemplified by: promotes kidney yang, balancing vital energy flows within the body, boost the yang, loss of energy, and reinforce the *Qi*.

Indications of use of seahorses could vary regionally, as indicated by their use to treat sexual disorders and as aphrodisiacs in Taiwan and Indonesia and to treat respiratory ailments in China and Hong Kong (Vincent 1996). Variations in use were also found within Latin America (e.g., Baum and Vincent 2005, 2011a, b; Baum et al. 2011; Rosa 2005; Rosa et al. 2005, 2010, 2011).

## 10.4 How are Seahorses Used in TM?

The main form of preparation of remedies reported in the literature involved toasting and grinding whole dried seahorses, or parts of them. The resulting powder can be consumed either as the sole ingredient or in combination with other products; other forms of use cited in the literature were: as prepackaged medicines, as tonic food, or as a basis for concoctions that may also contain other ingredients—plants and animals (see Table 10.2). Seahorses used in TCM are generally ground to powder which may be applied directly to wounds, or mixed with a liquid (e.g., warm water); additionally, whole seahorses may be steeped (and sometimes) fermented in a liquid (strong alcohol or another Chinese medicine) or even prepared in soups (Vincent 1996). In Latin America, they are typically roasted and ground to powder for administration in teas (Baum and Vincent 2005; Rosa et al. 2010), but could also be used in the preparation of concoctions or mixed with other ingredients (see Baum and Vincent 2011a, b; Baum et al. 2011; Rosa et al. 2010). Seahorse-based prepackaged medicines were used by TCM practitioners in Asia and North America (LaFrance and Vincent 2011; Lee 2000a; Vincent 1996;

**Table 10.1** Medicinal uses and properties of seahorses as quoted in the literature.

| ICD 10 | Indication of use and therapeutic properties | *N* |
|---|---|---|
| Mental and behavioral disorders | Aphrodisiac; alcoholism; diminished desire (also quoted as to: promote male sexual desire; promote sexual desire); dizziness; erectile dysfunction (also quoted as: impotence); general enervation; insomnia; amnesia (also quoted as: loss of memory); mental diseases (also quoted as: mental disorders); neurasthenia; to normalize sexual activity; overburdening of brain; poor memory; to prevent melancholy; to promote sexual function (also quoted as: to promote male sexual function; to promote vigour; to strengthen sexual functionality; to improve the sexual potency of older men; to increase male "strength"; to enhance the husband's performance; to increase sexual potency; powerful masculine stimulant; reduced potency; sexual stimulant); premature ejaculation; psychological disorders; sexual dysfunction (also quoted as: astysia; sexual disorders); as a sexual tonic; to strengthen weakened minds; trance; upsets; vertigo; virility treatment | 41 |
| Genitourinary system | Amenorrhoea; chronic nephritis; climacteric syndrome; delayed menstruation; dysfunction of sexual organs (also quoted as: sexual organs disorders); frequent urination at night; hypermenorrhea; incontinence (also quoted as: leakage; enuresis; urinary incontinence); infertility (as quoted as: to promote fertility; female infertility; female sterility); kidney diseases (also quoted as: kidney disorders; kidney problems; kidney trouble); kidney pain; leucorrhoea; low sperm count; menoxenia; painful and unsuccessful erections; weak erection; period cramps; to prevent the retention of urine; renal colic; scanty and pale menstruation; seminal emission; spermatorrhoea; strengthening the genital functions; strengthening reproductive organs; urinary infections; uterine bleeding; vaginal discharges; weak sperm | 38 |
| Skin and subcutaneous tissue | Baldness (also quoted as: hair loss; alopecia); clove sores (also quoted as: clove sores on the back); to dissipate skin nodules; eczema; boils; carbuncles; furuncles (also quoted as: furuncular infections in children with weak constitutions); malign sores on the back; pallor (also quoted as: paleness; pale face); persistent nodules (also quoted as: persistent nodules formation); to reduce nodulation; to reduce swelling; to remove lichens; severe acne; skin allergy; skin ailments (also quoted as: skin diseases; skin health); swelling due to sores and boils; toxic swellings (also quoted as: toxic swollen sores); to treat tenacious pimples; vitiligo | 30 |

(continued)

**Table 10.1** (continued)

| ICD 10 | Indication of use and therapeutic properties | *N* |
|---|---|---|
| Symptoms, signs and abnormal clinical and laboratory findings, not elsewhere classified | Chills; cold in the lower part of the abdomen; cold limbs; debility in the elderly; fevers (also quoted as: recurrent fever); fatigue (also quoted as: masculine fatigue; general fatigue; general lassitude; general debilitation; weariness); general lethargy; general malaise; general pain (also quoted as: pain); inflammation; loss of appetite (also quoted as: poor appetite; lack of appetite); lymph nodes disorders; swelling in lymph nodes; oedema; pains in the sides; sweating (also quoted as: day sweat; night sweat); vague pain in the abdomen | 28 |
| Respiratory system | Asthma (also quoted as: chronic asthma; asthma in children); bronchitis; certain forms of allergic conditions; chronic cough; consumptive coughs; dyspnoea (also quoted as: shortness of breath); flu; lung congestion; phlegm (also quoted as: excess throat phlegm); respiratory ailments (also quoted as: respiratory disorders; respiratory infections; respiratory problems); sinusitis; throat disorders (also quoted as: throat infections); to stop dry coughs; wheezing | 22 |
| Endocrine, nutritional, and metabolic diseases | As an anti-ageing tonic for the middle-aged and older; diabetes; for general health (also quoted as: for health-keeping); as a general tonic (also quoted as: tonics; general tonic and strengthener; strengthening organs); high cholesterol; physical weakness; poor nourishment; premature senility (also quoted as: signs of premature aging); thyroid disorders (also quoted as: goitre; throat goitres); weakness (also quoted as: general weakness; recovering from all kinds of weaknesses; weak constitutions in children) | 20 |
| Musculoskeletal system and connective tissue | Back pain (also quoted as: sores on back; backache); debility of limbs (also quoted as: weakness of the limbs); good for the bones; keep the back healthy; lumbago; osteoporosis; pains in the legs; rheumatism; to strengthen the bones and tendons; soreness in the lumbar area; soreness of the knees; sores on waist (also quoted as: soreness of the waist); weakness in the lumbar area (also quoted as: to strengthen the lumbar area); weakness of the knees; weakness of the waist | 20 |
| Certain infectious and parasitic diseases | Abscesses; amoebiasis; anti-malarial; diarrhoea; erysipelas; microbial pathogenic diseases; leprosy (also quoted as: to remove leprous spots); lymphatic tubercle; rabies; septic infections; tuberculosis; whooping cough in children | 13 |
| Digestive system | Gas pain; gastritis; hyperacidity; liver diseases; mass in the abdomen (also quoted as: abdominal accumulations and masses; to eliminate lumps; to remove lumps and masses in the lower abdomen of women; mobile and fixed abdominal masses); pale tongue with white fur; stomach ache (also quoted as: stomach pains); toothache | 13 |

(continued)

**Table 10.1** (continued)

| ICD 10 | Indication of use and therapeutic properties | *N* |
|---|---|---|
| Injury, poisoning, and certain other consequences of external causes | Bites from mad dogs; broken bones (also quoted as: fractures); injuries from falls; contusions; strains; wounds (also quoted as: open wounds); to neutralize the poison of the sea-hare; traumatic injuries; cuts (also quoted as: wound cuts) | 12 |
| Blood and blood-forming organs and certain disorders involving the immune mechanism | Anemia; bleeding (also quoted as: bleeding in women; loss of blood; bleeding with pain); hemorrhoids; poor immune system; slow and heavy pulse; spleen disorders; to clear toxins from blood; good for bone marrow | 11 |
| Circulatory system | Arteriosclerosis; cardiac stimulant; circulatory problems; heart diseases; to keep the heart healthy; increasing blood circulation; to active circulation of blood; lymphadenitis; palpitations; stroke; thrombosis | 11 |
| Pregnancy, childbirth, and the puerperium | Dystocia (also quoted as: facilitate parturition; difficult or delayed childbirth; as an aid in childbirth; difficult labour); to increase [women] milk; postpartum haemorrhage; problems postpartum (to speed postpartum recovery); to prevent abortion | 9 |
| Neoplasms | Cancer (also quoted as: tumours); breast cancer; cervical cancer; intestinal cancer; kidney cancer | 6 |
| Nervous system | Epilepsy (also quoted as: to prevent epilepsy or other similar disorders); fits; headache; stimulates the brain | 5 |
| Ear and mastoid process | Earache; tinnitus | 2 |
| Eye and adnexa | Pain in the eyes; poor eyesight | 2 |
| Certain conditions originating in the perinatal period | To decrease the navel of newborn | 1 |

Categories used follow the International Statistical Classification of Diseases and Related Health Problems (ICD-10)

Vincent et al. 2007a). Exceptionally, fresh or live seahorses were used in traditional medicine preparations, as exemplified by the use of a fresh seahorse boiled with pork and ginger and the addition of live seahorses to alcoholic beverages (Vincent 1996; Giles et al. 2006).

Examples of posology related to seahorse-based remedies include: Brazil—seahorses should be taken as a tea (in the treatment of asthma), three times a day, during three consecutive days, three days after the disease commenced (Bonifácio 2002; Silva et al. 2004), or once a day, in waning moon periods (Rosa et al. 2010); dosages depend on the users' age, being lower to children (Andrade and Costa-Neto 2006; Rosa et al. 2010)—for example, when taken mixed with honey or sugar and other ingredients in a syrup, the dosage should be one tablespoon for adults, and one teaspoon for children (Costa-Neto 1999a). In Asia, Vincent (1996) recorded the following posologies: Indonesia—two tablets of seahorse-based *jamu* pills (labelled *Health Ginjal Seahorse Ghenshen Kidney Pills*), indicated for a range of health conditions), taken twice daily, in the morning and at night, with

**Table 10.2** Use of seahorses in traditional medicines as described in the literature. (–) no record found; (?) dubious record

| Species | Diseases and conditions treated | Part used | Preparation | Source |
|---|---|---|---|---|
| *Hippocampus abdominalis* | – | – | – | HippocampusInfo (http://hippocampusinfo.org), Lourie et al. (1999, 2004), Woods et al. (2006) |
| *Hippocampus algiricus* (?) | – | – | – | CITES (2002) |
| *Hippocampus angustus* (originally cited as *H. erinaceus* by some authors) | – | Whole animal | – | Fan (2005), HippocampusInfo (http://hippocampusinfo.org), LaFrance and Vincent (2011) |
| *Hippocampus barbouri* | – | Whole animal | – | CITES (2002), HippocampusInfo (http://hippocampusinfo.org), LaFrance and Vincent (2011), Lourie et al. (1999, 2004), Martin-Smith and Vincent (2011a, b), May and Tomoda (2002), Perry et al. (2010), Project Seahorse (2002a) |
| *Hippocampus borboniensis* | – | – | – | CITES (2002), HippocampusInfo (http://hippocampusinfo.org), Lourie et al. (2004), Project Seahorse (2003a) |
| *Hippocampus camelopardalis* | – | – | – | CITES (2002), HippocampusInfo (http://hippocampusinfo.org), Lourie et al. (2004), Project Seahorse (2003b) |

(continued)

**Table 10.2** (continued)

| Species | Diseases and conditions treated | Part used | Preparation | Source |
|---|---|---|---|---|
| *Hippocampus comes* | – | Whole animal | – | CITES (2002), HippocampusInfo (http://hippocampusinfo.org), LaFrance and Vincent (2011), Lourie et al. (1999, 2004), Martin-Smith and Vincent (2011a), Perry et al. (2010), Project Seahorse (2002b), Sanders et al. (2008), Vincent (1996) |
| *Hippocampus coronatus* | Anemia; bleeding; difficult labour; dizziness; furuncular infections in children with weak constitutions; impotence; insomnia; kidney trouble; mobile and fixed abdominal masses; neurasthenia; overburdening of brain; pain; paleness; poor appetite; poor memory; premature senility; sores on waist and back; strengthening organs (especially reproductive); swelling due to sores and boils; sweating; swollen toxic sores; urinary frequency at night; urinary incontinence; vaginal discharge; weakness; wheezing | Whole animal | Pills containing seahorse and other ingredients; with Lycii Fructus (*Lycium barbarum*) and Jujubae Fructus (*Ziziphus jujuba*); with lean pork; soaked in yellow rice wine and roasted | Bensky et al. (2004), Fan (2005), Gaski and Johnson (1994), Hirschhorn (1982), Vincent (1996) |

(continued)

**Table 10.2** (continued)

| Species | Diseases and conditions treated | Part used | Preparation | Source |
|---|---|---|---|---|
| *Hippocampus erectus* | Alcoholism; asthma; bronchitis; cancer; diabetes; heart diseases; impotence; osteoporosis; rheumatism; thrombosis | Whole animal | Tea of roasted and ground seahorse | Alves et al. (2007), Dias et al. (2002), El-Deir et al. (2012), HippocampusInfo (http://hippocampusinfo.org), LaFrance and Vincent (2011), Lourie et al. (2004), Rosa (2005), Rosa et al. (2011), Project Seahorse (2003c) |
| *Hippocampus fuscus* | – | Whole animal | – | CITES (2002), HippocampusInfo (http://hippocampusinfo.org), LaFrance and Vincent (2011), Lourie et al. (1999, 2004), Project Seahorse (2003d) |
| *Hippocampus histrix* | Bleeding; debility in the elderly; difficult labour; dyspnoea; enuresis; external use for boils; external use for carbuncles; furuncular infections in children with weak constitutions; impotence; mass in the abdomen; mobile and fixed abdominal masses; pain; swelling due to sores and boils; swollen toxic sores; traumatic injuries; urinary frequency at night; urinary incontinence; vaginal discharge; wheezing | Whole animal | With Lycii Fructus (*Lycium barbarum*) and Jujubae Fructus (*Ziziphus jujuba*); with lean pork; soaked in yellow rice wine and roasted | Bensky et al. (2004), CITES (2002); Fan (2005), Goenka (2005), HippocampusInfo (http://hippocampusinfo.org), Lourie et al. (1999, 2004), Martin-Smith and Vincent (2011a, b), May and Tomoda (2002), Project Seahorse (2002c), Vincent (1996), Yao and Chang (1995), Zhang (2000), Zheng (2000) |

(continued)

**Table 10.2** (continued)

| Species | Diseases and conditions treated | Part used | Preparation | Source |
|---|---|---|---|---|
| *Hippocampus ingens* | Aphrodisiacs; asthma; kidney pain; masculine fatigue | Whole animal | Marinated with ginseng roots; in a tonic soup prepared with pork | Baum and Vincent (2005); Baum et al. (2011), CITES (2002), HippocampusInfo (http://hippocampusinfo.org), LaFrance and Vincent (2011), Lourie et al. (1999, 2004), Project Seahorse (2003e), Sanders et al. (2008), Vincent (1996) |
| *Hippocampus kelloggi* | Aphrodisiac; bleeding; debility in the elderly; difficult labour; dyspnoea; enuresis; external use for boils; external use for carbuncles; for male "strength"; furuncular infections in children with weak constitutions;impotence; mass in the abdomen; mobile and fixed abdominal masses; to normalize sexual activity; pain; to promote kidney function; swollen toxic sores; swelling due to sores and boils; traumatic injuries; urinary frequency at night; urinary incontinence; vaginal discharge; wheezing | Whole or halved animal | Ground into powder; in alcohol-based tonics; prepackaged pills; packaged with other herbs, for example, "Kuku Bima"; soups; with Lycii Fructus (*Lycium barbarum*) and Jujubae Fructus (*Ziziphus jujuba*); with lean pork; soaked in yellow rice wine and roasted | Bensky et al. (2004), CITES (2002), Fan (2005), Goenka (2005), HippocampusInfo (http://hippocampusinfo.org), Ho and Tan (2011), LaFrance and Vincent (2011), Lourie et al. (2004), Low and Tan (2007), May and Tomoda (2002), Perry et al. (2010), Project Seahorse (2002d), Tan et al. (2010), Vincent (1996), Wong and Dahlen (1999), Yao and Chang (1995), Zhang (2000), Zheng (2000) |

(continued)

**Table 10.2** (continued)

| Species | Diseases and conditions treated | Part used | Preparation | Source |
|---|---|---|---|---|
| *Hippocampus kuda* | Asthma; bleeding; debility in the elderly; difficult labour; dyspnoea; enuresis; external use for boils; external use for carbuncles; furuncular infections in children with weak constitutions; gas pain;hyperacidity; impotence; mass in the abdomen; mobile and fixed abdominal masses; pain; swelling due to sores and boils; swollen toxic sores; traumatic injuries; urinary frequency at night; urinary incontinence; vaginal discharge; wheezing; whooping cough in children; cut wounds | Whole animal | With Lycii Fructus (*Lycium barbarum*) and Jujubae Fructus (*Ziziphus jujuba*); with lean pork; soaked in yellow rice wine and roasted; dried seahorse powdered, roasted and mixed in honey; dried seahorse burnt and mixed with coconut oil (external use); drink of dried seahorse slightly burned, soaked in water; ready-made tonics of fermented seahorse in alcohol; live, for incorporation into alcohol-based tonics | Alino et al. (1990), Bensky et al. (2004), CITES (2002), Fan (2005), Giles et al. (2006), Goenka (2005), HippocampusInfo (http://hippocampusinfo.org), LaFrance and Vincent (2011), Lourie et al. (1999, 2004), Marichamy et al. (1993), May and Tomoda (2002), Perry et al. (2010), Project Seahorse (2003f), Qian et al. (2008), Vincent (1996), Yao and Chang (1995), Zhang (2000), Zheng (2000) |
| *Hippocampus mohnikei* (originally quoted as *H. japonicus* by some authors) | Bleeding; debility in the elderly; difficult labour; dyspnoea; enuresis; external use for boils; external use for carbuncles; furuncular infections in children with weak constitutions; impotence; mass in the abdomen; mobile and fixed abdominal masses; pain; swelling due to sores and boils; swollen toxic sores; traumatic injuries; urinary frequency at night; urinary incontinence; vaginal discharge; wheezing | Whole animal | With Lycii Fructus (*Lycium barbarum*) and Jujubae Fructus (*Ziziphus jujuba*); with lean pork; soaked in yellow rice wine and roasted | Bensky et al. (2004), CITES (2002), Fan (2005), HippocampusInfo (http://hippocampusinfo.org), Lee (2000b), May and Tomoda (2002), Vincent (1996), Yao and Chang (1995), Zhang (2000), Zheng (2000) |

(continued)

**Table 10.2** (continued)

| Species | Diseases and conditions treated | Part used | Preparation | Source |
|---|---|---|---|---|
| *Hippocampus reidi* | Alcoholism; asthma; bleeding; bleeding in women; bronchitis; cancer; diabetes; earache; epilepsy; erysipelas; gastritis; heart diseases; impotence; inflammation; leakage; oedema; osteoporosis; period cramps; physical weakness; postpartum hemorrhage; problems postpartum (to speed postpartum recovery); rheumatism; sinusitis; thrombosis; stroke; to prevent abortion; toothache; tuberculosis; vitiligo | Head, "stomach" or tail; whole animal; without head | Mix of seahorse, medicinal plants and "cachaça" (Brazilian sugar cane liquor); roasted seahorse with white onion (*Allium* sp.) and "juá" fruit (*Zizyphus joazeiro* Mart.); roasted with "cachaça"; concoction of seashore mixed with homeopathic products; concoction of seahorse powder mixed with white onion (*Allium* sp.), "arapuá" stone (scutellum), rosemary (*Rosmarinus officinalis*), garlic skin (*Allium sativum*), sugar and water; roasted and ground seahorse taken with food or coffee; tea of boiled seahorse; tea of roasted (or sun dried) and ground seahorse; tea of a mixture of alligator skin, starfish and seahorses; tea of a mixture of ground "arapuá" scutellum (*Trigona spinipes*), ground seahorse, starfish (*Echinaster brasiliensis*) and prehensile-tailed porcupine (*Coendou* sp); tea of roasted and ground seahorse and lavender (*Lavandula* sp.) | Alves et al. (2007, 2008a, 2009, 2010a), Alves and Rosa (2006, 2007b, c, 2010), Begossi (1992), Begossi and Figueiredo (1995), Costa-Neto (1999a, b, c, 2000, 2002, 2004), El-Deir et al. (2012), Ferreira et al. (2009, 2012), LaFrance and Vincent (2011), Lourie et al. (2004), Oliveira et al. (2010), Osório (2005), Project Seahorse (2003g), Rosa (2005; IL Rosa, unpublished data), Rosa et al. (2005, 2010, 2011), Seixas and Begossi (2001) |

(continued)

**Table 10.2** (continued)

| Species | Diseases and conditions treated | Part used | Preparation | Source |
|---|---|---|---|---|
| *Hippocampus spinosissimus* | – | Whole animal | – | Baum and Vincent (2005), CITES (2002), HippocampusInfo (http://hippocampusinfo.org), LaFrance and Vincent (2011), Lourie et al. (1999, 2004), Martin-Smith and Vincent (2011a), May and Tomoda (2002), Morgan and Panes (2008), Perry et al. (2010), Project Seahorse (2003h) |
| *Hippocampus trimaculatus* | Bleeding; debility in the elderly; difficult labour; dyspnoea; enuresis; external use for boils; external use for carbuncles; furuncular infections in children with weak constitutions; impotence; mass in the abdomen; mobile and fixed abdominal masses; pain; swelling due to sores and boils; swollen toxic sores; traumatic injuries; urinary frequency at night; urinary incontinence; vaginal discharge; wheezing | Whole animal | With Lycii Fructus (*Lycium barbarum*) and Jujubae Fructus (*Ziziphus jujuba*); with lean pork; soaked in yellow rice wine and roasted; ready-made tonics of fermented seahorse in alcohol; live, for incorporation into alcohol-based tonics | Baum and Vincent (2005), Bensky et al. (2004), CITES (2002), Fan (2005), Giles et al. (2006), Goenka (2005), HippocampusInfo (http://hippocampusinfo.org), LaFrance and Vincent (2011), Lourie et al. (2004), May and Tomoda (2002), Perry et al. (2010), Project Seahorse (2003i), Vincent (1995b, 1996), Yao and Chang (1995), Zhang (2000), Zheng (2000) |
| *Hippocampus whitei* | – | Whole animal | – | CITES (2002), LaFrance and Vincent (2011) |
| *Hippocampus zebra* | – | – | – | HippocampusInfo (http://hippocampusinfo.org) |

(continued)

**Table 10.2** (continued)

| Species | Diseases and conditions treated | Part used | Preparation | Source |
|---|---|---|---|---|
| *Hippocampus* sp. | Abscesses; abdominal accumulations and masses; active circulation of blood; alopecia; amenorrhoea; amnesia; amoebiasis; as an anti-ageing tonic for the middle-aged and older; antimalarial; aphrodisiacs; arteriosclerosis; as an aid in childbirth; asthma; asthma in children; astysia; backache; back pain; baldness; bites from mad dogs; bleeding with pain; broken bones; bronchitis; cancer; cardiac stimulant; certain forms of allergic conditions; chills and fevers; chronic asthma; chronic cough; chronic nephritis; circulatory problems; to clear toxins from blood; climacteric syndrome; clove sores; clove sores on the back; cold in the lower part of the abdomen; cold limbs; consumptive coughs; day sweat; to decrease the navel of newborn; debility of limbs; delayed menstruation; diabetes; difficult or delayed childbirth; | Whole or halved animal; pieces | Roasted and ground; ashes of seahorses mixed with liquid pitch, tallow, and oil of sweet marjoram; blended into a powder, and steeped in whisky; ground and consumed in a drink; ground and mixed with other ingredients and taken with liquor, tea or mixed with food; ground and taken with juice; dried in sand or left out in the night air, roasted and ground and taken as a tea; tea of roasted and ground seahorse mixed with the powder of a roasted starfish; powdered seahorse mixed with water, soup or juice; powdered seahorse added into capsules; powdered seahorse into coffee; burnt to charcoal and drunk with water; dry seahorse powder with honey; powdered and applied directly to wounds; a seahorse pair (fried) with *Manis pentadactyla* (fried in loess), cinnabar, realgar, borneol, musk, mixed in metallic mercury, ground and | Almeida and Albuquerque (2002), Andrade and Costa-Neto (2006), Araújo (2004), Ashwell and Walston (2008), Baum and Vincent (2005, 2011b), Baum et al. (2011), Barros (1977), Bostock and Riley (1855), Bruckner (2002), Camargo (1985), Cameron et al. (2004), Chen and Yan (2009), Christie et al. (2011), Clarke (2002), Costa-Neto and Mota (2010), Demunshi and Chugh (2010), Dharmananda (2002), Eastman (1915), Ewbank (1856), Fontenelle (1959), Fulder (1993), Gaski and Johnson (1994), Goenka (2005); Hoover (2003), Hopkins (2006), James (1747), Lages-Filho (1934), laFrance and Vincent (2011), Lee (2000a, b), Lee and Mills (2000), Lei et al. (1996), Lipton (2000), Liu et al. (2005), Lockyear (2000), Lourie et al. (1999, 2004), Martin-Smith and Vincent (2006, 2011a), May and Tomoda (2002), McPherson |

(continued)

**Table 10.2** (continued)

| Species | Diseases and conditions treated | Part used | Preparation | Source |
| --- | --- | --- | --- | --- |
| | difficult labour; diarrhoea; to dissipate skin nodules; dizziness; dysfunction of sexual organs; dystocia; eczema; to eliminate lumps; to enhance the husband's performance; erectile dysfunction; excess throat phlegm; to facilitate parturition; facilitating childbirth; fatigue; female infertility and diminished desire; fits; flu; for health-keeping; frequent urination at night; furuncles; gas pain; general debilitation; general enervation; general fatigue; for general health; general lassitude; general lethargy; general malaise; general pain; as a general tonic; as a general tonic and strengthener; general weakness; goitres; hair loss; headache; heart disease; hemorrhoids; high cholesterol; hypermenorrhoea; impotence; incontinence; to increase male "strength"; to increase [women] milk; increasing blood circulation; to increase sexual potency; infertility; | | applied externally; mixture of dried and burned seahorse and coconut oil and applied externally; sun dried and kept/worn/held in the hand; worn as a necklace and then thrown away; fresh or dried seahorse fastened to the left arm; blended with pegasids, pipefishes and sea snakes; seahorses and 9–10 other herbs, roots and medicines; ingested with other animal and plant ingredients; taken with rice and wine; with fruits or lean pork; medical dishes called *yakuzen*; broth of boiled seahorse; boiled seahorses in water for half an hour and the liquid drunk while a mantra is chanted; sachet of seahorse jamu remedy together with one spoonful of honey, one egg and orange juice mixed with boiling water; extract of roasted, boiled and ground seahorse; soaked in liquid; a male–female pair mixed with *Radix saussureae*, *Rhizoma rhei*, *Semen pharbitidis*, *Semen croton* seeds' skin, | and McPherson and Vincent (2004, 2011), Meenakarn and Leepayakoon (2005), Moreau et al. (2000), Nogueira et al. (2005), Pajaro (2000), Parry-Jones and Vincent (1998), Perry (2000), Pitrè (1896), Qian et al. (2008), Quijano (2000), IL Rosa (unpublished data), Rosa et al. (2010), Salin and Mohanakumaran (2006), Scales (2009), Sharlip (1981), Silva et al. (2004), Sreepada et al. (2002), Van and Tap (2008), Vincent (1995a, b, c, 1996, 1997, 2000, 2006), Vincent et al. (2007a, b, 2011a), Zhang (2000), Zheng (2000) |

(continued)

**Table 10.2** (continued)

| Species | Diseases and conditions treated | Part used | Preparation | Source |
|---|---|---|---|---|
| | injuries from falls, fractures, contusions and strains; insomnia; to keep the heart and back healthy; kidney and liver diseases; kidney disorders; kidney problems; lack of appetite; leprosy; leucorrhoea; loss of blood; loss of memory; low sperm count; lung congestion; lumbago; lymph node disorders; lymph node problems; lymphadenitis; lymphatic tubercle; malign sores on the back; mass in the abdomen; menoxenia; mental disorders; neurasthenia and other psychological disorders; night sweat; open wounds and cuts; pain; pain in the eyes; pains in the sides; painful and unsuccessful erections; pains in the legs; pale face; pale tongue with white fur; pallor; palpitations; persistent nodules; persistent nodules formation; phlegm; poor eyesight; poor immune system; poor memory; powerful masculine stimulant; premature ejaculation; poor nourishment; to prevent epilepsy or other similar disorders; to prevent | | *Pericarpium citri reticulatae viride*, soaked in boy's urine and roasted and powdered taken with water; in alcohol-based tonics; whole *Gekko gekko* and seahorses steeped in alcohol; deer foetus or seahorse mixed with Ginseng (*Panax ginseng*) in strong alcohol; drink of a mixture of seahorses, ginseng and urchin spines in whisky (could add guinea bark or mace); drink of decomposed seahorses (live seahorses left to die and rot) in 5-year-old whisky; "sea treasure" (seahorse, pipefish, sea bird, sea snake, and sea sparrow) roasted with honey, ground, and mixed with alcohols (dan ginseng and red olive into rice wine), or the ground powder stewed in chicken soup; roasted seahorse or Venus shell, gathered during the day on a Friday, crushed into powder and soaked in good wine; seahorses incorporated into strong spirits; spirits containing two seahorses or seahorse and several animal and plant ingredients; | |

(continued)

Table 10.2 (continued)

| Species | Diseases and conditions treated | Part used | Preparation | Source |
|---|---|---|---|---|
| | melancholy; to prevent the retention of urine; to promote sexual function; to promote sexual desire and strengthens sexual functionality; to promote vigour and fertility; rabies; recovering from all kinds of weaknesses; recurrent fever; to reduce nodulation; reduced potency; to reduce swelling; to remove lichens and leprous spots; to remove lumps and masses in the lower abdomen of women; renal colic; respiratory ailments; respiratory disorders; respiratory infections; respiratory problems; rheumatism; to neutralize the poison of the sea-hare; scanty and pale menstruation; seminal emission; septic infections; severe acne; sexual disorders; sexual dysfunctions; sexual stimulant; shortness of breath; signs of premature aging; skin allergy; skin ailments, skin disease; skin health; soreness and weakness in the lumbar area; slow and heavy pulse; soreness of the waist and the knees; spermatorrhoea; | | fermented in strong alcohol with other medicinal material/herbs/Chinese herbs, which could include pipefishes; two small seahorses steeped in strong alcohol; pieces of seahorses in bottles of alcoholic beverages; in a drink of a whole seahorse steeped in a liquid (strong alcohol or another Chinese medicine), sometimes fermented; drink of seahorses steeped in "wine" (strong alcohol) with plant and animal matter and fermented for months; drink of a mixture of seahorse with palm wine; seahorses put in bottles of cognac; seahorse boiled into a soup, cooked with lean pork and eaten directly; tonic soup with seahorse, pipefish and pork; seahorse boiled into a soup, cooked with lean pork and eaten directly; tonic soup with seahorse, pipefish and pork; cut the tail tip off a live seahorse and suck the stump; fresh seahorse boiled with pork and ginger and eaten; oil of roses in which a seahorse has been dipped and killed; | |

(continued)

**Table 10.2** (continued)

| Species | Diseases and conditions treated | Part used | Preparation | Source |
|---|---|---|---|---|
| | spleen disorders; stimulates the brain; stomach aches; stomach pains; to strengthen the bones and tendons; strengthening the genital functions; to strengthen the lumbar area; to strengthen the sinews; swelling due to sores and boils; swelling in lymph nodes; throat disorders; throat goitres; throat infections; thyroid disorders; tinnitus; to improve the sexual potency of older men; as tonics; to promote male sexual function or desire; to stop dry coughs; to strengthen weakened minds; toxic swollen sores; trance; to treat tenacious pimples; toxic swellings; tumours (e.g., breast, kidney, cervical and intestinal cancers); upsets; urinary incontinence; urinary infections; virility treatment; female sterility; uterine bleeding; vaginal discharges; vague pain in the abdomen; vertigo; weak erection; weakness of the limbs; weakness of the waist and the knees; weak constitutions in children; weak sperm; weariness; wheezing; wounds. | | prepackaged tablets or pills containing seahorse and other ingredients | |

warm water; or one sachet of another seahorse *jamu* remedy (*Jamu Ramuan (Traditional) "Tangkur Jaya"*), together with one spoonful of honey, one egg and orange juice mixed with boiling water, taken three times weekly; Philippines—one or two seahorses should be roasted and boiled in four cups of water up to the equivalent of one cupful of liquid, ground and the resulting extract should be drunk twice daily; Hong Kong SAR—a total of 113.4 g of whole seahorse boiled into a soup, cooked with lean pork for a few hours, then eaten directly; as a tonic remedy—adults would usually consume 11.4–19.0 g of seahorse and pipefish once every 2–3 days, while children would take doses of 7.6 g. China—Yao and Chang (1995) cited the use of 3–9 g (appropriate quantity to be ground into powder and applied topically) of a seahorse-based remedy for external use to treat carbuncles and boils.

Prescriptions of prepackaged medicines containing seahorses to treat a range of illnesses were found in Zhu (1989), as follows: *Hai Ma Bu Shen Wan* (*Seahorse Kidney-Supplementing Pills*)—ten pills taken twice daily; *Gui Ling Ji* (*Tortoise Age Conglomerate*)—two or three capsules a day; *Zhi Bao San BianWan* (*Supreme Gem Penis Trio Pills*)—one pill, once a day. Hirschhorn (1982) mentioned that one pill of *Tzepao sanpien tablets*, a remedy containing seahorse and other 39 ingredients, should be taken with warm, boiled water before breakfast.

Additionally, Lei et al. (1996) described the preparation of various infusions termed *spirit of seahorse(s)*,with their respective ways of administration, and May and Tomoda (2002) recorded the posology of two seahorse-based remedies, which according to them are no longer in use: *Hai ma tang* and *Hai ma ba du san*.

Reports of unwanted effects related to the use of seahorses as remedies, although uncommon in the literature, are not neglectable. Examples of contraindications and side-effects found in the literature include: not to be used during pregnancy (Bensky et al. 2004; Zhu 1989), should not be used by those with renal insufficiency and weak digestion; in patients with chronic renal failure, it can aggravate uremia and lead to more severe renal damage (Bensky et al. 2004); if not properly administered, could lead to death (Alves and Rosa 2006).

Folkloric uses of seahorses have also been recorded, as follows: holding a seahorse in the hand to facilitate labour (Bensky et al. 2004; May and Tomoda 2002; Vincent 1996), cutting the tail tip off a live seahorse and sucking the stump to extract the juices to promote vigor (Vincent 1996), wearing a seahorse as a necklace and then throwing it away to treat asthma, thus transferring the disease to whoever picked up the seahorse (Baum and Vincent 2011a), keeping a seahorse (either dried of fresh) fastened to the left arm as an amulet against recurrent fever (Pitrè 1896); a female patient should prepare the concoction with a male seahorse and vice versa; the number of seahorse bony rings used to prepare the remedy must be similar to the age of the child receiving it (Rosa et al. 2010).

The use of folk seahorse-based remedy was sometimes connected to a belief system which requires that people receiving the remedies must not know what they are taking. As an example, in Brazil, users of seahorse-based medicines for treating asthma should not know what they are taking, otherwise they will not be cured (Rosa et al. 2010).

In addition to the aforementioned medicinal uses, recent studies have indicated the use of seahorses in homeopathy (*Hippocampus kuda*—Gray 2005; Shukla 2008; Sonz et al. 2003). Another poorly known aspect of the medicinal use of seahorses is the prescription of these animals in the context of ethnoveterinary therapies. This particular use was mentioned by Bennet (1834), who reported the use of dried seahorses by the Malays as an excellent medicine and tonic for horses, and by Xie and Preast (2010) who also described the use of seahorse-based remedies in Chinese ethnoveterinary practices.

## 10.5 Implications for Conservation

Traditional medicines can provide valuable clues for the development of novel compounds or drugs isolated or derived from marine life forms, and as pointed out by Costa-Neto (2001), the empirical knowledge on which the use of animal products in folk medicines is based is not necessarily devoid of scientific grounding. In fact, different authors have shown that many animal-based pharmaceutical products are used by allopathic medicine to treat several ailments (Alves and Rosa 2007a; Bisset 1991; Chivian 2002; Trowell 2003). Furthermore, Demunshi and Chugh (2010) have pointed out that substances obtained from marine organisms can have potential pharmaceutical, therapeutic, and diagnostic value.

Pharmacological results, however, cannot overlook conservation issues, and must be assessed with much care. As discussed by Demunshi and Chugh (2010), as the interest of pharmaceutical and allied industries to accrue marine life based products in their market profile is increasing rapidly, the conservation issues of the marine catch are also rising.

Seahorses are targeted by both direct and incidental fisheries, which together remove millions of specimens from their habitats yearly, to supply the demands of a local and/or international trade that has involved 93 countries from 1996 to 2008 (Vincent et al. 2011a); additionally, their populations have been affected by habitat degradation (Lourie et al. 1999, 2004; Foster and Vincent 2004; Vincent et al. 2011a).

As a result of those pressures, from the 48 species recognized as valid by Project Seahorse (2011), 37 figure in the Red List of Threatened Species published by the International Union for Conservation of Nature (IUCN 2011), most of them (28; 75.7 %) as Data Deficient, a category that denotes the need for more research. Regionally speaking, seahorses are also included in lists of threatened species in many countries (Vincent et al. 2011a).

Additionally, the entire genus *Hippocampus* was included in the Appendix II of CITES, to ensure that trade is non-detrimental to wild populations of seahorses.

Trade related to traditional medicinal uses of seahorses certainly represents a major focal point within the global seahorse trade, both in volume and in the number of species traded. In fact, Vincent et al. (2011a) showed that ca. 95 % of the seahorses in trade are sold for use in traditional medicines, and that of the 28 species of *Hippocampus* for which CITES trade data were available, 18 were reportedly used in the dried trade, probably mostly for TM.

The use of seahorse as remedies, especially for TCM, has spread globally, and as discussed by Vincent et al. (2007a), China's economic growth (which led to higher incomes), and the global diaspora of Chinese people apparently contributed to a greater demand for seahorses, and thus to direct fishing pressure. Moreover, Vincent (1996) reported an increase in popularity of TCM prepackaged medicines, which came about partly because they are faster to administer. The rapid growth of manufactured seahorse medicines can be exemplified by the number of patented remedies containing seahorses: 24 in 1994 (Gaski and Johnson 1994), and 43 in 2004, among pills/capsules, plasters, and liquids (Cameron et al. 2004).

The use of prepackaged medicines is a cause of growing concern in seahorse conservation, since such remedies may incorporate small seahorse species and juveniles, which were previously undesirable when consumers chose seahorses individually (Lourie et al. 1999; Vincent 1996; Vincent et al. 2007a). It should be noted that prepackaged is not favored in some TM systems, as exemplified by Brazil, where healers generally are not in favour of ready-to-use seahorse remedies because they are afraid they cannot guarantee the quality or even the composition of the product (Rosa et al. 2010).

There is a need to further increase awareness of the plight of endangered species among Chinese medicine practitioners and consumers, as well as to promote alternative treatments that use sustainably harvested, and often herbal, ingredients. In the case of seahorses, Vincent (1996) has pointed out that traders and practitioners in both China and Taiwan regard seahorses as very useful, but not vital ingredients; May and Tomoda (2002) mentioned that seahorses are not a crucial item in Chinese medicine practice; and Call (2006), based on a survey conducted in the United States, pointed out that the use of seahorses was recorded as of "minimum importance" by 61.4 % of 145 Chinese herbology practitioners. In this sense, seahorses substitution could be acceptable, and consumption adjusted to relieve pressure on certain species and size classes (Lourie et al. 1999). In fact, according to Call (2006), the Hong Kong Chinese Medicinal Merchants Association (HK—CMMA) has recommended the pursuit of alternatives to seahorses, seeking their conservation.

Scales (2009) cited some ingredients that could potentially be used as seahorse replacements, among which Chinese dodder seed. In Brazil, where seahorses are mainly used in the treatment of asthma in folk medicines, 13 plant-derived alternative remedies have been mentioned as substitutes for the same purpose by TM traders (Rosa et al. 2010), among which *Abelmoschus esculentus*, which was cited as one of the most efficient remedies to treat asthma (Osório 2005), and *Uncaria tomentosa*, sometimes referred to as the definite remedy for that ailment (personal observation).

Replacements in traditional medicine recipes, where suitable, can reduce the pressure on threatened species (Sodeinde and Soewu 1999). Nevertheless, precaution should be taken when suggesting the replacement of animal products with plants which may also be under conservation concerns (Alves and Rosa 2007a). The use of substitutes should be carefully assessed prior to recommendation, especially in relation to how much can be explored sustainably (Sodeinde and Soewu 1999). Such approach should also take into account the socio-cultural contexts of the trade and use of seahorses (Rosa et al. 2010).

According to Call (2006), neither the replacement of ingredients in prescriptions is a new phenomenon within the context of TCM, nor the idea of questioning Chinese practitioners on the subject of substitutions for endangered species. In her view, what is new is the willingness of the Chinese community to pursue its own answers to concerns about the use of endangered species.

In fact, in recent years, concern over the medicinal uses of threatened has been expressed in some publications focusing on TCM. As an example, the book "Chinese Herbal medicine—Materia Medica" (Bensky et al. 2004), highlights the relevance of CITES to TCM and dedicates a chapter to Appendix I CITES species. Additionally, Appendix II CITES species, including seahorses, are identified as such in the "additional product information" section.

In an ethnoveterinary context, Xie and Preast (2010) also expressed a growing concern over seahorse declining populations, by emphasizing that seahorses are endangered, prohibited and no longer in use in ethnoveterinary practices, and by recommending that *Hai Ma* (*Hippocampus*) in the original formula should be replaced with the plant-derived medicines *Rou Cong Rong* (*Cistanche*) and *Tu Si Zi* (*Cuscuta*).

Aquaculture is also cited in the literature as a possible avenue to diminish pressure on wild seahorse populations (see Koldewey and Martin-Smith 2010). Nonetheless, in relation to TM, in addition to an existing view that wild seahorses may be more efficacious than cultivated ones ("wild is better") (Moreau et al. 2000), aquacultures developed so far have not proven commercially viable to produce enough volumes of seahorses to reduce the demands of the TCM trade on wild populations (Koldewey and Martin-Smith 2010).

In general terms, possible measures towards seahorse conservation include: fisheries management, the inclusion of marine protected areas, select aquaculture ventures, trade regulation, improved governance (particularly), and consumer engagement (see Vincent et al. 2011a). Specifically in relation to the trade in seahorses for medicinal purposes, measures that could be explored (based on actions proposed for other animal species traded as medicines) include: further international cooperation; enhanced labeling of traditional remedies; public education; search for feasible approaches to replace ingredients; enhanced trade monitoring; sustained dialog with traders and users of TM, particularly of TCM; exploring effective channels of communication to convey the issue of the relationship of TCM and wildlife conservation to the members of the TCM community (e.g., Alves et al. 2008b, 2010b; Ashwell and Walston 2008; Kang and Phipps 2003; Lee et al. 1998; Marshall 1998).

One unifying aspect of the possible avenues that can be explored to address the conservation concerns related to the medicinal uses of seahorses is the need to consider, among other aspects, the socioeconomic and cultural issues related to that activity, given that (1) seahorses account for a part of fishers' incomes in some countries (Pajaro et al. 1997; Rosa et al. 2005; Vincent et al. 2007b), and constitute a part of their living following the decline of traditional fishing resources; (2) seahorses have a perceived importance in traditional medicines (Lourie et al. 1999; Rosa et al. 2010; Vincent 1996), and their use in zootherapeutic and magic ritualistic practices constitutes a relevant transcultural phenomenon which implies additional pressure to their populations (Andrade and Costa Neto 2006); such aggregated symbolic and spiritual values should also be addressed in the context of seahorse conservation. As pointed out by Lourie et al. (1999), attempts to stigmatize traditional medicine as "mere superstition" overlook its considerable successes in medical treatment and create a cultural conflict that does nothing to promote wise resource management.

Lastly, the conservation of seahorse wild populations cannot be dissociated from further actions and policies to address the indirect seahorse fisheries. Globally, incidental captures in trawling nets constitutes the main source of dried seahorses (see Vincent 1996; McPherson and Vincent 2004; Baum and Vincent 2005; Giles et al. 2006; Perry et al. 2010; Rosa et al. 2011; Vincent et al. 2011a), and as remarked by Vincent et al. (2011a), the total extraction is large (tens of millions of animals annually), and unsustainable.

## 10.6 Final Remarks

Seahorses are charismatic animals, and serve as flagship species which may provide support for a number of general concerns, such as the need of marine protected areas, the lack of management of subsistence fisheries, the sequential overexploitation of resources, the degradation of coastal environments, and the problems arising from excessive and destructive by-catch (Lourie et al. 1999; Vincent et al. 2011a).

A greater understanding of the processes and stakeholders involved in the commercial exploitation and multiple uses of seahorses may potentially benefit a wider range of species and environments, as well as further promote integrative approaches among conservation needs, socioeconomic aspects and cultural practices.

## References

Alakbarli F (2006) Medical manuscripts of Azerbaijan. Heydar Aliyev Foundation, Baku

Alino PM, Cajipe GJB, Ganzon-Fortes ET, Licuanan WRY, Montano NE, Tupas LM (1990) The use of marine organisms in folk medicine and horticulture: a preliminary study. Seaweed Information Center (SICEN), University of the Philippines, Quezon City

Almeida CFCBR, Albuquerque UP (2002) Uso e conservação de plantas e animais medicinais no Estado de Pernambuco (Nordeste do Brasil): Um estudo de caso. Interciencia 27(6):276–285

Alves RRN, Rosa IL (2006) From cnidarians to mammals: the use of animals as remedies in fishing communities in NE Brazil. J Ethnopharmacol 107:259–276
Alves RRN, Rosa IL (2007a) Biodiversity, traditional medicine and public health: where do they meet? J Ethnobiol Ethnomed 3(14):9
Alves RRN, Rosa IL (2007b) Zootherapeutic practices among fishing communities in North and Northeast Brazil: a comparison. J Ethnopharmacol 111:82–103
Alves RRN, Rosa IL (2007c) Zootherapy goes to town: the use of animal-based remedies in urban areas of NE and N Brazil. J Ethnopharmacol 113:541–555
Alves RRN, Rosa IL (2010) Trade of animals used in Brazilian traditional medicine: trends and implications for conservation. Hum Ecol 38(5):691–740
Alves RRN, Rosa IL, Santana GG (2007) The role of animal-derived remedies as complementary medicine in Brazil. Bioscience 57(11):949–955
Alves RRN, Lima HN, Tavares MC, Souto WMS, Barboza RRD, Vasconcellos A (2008a) Animal-based remedies as complementary medicines in Santa Cruz do Capibaribe, Brazil. BMC Complement Altern Med 8:44
Alves RRN, Vieira WLS, Santana GG (2008b) Reptiles used in traditional folk medicine: conservation implications. Biodiversity Conserv 17(1):2037–2049
Alves RRN, Neto NAL, Brooks SE, Albuquerque UP (2009) Commercialization of animal-derived remedies as complementary medicine in the semi-arid region of Northeastern Brazil. J Ethnopharmacol 124(3):600–608
Alves RRN, Oliveira MGG, Barboza RRD, Lopez LCS (2010a) An ethnozoological survey of medicinal animals commercialized in the markets of Campina Grande, NE Brazil. Hum Ecol Rev 17(1):11–17
Alves RRN, Souto WMS, Barboza RRD (2010b) Primates in traditional folk medicine: a world overview. Mamm Rev 40(2):155–180
Andrade JN, Costa Neto EM (2006) O comércio de produtos zooterápicos na cidade de Feira de Santana, Bahia, Brasil. Sitientibus Série Ciências Biológicas (Etnobiologia) 6:37–43
Araújo AM (2004) Medicina rústica. Martins Fontes, São Paulo
Ashwell D, Walston N (2008) An overview of the use and trade of plants and animals in traditional medicine systems in Cambodia, 1st edn. TRAFFIC Southeast Asia, Greater Mekong Programme, Ha Noi, Vietnam
Barros S (1977) Arte, folclore, subdesenvolvimento. Civilização Brasileira, Rio de Janeiro
Baum JK, Vincent ACJ (2005) Magnitude and inferred impacts of the seahorse trade in Latin America. Environ Conserv 32(04):305–319
Baum JK, Vincent ACJ (2011a) Seahorse trade in Central America. In: Vincent ACJ, Giles BG, Czembor CA, Foster SJ (eds) Trade in seahorses and other syngnathids in countries outside Asia (1998–2001). Fisheries Centre Research Reports 19(1). Fisheries Centre, University of British Columbia, pp 78–101
Baum JK, Vincent ACJ (2011b) Seahorse trade in Mexico. In: Vincent ACJ, Giles BG, Czembor CA, Foster SJ (eds) Trade in seahorses and other syngnathids in countries outside Asia (1998–2001). Fisheries Centre Research Reports 19(1). Fisheries Centre, University of British Columbia, pp 57–77
Baum JK, Rosa IL, Vincent ACJ (2011) Seahorse trade in South America. In: Vincent ACJ, Giles BG, Czembor CA, Foster SJ (eds) Trade in seahorses and other syngnathids in countries outside Asia (1998–2001). Fisheries Centre Research Reports 19(1). Fisheries Centre, University of British Columbia, pp 102–128
Begossi A (1992) Food taboos at Búzios Island (SE Brazil): their significance and relation to folk medicine. J Ethnobiol 12(1):117–139
Begossi A, Braga F (1992) Food taboos and folk medicine among fishermen from the Tocantins river. Amazoniana 12:341–352
Begossi A, Figueiredo JL (1995) Ethnoichthyology of southern coastal fishermen: cases from Búzios Island and Sepetiba Bay (Brazil). Bull Mar Sci 56(2):710–717
Bennet G (1834) Wanderings in New South Wales, Batavia, Pedir Coast, Singapore, and China: being the journal of a naturalist in those countries during 1832, 1833, and 1834. Richard Bentley, London

Bensky D, Clavey S, Stöger E (2004) Chinese herbal medicine: materia medica. Eastland Press, Seatle
Bisset NG (1991) One man's poison, another man's medicine? J Ethnopharmacol 32(1–3):71–81
Bonifácio KM (2002) Estudo etnoictiológico e de comércio de cavalos-marinhos (Teleostei: Syngnathidae: *Hippocampus*) por pescadores artesanais do Estuário do Rio Mamanguape—PB. BSc monograph, Universidade Federal da Paraíba
Bostock J, Riley HT (1855) The natural history. Pliny the Elder, Taylor and Francis, London
Brautigam AJ, Howes T, Hutton H, Hutton J (1994) Analysis of the market for tigers, bears and musk deer in the Russian Far East. Traffic Bull 15(1):23–30
Bruckner AW (2002) Life-saving products from coral reefs. Issue Sci Technol 18(3):39–44
Bruckner AW, Field JD, Daves NK (2005) The proceedings of the international workshop on CITES implementation for seahorse conservation and trade. NOAA Technical Memorandum NMFS-OPR-36, Silver Spring
But PP, Tam YK, Lung LC (1991) Ethnopharmacology of rhinoceros horn. II: Antipyretic effects of prescriptions containing rhinoceros horn or water buffalo horn. J Ethnopharmacol 33(1–2): 45–50
Call E (2006) Mending the web of life—Chinese medicine & species conservation. IFAW/AHPA-ERB Foundation, Yarmouth Port/Silver Spring
Camargo MTLA (1985) Medicina popular: aspectos metodológicos para pesquisa: garrafada, objeto de pesquisa: componentes medicinais de origem vegetal, animal e mineral. Almed, São Paulo
Cameron G, Pendry S, Allan C, Wu J (2004) Traditional Asian medicine identification guide for law enforcers: version II. Her Majesty's Customs and Excise, London, and TRAFFIC International, Cambridge
Chivian E (2002) Biodiversity: its importance to human health. Center for Health and the Global Environment, Harvard Medical School, Boston
Christie P, Oracio EG, Eisma-Osorio L (2011) Impacts of the CITES listing of seahorses on the status of the species and on human well-being in the Philippines: a case study. FAO Fisheries and Aquaculture Circular, vol 1058. FAO, Rome
Chen TM, Yan M (2009) A tradition of soup: flavors from China's Pearl River Delta. North Atlantic Books, Berkeley
CITES (2002) CoP12 Prop. 12.37—consideration of proposals for amendment of Appendices I and II. Inclusion of all species of the genus *Hippocampus* (*Hippocampus* spp.) in Appendix II of CITES
Clarke S (2002) Trade in Asian dried seafood: characterization, estimation and implications for conservation. Wildl Conserv Soc, New York
Costa-Neto EM (1999a) Barata é um santo remédio: introdução a zooterapia popular no Estado da Bahia. Editora Universitária da UEFS, Feira de Santana
Costa-Neto EM (1999b) Healing with animals in Feira de Santana City, Bahia, Brazil. J Ethnopharmacol 65:225–230
Costa-Neto EM (1999c) Traditional use and sale of animals as medicines in Feira de Santana City, Bahia, Brazil. Indig Knowl Dev Monit 7(2):6–9
Costa-Neto EM (2000) Zootherapy based medicinal traditions in Brazil. Honey Bee 11(2):2–4
Costa-Neto EM (2001) A cultura pesqueira do litoral Norte da Bahia. Salvador/Macéio, EDUFBA/EDUFAL, Salvador/Macéio, Brazil
Costa-Neto EM (2002) The use of insects in folk medicine in the State of Bahia, Northeastern Brazil, With notes on insects reported elsewhere in Brazilian folk medicine. Hum Ecol 30(2):245–263
Costa-Neto EM (2004) Implications and applications of folk zootherapy in the state of Bahia, Northeastern Brazil. Sustain Dev 12(3):161–174
Costa-Neto EM, Marques JGW (2000) Conhecimento ictiológico tradicional e a distribuição temporal e espacial de recursos pesqueiros pelos pescadores de Conde, Estado da Bahia, Brasil. Etnoecológica 4(6):56–68

Costa-Neto EM, Mota PC (2010) Animal species traded as ethnomedicinal resources in the Federal District, central west region of Brazil. Open Complementary Med J 2:24–30

Demunshi Y, Chugh A (2010) Role of traditional knowledge in marine bioprospecting. Biodivers Conserv 19:3015–3033

Dharmananda S (2002) Treatment of diabetes with Chinese herbs and acupuncture. http://www.itmonline.org/journal/arts/diabetes.htm. Accessed 28 Aug 2011

Dias TL, Rosa IL, Baum JK (2002) Threatened fishes of the world: *Hippocampus erectus* Perry, 1810 (Syngnathidae). Environ Biol Fishes 65(3):326–326

Eastman CR (1915) Olden time knowledge of *Hippocampus*. Ann Rep Board Regents Smithson Inst 1915:349–357

El-Deir ACA, Collier CA, Neto MSA, Silva KMS, Policarpo IS, Araújo TAS, Alves RRN, Albuquerque UP, Moura GJB (2012) Ichthyofauna used in traditional medicine in Brazil. Evidence-based complementary and alternative medicine (in press)

El-Kamali HH (2000) Folk medicinal use of some animal products in Central Sudan. J Ethnopharmacol 72:279–282

Ellis R (2005) Tiger bone & rhino horn: the destruction of wildlife for traditional Chinese medicine. Island Press, Washington D.C

Ewbank T (1856) Life in Brazil; or a journal of a visit to the land of the cocoa and the palm. Harper & Brothers, New York

Fan Z (2005) National report—China. In: Bruckner AW, Field JD, Daves NK (eds) The proceedings of the international workshop on CITES implementation for seahorse conservation and trade. NOAA Technical Memorandum NMFS-OPR-35, vol 1. Silver Spring, pp 54–61

Ferreira FS, Albuquerque UP, Coutinho HDM (2012) The trade in medicinal animals in Northeastern Brazil. Evid Based Complement Alternat Med. doi: 10.1155/2012/126938

Ferreira FS, Brito S, Ribeiro S, Saraiva A, Almeida W, Alves RRN (2009) Animal-based folk remedies sold in public markets in Crato and Juazeiro do Norte, Ceara, Brazil. BMC Complement Altern Med 9(1):17

Fontenelle LFR (1959) Aimorés: análise antropológica de um programa de saúde. Dasp/Serviço de Documentação, Rio de Janeiro

Foster SJ, Vincent ACJ (2004) Life history and ecology of seahorses: implications for conservation and management. J Fish Biol 65(1):1–61

Fulder S (1993) The book of ginseng: and other Chinese herbs for vitality. Healing Arts Press, Rochester

Gaski AL, Johnson KA (1994) Prescription for extinction: endangered species and patented Oriental medicines in trade. TRAFFIC USA and TRAFFIC International, Whasington D.C./Cambridge

Giles BG, Ky TS, Hoang H, Vincent ACJ (2006) The catch and trade of seahorses in Vietnam. Hum Exploitation Biodiversity Conserv:157–173

Goenka D (2005) National report—India. In: Bruckner AW, Field JD, Daves NK (eds) The proceedings of the international workshop on CITES implementation for seahorse conservation and trade. NOAA Technical Memorandum NMFS-OPR-35, vol 1. Siler Spring, pp 62–68

Gray AC (2005) Experience of medicine 2—clinical manuals in Homeopathic medicine, vol Two. Alastair Gray, Sidney

Gunther RT (1959) The Greek Herbal of Dioscorides. Hafner, New York

Hirschhorn HH (1982) Natural substances in currently available Chinese herbal and patent medicines. J Ethnopharmacol 6(1):109–119

Ho CCK, Tan HM (2011) Rise of herbal and traditional medicine in erectile dysfunction management. Curr Urol Rep:1–9

Hopkins J (2006) Asian aphrodisiacs. Periplus Editions (HK) Ltd., Singapore

Hoover C (2003) Response to 'Sex, drugs and animal parts: will Viagra save threatened species?'. Environ Conserv 30(04):317–318

Huxtable RJ (1992) The pharmacology of extinction. J Ethnopharmacol 37(1):1–11

IUCN (2011) Red list of threatened species. Version 2011.2. www.iucnredlist.org. Accessed 26 Oct 2011
Jacobs F (1832) Aeliani de natura animalium: libri XVII. Impensis Friderici Frommanni, Jena
James R (1747) Pharmacopoeia universalis: or, A new universal English dispensatory. J. Hodges and J. Wood, London
Kang S, Phipps M (2003) A question of attitude: South Korea's traditional medicine practitioners and wildlife conservation, 1st edn. TRAFFIC East Asia, Hong Kong
Koldewey HJ, Martin-Smith KM (2010) A global review of seahorse aquaculture. Aquaculture 302(3–4):131–152
Lages-Filho J (1934) A Medicina popular em Alagoas. Liv. Scientifica, Salvador
LaFrance P, Vincent ACJ (2011) Seahorse trade in the USA and Canada. In: Vincent ACJ, Giles BG, Czembor CA, Foster SJ (eds) Trade in seahorses and other syngnathids in countries outside Asia (1998–2001). Fisheries Centre Research Reports 19(1). Fisheries Centre, University of British Columbia, pp 39–56
Lee SKH (1999) Trade in traditional medicine using endangered species: an international context. In: Proceedings of the second Australian symposium on traditional medicine and wildlife conservation, Melbourne, 1999. Environment Australia, Canberra, pp 15–19
Lee J (2000a) TCM use of marine medicinals: content and context. In: Moreau M-A, Hall HJ, Vincent ACJ (eds) Proceedings of the first international workshop on the management and culture of marine species used in traditional medicines, Cebu City, 1998. Project Seahorse, Montreal pp 81–83
Lee Y-J (2000b) The state of seahorses as herbs in Korean oriental medicine. In: Moreau M-A, Hall HJ, Vincent ACJ (eds) Proceedings of the first international workshop on the management and culture of marine species used in traditional medicines, Cebu City, 1998. Project Seahorse, Montreal pp 92–94
Lee SK, Mills J (2000) How conservationists can work with traditional medicine communities. In: Moreau M-A, Hall HJ, Vincent ACJ (eds) Proceedings of the first international workshop on the management and culture of marine species used in traditional medicines, Cebu City, 1998. Project Seahorse, Montreal pp 87–89
Lee S, Hoover C, Gaski A, Mills J (1998) A world apart? Attitudes toward traditional Chinese medicine and endangered species in Hong Kong and the United States. TRAFFIC East Asia, TRAFFIC North America, and World Wildlife Fund-US, Washington D.C
Lei L, Yangbo Y, Shuyun L, Hongjun L (1996) Preparations and uses of Chinese medicated spirits and wine. Shandong Friendship Press, Jinan
Lev E (2003) Traditional healing with animals (zootherapy): medieval to present-day levantine practice. J Ethnopharmacol 85(1):107–118
Lipton AP (2000) Fishery for marine medicinal species in India: problems and responses. In: Moreau M-A, Hall HJ, Vincent ACJ (eds) Proceedings of the first international workshop on the management and culture of marine species used in traditional medicines, Cebu City, 1998. Project Seahorse, Montreal pp 75–77
Liu C, Tseng A, Yang S (2005) Chinese herbal medicine: modern applications of traditional formulas. Alternative medicine, CRC Press, Boca Raton
Lockyear J (2000) South African seahorses: consumption, threats and research problems In: Moreau M-A, Hall HJ, Vincent ACJ (eds) Proceedings of the First International Workshop on the Management and Culture of Marine Species Used in Traditional Medicines, Cebu City, 1998. Project Seahorse, Montreal pp 99-101
Lourie SA, Vincent ACJ, Hall HJ (1999) Seahorses: an identification guide to the world's species and their conservation. Project Seahorse, London
Lourie SA, Vincent ACJ, Hall HJ (2004) A guide to the identification of seahorses. project seahorse and TRAFFIC North America. University of British Columbia and World Wildlife Fund, Washington D.C
Low WY, Tan HM (2007) Asian traditional medicine for erectile dysfunction. J Men's Health Gend 4(3):245–250

Mainka SA, Mills JA (1995) Wildlife and traditional Chinese medicine: supply and demand for wildlife species. J Zoo Wildl Med:193–200

Marichamy R, Lipton AP, Ganapathy A, Ramalingam JR (1993) Large scale exploitation of sea horse (*Hippocampus kuda*) along the palk bay coast of Tamil Nadu. Marine Fish Inform Serv Tech Ext Ser 119:17–20

Marques JGW (1997) Fauna medicinal: recurso do ambiente ou ameaça à biodiversidade? Mutum 1(4):4

Marshall NT (1998) Searching for a cure: conservation of medicinal wildlife resources in East and Southern Africa. TRAFFIC International, Cambridge

Martin-Smith KM, Vincent ACJ (2006) Exploitation and trade of Australian seahorses, pipehorses, sea dragons and pipefishes (family Syngnathidae). Oryx 40(02):141–151

Martin-Smith KM, Vincent ACJ (2011a) Syngnathid trade in Australia. In: Vincent ACJ, Giles BG, Czembor CA, Foster SJ (eds) Trade in seahorses and other syngnathids in countries outside Asia (1998–2001). Fisheries Centre Research Reports 19(1). Fisheries Centre, University of British Columbia, pp 138–165

Martin-Smith KM, Vincent ACJ (2011b) Syngnathid trade in New Zealand. In: Vincent ACJ, Giles BG, Czembor CA, Foster SJ (eds) Trade in seahorses and other syngnathids in countries outside Asia (1998–2001). Fisheries Centre Research Reports 19(1). Fisheries Centre, University of British Columbia, pp 166–175

May B, Tomoda T (2002) Seahorses in the Ben cao gang mu and in contemporary Chinese medicine. J Australian Chinese Med Education Res Council 7(1):2–13

McPherson JM, Vincent ACJ (2004) Assessing East African trade in seahorse species as a basis for conservation under international controls. Aquatic Conserv Marine Freshw Ecosyst 14(5):521–538

McPherson JM, Vincent ACJ (2011) Trade in seahorses and other syngnathids in Africa. In: Vincent ACJ, Giles BG, Czembor CA, Foster SJ (eds) Trade in seahorses and other syngnathids in countries outside Asia (1998–2001). Fisheries Centre Research Reports 19(1). Fisheries Centre, University of British Columbia, pp 7–38

Meenakarn S, Leepayakoon P (2005) National report—Thailand. In: Bruckner AW, Field JD, Daves NK (eds) The proceedings of the international workshop on CITES implementation for seahorse conservation and trade. NOAA Technical Memorandum NMFS-OPR-35, vol 1. Silver Spring, pp 89–92

Mills J, Servheen C (1994) The Asian trade in bears and bear parts: impacts and conservation recommendations. Bears: Their Biology and Management, pp 161–167

Moreau M-A, Hall HJ, Vincent ACJ (2000) Proceedings of the first international workshop on the management and culture of marine species used in traditional medicines. Project Seahorse, Montreal

Morgan SK, Panes HM (2008) Threatened fishes of the world: *Hippocampus spinosissimus* Weber 1913 (Syngnathidae). Environ Biol Fishes 82(1):21–22

Nogueira AJ, Vargas A, Brayner F, Campos M (2005) Medicina popular. Prefeitura Municipal, Rio de Janeiro

Oliveira ES, Torres DF, Brooks SE, Alves RRN (2010) The medicinal animal markets in the metropolitan region of Natal City, Northeastern Brazil. J Ethnopharmacol 130(1):54–60

Osório FM (2005) Análise do comércio de cavalos-marinhos no Ceará. Universidade Federal do Ceará, BSc monograph

Pajaro MG (2000) The importance of trade in marine medicinals to local communities In: Moreau M-A, Hall HJ, Vincent ACJ (eds) Proceedings of the first international workshop on the management and culture of marine species used in traditional medicines, Cebu City, 1998. Project Seahorse, Montreal, pp 68–71

Pajaro MG, Vincent ACJ, Buhat DY, Perante NC (1997) The role of seahorse fishers in conservation and management. In: Lunn KE, Boehm JR, Hall HJ, Vincent ACJ (eds) Proceedings of the first international symposium in marine conservation, 1997. Hong Kong Marine Conservation Society, Hong Kong, pp 118–126

Perry A (2000) Global survey of marine medicinals. In: Moreau M-A, Hall HJ, Vincent ACJ (eds) Proceedings of the first international workshop on the management and culture of marine species used in traditional medicines. Project Seahorse, Montreal, pp 35–43
Perry AL, Lunn KE, Vincent ACJ (2010) Fisheries, large-scale trade, and conservation of seahorses in Malaysia and Thailand. Aquatic Conserv Marine Freshw Ecosyst 20(4):464–475
Parry-Jones R, Vincent A (1998) Can we tame wild medicine? New Scientist 157(2115):26–29
Pitrè G (1896) Medicina popolare siciliana. Carlo Clausen, Torino-Palermo
Project Seahorse (2002a) *Hippocampus barbouri*. In: 2011 IUCN Red List of Threatened Species. IUCN. Accessed 28 Oct 2011
Project Seahorse (2002b) *Hippocampus comes*. In: 2011 IUCN Red List of Threatened Species. IUCN. Accessed 28 Oct 2011
Project Seahorse (2002c) *Hippocampus histrix*. In: 2011 IUCN Red List of Threatened Species. IUCN. Accessed 28 Oct 2011
Project Seahorse (2002d) *Hippocampus kelloggi*. In: 2011 IUCN Red List of Threatened Species. IUCN. Accessed 28 Oct 2011
Project Seahorse (2003a) *Hippocampus borboniensis*. In: 2011 IUCN Red List of Threatened Species. IUCN. Accessed 28 Oct 2011
Project Seahorse (2003b) *Hippocampus camelopardalis*. In: 2011 IUCN Red List of Threatened Species. IUCN. Accessed 28 Oct 2011
Project Seahorse (2003c) *Hippocampus erectus*. In: 2011 IUCN Red List of Threatened Species. IUCN. Accessed 28 Oct 2011
Project Seahorse (2003d) *Hippocampus fuscus*. In: 2011 IUCN Red List of Threatened Species. IUCN. Accessed 28 Oct 2011
Project Seahorse (2003e) *Hippocampus ingens*. In: 2011 IUCN Red List of Threatened Species. IUCN. Accessed 28 Oct 2011
Project Seahorse (2003f) *Hippocampus kuda*. In: 2011 IUCN Red List of Threatened Species. IUCN. Accessed 28 Oct 2011
Project Seahorse (2003g) *Hippocampus reidi*. In: 2011 IUCN Red List of Threatened Species. IUCN. Accessed 28 Oct 2011
Project Seahorse (2003h) *Hippocampus spinosissimus*. In: 2011 IUCN Red List of Threatened Species. IUCN. Accessed 28 Oct 2011
Project Seahorse (2003i) *Hippocampus trimaculatus*. In: 2011 IUCN Red List of Threatened Species. IUCN. Accessed 28 Oct 2011
Project Seahorse (2011) Why seahorses? Essential facts about seahorses. http://seahorse.fisheries.ubc.ca/why-seahorses/essential-facts. Accessed 28 Oct 2011
Qian ZJ, Ryu BM, Kim MM, Kim SK (2008) Free radical and reactive oxygen species scavenging activities of the extracts from seahorse, *Hippocampus kuda* Bleeler. Biotechnol Bioprocess Eng 13(6):705–715
Quijano SB (2000) Role of women in the production, trade and use of marine species as medicine. In: Moreau M-A, Hall HJ, Vincent ACJ (eds) Proceedings of the first international workshop on the management and culture of marine species used in traditional medicines, Cebu City, 1998. Project Seahorse, Montreal, pp 72–74
Rasteiro A (2003) Religião, medicação e informação/desinformação na época de Amato Lusitano. In: Sallvado A (ed) Medicina na Beira do Interior da Pré-História ao Século XXI. Cadernos de Cultura, n.17– Século XVI. Quarteto Editora, Coimbra
Rocha LA (1960) História da Medicina em Pernambuco: Séculos XVI, XVII e XVIII. Arquivo Público Estadual, Recife
Robinson MM, Zhang X (2011) Traditional medicines: global situation, issues and challenges. World health organization. http://www.who.int/medicines/areas/policy/world_medicines_situation/WMS_ch18_wTraditionalMed.pdf
Rosa IL (2005) National report—Brazil. In: Bruckner AW, Field JD, Daves NK (eds) The proceedings of the international workshop on CITES implementation for seahorse conservation and trade. NOAA Technical Memorandum NMFS-OPR-35, vol 1. Silver Spring, pp 46–53

Rosa IML, Alves R, Bonifácio KM, Mourão JS, Osório FM, Oliveira TPR, Nottingham MC (2005) Fishers' knowledge and seahorse conservation in Brazil. J Ethnobiol Ethnomed 1(1):12

Rosa IL, Oliveira TPR, Alves RRN (2010) Entre o corpo e o espírito: uso medicinal de cavalos-marinhos no Brasil. In: Alves RRN, Souto WMS, Mourão JS (eds) A Etnozoologia no Brasil: importância, status atual e perspectivas. NUPEEA, Recife, pp 323–346

Rosa IL, Oliveira TPR, Osório FM, Moraes LE, Castro ALC, Barros GML, Alves RRN (2011) Fisheries and trade of seahorses in Brazil: historical perspective, current trends, and future directions. Biodivers Conserv 20(9):1951–1971

Salin KR, Mohanakumaran NC (2006) Resources and biodiversity of seahorses and the need for their conservation in India. Aquac Asia 11(3):3

Sanders JG, Cribbs JE, Fienberg HG, Hulburd GC, Katz LS, Palumbi SR (2008) The tip of the tail: molecular identification of seahorses for sale in apothecary shops and curio stores in California. Conserv Genet 9(1):65–71

Scales H (2009) Poseidon's steed: the story of seahorses, from myth to reality. Gotham, New York

Scarborough J (1996) Drugs and medicines in the Roman world. Expedition Philadelphia 38:38–51

Seixas C, Begossi A (2001) Ethnozoology of caiçaras from Aventureiro, Ilha Grande. J Ethnobiol 21(1):107–135

Servheen C (1999) The trade in bears and bear parts. In: Servheen C, Herrero S, Peyton B (eds) Bears: status survey and conservation action plan. IUCN, Gland/Cambridge, pp 33–38

Sharlip ID (1981) Penile revascularization in the treatment of impotence. West J Med 134(3):206

Shukla C (2008) The quintessence of homeopathic remedies—provings & cases of Hawthorn, Mango, Oxygen, Sea Horse, Python & Crow. Homeolinks Publishers, Haren

Silva MLV, Alves ÑGC, Almeida AV (2004) A zooterapia no Recife (Pernambuco): uma articulação entre as práticas e a história. Biotemas 17(1):95–116

Sodeinde OA, Soewu DA (1999) Pilot study of the traditional medicine trade in Nigeria. Traffic Bulletin 18(1):35–40

Solavan A, Paulmurugan R, Wilsanand V, Ranjith Sing AJA (2004) Traditional therapeutic uses of animals among tribal population of Tamil Nadu. Indian J Tradit Knowl 3(2):206–207

Sonz S, Stewart R, Kushner S (2003) *Hippocampus kuda*, a proving of seahorse. Am Homeopath 9:100–123

Sreepada RA, Desai UM, Naik S (2002) The plight of Indian sea horses: need for conservation and management. Curr Sci 82(4):377–378

Tan BKH, Ong VYC, Lim PHC (2010) Clinical applications of modern and traditional pharmacopoeias for sexual health in Singapore and Asia. In: Park NC, Okuyama A, Xin Z-C, Park JH, Earle C, Adaikan PG (eds) Modern oriental phytotherapy in sexual medicine. Koonja Publishing Inc., Seoul

Trowell S (2003) Drugs from Bugs: The promise of pharmaceutical entomology. Futurist 37(1):17–19

Urban S (1753) The Gentleman's magazine and historical chronicle, vol 23. Edward Cave, London

Van NDN, Tap N (2008) An overview of the use of plants and animals in traditional medicine systems in Viet Nam, 1st edn. Greater Mekong Programme, Ha Noi, Viet Nam, TRAFFIC Southeast Asia

Vasconcelos JL (1985) Etnografia portuguesa, vol IX. Imprensa Nacional Casa da Moeda, Lisboa

Vincent ACJ (1995a) Seahorse conservation and traditional Chinese medicine. Abstr Chin Med 6(2):272–281

Vincent ACJ (1995b) Trade in seahorses for traditional Chinese medicines, aquarium fishes and curios. Traffic Bulletin 15(3):125–128

Vincent ACJ (1995c) Exploitation of seahorses and pipefishes. Naga Iclarm Quartely 18(1):18–19

Vincent ACJ (1996) The international trade in seahorses. TRAFFIC International, Cambridge

Vincent ACJ (1997) Sustainability of seahorse fishing. In: Lessios HA, Macintyre IG (eds) Eighth international coral reef symposium, Panama, 1997. Proceedings of the eighth international coral reef symposium. smithsonian tropical research institute, pp 2045–2050

Vincent ACJ (2000) Setting the scene for marine medicinal species. In: Moreau M-A, Hall HJ, Vincent ACJ (eds) Proceedings of the first international workshop on the management and culture of marine species used in traditional medicines, Cebu City, 1998. Project Seahorse, Montreal, pp 61–65

Vincent ACJ (2006) Live food and non-food fisheries on coral reefs, and their potential management. In: Côté IM, Reynolds JD (eds) Coral Reef Conservation. Cambridge University Press, Cambridge, pp 183–236

Vincent ACJ, Marsden AD, Sumaila UR (2007a) Possible contributions of globalization in creating and addressing sea horse conservation problems. In: Taylor WW, Schechter MG, Wolfson LG (eds) Globalization: Effects on Fisheries Resources. Cambridge University Press, New York, pp 185–214

Vincent ACJ, Meeuwig JJ, Pajaro MG, Perante NC (2007b) Characterizing a small-scale, data-poor, artisanal fishery: seahorses in the central Philippines. Fish Res 86(2–3):207–215

Vincent ACJ, Foster SJ, Koldewey HJ (2011a) Conservation and management of seahorses and other Syngnathidae. J Fish Biol 78(6):1681–1724

Vincent ACJ, Giles BG, Czembor CA, Foster SF (2011b) Trade in seahorses and other syngnathids in countries outside Asia. Fisheries Centre, University of British Columbia, Vancouver

Wemmer C (1998) Deer: status survey and conservation action plan, vol 43. IUCN, Gland/Cambridge

Wong K, Dahlen M (1999) Streetwise guide: Chinese herbal medicine. Red Mansions, San Francisco

Woods CMC, Morgan SK, Martin-Smith K, Pogonoski JJ, Paxton JR, Pollard DA, Morgan AJ (2006) *Hippocampus abdominalis*. In: 2011 IUCN Red List of Threatened Species. IUCN. Accessed 28 Oct 2011

Wu S, Liu NF, Zhang YM, Ma GZ (2004) Assessment of threatened status of Chinese pangolin (Manis pentadactyla). Chin J Appl Environ Biol 10(4):456–461

Xie H, Preast V (2010) Xie's Chinese veterinary herbology. Wiley-Blackwell, Ames

Yao DM, Chang CP (1995) A colored atlas of the Chinese materia medica specified in pharmacopoeia of the People's Republic of China. Guangdong Science and Technology Press, Guangdong

Zhang E (2000) Preliminary report on seahorse trade in Shanghai and Tianjin. In: Moreau M-A, Hall HJ, Vincent ACJ (eds) Proceedings of the first international workshop on the management and culture of marine species used in traditional medicines, Cebu City, 1998. Project Seahorse, Montreal, pp 84–86

Zheng H (2000) Marine species in traditional Chinese medicines. In: Moreau M-A, Hall HJ, Vincent ACJ (eds) Proceedings of the first international workshop on the management and culture of marine species used in traditional medicines, Cebu City, 1998. Project Seahorse, Montreal, pp 90–91

Zhu CH (1989) Clinical handbook of Chinese prepared medicines, Paradigm Title Series. Paradigm Publications, Brookline

# Chapter 11
# Aquatic Mammals Used in Traditional Folk Medicine: A Global Analysis

**Rômulo Romeu Nóbrega Alves, Wedson Medeiros Silva Souto, Ronnie Enderson Mariano Carvalho Cunha Oliveira, Raynner Rilke Duarte Barboza and Ierecê Lucena Rosa**

**Abstract** Aquatic mammals are medium- to large-sized animals, which have been the source of large amounts of subproducts—among which medicines—to humans since ancient times. They are also among the most threatened animal taxa worldwide. This chapter provides an overview of the global use of aquatic mammals in traditional folk medicine and the implications for conservation, with emphasis on species of the orders Cetacea, Sirenia, and Carnivora. At least 24 species of aquatic mammals are used in traditional medicine worldwide, although all these species are listed in the IUCN Red List, including one species categorized as Critically Endangered, three as Endangered, eight as Vulnerable. The medicinal use probably does not pose a major threat to most aquatic mammals; however, their exploitation for medicinal purposes is troubling, as it constitutes an additional and persistent impact for a group of animals that has suffered major direct impacts, especially hunting (for food and fur) and habitat loss, as well as impacts related to global warming.

**Keywords** Aquatic mammals · Traditional folk medicine · Conservation

R. R. N. Alves (✉) · R. E. M. C. C. Oliveira
Departamento de Biologia, Universidade Estadual da Paraíba, Av. das Baraúnas, 351/Campus Universitário, Bodocongó, Campina Grande-PB 58109-753, Brazil
e-mail: romulo_nobrega@yahoo.com.br

W. M. S. Souto · R. R. D. Barboza
Departamento de Sistemática e Ecologia,
Programa de Pós-Graduação em Ciências Biológicas (Zoologia),
Universidade Federal da Paraíba,
João Pessoa-PB 58059-970, Brazil

I. L. Rosa
Departamento de Sistemática e Ecologia, Universidade Federal da Paraíba,
João Pessoa-PB 58059-970, Brazil

R. R. N. Alves and I. L. Rosa (eds.), *Animals in Traditional Folk Medicine*,
DOI: 10.1007/978-3-642-29026-8_11, © Springer-Verlag Berlin Heidelberg 2013

## 11.1 Introduction

Aquatic mammals are animals that occupy a variety of habitats, from pelagic continental shelf and coastal marine waters to estuarine, riverine, and lacustrine areas (Vidal 1993); some may spend their entire life in the water, while others move between water and land. Although these animals show some common basic characteristics, they exhibit a wide variety of ecological characteristics such as size, behavior, diet, and habitat (Crespo and Hall 2002). Generally, they are medium to large sized and are therefore a potential source of large amounts of subproducts, which have been utilized by humans since ancient times, as attested by archeological evidence and historical accounts (Colten and Arnold 1998; Crespo and Hall 2002; Porcasi and Fujita 2000; Romero et al. 2002; Romero and Creswell 2005). Aquatic mammals have a history of interaction with humans. For instance, in the cold regions of the planet, the scarcity of animal protein of terrestrial origin and of alternative food sources, led their inhabitants to the coastal zone or the sea, where mammals were especially important (Colten and Arnold 1998). Besides their use as food, aquatic mammals, cetaceans especially, are utilized as artifacts, adornments, fuel, and for other purposes (Perrin et al. 2008; Ripple and Perrine 1999; Roman 2006; Toledo et al. 2010). In addition, various species of aquatic mammals have been a source of fascination for peoples of different communities throughout the world (Alie 2008; Cravalho 1999). Dolphins, whales, sea lions, and porpoises, among others, appeared since early times in the myths and the magic symbolisms of many cultures, and were integrated into rituals and ceremonies (Alves and Rosa 2008; Cravalho 1999; Crespo and Hall 2002; Eason 2008; McKillop 1985)

From an environmental perspective, the interactions between humans and aquatic mammals can have negative impacts on the latter, when hunting by humans implies an additional pressure on the species, or positive impacts, when cultural aspects impede their capture and killing. Because of hunting and use by humans, combined with loss of suitable habitat, conflicts with people, and other causes, many species of aquatic mammals are now considered to be endangered in the wild. Considering that many species are captured for use in traditional medicine, ethnozoological studies constitute an important asset to conservation and environmental protection efforts. This chapter focuses on the global use of aquatic mammals in traditional folk medicine around the world, with special emphasis on species of the orders Cetacea, Sirenia, and Carnivora. Considering that dolphins are particularly used in folk medicine, their use in Brazilian popular medicine is discussed in more detail, exemplifying thus the importance of these animals in this context.

## 11.2 Background

The relationship between people and aquatic mammals has been around since ancient times, and has been expressed in different ways. Examples are the 9,000-year-old drawings carved into rocks in northern Norway of killer whales

(*Orcinus orca*) and other local animals, and the Greek and Minoan (Crete 3000–1500 BC) portrayal of dolphins on frescoes, mosaic floors, coins and vases, and in sculpture (Perrin et al. 2008). Various authors have highlighted that human interaction with seals, sea otters, and other marine mammals spans millennia and the entire globe (Hildebrandt and Jones 1992; Klein and Cruz-Uribe 1996; McNiven and Bedingfield 2008; Monks 2005; Rick et al. 2011; Stringer et al. 2008), and that long history of interactions between humans and aquatic mammals is reflected in various cosmological and supernatural traditions, through many folktales and fables (Alves and Rosa 2008; Cravalho 1999; Perrin et al. 2008; Ripple and Perrine 1999).

In this sense, it is not surprising that these animals had an important role in the traditional medicine of different cultures, standing out among the wild animals commonly used in folk medicine (Alves and Rosa 2008; Cravalho 1999; Perrin et al. 2008; Ripple and Perrine 1999).

## 11.3 Procedures

The diversity of aquatic mammals used in traditional medicine was assessed using published articles and reports; only taxa that could be identified to species level were included in the database. Scientific names provided in publications were updated according to the ITIS Catalogue of Life: 2011 Annual Checklist. The conservation status of the species follows IUCN (2011) and CITES (convention on international trade in endangered species of fauna and flora) (CITES 2011).

## 11.4 Results and Discussion

Our survey indicated that at least 24 aquatic mammals are used in traditional medicine worldwide, represented by cetaceans (12 species), Carnivora (8), and Sirenia (4). The families with the highest number of species used as remedies were Delphinidae, Otariidae, and Phocidae (with 4 species each).

We found records of the use of aquatic mammals for medicinal purposes pertaining to 45 countries; however, the use of some species is widespread and cross-border. For instance, *Sotalia fluviatilis* is used for medicinal and magical religious purposes in Brazil, Bolivia, and Peru; *Trichechus innuguis* in Brazil, Colombia, Ecuador, Guyana, and Venezuela.

Among other aspects, the substantial number of species of aquatic mammals used in traditional folk medicine is related to their large body size. As remarked by Perrin et al. (2008), these animals generally have substantial amounts of fatty tissue, thus furnishing a large amount of fat, an ingredient widely used in popular medicine worldwide (Alves and Alves 2011; Lev 2006; MacKinney 1946; Quave et al. 2010).

Peoples' access to aquatic mammals is potentiated by the fact that some species are docile, and thus easy to capture; others frequently beach on the coast, and others are prey and die in fishing nets. All these factors contribute to a greater access to the animals, despite many species being protected by laws in different countries. For instance, the large terrestrial breeding colonies of some pinnipeds have made them highly susceptible to hunting in the distant and near past. As discussed by Rick et al. (2011) due to their large amounts of meat, oil, ivory, and other important raw material and dietary resources, pinnipeds (seals, sea lions, and walruses) and sea otters have been hunted and scavenged by people in the northeastern Pacific for much of the Holocene or earlier.

In more recent times, during the fur and oil trade of the eighteenth and nineteenth centuries, human hunting of pinnipeds and sea otters decimated marine mammal populations, causing the extirpation of local populations of sea otters (*Enhydra lutris*), Guadalupe fur seals (*Arctocephalus townsendi*), and northern elephant seals (*Mirounga angustirostris*) and the extinction of the Steller's sea cow (*Hydrodamalis gigas*) (Ellis 2003; Ogden 1941; Scammon 1968).

Considering that traditional medicine is strongly associated with religious magical elements, cultural factors certainly influence the use of aquatic mammals in traditional folk medicine. In fact, the role of aquatic mammals in folk practices related to magic and to the healing and/or prevention of illnesses has been recorded in different social-cultural contexts throughout the world.

Examples are varied and involve different taxa: amulets derived from aquatic mammals are utilized in the prevention of diseases of spiritual cause (Alves and Rosa 2008); Maori folklore of the Ngati Porou people tells of their ancestors being carried safely across the Pacific to New Zealand on the back of a whale (Ihimaera 1987); the Ngai Tahu people consider the sperm whales off the coast of the South Island as taonga (treasures); if a whale strands, prayers are said to return its spirit to Tangaroa, the Maori god of the sea. After this, the lower jawbone is removed for ceremonial carving and placement on the marae, the tribes' traditional meeting grounds (Perrin et al. 2008). The Rama Indians of coastal Nicaragua believe that if a hunter fails to return a manatee's bones to where it was killed, he will be unable to find manatees on future hunts. Near the Stewart River on Cape York Peninsula of northern Australia, the natives used to believe that any dugong remains must be discarded into the river (rather than burned) to avoid horrible luck (Ripple and Perrine 1999).

Various parts and products of aquatic mammals are used in the preparation of folk medicines, including, skin, fat, oil, eyes, bones, penises, vagina, brain, embryo, teeth, flippers, tusks, milk, wax, and meat. The influence of Westernization is reflected in the presentation of some zootherapeutic products, which are either manufactured or pre-packaged. Examples are the fat extracted from the manatee (*Trichecus* sp.), sold as tablets and as a manufactured soap in Brazil (Alves and Rosa 2007a).

It should be noted that in some cases adulteration occurs, and medicinal products of domestic animals are sold as if they are derived from aquatic mammals. A study by Malik et al. (1997) exemplifies this situation well: their genetic research on Pinniped (seals, sea lions, fur seals, and walrus) penises purchased in traditional Chinese medicine shops in Asia and North America revealed that of 21 samples marketed specifically as seal penises, seven were not derived from pinnipeds; one was identified as originating from domestic cattle (*Bos taurus*) and six could not be identified to species due to a lack of relevant published sequence. Similarly, field surveys conducted in Brazil by Alves and Rosa (2007a, 2010) found evidence that cattle lard (*Bos taurus*) is sometimes sold as being from manatee (*T. manatus*), a threatened species and protected by Brazilian law, but extremely important in the popular pharmacopeia (Alves and Rosa 2007a, 2010; Alves et al. 2007).

Aquatic mammals are used to treat approximately 60 diseases. Besides their role in the treatment of clinical diseases, medicinal products often have magical-religious significances, where they are used in the treatment and prevention of diseases considered of spiritual cause or as amulets to prevent these diseases, for good luck and to attract sex partners. The same species can also furnish products used for different purposes. In Indonesia, for instance, the siren's meat is believed to give strength, the fat is used in obstetrics and cooking, the siren's tears are a love potion, and the incisors are thought to render their owner invulnerable to harm (Ripple and Perrine 1999).

Patterns of species exploitation by humans can be influenced by both environmental conditions and culture (e.g., religion) (Adeola 1992; Alves et al. 2010; Léo Neto and Alves 2010; Richerson et al. 1996), and depending on the proposed use and on the associated cultural aspects, aquatic mammals can be utilized live or dead. While dead specimens provide different parts used as remedies, such as meat, fat, and oil the use of the live animal has also been recorded, generally in a ritualistic context. A good example was recorded by Perrin et al. (2008) in the Gulf of Carpentaria, northern Australia, where a tribe of Aborigines has been in direct communication with Indian Ocean bottlenose dolphins (*Tursiops aduncus*) for thousands of years. In that specific case, a medicine man calls the dolphins and "speaks" to them telepathically. Through these communications, he ensures that the tribes' fortune and happiness are maintained.

All species recorded in our survey are included on the IUCN's 2011 Red List of Threatened Species (one species categorized as Critically Endangered, three as endangered, eight as Vulnerable, and one as Near Threatened), and 15 are CITES-listed (see Table 11.1). For some of those species, medicinal use represents an additional pressure that along with other issues has been contributing to the decline of natural populations. A decade ago, Beasley and Davidson (2001) drew attention to the fact that, the dugong *Dugong dugon* and the Mekong River Irrawaddy dolphin *Orcaella brevirostri* (globally threatened species which are highly valued in traditional medicine) were species of particular concern. While dugongs' tusks are valued for their medicinal and aphrodisiac properties, their skull and postcranial are cut into small pieces and sold for medicinal use. Additionally, anecdotal evidence suggests that most parts of dugongs can be used in traditional medicine—the oil is used for rheumatism, the bones for fever in children, the meat

**Table 11.1** Aquatic mammals used in traditional medicine

| Family/Species | IUCN Red List | CITES | Country(-ies) | Use(s) | Reference(s) |
|---|---|---|---|---|---|
| **Balaenopteridae** | | | | | |
| *Balaenoptera acutorostrata* Lacépède, 1804 | LC | I | Brazil | Menstrual cramps, rheumatism, sore throat, wounds | 1–8 |
| *Megaptera novaeangliae* (Borowski, 1781) | LC | I | Brazil, Mexico | Unspecified | 9, 10 |
| **Delphinidae** | | | | | |
| *Delphinus delphis* Linnaeus, 1758 | LC | | Africa, Brazil, China | In Brazilian Amazon, local people use dolphins in love fetishes; In other areas, uses of *D. delphis* are not mentioned | 11–16 |
| *Orcaella brevirostris* (Owen in Gray, 1866) | VU | I | Cambodia, Philippines | For sick livestock or humans, to keep away evil spirits that cause chicken sickness, for sick buffaloes, for sick children | 17–19 |
| *Sotalia fluviatilis* (Gervais and Deville, 1853) | DD | I | Bolivia, Brazil, Colombia, Peru | Arthritis, arthrosis, asthma, athlete's foot, bloating, cancer, chest ailments, chilblains, dermatophytosis, earache, erysipelas, hemorrhoids and their inflammation, headache, hernia, impiger, inflammation, injuries caused by the spines of the 'arraia', pityriasis, rheumatism, schistosomiasis, sore throat, swelling, thrombosis, tumor, womb disorders, wounds; attract money. It is used also as love charms or aphrodisiacs when prepared in a special manner | 1–3, 5, 6, 16, 20–33 |
| *Sotalia guianensis* (van Bénéden, 1864) | DD | I | Bolivia, Brazil, Peru | Arthritis, arthrosis, asthma, athlete's foot, bloating, cancer, chest ailments, chilblains, dermatophytosis, earache, erysipelas, hemorrhoids and their inflammation, headache, hernia, impiger, inflammation, injuries caused by the spines of the 'arraia', pityriasis, rheumatism, schistosomiasis, sore throat, swelling, thrombosis, tumor, womb disorders, wounds, sexual attraction; attract money | 1–3, 5, 6, 16, 27, 28, 31–35 |
| **Dugongidae** | | | | | |
| *Dugong dugon* (Müller, 1776) | VU | I | Cambodia and SE Asia | Aphrodisiac, fever in children, rheumatism, to protect households from evil spirits | 12, 19, 36 |

(continued)

**Table 11.1** (continued)

| Family/Species | IUCN Red List | CITES | Country(-ies) | Use(s) | Reference(s) |
|---|---|---|---|---|---|
| Iniidae | | | | | |
| *Inia geoffrensis* (Blainville, 1817) | DD | | Brazil, Bolivia, Colombia, Ecuador, Peru, Venezuela | Asthma, athlete's foot, cancer, chest ailments, chilblains, convulsions in children, dermatophytosis, earache, effusion, epilepsy, erysipelas, headache, hemorrhoids and their inflammation, hernia, impiger, inflammation, injuries caused by the spines of the 'arraia', pityriasis, rheumatism, schistosomiasis, sore throat, stroke, swelling, thrombosis, tumor, used to increase sexual desire, used for sexual attraction, used to make money, used in making *"patuás"* (a kind of amulet that are seeking good luck, love, and financial success), womb disorders, wounds | 1–3, 5, 6, 16, 25, 31–35, 37, 38 |
| *Lipotes vexillifer* Miller, 1918 | CR | I | India | unspecified | 39 |
| **Monodontidae** | | | | | |
| *Monodon monoceros* Linnaeus, 1758 | NT | | Canada, France, Germany, Greenland, Italy, Norway | Not mentioned | 11, 12 |
| **Otariidae** | | | | | |
| *Arctocephalus australis* (Zimmermann, 1783) | LC | II | Not mentioned | Not mentioned | 12, 40 |
| *Arctocephalus pusillus* (Schreber, 1775) | LC | II | Australia, Canada, Chile, China, Falkland Islands, Hong Kong, Japan, Malaysia, Namibia, Peru, South Africa, Vietnam | Chest and ear infections, protective charm | 11, 12, 14, 15, 37, 40, 41 |
| *Callorhinus ursinus* (Linnaeus, 1758) | VU | | China | Liver and kidney problems, used as substitute for *Phoca vitulina* products | 11, 13, 40, 42, 43 |
| *Eumetopias jubatus* (Schreber, 1776) | EN | | E Asia, SE Asia | Used as substitute for Tiger penis in Asian Traditional Medicine systems | 44 |

(continued)

**Table 11.1** (continued)

| Family/Species | IUCN Red List | CITES | Country(-ies) | Use(s) | Reference(s) |
|---|---|---|---|---|---|
| **Phocidae** | | | | | |
| *Cystophora cristata* (Erxleben, 1777) | VU | | Not mentioned | Not mentioned | 45 |
| *Pagophilus groenlandicus* (Erxleben, 1777) | LC | | Not mentioned | Not mentioned | 45 |
| *Phoca vitulina* Linnaeus, 1758 | LC | | China | Not mentioned | 12, 13 |
| *Pusa hispida* Schreber, 1775 | LC | | Not mentioned | Not mentioned | 45 |
| **Physeteridae** | | | | | |
| *Physeter macrocephalus* Linnaeus, 1758 | VU | I | Australia, Belgium, Brazil, Canada, Denmark, Falkland Islands, Fiji, France, Greenland, Iceland, Israel, Japan, Jordan, Lao PDR, Netherlands, New Zealand, Norway, Russia, Sudan, Switzerland, Tonga, United Kingdom, the United States of America | Used to reinforce potency, kidneys and for general health, blood tonic, asthma, backache, blood tonic, cough, general health, heart diseases, hysteria, kidney diseases, menstrual cramps, paralysis, reinforces potency, rheumatism, sore throat, wounds | 1–6, 11–13, 34, 46–52 |
| **Platanistidae** | | | | | |
| *Platanista gangetica* (Roxburgh 1801) | EN | I | Nepal | Gastric problems | 37 |
| *Platanista minor* (Owen 1853) | EN | I | Pakistan | Unspecified | 53, 54 |

(continued)

**Table 11.1** (continued)

| Family/Species | IUCN Red List | CITES | Country(-ies) | Use(s) | Reference(s) |
|---|---|---|---|---|---|
| **Trichechidae** | | | | | |
| *Trichechus inunguis* (Natterer 1883) | VU | I | Brazil, Colombia, Ecuador, Guyana, Venezuela | Allergy, arthritis, arthrosis, asthma, backache, burns, bursitis, 'estrepes' (suck a splinter out of skin), hernia, inflammation, injuries caused by bang, insect bites, itch, luxation, menstrual cramps, muscle strain, osteoarthritis, osteoporosis, pain, rheumatism, sore throat, sprains, tumor, vaginal discharge, wounds, wrench | 1–3, 5, 6, 12, 25, 33, 34, 46, 55–62 |
| *Trichechus manatus* Linnaeus 1758 | VU | I | Brazil | Arthrosis, asthma, backache, burns, hernia, injuries caused by bang, insect bites, luxation, menstrual cramps, muscle strain, rheumatism, sore throat, sprains, suck a splinter out of skin or fresh, tumor, vaginal discharge, wounds | 1–3, 5, 6, 34, 55–58, 62, 63 |
| *Trichechus senegalensis* Link 1795 | VU | II | Nigeria and Weast Africa | This species is taken for food and traditional medicinal products throughout their range | 64–67 |

*IUCN red list categories*: DD Data deficient, LC Least Concern, NT Near Threatened, VU Vulnerable, EN Endangered; CITES Appendices I or II. *References*: 1. Alves et al. (2007), 2. Alves and Rosa (2007b), 3. Alves and Rosa (2006), 4. Costa-Neto and Alves (2010), 5. Alves et al. (2009a), 6. Alves and Rosa (2007a), 7. Alves and Alves (2011), 8. Alves and Rosa (2010), 9. Bayúgar (2007), 10. Costa-Neto et al. (2002), 11. CITES (2001), 12. CITES (2002a), 13. Jiangsu New Medical College (2002), 14. Marshall (1998), 15. Pujol (1993), 16. Gravena et al. (2008), 17. Smith et al. (2004), 18. Reeves et al. (2008), 19. Ashwell and Walston (2008), 20. Secchi (2010), 21. Silva and Best (1994), 22. Silva and Best (1996b), 23. Trujillo and Diazgranados (2002), 24. Alves and Rosa (2008), 25. Tratado de Cooperacion Amazonica (1999), 26. Dourojeanni (1990), 27. Siciliano (1994), 28. Reeves et al. (2003), 29. Salter (1994), 30. Smith (1996), 31. Silva and Best (1989), 32. Cravalho (1999), 33. Alves (2006), 34. Alves (2009), 35. Silva (1990), 36. Hines et al. (2004), 37. Nilsson (1990), 38. Moura and Marques (2008), 39. Pompa et al. (2011), 40. Malik et al. (1997), 41. Simelane and Kerley (1998), 42. Bensky et al. (1993), 43. Enqin (1991), 44. Yates (2005), 45. Perry (2000), 46. CITES (2002b), 47. Lev and Amar (2000), 48. El-Kamali (2000), 49. Costa-Neto (2000), 50. Lev (2003), 51. Lev (2006b), 52. Lev (2006a), 53. Perveen (2009), 54. Sinha and Sharma (2003), 55. Costa-Neto (1999a), 56. Costa-Neto (1999b), 57. Costa-Neto (1999c), 58. Costa-Neto (1999d), 59. Begossi and Braga (1992), 60. Alves et al. (2011), 61. Alves et al. (2009b), 62. Costa-Neto (2004), 63. Seixas and Begossi (2001), 64. Powell (1996), 65. Powell and Kouadio (2008), 66. Adeola (1992), 67. Costa-Neto (2005) for consumption, the hair for toothache, and the tusks and flippers can be used to protect households from evil spirits (Hines et al. 2004).

In face of the multiple uses given to aquatic mammals, it is difficult to assess the impact of the medicinal use alone on wild populations. For example, the meat of manatees has been used as food, their oil for cooking, and their skin for the manufacture of whips. These products, as well as ear bones, have also been used for medicinal purposes in different forms. In addition, ear bones have been used as amulets (Ripple and Perrine 1999). In this sense, although manatees are utilized in traditional medicine, the animal can be killed for other reasons or can even be found dead when beached, and on these occasions, their products are utilized. An example comes from the Lao PDR, where Baird (1995) noted that although dolphin bones and oil are used in traditional medicine the animals are apparently not killed specifically for that purpose.

Moreover, Amazonian tribes believe that those who drown or eat manatees become manatees themselves (Lefebvre et al. 2011; Romero and Creswell 2005), it should be noted that in some cases, myths and legends associated with these animals—although not always sufficient to stop their exploitation—can provide them with some level of protection (Alves and Rosa 2008). In Cambodia, for example, it is generally considered unlucky to deliberately kill a dolphin, but if a dead dolphin is found, the body parts are then often used for traditional medicine, for both livestock and humans (Ashwell and Walston 2008). Examples include: dolphin bones ground into fine powder and placed in water for sick livestock or humans to drink; dried flesh hung in chicken coops to keep away evil spirits that cause chicken sickness; dolphin teeth placed in holes made in the horns of sick buffaloes; and dolphin teeth made into a necklace to be hung round the neck of a sick child (Ashwell and Walston 2008; Beasley and Davidson 2001). These examples highlight that the use of aquatic mammals in a magical context should be addressed within a perspective of conservation, since the complex system of beliefs associated with this form of interaction can be a source of impact, as well as a factor that may offer a certain level of protection to some species.

## 11.5 Use of Dolphin in Brazilian Traditional Medicine

Dolphins are among the animal species most frequently used in traditional folk practices in different social-cultural contexts throughout Brazil (Alves and Rosa 2008). They are the subject of numerous myths and legends in South America, and are an integral form of Brazilian folklore. For instance, there is a belief that a person who kills a pink river dolphin (*Inia geofreensis*) will not succeed in killing anything else afterwards, and will always be punished. There are also stories of dolphins taking paddles away from lone canoeists, and of dolphins helping people whose boats have capsized by pushing them ashore. There are even stories of fishermen training dolphins to help them fish by steering fish toward them (Lamb 1967). Many people purportedly believe that dolphins seduce girls and are the father of all children of unknown paternity, and some women name dolphins as legally responsible for their illegitimate children. When women give birth to children with spina bifida, they blame it on dolphins because the defect leaves an opening in the head that resembles a dolphin's blowhole (Morell 1997). Dolphins are often believed to transform themselves in the early hours of the night into handsome, young, tall, white men, who are talented dancers, who, before dawn, jump into the water and turn into dolphins again (Cravalho 1999). During their transformation into humans, dolphins are believed to enter households and paralyze their occupants. At such times, they are said to engage in sexual intercourse with men or women, depending on the gender of the dolphin. They often return to the same household, and the victims become ill, suffering from loss of appetite, body stiffness, and vocal aberration. Some people are believed to die from it as a consequence and their soul is taken to the bottom of the sea to serve as the dolphin's consort (Cravalho 1999).

To defend themselves from a dolphin's attack people are advised to draw a cross with garlic under their hammocks, on the back of their necks, and on their foreheads, in the belief that dolphins do not like garlic. To a certain extent, such myths have given dolphins some protection for many generations, and they are generally left unmolested by the local people. As noted by Culik (2004), freshwater dolphins have been protected by such beliefs from Colombia to southern Brazil as well as in the Amazon. Nevertheless, myths alone have not sufficed to preserve the dolphins. Despite the lack of records of past or recent commercial fisheries, gillnets deployed in the dolphins' favored habitats clearly have the potential to cause significant damage to these species at the population level, as they have to cetaceans worldwide (Northridge and Hofman 1999). As pointed by Martin et al. (2004), Amazonian dolphins selectively occur in areas known to be favored for gillnet deployment by local fishermen, and this may explain why entanglement is apparently a common cause of mortality. Increasing human populations and rapid economic development, particularly throughout Brazil, cause additional threats (Dudgeon et al. 2006). Freshwater dolphins are often in direct competition with humans for space (Plagányi and Butterworth 2002), and as a result many freshwater dolphin populations are facing threats that coastal or oceanic cetaceans do not encounter (Earle 1996). Following the recent increase in fishing activities in Brazil, negative interactions between dolphins and fisheries are on the rise (Silva and Best 1996a). This occurs by accidental entanglement in fishing gear (Menegaldo 2008; Silva and Best 1996a) and by direct conflict with fishermen, who consider the dolphins as direct competitors for some species of fish (Silva and Best 1996a). Similar conflicts have also been recorded in the Colombian and Peruvian Amazon (Leatherwood and Reeves 1997; Trujillo et al. 2002). Additionally, some are deliberately hunted for their meat and oil, which are used as fish bait and as an emulsion to protect boats from water. They are also occasionally hunted for their body parts, which are used in traditional medicine.

In Brazil there is a market for the eyes, teeth, and genital organs of the dolphins, which are used as magical charms. Dolphins products were commonly sold in outdoor markets and stores that sell religious articles in the Brazilian cities (Fig. 11.1).

**Fig. 11.1** Examples of dolphin products used in Brazilian folk medicine: **a** Eyeball, *Sotalia guianensis*; **b** Eyes, penis, perfume and vagina, *Sotalia fluviatilis*; **c** and **d** Product locally known as "Água da bota" (water of the female dolphin), containing a vagina of *S. fluviatilis*; **e** Product locally known as "Água de caranguejo" (water of the crab), also containing a dolphin's embryo; **f** Snout, *S. fluviatilis*. Teeth sold for medicinal and magic religious purposes in Belém city, N Brazil. Photo credits: **a** Gustavo Toledo; **b**, **c**, **d**, **e** and **f** Rômulo Alves

Traders obtain specimens or parts of dolphins through sellers that periodically bring these products to them or directly from fishers that come from rural areas which have accidently captured dolphins (Alves and Rosa 2008). The species traded are *Sotalia fluviatilis, Sotalia guianensis* and *Inia geofreensis*. Traders reported that dolphins are harpooned, or killed by blows to their heads while entangled in nets. Incidental captures seem to be common, as well as mortality due to collisions with boat engines. In fact, part of these products are derived from incidental catches in fishing gear (Alves and Rosa 2008), but Best and Silva (1989) confirmed the intentional death by harpoon of at least four tucuxis and one boto, which represented 7.5 % of their sample ($n = 67$). Cases of aggression and intentional killing of aquatic mammals are not uncommon in the Brazilian territory. Simões-Lopes and Ximenez (1990) reported two cases of human aggression toward *Sotalia guianensis* in the Baía Norte of Santa Catarina Island, where knife cuts and strikes were also the probable cause of death. Such retaliatory measure is usually related to competition for prey or damage to fishing gear (Loch et al. 2009).

The fat and oil of the skin of dolphins also are used commercially and in traditional medicine (Alves and Rosa 2007a; Alves et al. 2007). Despite intensive use and commercialization, there is a general lack of information, which renders difficult any evaluation of the magnitude and impact of the harvesting and trade on the natural populations. For conservation of wild species to be effective, both biological and social elements of the conservation process must be considered as a basis for minimizing threats.

They use fat and oil of the skin of dolphins, generally as an ointment rubbed on a wound or sore place, or swallowed, depending on the illness to be treated. The fat of dolphins is used for treating at least 15 diseases: Asthma, headache, rheumatism, hernia, womb disorders, sore throat, injuries caused by the spines of the 'arraia', swelling, hemorrhoids inflammation, wounds, earache, erysipelas, athlete's foot, tumor, cancer (Alves and Rosa 2008). Dolphins' teeth were also reportedly used in a concoction for treating asthma, after being sun-dried, grated, and crushed to powder. Alves and Rosa (2006, 2007b) documented the use of the fat of *S. fluviatilis* and *S. guianensis* for therapeutic purposes in fishing communities in the states of Piauí, Paraíba, and Maranhão, and these results suggest a geographic continuum in the use of dolphins as medicine in fishing communities located in coastal Brazil, which needs to be further examined in the context of their conservation and management.

Beyond their utilization as medicines, parts of dolphins are traded for their purported magical/religious attributes, being popular mainly among adepts of the Afro-Brazilian religions. Dolphins' products traded are: eyes, teeth, brain, embryo, penis, and vagina. Additionally, these parts are also used in the preparation of purifying "baths" which are indicated for attracting sexual partners. For the "baths", interviewees stated that the "water" of the dolphin is prepared using parts of dolphin (vagina or penis, according to gender of user) immersed in alcohol or patchouly oil (*Pogostemon* sp.). The "water" can also be produced using parts of dolphin (penis or vagina) and other animal species. In one shop in the city of Belém, a dolphin's embryo and the crab *Ucides cordatus* were used as ingredients for making the "water". The extracts are subsequently mixed with water during baths, or applied as a perfume. According to traders, this procedure will ensure that the user is successful in love-related matters. Another product indicated to assure the user success in love is the "perfume do boto" (dolphin's perfume), also called "asseio do(a) boto(a)", which is produced using a penis or vagina of *S. fluviatilis*. This product is either used as a perfume or is rubbed on the genital organ prior to intercourse, observing gender distinction: women should use the product made with the sexual organ of a female dolphin, while men should use the infusion containing the sexual organ of a male dolphin (Alves and Rosa 2008).

Various studies have demonstrated that the dolphin's sexual organs are believed to have special powers. It is the sexual nature of this folklore that leads to the perceived value of boto body parts as love charms. The most important of these are the eyes and genitalia (Silva 1990) valued particularly by Amazonian city dwellers (Smith 1996). For example, holding the eye of the *boto* while conversing with a member of the opposite sex is thought to make one highly sexually attractive because no one can resist the boto's gaze (Shao-Ke et al. 2010). *Boto* genitalia are considered even more powerful (Best and Silva 1989). Grinding the dry penis of the *boto*, mixing it with talcum powder or perfume, and applying it to the genital area is believed to increase the pleasure that a woman can give a man. Similarly, the application of a ground dolphin's vagina to a man's penis is believed to increase his potency (Cravalho 1999).

Another very popular product derived from dolphins is the "patuá", a kind of amulet that is hung around the neck glued on a piece of cloth, or kept in one's pocket or wallet. Parts of the dolphins that are used in the "patuá" are: eyes, or pieces of the penis or vagina. "Patuás" can be produced using one or more animal parts derived from one or more species of animal. For example, the same "patuá" can contain a dolphin eye and a dried seahorse (*Hippocampus* spp.). According to the shop owners where they are sold, these amulets are very popular among customers seeking good luck, love, and financial success (Alves and Rosa 2008). Loch et al. (2009) pointed that the "boto" amulets sold in markets of main Amazonian cities are not derived from the true boto (*Inia geoffrensis*). All amulets, if they are of dolphin origin at all, are unambiguously derived from the marine species *Sotalia guianensis*. This implies that the "boto" fetishes most likely originate in the coastal areas of North Brazil, and are then exported to the central Amazon cities for sale. In distant inland regions such as the city of Porto Velho, which is located some 4,000 km inland from Belém, a surprising 90 % of the samples were either pig or sheep eyes. The fetishes in Porto Velho were also the most expensive (~US $7.50/piece), approximately thrice the price in Belém (~US $2.50/piece) and more than twice the sale price in Manaus (~US $4.00/piece). The high price of fetishes, and the use of domestic animal eyeballs do not reflect regional scarcity of the boto, *Inia geoffrensis*, or the tucuxi (*Sotalia fluviatilis*), both of which are abundant near Porto Velho.

The use of body parts of dolphins for their magical power is clearly of some antiquity (Agassiz and Agassiz 1868; Cascudo 1962). A nineteenth century source mentions that "in spite of the respect they have for him as sorcerer (who at will changes from dolphin to human and from human to dolphin), still they kill him, to take out the eyes, the teeth, and the penis, things to all of which they attribute extraordinary virtues" (Cascudo 1962). Beliefs about dolphins apparently derive from African, European, and indigenous cultures (Slater 1994), and Brazilian stories of humans encountering enchanted dolphins have been told for decades (Cravalho 1999).

Demand for boto-derived fetishes is high and supposed boto body parts are sold in virtually all Amazonian city markets and are used mostly by city dwellers (Smith 1996). The indigenous and Caboclo (riverine peoples of mixed ancestry) societies appear not to use boto body parts, and largely appear to respect cultural taboos against killing these magical animals (Best and Silva 1989). The ritualistic use of dolphin body parts has also been reported in the coastal regions of northern and northeastern Brazil (Alves et al. 2007; Reeves et al. 2003; Siciliano 1994), where genital organs and eyes of both the estuarine dolphin (*Sotalia guianensis*) and the freshwater tucuxi (*Sotalia fluvitalis*) are reportedly sold as amulets. The northern and northeastern coastal region of Brazil has a strong West African cultural heritage, including Afro-Brazilian religions such as Candomblée and

Macumba. Unlike the case of folklore and traditional practices of the Amazon interior, these coastal populations and religions have no taboos against killing dolphins or using their body parts in rituals.

The demand for dolphin products in local markets may impact dolphins' populations. However the magnitude of this impact needs to be further investigated and better understood. The combination of direct catch, incidental catch in trawl nets (bycatch), and habitat destruction has placed dolphin populations at risk. The demand for dolphin products for use in traditional medicine and for magic/religious is an additional pressure and should be considered in conservation and management strategies for those species (Alves and Rosa 2008). River dolphins are 'flagship' species for their habitats—charismatic representatives of the biodiversity within the complex ecosystems they inhabit (Karczmarski 2000; Walpole and Leader-Williams 2002; WWF 2006). Efforts to safeguard these cetaceans will not only help save many other species, but will directly contribute to human development and survival by ensuring the availability of adequate and clean freshwater (WWF 2006).

Dolphin parts, such as eyeballs, genitalia, and oil, are increasingly being sold as aphrodisiacs or medicines throughout their range, especially in areas with a large influx of new settlers (Alves and Rosa 2007a; Smith 1996). Traditional Amazonian folklore includes a host of supernatural beings believed to protect its forests and rivers. One of the most powerful and widely recognized of these beings is the Amazon River dolphin *Inia geoffrensis*, locally known as boto. The *boto* is traditionally viewed as a mischievous and tempestuous being, both feared and respected. The most sensational folklore concerning botos is that they transform themselves at dusk into handsome Caucasian men, who are skilled at dancing and seducing young women. *Botos* are also believed to at times enter boats or households, paralyzing their occupants to engage in sexual intercourse. Before first light, the boto returns back into water, reverting into its dolphin form (Cravalho 1999). Indeed, unexpected teenage pregnancies in the Amazon region have been traditionally attributed to seduction by the boto (Cravalho 1999).

The use of the products or parts of dolphins is widespread in northern and northeastern Brazil, both in urban and rural areas, reflecting the cultural importance of the cetaceans in the country. Conservation and management plans are urgently required, but these will have to recognize the cultural aspects of human communities that use dolphins for food, medicines, or for magic/religious purposes (Alves and Rosa 2008). Understanding the socioeconomic aspects of use and trade of dolphins is also important for the development of any successful management plan.

## 11.6 Final Considerations

From the tropical areas to the poles, in the different regions where aquatic mammals occur, their interactions with humans are commonplace. These animals have a substantial utilitarian value and have been the source of inspiration for myths and stories in different localities. Use in traditional folk medicine involves at least 24 species, which supply a variety of products that are used for the treatment of various diseases, some treated through magical-religious rituals.

In some cases, myths impede the capture and use of these animals, nevertheless, the collection of aquatic mammals from the wild for medicinal purposes has been invoked as a factor contributing to the decline of certain species. Of the species cataloged in this review, many are included in the IUCN Red List of Threatened Species, although the reasons for their inclusion are not necessarily related to medicinal use. The trade of aquatic mammals for medicinal purposes can be an important factor for their overexploitation, as exemplified by countries such as Brazil and Cambodia, where products derived from dolphins and manatees are commonly sold in public markets in various cities.

The need to further assess the implications of the use and trade of aquatic mammals parts for traditional medicine on their wild populations is evident, as well as the need for including such uses in discussions of species conservation. The conservation and management of aquatic mammals is a complex matter, and should include a variety of alternative solutions, taking into account ecological, economic, ethical, cultural, and social variables.

## References

Adeola MO (1992) Importance of wild Animals and their parts in the culture, religious festivals, and traditional medicine, of Nigeria. Environ Conserv 19(2):125–134

Agassiz L, Agassiz ECC (1868) A journey in Brazil, 1st edn. Ticknor and Fields, Boston

Alie K (2008) Whales: more valuable alive than dead? A question for decision makers in eastern caribbean whale-watching destinations. Bus finance Econ Emerg Econ 3(1):177–190

Alves RRN (2006) Uso e comércio de animais para fins medicinais e mágico-religiosos no Norte e Nordeste do Brasil. DSc. Thesis, Universidade Federal da Paraíba, João Pessoa, Paraíba

Alves RRN (2009) Fauna used in popular medicine in Northeast Brazil. J Ethnobiol Ethnomed 5(1):1–30. doi:10.1186/1746-4269-5-1

Alves RRN, Alves H (2011) The faunal drugstore: animal-based remedies used in traditional medicines in Latin America. J Ethnobiol Ethnomed 7(1):1–9. doi:10.1186/1746-4269-7-9

Alves RRN, Rosa IL (2006) From cnidarians to mammals: the use of animals as remedies in fishing communities in NE Brazil. J Ethnopharmacol 107:259–276

Alves RRN, Rosa IL (2007a) Zootherapy goes to town: the use of animal-based remedies in urban areas of NE and N Brazil. J Ethnopharmacol 113:541–555. doi:10.1016/j.jep2007.07.015

Alves RRN, Rosa IL (2007b) Zootherapeutic practices among fishing communities in North and Northeast Brazil: a comparison. J Ethnopharmacol 111:82–103. doi:10.1016/j.jep2006.10.033

Alves RRN, Rosa IL (2008) Use of Tucuxi Dolphin Sotalia fluviatilis for medicinal and magic/religious purposes in North of Brazil. Human Ecology 36:443–447. doi:10.1007/s10745-008-9174-5

Alves RRN, Rosa IL (2010) Trade of animals used in Brazilian traditional medicine: trends and implications for conservation. Hum Ecol 38(5):691–704

Alves RRN, Rosa IL, Santana GG (2007) The role of animal-derived remedies as complementary medicine in Brazil. Bioscience 57(11):949–955

Alves RRN, Silva CC, Barboza RRD, Souto WMS (2009a) Zootherapy as an alternative therapeutic in South America. J Altern Med Res 1(1):21–47

Alves RRN, Léo Neto NA, Brooks SE, Albuquerque UP (2009b) Commercialization of animal-derived remedies as complementary medicine in the semi-arid region of Northeastern Brazil. J Ethnopharmacol 124(3):600–608

Alves RRN, Souto WMS, Barboza RRD (2010) Primates in traditional folk medicine: a world overview. Mamm Rev 40(2):155–180. doi:10.1111/j.1365-2907.2010.00158.x

Alves RRN, Barbosa JAA, Santos SLDX, Souto WMS, Barboza RRD (2011) Animal-based remedies as complementary medicines in the semi-arid region of Northeastern Brazil. Evidence-based complementary and alternative medicine 2011 (ID 179876):1–15. doi:10.1093/ecam/nep134

Ashwell D, Walston N (2008) An overview of the use and trade of plants and animals in traditional medicine systems in Cambodia, 1st edn. TRAFFIC Southeast Asia, Greater Mekong Programme, Ha Noi, Vietnam

Baird IG (1995) Lao PDR: an overview of traditional medicines derived from wild animals and plants. TRAFFIC Southeast Asia, Kuala Lumpur, Malaysia

Bayúgar AC (2007) Conocimiento y comparación del uso de la fauna silvestre en dos comunidades Ejidales del municipio de Hueytamalco, Puebla, Mexico. MSc. dissertation, Instituto de Ecologia, A.C., Xalapa, Veracruz, Mexico

Beasley I, Davidson P (2001) Abundance, distribution and conservation management of marine mammals in Cambodia's coastal waters. Wildlife Conservation Society, Phnom Penh, Camboja

Begossi A, Braga FMS (1992) Food taboos and folk medicine among fishermen from the Tocantins River (Brazil). Amazoniana 12:101–118

Bensky D, Gamble A, Kaptchuk T (1993) Chinese herbal medicine materia medica revised edition, 1st edn. Eastland Press, Seattle

Best RC, Silva VMF (1989) Biology, status and conservation of Inia geoffrensis in the Amazon and Orinoco River basins. In: Perrin WF, Brownell RL Jr, Zhou K, Liu J (eds) Biology and conservation of the river dolphins, 1st edn. IUCN, Gland, Switzerland

Cascudo CL (1962) Dicionário do folclore brasileiro, 2nd edn. Ed. do Instituto Nacional do Livro, Rio de Janeiro, Brazil

CITES (2001) List of animal species used in traditional medicine. CITES. http://www.cites.org/eng/com/AC/17/E17i-05Rev.doc. Accessed 14 Jan 2011

CITES (2002a) List of species traded for medicinal purposes. http://www.cites.org/eng/com/ac/18/E18-13-1.pdf

CITES (2002b) Trade in products, possibly used as medicinals, of the species listed in document AC18 Doc. 13.1. CITES. http://www.cites.org/common/com/ac/18/E18i-08.doc

CITES (2011) CITES appendix. http://www.cites.org/eng/resources/species.html

Colten RH, Arnold JE (1998) Prehistoric marine mammal hunting on California's northern Channel Islands. Am Antiq 63(4):679–701

Costa-Neto EM (1999a) Healing with animals in Feira de Santana City, Bahia, Brazil. J Ethnopharmacol 65:225–230

Costa-Neto EM (1999b) Barata é um santo remédio : introdução à zooterapia popular no estado da Bahia, 1st edn. EdUEFS, Feira de Santana, Brazil

Costa-Neto EM (1999c) Recursos animais utilizados na medicina tradicional dos índios Pankararé que habitam o Nordeste do Estado da Bahia, Brazil. Actual Biol 21(70):69–79

Costa-Neto EM (1999d) Traditional use and sale of animals as medicines in Feira de Santana City, Bahia, Brazil. Indig Knowl Dev Monit 7(2):6–9

Costa-Neto EM (2000) Zootherapy Based Medicinal Traditions in Brazil. Honey Bee 11 (2)

Costa-Neto EM (2004) Implications and applications of folk zootherapy in the state of Bahia, Northeastern Brazil. Sustain Dev 12(3):161–174

Costa-Neto EM (2005) Animal-based medicines: biological prospection and the sustainable use of zootherapeutic resources. Anais da Academia Brasileira de Ciências 77(1):33–43

Costa-Neto EM, Alves RRN (2010) Estado da arte da zooterapia popular no Brasil. In: Costa-Neto EM, Alves RRN (eds) Zooterapia: os Animais na medicina popular brasileira, vol 2., 1NUPEEA, Recife, Brazil, pp 13–54

Costa-Neto EM, Dias CV, Melo MN (2002) O conhecimento ictiológico tradicional dos pescadores da cidade de Barra, região do Médio São Francisco, Estado da Bahia, Brasil. Acta Scientiarum 24(2):561–572
Cravalho MA (1999) Shameless creatures: an ethnozoology of the Amazon River dolphin. Ethnology 38:47–58
Crespo EA, Hall MA (2002) Interactions between aquatic mammals and humans in the context of ecosystem management. In: Evans PGH, Raga JA (eds) Marine mammals: biology and conservation, 1st edn. Kluwer Academic/Plenum Publishers, New York, pp 463–490
Culik BM (2004) Review of small cetaceans: distribution, behaviour, migration and threats. Marine mammal action plan/regional seas reports and studies.
Dourojeanni M (1990) Amazonia, Qué hacer?, 1st edn. Centro de Estudios Teológicos de la Amazonia, Iquitos, Peru
Dudgeon D, Arthington AH, Gessner MO, Kawabata ZI, Knowler DJ, Lévêque C, Naiman RJ, Prieur-Richard AH, Soto D, Stiassny MLJ (2006) Freshwater biodiversity: importance, threats, status and conservation challenges. Biol Rev Camb Philos Soc 81(2):163–182
Earle M (1996) Ecological interactions between cetaceans and fisheries. In: Simmonds MP, Hutchinson JD (eds) The conservation of whales and dolphins: science and practice, vol 1. Wiley, Chichester, pp 167–204
Eason C (2008) Fabulous creatures, mythical monsters, and animal power symbols: a handbook, 1st edn. Greenwood Publishing Group, Westport
El-Kamali HH (2000) Folk medicinal use of some animal products in Central Sudan. J Ethnopharmacol 72:279–282
Ellis R (2003) The empty Ocean. Island Press/Shearwater Books, Washington
Enqin Z (1991) Rare chinese materia medica, 1st edn. Science Press, Shanghai, China
Gravena W, Hrbek T, Silva VMF, Farias IP (2008) Amazon River dolphin love fetishes: from folklore to molecular forensics. Marine Mamm Sci 24(4):969–978. doi:10.1111/j.1748-7692.2008.00237.x
Hildebrandt WR, Jones TL (1992) Evolution of marine mammal hunting: a view from the California and Oregon coasts. J Anthropol Archaeol 11(4):360–401
Hines E, Adulyanukosol K, Charuchinda M, Phay S, Leng SA (2004) Conservation of the Dugong (Dugong dugon) on the Eastern Coast of the Gulf of Thailand (Thailand and Cambodia): final report, 1st edn. Ocean Park Conservation Foundation (Hong Kong)/Project Aware (Austrália), Aberdeen, Hong Kong/Sydney, Australia
Ihimaera W (1987) The whale rider reed books. Auckland, New Zealand
IUCN (2011) IUCN red list of threatened species. http://www.iucnredlist.org. Accessed 30 Sept 2011
Jiangsu New Medical College (2002) The chinese materia medica dictionary, 2nd edn. Jiangsu New Medical College, Shanghai, China
Karczmarski L (2000) Conservation and management of humpback dolphins: the South African perspective. Oryx 34(3):207–216
Klein RG, Cruz-Uribe K (1996) Exploitation of large bovids and seals at middle and later stone age sites in South Africa. J Hum Evol 31(4):315–334
Lamb FB (1967) The fishermen's porpoise. In: Eleanore D, Clark M (eds) The dolphin smile: twenty-nine centuries of dolphin lore, 1st edn. MacMillan, New York
Leatherwood S, Reeves RR (1997) Conservación de los delfines de rio, Inia geoffrensis y Sotalia fluviatilis, en la Amazonia Peruana. In: Fang TG, Bodmer RE, Aquino R, Valqui MH (eds) Manejo de Fauna Silvestre en la Amazonia, vol 1. UNAP, La Paz, Bolivia
Lefebvre LW, Marmontel M, Reid JP, Rathbun GB, Domning DP (2011) Status and biogeography of the West Indian manatee. In: Woods CA, Sergile FE (eds) Biogeography of the West Indies, 1st edn. CRC Press, Boca Raton, pp 425–474
Léo Neto NA, Alves RRN (2010) A Natureza sagrada do Candomblé: análise da construção mística acerca da Natureza em terreiros de candomblé no Nordeste de Brasil. Interciencia 35(8):568–574

Lev E (2003) Traditional healing with animals (zootherapy): medieval to present-day Levantine practice. J Ethnopharmacol 85:107–118. doi:10.1016/S0378-8741(02)00377-X
Lev E (2006a) Healing with animals in the Levant from the 10th to the 18th century. J Ethnobiol Ethnomed 2(11):9. doi:10.1186/1746-4269-2-11
Lev E (2006b) Ethno-diversity within current ethno-pharmacology as part of Israeli traditional medicine a review. J Ethnobiol Ethnomed 2(4):1–12. doi:10.1186/1746-4269-2-4
Lev E, Amar Z (2000) Ethnopharmacological survey of traditional drugs sold in Israel at the end of the 20th century. J Ethnopharmacol 72:191–205
Loch C, Marmontel M, Simões-Lopes PC (2009) Conflicts with fisheries and intentional killing of freshwater dolphins (Cetacea: Odontoceti) in the Western Brazilian Amazon. Biodivers Conserv 18(14):3979–3988
MacKinney LC (1946) Animal substances in materia medica. J Hist Med Allied Sci 1(1):149–170
Malik S, Wilson PJ, Smith RJ, Lavigne DM, White BN (1997) Pinniped penises in trade: a molecular-genetic investigation. Conserv Biol 11(6):1365–1374
Marshall NT (1998) Searching for a cure, conservation of medicinal wildlife resources in East and Southern Africa, 1st edn. TRAFFIC International, Cambridge
Martin AR, Silva VMF, Salmon DL (2004) Riverine habitat preferences of botos (Inia geoffrensis) and tucuxis (Sotalia fluviatilis) in the central Amazon. Marine Mamm Sci 20(2):189–200
McKillop HI (1985) Prehistoric exploitation of the manatee in the Maya and circum-Caribbean areas. World Archaeol 16(3):337–353
McNiven IJ, Bedingfield AC (2008) Past and present marine mammal hunting rates and abundances: dugong (Dugong dugon) evidence from Dabangai Bone Mound, Torres Strait. J Archaeol Sci 35(2):505–515
Menegaldo LR (2008) Etnoconhecimento como parte e instrumento de gestão ambiental dos mamíferos aquáticos no Projeto PIATAM (Amazônia Central). Monografia de Especialização em Gestão Ambiental, Universidade do Estado do Amazonas, Manaus, Brazil
Monks GG (2005) The exploitation and cultural importance of sea mammals, 1st edn. Oxbow Books, Oakville
Morell V (1997) Looking for Big Pink. Int Wildl 27:26–31
Moura FBP, Marques JGW (2008) Zooterapia popular na Chapada Diamantina: uma Medicina incidental? Ciência & Saúde Coletiva 13:2179–2188
Nilsson G (1990) The endangered species handbook, 1st edn. The Animal Welfare Institute, Washington
Northridge SP, Hofman RJ (1999) Marine mammal interactions with fisheries. In: Twiss JR, Reeves RR (eds) Conservation and management of marine mammals, 1st edn. Smithsonian Institution Press, Washington, pp 99–119
Ogden A (1941) The california sea otter trade 1784–1848. University of california publications in history 26. University of California Press, Berkeley
Perrin WF, Wursig B, Thewissen JGM (2008) Encyclopedia of marine mammals, 1st edn. Academic Press, San Diego
Perry A (2000) A Global survey of marine medicinals. In: Moreau M-A, Hall HJ, Vincent ACJ (eds) Proceedings of the first international workshop on the management and culture of marine species used in traditional medicines, 1st edn. Project Seahorse, Montreal, Canada, pp 35–43
Perveen F (2009) A beauty of Indus River: the Indus dolphin, its protection in Pakistan—a beauty of Indus River: the Indus River dolphin, Platanista minor Owen, its threats and protection in N-WFP, Pakistan, 1st edn. VDM Verlag, Saarbrücken, Germany
Plagányi ÉE, Butterworth DS (2002) Competition with fisheries. In: Perrin WF, Würsig B, Thewissen H (eds) Encyclopedia of marine mammals, 1st edn. Academic Press, San Diego, pp 268–273
Pompa S, Ehrlich PR, Ceballos G (2011) Global distribution and conservation of marine mammals. Proc Nat Acad Sci 108(33):13600–13605

Porcasi JF, Fujita H (2000) The dolphin hunters: a specialized prehistoric maritime adaptation in the Southern California Channel Islands and Baja California. Am Antiq 65(3):543–566

Powell JA (1996) The distribution and biology of the West African manatee (Trichechus senegalensis Link, 1795), 1st edn. United Nations Environment Programme, Regional Seas Programme, Oceans and Coastal Areas, Nairobi, Kenya

Powell JA, Kouadio A (2008) Trichechus senegalensis. In: IUCN Red list of threatened species (2011). IUCN. http://www.iucnredlist.org/apps/redlist/details/22104/0. Accessed 10 Aug 2011

Pujol J (1993) Natur Africa: The herbalist handbook, 1st edn. Jean Pujol Natural Healers Foundation, Durban, South Africa

Quave CL, Lohani U, Verde A, Fajardo J, Rivera D, Obón C, Valdes A, Pieroni A (2010) A comparative assessment of zootherapeutic remedies from selected areas in Albania, Italy, Spain and Nepal. J Ethnobiol 30(1):92–125

Reeves RR, Smith BD, Crespo A, di Sciara GN (2003) Dolphins, whales and porpoises: 2002–2010 conservation action plan for the world's cetaceans, 1st edn. IUCN/SSC Cetacean Specialist Group, Gland, Switzerland/Cambridge

Reeves RR, Jefferson TA, Karczmarski L, Laidre K, O'Corry-Crowe G, Rojas-Bracho L, Secchi ER, Slooten E, Smith BD, Wang JY, Zhou K (2008) Orcaella brevirostris. In: IUCN red list of threatened species 2011. IUCN. http://www.iucnredlist.org/apps/redlist/details/15419/0. Accessed 8 Sept 2011

Richerson PJ, Mulder MB, Vila BJ (1996) Principles of human ecology, 1st edn. Pearson Custom Pub, London

Rick TC, Braje TJ, DeLong RL (2011) People, pinnipeds, and Sea Otters of the Northeast Pacific. In: Braje TJ, Rick TC (eds) Human impacts on seals, Sea Lions, and Sea Otters: Integrating archaeology and ecology in the Northeast Pacific. University of California Press, Berkeley/Los Angeles/London, pp 1–17

Ripple J, Perrine D (1999) Manatees and dugongs of the world, 1st edn. Voyageur Press, Stillwater

Roman J (2006) Whale. Animal series, 1st edn. Reaktion books Ltd, London

Romero A, Baker R, Cresswell JE, Singh A, McKie A, Manna M (2002) Environmental history of marine mammal exploitation in Trinidad and Tobago, WI, and its ecological impact. Environ Histor 8(3):255–274

Romero A, Creswell J (2005) In the land of the Mermaid: how culture, not ecology, influenced marine mammal exploitation in the Southeastern Caribbean. In: Romero A, West SE (eds) Environmental issues in Latin America and the Caribbean, 1st edn. Springer, Dordrecht, pp 3–30

Scammon CM (1968) The marine mammals of the Northwestern coast of North America. Originally published in 1874, Dover, New York

Secchi E (2010) Sotalia fluviatilis In: IUCN red list of threatened species 2011. IUCN. http://www.iucnredlist.org/apps/redlist/details/190871/0/print. Accessed 10 Aug 2011

Seixas CS, Begossi A (2001) Ethnozoology of fishing communities from Ilha Grande (Atlantic forest coast, Brazil). J Ethnobiol 21:107–135

Shao-Ke L, Ka-Lok W, Ming L, Paul B, Stephen T, Pang-Chui S (2010) An integrated web medicinal materials DNA database: MMDBD (Medicinal materials DNA Barcode database). BMC Genomics 11:1–8

Siciliano S (1994) Review of small cetaceans and fishery interactions in coastal waters of Brazil. Rep Int Whal Comm Special Issue 15:241–250

Silva VMF (1990) Botos, mitológicos hóspedes da Amazônia. Ciência Hoje 11:14–18

Silva VMF, Best RC (1994) Tucuxi Sotalia fluviatilis (Gervais, 1853). In: Ridgway SH, Harrison R (eds) Handbook of marine mammals, 1st edn. Academic Press, London, pp 43–69

Silva VMF, Best RC (1996a) Freshwater dolphin/fisheries interaction in the Central Amazon. Amazoniana 14(1/2):165–175

Silva VMF, Best RC (1996b) Sotalia fluviatilis. mammalian species 527:1–7

Simelane TS, Kerley GIH (1998) The conservation implications of the use of indigenous vertebrates by traditional healers in Xhosa communities. S Afr J Wildl Res 28:121–126

Simões-Lopes PC, Ximenez A (1990) O impacto da pesca artesanal em áreas de nascimento do boto cinza, Sotalia fluviatilis (Cetacea, Delphinidae), Santa Catarina, Brasil. Biotemas 3(1): 67–72

Sinha RK, Sharma G (2003) Current status of the ganges river dolphin, Platanista gangetica in the rivers Kosi and Son, Bihar, India. J Bombay Nat Hist Soc 100(1):27–37

Slater C (1994) Dance of the dolphin: transformation and disenchantment in the Amazonian imagination, 1st edn. University of Chicago Press, Chicago

Smith A (1996) The river dolphins: the road to extinction. In: Simmonds MP, Hutchinson JD (eds) The conservation of whales and dolphins: science and practice. John Wiley and Sons, Chichester, pp 339–373

Smith BD, Beasley I, Buccat M, Calderon V, Evina R, Lemmuel de Valle J, Cadigal A, Tura E, Vistacion Z (2004) Status, ecology and conservation of Irrawaddy dolphins (Orcaella brevirostris) in Malampaya Sound, Palawan, Philippines. J Cetacea Res Manag 6(1):41–52

Stringer CB, Finlayson JC, Barton RNE, Fernández-Jalvo Y, Cáceres I, Sabin RC, Rhodes EJ, Currant AP, Rodríguez-Vidal J, Giles-Pacheco F (2008) Neanderthal exploitation of marine mammals in Gibraltar. Proc Nat Acad Sci 105(38):14319–14324

Toledo GAC, Campos BATP, Feitosa ICS, Souto AS, Alves RRN (2010) Interações entre pescadores artesanais e o boto-cinza (Sotalia guianensis—Van Bénéden, 1864) na região de Baía Formosa, Rio Grande do Norte—Brasil. In: Alves RRN, Souto WMS, Mourão JS (eds) A Etnozoologia no Brasil: Importância, Status atual e Perspectivas, vol. 7, 1st edn. NUPEEA, Recife, PE, Brazil, pp 277–296

Tratado de Cooperacion Amazonica (1999) Conservacion y uso de la fauna silvestre en areas protegidas de la Amazonia, 1st edn. Secretaria Pro Tempore (Venezuela)/FAO, Caracas, Venezuela

Trujillo F, Diazgranados MC, Omacha F (2002) Delfines de rıo en la Amazonia e Orinoquia: Ecologıa e Conservacion, 1st edn. Fundación Omacha, Bogotá, Colombia

Vidal O (1993) Aquatic mammal conservation in Latin America: problems and perspectives. Conserv Biol 7(4):788–795

Walpole MJ, Leader-Williams N (2002) Tourism and flagship species in conservation. Biodivers Conserv 11(3):543–547

WWF (2006) Species/Freshwater fact sheet: river dolphins. http://assets.wwf.ch/downloads/riverdolphinsw.pdf. Accessed 2 Oct 2011

Yates BC (2005) Distinguishing real vs fake Tiger Penises. Identif Guides Wildl Law Enforc 6:1–8

# Chapter 12
# Marine Invertebrates in Traditional Medicines

**Rômulo Romeu Nóbrega Alves, Tacyana Pereira Ribeiro Oliveira, Ierecê Lucena Rosa and Anthony B. Cunningham**

**Abstract** This chapter reviews the uses of marine invertebrates for medicinal purposes, based on an extensive literature survey. We identified at least 266 species which are prescribed to treat a number of diseases and conditions categories, especially those of the digestive and genitourinary systems. Among the medicinal species, 19 species feature on the IUCN (International Union for Conservation of Nature) Red List and/or are CITES-listed (Convention on International Trade in Endangered Species of Wild Fauna and Flora). This highlights the need to assess the impacts of zootherapy in the context of marine biodiversity conservation. Conservation efforts should also be directed to unprotected species, whose medicinal use is widespread. In addition to biological aspects, economical and sociocultural factors influence the relationship between people and zootherapeutical resources. Therefore, studies bridging ecological, anthropological, and pharmacological aspects of the medicinal use of animals should be fostered.

*As an old saying in Western countries goes* 'An apple a day keeps the doctor away'*, so now there is a new saying in China, 'A sea cucumber a day keeps the doctor away'*

(Chen 2003).

R. R. N. Alves (✉)
Ethnozoology, Conservation and Biodiversity Research Group, Departamento de Biologia, Universidade Estadual da Paraíba, Campina Grande-PB 58109-753, Brazil
e-mail: romulo_nobrega@yahoo.com.br

T. P. R. Oliveira · I. L. Rosa
Departamento de Sistemática e Ecologia, Universidade Federal da Paraíba, João Pessoa-PB 58059-970, Brazil

A. B. Cunningham
Forests and Livelihoods Programme, Centre for International Forestry Research, PO Box 0113 BOCBD, Bogor Indonesia

R. R. N. Alves and I. L. Rosa (eds.), *Animals in Traditional Folk Medicine*,
DOI: 10.1007/978-3-642-29026-8_12, © Springer-Verlag Berlin Heidelberg 2013 

## 12.1 Introduction

More than 70 % of our planet's surface is covered by the oceans, and perhaps the potentially available biodiversity on the deep sea floor or coral reefs is higher than that existing in the rainforests (Haefner 2003). The oceans are the world's largest reservoirs of natural resources, which have been used in different ways since the earliest humans have existed. Food is one of the most important of these resources, but others include water, minerals, petroleum, building materials, chemicals, and energy (Walker and Wood 2005).

Humans have used marine animals since prehistoric times as food sources, for personal adornment, medicines, souvenirs, as pets, among other uses (Dias et al. 2011; Gössling et al. 2004; Marean et al. 2007; Morwood et al. 1998; Pajaro et al. 2004; Rosa et al. 2011; Stringer et al. 2008). In general, attention has traditionally focused on their capture, use, and trade for food, while other forms of exploitation were comparatively considered minor, and the use of non-food marine species (e.g., in traditional medicines, for aquarium display, as curiosities), although widespread, has not been well studied or documented (Pajaro et al. 2004; Vincent 2006).

A number of marine resources are used for medicinal purposes, including algae and several taxa of animals—from sponges to marine mammals (Alino et al. 1990; Alves and Alves 2011; Alves et al. 2007; Perry 2000). For thousands of years people have believed in the healing power of the sea. As the Greek dramatist Euripides says in one of his plays, "the sea washes away the stains and wounds of the world". The ancient Egyptians and Greeks examined the effects of seawater on human health, and credited the sea and the substances it contained with healing properties (Schröder 2010).

Marine products have for centuries been an integral part of folk medicine all around the world, as exemplified by sea salt–traditionally used to treat skin diseases—and by algae to treat parasitic worms (Schröder 2010). Among the marine organisms, animals play a key role in folk medicine, providing a range of traditional natural remedies (Alves and Dias 2010; Herbert et al. 2003; Hu 1980; Prabhakar and Roy 2009), and their use as remedies is a common practice in many cultures.

Despite the scarcity of information about the historical medicinal uses of marine organisms (Benoist 1951), some authors have pointed that historical documents and archeological research indicate that these animals have been used in traditional medicines since ancient times (Lev 2003; Voultsiadou 2010). In China, for example, sea cucumbers are regarded as a folk remedy and their utilization has been recorded since the Ming Dynasty (Conand 2006; Voultsiadou 2010) pointed out that 38 marine invertebrates were recorded in the classical Greek texts for their pharmaceutical properties, which corresponded, in terms of current taxonomic groups, to 7 different Phyla and 11 Classes of the animal kingdom, the most diverse taxon being by far mollusks (55% of the recorded invertebrates), followed by crustaceans.

Nonetheless, the exact nature and extent of the medicinal use of marine species are still far from being known, given the paucity of published baseline data. As remarked by Perry (2000), the therapeutic use of medicinal species has effectively not been studied in the past. On the other hand, the existing literature records clearly indicate the importance of medicinal invertebrates in different countries and regions (Alves and Dias 2010; Alves et al. 2007; Herbert et al. 2003; Hu 1980; Prabhakar and Roy 2009; Whiting et al. 2011).

Further studies on the topic are much needed, especially in face of the global magnitude of the use of traditional medicines—an aspect that can drive unsustainable harvesting, especially for high demand or high priced species. Particularly in the case of marine taxa, there is evidence that some of the species used in traditional medicine are threatened with extinction (Alves and Dias 2010; Perry 2000).

This chapter provides an overview of the medicinal use of marine invertebrates, aiming to document their uses worldwide. It also provides insight into the following questions: Which are the species used in traditional medicines? Which are the main taxonomic groups used? What are the indications of use of these animals? How to convey information about the medicinal use for conservation actions directed at the exploited species?

## 12.2 Methods

The relevant information and data were gathered through an extensive literature review. Our main sources were: published articles, books and book chapters, as well as reports, available at international online databases such as Web of Science, Scopus, Google Scholar, or journals' web sites.

The resulting database encompasses information about species, family names, and conditions to which remedies are prescribed. Only taxa that were identified to species level were included in the database. Scientific names provided in publications were updated according to the World Register of Marine Species (WoRMS) (http://www.marinespecies.org/index.php). The following sources were consulted: Alino et al. (1990), Almeida (2007), Almeida et al. (2005), Almeida and Albuquerque (2002), Alves (2009), Alves et al. (2007, 2008a, 2009a, c), Alves and Alves (2011), Alves and Dias (2010), Alves and Rosa (2006, 2007b, c), Barajas (1961), Begossi (1992), Bensky et al. (2004), Chen (2005), Choo (2004, 2008), Chopra et al. (2006), CITES (2001); Conand (1990), Costa-Neto (1999a, b, c, 2000a, b, 2001), Costa-Neto and Marques (2000), Cunningham and Zondi (1991), Ferreira et al. (2009), Figueiredo (1994), Fredalina et al. (1999), Gopal et al. (2008), Hamidah et al. (2009), Herbert et al. (2003), Lev (2003, 2007), Lev and Amar (2002), Liu et al. (2005), Lovatelli et al. (2004), Mallmann (1996), Marshall et al. (2001), Moura and Marques (2008), Nadkarni (1994), Niir Board of Consultants and Engineers (2003), Padmanabhan and Sujana (2008), Pereira (1842), Perezrul (2006), Read et al. (1937), Salamanca and Pajaro (1996), Seraj et al. (2011), Silva et al. (2004), Sousa (2010), Tang (1987), Titcomb et al. (1978), Traffic Europe-Russia (2006), Van and

Tap (2008), Vohora and Khan (1978), Voultsiadou (2010), Whiting et al. (2011), Xiyin et al. (2004), Yinfeng et al. (1997) and Zaidnuddin and Ibrahim (2006).

Diseases and health conditions to which marine invertebrates were prescribed were categorized according to the 10th revision (version 2010) of the International Statistical Classification of Diseases and Related Health Problems (ICD-10) (WHO 2011), while the conservation status of the medicinal invertebrate species followed the International Union for Conservation of Nature (IUCN) Red List (http://www.iucnredlist.org/) and the Convention on International Trade in Endangered Species of Wild Fauna and Flora (CITES) (http://www.cites.org/eng/resources/species.html).

## 12.3 Medicinal Marine Invertebrates

The use of medicinal fauna has been increasingly examined over the past two decades, particularly in countries of Latin America, Asia, and Africa (Adeola 1992; Alves and Alves 2011; Ashwell and Walston 2008; Quave et al. 2010; Sodeinde and Soewu 1999; Van and Tap 2008). In these studies, the predominant use of vertebrates as medicinal fauna in different medical systems worldwide has been highlighted, and as remarked by Perry (2000), this is an expected trend, considering the frequent interactions between people and vertebrates—typically, large-bodied animals, which may provide a wide range of medicinal products.

In the marine realm, invertebrates, along with fishes, and to a lesser extent, some reptile and mammal species, comprise the therapeutic "tool-box" used in traditional medicines by different human communities (Alves and Rosa 2006, 2007b). The relative dominance of invertebrates may reflect the high proportional abundance of these animals in the marine environment (Ruppert et al. 1994).

A decade ago, Perry (2000) estimated that approximately 185 species of invertebrates were used globally for their medicinal value. In this review that figure has risen to at least 266 marine invertebrate species, belonging to 119 families (Table 12.2). The species cataloged comprised 10 taxonomic categories, the groups with the largest number of medicinal species being mollusks (56.8% of the recorded species; $n = 151$), echinoderms (16.2%; 43), crustaceans (15.4%; 41), and cnidarians (8.3%; 22). Annelids, chordates, sponges, chelicerates, and bryozoans together accounted for 3.4% of recorded species ($n = 9$) (Fig. 12.1). Some examples of medicinal invertebrates are shown in Fig. 12.2.

Zootherapeutic products can be consumed in several forms: marine medicinal ingredients may be cooked and boiled, eaten or taken as drink, taken alone or in combination with other herbal ingredients, eaten raw, aged, or dried, ground, or applied externally; reported ingredients were flesh, skin, shell, skeleton (coral, crustacean), teeth, head, heart, eggs/gonads, ink and bone (cuttlefish), sucking dish, eye, pearls, foot, operculum and hypobranchial gland (murex snail), or whole animals (e.g., Alves and Alves 2011; Alves et al. 2007; Lev 2003; Nadkarni 1994; Perry 2000; Voultsiadou 2010).

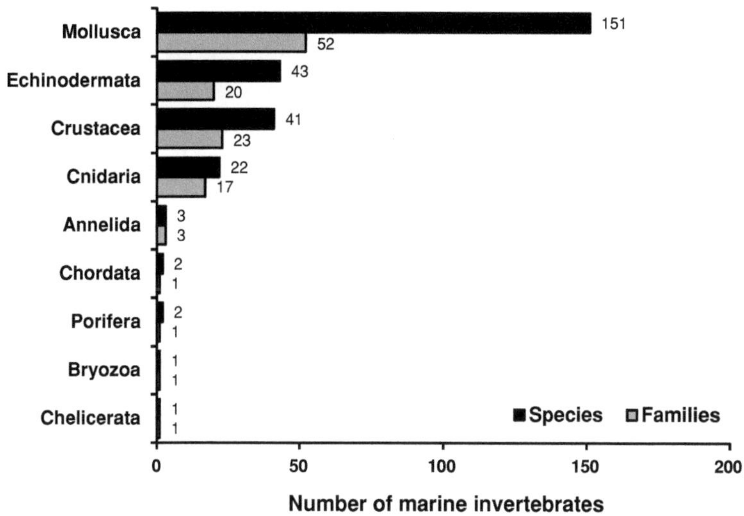


**Fig. 12.1** Number of marine invertebrate families and species used in traditional medicine per taxonomic group

Given the limited amount of specific studies on medicinal invertebrates, and the predominant descriptive nature of data sources on marine medicinals—based simply on lists of species, which are often taxonomically incorrect or are restricted to the common names of the animals (Perry 2000)—it is very likely that the already expressive number of taxa listed in this review represents an underestimate of the total spectrum of invertebrate species used as remedies worldwide. Hence, we further highlight the need for planning and preparing studies with greater scientific rigor, especially in terms of taxonomic accuracy, and for expanding the data sources so that they can increasingly serve as a basis for different types of analyses, as well as for conservation and management strategies.

The rationale of use of some marine medicinals is not readily apparent (Perry 2000), nonetheless there is usually a cultural element associated with the use of animals in traditional medicines, and this includes both marine vertebrates and invertebrates (Alves and Rosa 2007b; Alves et al. 2007). In some cultures, organisms are used on the basis of appearance or behavior (Alves 2008; Alves et al. 2007, 2009b; Descola 1996; Groark 1997; Perry 2000), as exemplified by the use of flatfish in southern Africa to treat strokes on the basis of their resemblance to a paralyzed stroke victim (Perry 2000). This observation is in line with Radbill (1976), who pointed out that in homeopathic or imitative magic, it is assumed that certain qualities attributed to animals can be transferred to humans, and that this transfer can occur by inhalation, ingestion, or application of the body parts of those animals.

Our compilation of the literature records on medicinal invertebrates revealed at least 394 therapeutic properties and uses. From the 17 health condition categories (according to the ICD-10) listed in Table 12.1, those related to the digestive

**Fig. 12.2** Examples of marine invertebrates used in traditional medicines. **a** Dried starfish (*left*: *Oreaster reticulatus*; *right*: *Astropecten* sp.); **b** Several specimens of *O. reticulatus* for sale at popular markets in northeast Brazil; **c** Starfish for sale in Chengdu, Sichuan, China, for use in Chinese traditional medicine; and **d** Abalone shells (*Haliotis* spp.) traded in China. Photos: **a** Rômulo Alves; **b** Ierecê Rosa; **c**–**d** Anthony Cunningham

system were the most quoted, followed by diseases of the genitourinary system and skin disorders (see Table 12.1). In addition, eight other uses or therapeutic properties, which could not be classified according to ICD-10, were also reported in the literature: absorbent; absorbing liquids; acrid; alleviative of vayu; blue eye cosmetic; caustic in various ointments; dilating cavities; healing baths.

## 12.4 Conservation

Various authors have discussed the use of medicinal products derived from wild animal species under a conservationist perspective (Alves et al. 2008b; Ashwell and Walston 2008; Costa-Neto 2000a; Kang and Phipps 2003; Mahawar and Jaroli 2008). Nonetheless, it is well known that higher attention has been given to large carnivores, such as tigers, bears, and rhinos, threatened by the trade for traditional medicine, especially the Traditional Chinese Medicine (Lee 1999). Despite that

**Table 12.1** Medicinal uses of marine invertebrates as quoted in the literature, according to the categories listed in the international statistical classification of diseases and related health problems (ICD-10)

| ICD 10 | Indication of use and therapeutic properties | *N* |
|---|---|---|
| Digestive system | Acid regurgitation; abdominal bloating; abdominal pain (also quoted as: lower abdominal pain); anorectal fistulae; bilious affections (also quoted as: biliousness; antibilious); bowel complaints; chronic, non-healing ulcers or macerated areas, usually gums; chronicle intestinal disorders; clean and polish teeth (also quoted as: teeth cleaning; dentifrice); clyster therapy; colic; constipation (also quoted as: causes constipation; intestinal obstruction prevention; soften the bowels); 'consumed in excess it causes nausea, vomiting and diarrhea' (also quoted as: vomiting agent); dental problems in children; digestive; digestive impairment; duodenal ulcer; dysentery; dyspepsia; emetic; enlarged liver; epigastric pain; excessive gastric pain; 'fat is placed in the decaying teeth'; fixed and mobile abdominal mass; flatulence (also quoted as: distasteful belching; carminative; causes intestinal gas production); gastric ulcer; gastritis; heartburn; hepatomegaly; hyperacidity (also quoted as: antacid; reduces gastric acidity; harmonizes stomach acidity; increased gastric acidity); increases salivary secretion; indigestion; induces or worsens gastroenteritis symptoms; irritability of the intestines as in diarrhea and chronic dysentery (also quoted as: dysenteric disorders with pain in the region of the navel); irritates stomach; intestinal pains; jaundice; keep gums firm and healthy; laxative; liver ailments; malabsorption syndrome; organomegaly; parotid gland swelling; peptic ulcer; rectal prolapse; sores in the oral cavity (also quoted as: sores about the mouths of children; sores affecting the teeth); sprue; stomach ache (also quoted as: pain due to excess of stomach acid); stomach disorders; strengthening the liver; strengthening the stomach; stabilizing large and small intestines; teething; tongue diseases; toothache; tooth disinfection; ulcer pain; ulceration of gums; vomiting (also quoted as: vomiting of blood). | 81 |

(continued)

Table 12.1 (continued)

| ICD 10 | Indication of use and therapeutic properties | *N* |
|---|---|---|
| Genitourinary system | Abdominal swelling due to increase in uterus size; acid urine; ardor urine; bladder disorders; breast pain; cracks in the perineum; diuretic; disorders with erosion and seepage involving the genitals; dysmenorrhea; dysuria (also quoted as: dysuria symptoms); fertility problems; genital lesions; gonorrhea; helps menstrual blood flow; hemorrhage in women; kidney cleansing; kidney disorders (also quoted as: kidney problems); kidney stone; leucorrhoea; menorrhagia (also quoted as: excessive uterine bleeding); menstrual cycle abnormalities (also quoted as: irregular uterine bleeding; continuous uterine bleeding); metrorrhagia; nocturnal emission; obstruction of the urinary tract; oligomenorrhea; painful urinary dribbling (also quoted as: turbid painful urinary dribbling with blood in the urine); seminal weakness; spermatohrroea; urinary blockage; urinary frequency; urinary incontinence; urinary diseases (also quoted as: urinary dysfunction); uterovaginal dropsy; uterovaginal prolapse; uterus infection; uterus pain; vaginal bleeding; vaginal discharge (also quoted as: red and white vaginal discharge); venereal disease; womb disorders. | 48 |
| Skin and subcutaneous tissue | Alopecia; antiperspirant (also quoted as: increases perspiration); anti-wrinkle; astringent; boils; carbuncle; causes itching (also quoted as: skin itching); causes swelling (also quoted as: heals of various types of skin swelling; painful swelling); causes ulceration of the oral tissue; cracked skin (also quoted as: fissure; cracks in the feet); chloasma; damp rashes of long duration; dandruff; darkish discolorations on the face; demulcent; depilatory (also quoted as: eyebrow depilation–prevents eyebrow hair from growing again; reduces body hair growth); inhibits excessive tissue growth (also quoted as: inhibits excessive tissue growth during sore healing); macular rash; prickly heat; psoriasis; rough cheeks; scars; siccative; skin cleanser; skin diseases; skin spots (also quoted as: facial dark spots; sunspots); sores (also quoted as: chronic, non-healing sores; chronically open moist sores; sores with profuse suppuration; sores in the face); suppurating prolapsed parts; ulcers (also quoted as: chronic, non-healing skin ulcers; severe, chronic, non-healing ulceration anywhere on the body). | 46 |

(continued)

**Table 12.1** (continued)

| ICD 10 | Indication of use and therapeutic properties | *N* |
|---|---|---|
| Respiratory system | Asthma; bronchitis; catarrh; chest and rib pain; cough (also quoted as: chesty cough; causes cough; paroxysmal cough; coughing with blood); discharges from nose; expectorant; flu; irritates mouth cavity and throat; lung trouble; nasal congestion with fever; nosebleed; nose polyp removal; phlegm (also quoted as: accumulation of phlegm; antiphlegmonous; phlegm nodules; dissolves old phlegm; reduces old phlegm; phlegm lumps; stubborn phlegm; viscous yellow sputum difficult to expectorate; sensation of fullness and obstruction in the chest caused by phlegm); pneumonia; respiratory problems; shortness of breath; sinus; sore throat (also quoted as: ulceration of the throat; chronic, non-healing ulcers or macerated areas, usually throat); wheezing. | 35 |
| Symptoms, signs, and abnormal clinical and laboratory findings, not elsewhere classified | All types of nodules and masses around the body (also quoted as: dissipates nodules; various kinds of lumps); anti-inflammatory; convulsive attacks (also quoted as: childhood convulsions; febrile convulsions in children); cataplasm therapy; causes voice disorders (also quoted as: causes hoarseness); fevers (also quoted as high fever; low fever); loss of appetite; night sweats; pain (also quoted as: anodyne, pain killer; pain relief; local sedative; severe pain); reduces edema (also quoted as: superficial edema); seizures; spontaneous sweating; stimulant; swollen glands; wasting disease (also quoted as: wasting away of body). | 29 |
| Eye and adnexa | Blurred vision; cataract; cloudy vision; color blindness; conjunctivitis; diseases of the cornea; dry eyes; eye diseases (also quoted as: eye problems; eye troubles); eye inflammation; eye leucoma; itching of eyelids; keratitis; night blindness; night vision problems; ocular infections; opacity of cornea; painful eyes; photophobia; red eyes; remove pterygium; superficial disorders of the eyes (also quoted as: superficial visual obstructions); swollen eyes. | 25 |
| Blood and blood-forming organs and certain disorders involving the immune mechanism | Alterative; anemia; bleeding (also quoted as: causes bleeding; haemostatic; hemorrhage; restrain intense blood flow; stops blood flow; stops prolonged blood flow; stops bleeding; bleeding due to trauma, sores or eczema; bleeding form the upper parts of the body); blood deficiency; debility (also quoted as: general debility; weakness); enlarged spleen (also quoted as: swollen spleen); haematinic properties; hemorrhoids; illness recovery; increase blood viscosity and mucus production; reduces clotting in women; strengthen body's immune system. | 24 |

(continued)

**Table 12.1** (continued)

| ICD 10 | Indication of use and therapeutic properties | *N* |
|---|---|---|
| Injury, poisoning and certain other consequences of external causes | Alleviate the symptoms of intoxication with poison of *Colomesus psittacus*; alleviate the symptoms of intoxication with poison of 'niquim' (Pisces, Batrachoididae); antidote to poisons (also quoted as: antidote to poison from sea horse; poisoning treatment; poisonous secretion); black eye; burns; chilblains; cuts; dog bite (also quoted as: bite of a mad dog); injuries; pain relief in injuries caused by the dorsal fin spine of a species of catfish; snake bite; soothing of urchin wounds; strikes of harvest spiders; strikes of scorpions; wounds; scalds. | 20 |
| Mental and behavioral disorders | Anxiety; aphrodisiac; bad temper; depression; disorientation; dizziness; dream-disturbed sleep; dyspnea; fright mania; insomnia; irritability; mania; night terrors; premature ejaculation; restlessness; sedative; sexual impotence (also quoted as: impotence); vertigo; withdrawal mania. | 20 |
| Circulatory system | Cardiac stimulant (also quoted as: heart stimulant); chest pain; heart ailments (also quoted as: heart diseases); hypertension; pain in the heart; palpitation (also quoted as: continuous palpitations); rapid heartbeat; strengthen the heart (also quoted as: tonic to strengthen cardiac muscles). | 12 |
| Ear and mastoid process | Discharges from ears; ear diseases; earache (also quoted as: ear pain; pain of earache; pain in otorrhoea); hear loss; edema around auditory meatus; otorrhoea; sores affecting the ears; tinnitus; tympanitis. | 12 |
| Certain infectious and parasitic diseases | Leprosy; diarrhea; 'when poxes in children enter the eyes'; scrofula (also quoted as: scrofulous affections; scrofula on the lower nape); tuberculosis; typhus; whooping cough; worms. | 10 |
| Nervous system | Epilepsy; headache (also quoted as: chronic headache; hemicranias); nervine tonic (also quoted as: nervine tonic in giddiness; nervine tonic in headache; nervine tonic in vertigo; tonic to strengthen the central nervous system); nervous diseases. | 10 |
| Pregnancy, childbirth, and the puerperium | Bleeding during pregnancy; helps to peel away the placenta and membranes after labor (also quoted as: placenta removal; removes placenta retained after labor); increasing lactation; nipple inflammation; prevent abortion; remove a dead fetus. | 8 |
| Musculoskeletal system and connective tissue | Arthrosis; gout; lower back pain; osteoporosis; rheumatism; subcostal pain. | 6 |
| Endocrine, nutritional, and metabolic diseases | Diabetes; goiter; facilitates fluid metabolism; 'nutritious' (also quoted as: 'nutritive'); tonic. | 6 |
| Neoplasms | Abdominal tumors; cancer. | 2 |

**Table 12.2** Marine invertebrates used in traditional medicines

| Family/species |
|---|
| **PORIFERA** |
| Spongiidae: *Hippospongia communis* (Lamarck, 1814), *Spongia officinalis* Linnaeus, 1759 |
| **CNIDARIANS** |
| Actiniidae: *Anemonia viridis* (Forskål, 1775), *Anthopleura nigrescens* (Verrill, 1928), *Anthopleura xanthogrammica* (Brandt, 1835) |
| Antipathidae: *Antipathes dichotoma* Pallas, 1766[AppII], *Antipathes grandis* Verrill, 1928[AppII] |
| Coralliidae: *Corallium rubrum* (Linnaeus, 1758), *Paracorallium japonicum* (Kishinouyi, 1903)[AppIII] |
| Diadumenidae: *Diadumene lineata* (Verrill, 1869) |
| Gorgoniidae: *Gorgonia flabellum* Linnaeus, 1758 |
| Melithaeidae: *Melithaea ochracea* (Linnaeus, 1785), *Melithaea squamata* (Nutting, 1911) |
| Mussidae: *Mussismilia harttii* (Verril, 1868)[DD/AppII] |
| Oculinidae: *Galaxea fascicularis* (Linnaeus, 1767)[NT/AppII] |
| Parazoanthidae: *Savalia savaglia* (Bertoloni, 1819) |
| Pelagiidae: *Pelagia noctiluca* (Forsskål, 1775) |
| Physaliidae: *Physalia physalis* (Linnaeus, 1758) |
| Plexauridae: *Euplexaura erecta* Kükenthal, 1908 |
| Rhizostomatidae: *Rhopilema esculentum* Kishinouye, 1891 |
| Stylasteridae: *Stylaster scabiosus* Broch, 1935[AppII] |
| Subergorgiidae: *Subergorgia suberosa* (Pallas, 1766) |
| Tubiporidae: *Tubipora musica* Linnaeus, 1758[NT/AppII] |
| Veretillidae: *Cavernularia habereri* Moroff, 1902 |
| **ANNELIDA** |
| Amphinomidae: *Hermodice carunculata* (Pallas, 1766) |
| Arenicolidae: *Arenicola cristata* Stimpson, 1856 |
| Terebellidae: *Lanice conchilega* Pallas, 1766 |
| **MOLLUSCS** |
| Aplysiidae: *Aplysia depilans* Gmelin, 1791 |
| Arcidae: *Anadara broughtonii* (Schrenck, 1867), *Anadara sativa* Bernard, Cai & Morton, 1993, *Arca noae* Linnaeus, 1758, *Arca granosa* (Linnaeus, 1758) |
| Asteropseidae: *Asteropsis carinifera* (Lamarck, 1816) |
| Babyloniidae: *Babylonia lutosa* (Lamarck, 1816), *Babylonia japonica* (Reeve, 1842) |
| Buccinidae: *Neptunea cumingii* Crosse, 1862 |
| Bursidae: *Bursa granularis* (Röding, 1798) |
| Cardiidae: *Cerastoderma glaucum* (Bruguière, 1789), *Dallocardia muricata* (Linnaeus, 1758), *Hippopus hippopus* (Linnaeus, 1758)[LR/CD*–AppII], *Tridacna maxima* Röding, 1798[LR/CD*–AppII], *Tridacna squamosa* Lamarck, 1819[LR/CD*–AppII] |
| Cassidae: *Cassis tuberosa* (Linnaeus, 1758) |
| Cerithiidae: *Cerithium vulgatum* Bruguière, 1792 |
| Chitonidae: *Chiton salihafui* Bullock, 1972, *Onithochiton literatus* (Krauss, 1848) |
| Conidae: *Conus textile* Linnaeus, 1758 |
| Cypraeidae: *Arestorides argus* (Linnaeus, 1758), *Cypraea tigris* Linnaeus, 1758, *Erosaria citrina* (Gray, 1825), *Erronea pulchella* (Swainson, 1823), *Lyncina lynx* (Linnaeus, 1758), *Lyncina vitellus* (Linnaeus, 1758), *Mauritia arabica* (Linnaeus, 1758), *Mauritia mauritiana* (Linnaeus, 1758), *Monetaria annulus* (Linnaeus, 1758), *Monetaria caputserpentis* (Linnaeus, 1758), *Monetaria moneta* (Linnaeus, 1758), *Notocypraea angustata* (Gmelin, 1791) |

(continued)

**Table 12.2** (continued)

| Family/species |
|---|
| Donacidae: *Donax trunculus* Linnaeus, 1758, *Iphigenia brasiliensis* (Lamarck, 1818) |
| Fasciolariidae: *Fusinus ocellifer* (Lamarck, 1816), *Pleuroploca trapezium* (Linnaeus, 1758) |
| Haliotididae: *Haliotis asinina* Linnaeus, 1758, *Haliotis clathrata* Reeve, 1846, *Haliotis corrugata* Wood, 1828, *Haliotis cracherodii* Leach, 1814[CE], *Haliotis discus* Reeve, 1846, *Haliotis diversicolor* Reeve, 1846, |
| *Haliotis gigantea* Gmelin, 1791, *Haliotis laevigata* Donovan, 1808, *Haliotis midae* Linnaeus, 1758 *Haliotis ovina* Gmelin, 1791, *Haliotis rubra* Leach, 1814, *Haliotis tuberculata* Linnaeus, 1758, *Haliotis varia* Linnaeus, 1758 |
| Haminoeidae: *Bullacta exarata* (Philippi, 1849) |
| Ischnochitonidae: *Ischnochiton (Ischnochiton) oniscus* (Krauss, 1848) |
| Littorinidae: *Littoraria angulifera* (Lamarck, 1822) |
| Loliginidae: *Loligo vulgaris* Lamarck, 1798, *Uroteuthis chinensis* (Gray, 1849) |
| Lucinidae: *Phacoides pectinata* (Gmelin, 1791) |
| Mactridae: *Mactra antiquata* Spengler, 1802, *Mactra chinensis* Philippi, 1846, *Mactra quadrangularis* Reeve, 1854 |
| Malleidae: *Malleus albus* Lamarck, 1819 |
| Melongenidae: *Hemifusus ternatanus* (Gmelin, 1791), *Hemifusus tuba* (Gmelin, 1791), *Pugilina morio* (Linnaeus, 1758) |
| Muricidae: *Bolinus brandaris* (Linnaeus, 1758), *Hexaplex (Trunculariopsis) trunculus* (Linnaeus, 1758), *Murex (Murex) pecten* Lightfoot, 1786, *Purpura bufo* Lamarck, 1822, *Rapana venosa* (Valenciennes, 1846), *Reishia clavigera* (Küster, 1860), *Stramonita haemastoma* (Linnaeus, 1767), *Thaisella gradata* (Jonas, 1846) |
| Mytilidae: *Crenomytilus grayanus* (Dunker, 1853), *Lithophaga lithophaga* (Linnaeus, 1758)[AppII], *Modiolus barbatus* (Linnaeus, 1758), *Mytella charruana* (Orbigny, 1842), *Mytella guyanensis* Lamarck (1819), *Mytilus edulis* Linnaeus, 1758, *Mytilus galloprovincialis* Lamarck, 1819, *Perna viridis* (Linnaeus, 1758) |
| Nautilidae: *Nautilus pompilius* Linnaeus, 1758 |
| Octopodidae: *Amphioctopus fangsiao* (d'Orbigny, 1839), *Callistoctopus ornatus* (Gould, 1852), *Eledone moschata* (Lamarck, 1798), *Octopus variabilis* (Sasaki, 1929), *Octopus vulgaris* (Cuvier, 1799) |
| Olividae: *Oliva caroliniana* Duclos, 1835, *Oliva mustelina* Lamarck, 1811 |
| Ostreidae: *Crassostrea ariakensis* (Fujita, 1913), *Crassostrea belcheri* (G.B. Sowerby II, 1871), *Crassostrea gigas* (Thunberg, 1793), *Crassostrea nippona* (Seki, 1934), *Crassostrea rhizophorae* (Guilding, 1828), *Crassostrea virginica* (Gmelin, 1791), *Ostrea denselamellosa* Lischke, 1869, *Ostrea edulis* Linnaeus, 1758, *Planostrea pestigris* (Hanley, 1846), *Saccostrea echinata* (Quoy & Gaimard, 1835), *Saccostrea scyphophilla* (Peron & Lesueur, 1807) |
| Patellidae: *Patella caerulea* Linnaeus, 1758 |
| Pharidae: *Ensis minor* (Chenu, 1843), *Sinonovacula constricta* (Lamarck, 1818) |
| Pholadidae: *Pholas dactylus* Linnaeus, 1758 |
| Pinnidae: *Pinna nobilis* Linnaeus, 1758 |
| Placunidae: *Placuna placenta* (Linnaeus, 1758) |
| Potamididae: *Cerithidea microptera* (Kiener, 1842) |
| Pteriidae: *Pinctada imbricata* Röding, 1798, *Pinctada chemnitzii* (Philippi, 1849), *Pinctada margaritifera* (Linnaeus, 1758) |
| Ranellidae: *Charonia tritonis* (Linnaeus, 1758), *Linatella caudata* (Gmelin, 1791), *Ranella olearium* (Linnaeus, 1758)[LR/NT*], *Septa hepatica* (Röding, 1798) |

(continued)

**Table 12.2** (continued)

| Family/species |
|---|
| Sepiidae: *Sepia elegans* Blainville, 1827, *Sepia esculenta* Hoyle, 1885, *Sepia officinalis* Linnaeus, 1758, *Sepia pharaonis* Ehrenberg, 1831, *Sepiella inermis* (Van Hasselt, 1835 in Férussac and D'Orbigny, 1834–1848) |
| Siphonariidae: *Siphonaria normalis* Gould, 1846 |
| Solenidae: *Solen grandis* Dunker, 1862, *Solen marginatus* Pulteney, 1799, *Solen strictus* Gould, 1861 |
| Spirulidae: *Spirula spirula* (Linnaeus, 1758) |
| Spondylidae: *Spondylus imperialis* Chenu, 1844 |
| Strepsiduridae: *Melapium elatum* (Schubert & Wagner, 1829) |
| Strombidae: *Strombus pugilis* Linnaeus, 1758 |
| Teredinidae: *Neoteredo reynei* (Bartsch, 1920), *Lyrodus pedicellatus* (Quatrefages, 1849) |
| Trochidae: *Monodonta australis* (Lamarck, 1822), *Oxystele sinensis* (Gmelin, 1791), *Trochus incrassatus* Lamarck, 1822 |
| Turbinellidae: *Turbinella laevigata* (Anton, 1839), *Turbinella pyrum* (Linnaeus, 1767) |
| Turbinidae: *Turbo articulatus* Reeve, 1848, *Turbo chrysostomus* Linnaeus, 1758, *Turbo cornutus* Lightfoot, 1786, *Turbo coronatus* Gmelin, 1791, *Turbo petholatus* Linnaeus, 1758 |
| Turritellidae: *Turritella bacillum* Kiener, 1843 |
| Veneridae: *Anomalocardia brasiliana* (Gmelin, 1791), *Callista chione* (Linnaeus, 1758), *Chamelea gallina* (Linnaeus, 1758), *Circe scripta* (Linnaeus, 1758), *Cyclina sinensis* (Gmelin, 1791), *Mercenaria campechiensis* (Gmelin, 1791), *Meretrix casta* (Gmelin, 1791), *Meretrix petechialis* (Lamarck, 1818), *Meretrix meretrix* (Linnaeus, 1758), *Venerupis aspera* (Quoy & Gaimard, 1835), *Venerupis philippinarum* (A. Adams & Reeve, 1850), *Venus verrucosa* Linnaeus, 1758 |
| Volutidae: *Melo melo* (Lightfoot, 1786) |
| Xanthidae: *Etisus splendidus* Rathbun, 1906 |
| **CHELICERATA** |
| Limulidae: *Tachypleus tridentatus* (Leach, 1819)[DD*] |
| **CRUSTACEANS** |
| Balanidae: *Balanus amphitrite* Darwin, 1854 |
| Calappidae: *Calappa ocellata* Holthuis, 1958 |
| Cancridae: *Cancer pagurus* Linnaeus, 1758 |
| Crangonidae: *Crangon crangon* (Linnaeus, 1758) |
| Diogenidae: *Diogenes edwardsii* (DeHaan, 1849) |
| Epialtidae: *Doclea ovis* (Fabricius, 1787) |
| Galenidae: *Galene bispinosa* (Herbst, 1783) |
| Gecarcinidae: *Cardisoma guanhumi* Latreille, 1825 |
| Grapsidae: *Goniopsis cruentata* (Latreille, 1802), *Grapsus tenuicrustatus* (Herbst, 1783), *Plagusia depressa* (Fabricius, 1775) |
| Hippidae: *Emerita portoricensis* Schmitt, 1935 |
| Macrophthalmidae: *Macrophthalmus dilatatus* (de Hann, 1835) |
| Matutidae: *Matuta planipes* Fabricius, 1798 |
| Mithracidae: *Mithrax hispidus* (J. F. W. Herbst, 1790) |
| Nephropidae: *Homarus gammarus* (Linnaeus, 1758)[LC] |
| Ocypodidae: *Ocypode quadrata* (JC Fabricius, 1787), *Uca maracoani* (Latreille, 1802), *Ucides cordatus* (Linnaeus, 1763), *Uca (Minuca) pugnax* (Smith, 1870) |
| Paguridae: *Pagurus ochotensis* Brandt, 1851 |

(continued)

**Table 12.2** (continued)

| Family/species |
|---|
| Palaemonidae: *Macrobrachium malcolmsonii* (H. Milne Edwards, 1844) |
| Palinuridae: *Palinurus elephas* (Fabricius, 1787), *Panulirus homarus* (Linnaeus, 1758)[LC], *Panulirus ornatus* (Fabricius, 1798)[LC], *Panulirus stimpsoni* Holthuis, 1963[DD] |
| Penaeidae: *Fenneropenaeus chinensis* (Osbeck, 1765), *Xiphopenaeus kroyeri* (Heller, 1862), *Litopenaeus schmitti* (Burkenroad, 1936), *Trachysalambria curvirostris* (Stimpson, 1860) |
| Pollicipedidae: *Pollicipes mitella* (Linnaeus, 1758) |
| Portunidae: *Callinectes bocourti* A. Milne- Edwards, 1879, *Callinectes exasperatus* Gerstaecker, 1856, *Portunus (Portunus) trituberculatus* (Miers, 1876), *Scylla serrata* (Forskål, 1775) |
| Pseudosquillidae: *Cloridopsis dubia* (H. M. Edwards, 1837) |
| Sesarmidae: *Aratus pisoni* (H. Milne Edwards, 1837), *Sesarma dehaani* H. Milne Edwards, 1853 |
| Squillidae: *Harpiosquilla harpax* (de Haan, 1844), *Oratosquilla oratoria* (De Haan, 1844) |
| Upogebiidae: *Upogebia major* (De Haan, 1841) |
| **BRYOZOANS** |
| Celleporidae: *Celleporina costazii* (Audouin, 1826) |
| **ECHINODERMS** |
| Acanthasteridae: *Acanthaster planci* (Linnaeus, 1758) |
| Asteriidae: *Asterias amurensis* Lutken, 1871, *Asterias rollestoni* Bell, 1881 |
| Asterinidae: *Patiria pectinifera* (Muller & Troschel, 1842) |
| Astriclypeidae: *Echinodiscus bisperforatus* Leske, 1778 |
| Astropectinidae: *Astropecten bispinosus* (Otto, 1823), *Astropecten velitaris* von Martens, 1865, *Craspidaster hesperus* (Muller & Troschel, 1840) |
| Caudinidae: *Acaudina molpadioides* (Semper, 1867) |
| Cidaridae: *Prionocidaris australis* (Ramsay, 1885), *Prionocidaris pistillaris* (Lamarck, 1816) |
| Diadematidae: *Diadema setosum* (Leske, 1778) |
| Echinasteridae: *Echinaster (Othilia) brasiliensis* Müller & Troschel, 1842, *Echinaster (Othilia) echinophorus* (Lamarck, 1816) |
| Echinometridae: *Echinometra lucunter (Linnaeus,* 1758), *Echinometra mathaei* (Blainville, 1825), *Heliocidaris crassispina* (A. Agassiz, 1863), *Heterocentrotus mamillatus* (Linnaeus, 1758) |
| Holothuriidae: *Actinopyga echinites* (Jaeger, 1833), *Actinopyga lecanora* (Jaeger, 1833), *Actinopyga mauritiana* (Quoy & Gaimard, 1834), *Actinopyga miliaris* (Quoy & Gaimard, 1834), *Bohadschia argus* Jaeger, 1833, *Bohadschia marmorata* Jaeger, 1833, *Holothuria (Halodeima) atra* Jaeger, 1833, *Holothuria (Mertensiothuria) leucospilota* (Brandt, 1835), *Holothuria (Microthele) whitmaei* Bell, 1887 |
| Luidiidae: *Luidia senegalensis* Lamarck, 1916 |
| Mellitidae: *Leodia sexiesperforata* (Leske, 1778), *Mellita quinquiesperforata* (Leske, 1778) |
| Ophiactidae: *Ophiopholis mirabilis* (Duncan, 1879) |
| Oreasteridae: *Oreaster reticulatus* (Linnaeus, 1758), *Protoreaster lincki* (Blainville, 1830) |
| Parechinidae: *Paracentrotus lividus* (Lamarck, 1816) |
| Solasteridae: *Solaster dawsoni* Verrill, 1880 |
| Stichopodidae: *Apostichopus japonicus* (Selenka, 1867), *Isostichopus fuscus* (Ludwig, 1875)[AppIII], *Stichopus herrmanni* Semper, 1868, *Stichopus horrens* Selenka, 1867, *Thelenota ananas* (Jaeger, 1833) |
| Temnopleuridae: *Temnopleurus toreumaticus* (Leske, 1778) |

(continued)

**Table 12.2** (continued)

| Family/species |
|---|
| Toxopneustidae: *Lytechinus variegatus* (Lamarck, 1816), *Tripneustes gratilla* (Linnaeus, 1758) |
| **CHORDATA** |
| Pyuridae: *Halocynthia papillosa* (Linnaeus, 1767), *Pyura stolonifera* (Heller, 1878) |

IUCN categories
*CE* critically endangered; *DD* data deficient; *NT* near threatened; *LR/CD* lower risk/conservation dependent; *LC* least concern; *LR/NT* lower risk/near threatened
*needs updating
CITES categories
AppII–listed in appendix II; AppIII–listed in appendix III

fact, some invertebrate species exploited for medicinal purposes were also evaluated in terms of threatening and figure in lists of threatened species. According to our review, only 19 species (7.1%) figure in the IUCN and/or CITES databases (see Table 12.2), despite the well known use of invertebrates as folk remedies worldwide. On the other hand, some species are recorded in regional threatened lists. In Brazil, for example, the starfishes *Luidia senegalensis*, *Echinaster brasiliensis*, *Echinaster echinophorus* and *Oreaster reticulatus* figure in the Official List of Brazilian Fauna Threatened with Extinction (Machado et al. 2008). Major threats to these species include: incidental captures, such as bycatch in fishing trawls; habitat degradation; overfishing for aquarium trade; pollutants, such as oil, domestic and industrial sanitary wastes; and excessive tourism in their occurrence areas (Machado et al. 2008). Besides, they are traded for medicinal and religious use, as well as for decoration and curios (Alves and Dias 2010; Léo Neto et al. 2009; Machado et al. 2008). The case of the Brazilian starfishes exemplifies the situation of many other marine invertebrates: although medicinal use may imply an additional pressure to their populations, it is generally disregarded in most cases.

It must be highlighted that several medicinal species, especially mollusks, crustaceans, and echinoderms are collected for multiple purposes, mainly for food (Dias et al. 2011; Nishida et al. 2006; Purcell et al. 2010). In this sense, the access to medicinal animals (both vertebrates and invertebrates) and their medicinal by-products are associated to the fishing activities that routinely take place in human communities. As an example, recent studies in Brazil have shown that in coastal areas the majority of species used in traditional medicine has a marine and estuarine origin, many of them considered of fishing importance (Alves and Rosa 2006, 2007b, c; Alves et al. 2007). On the other hand, the use of invertebrates for medicinal purposes is not restricted to isolated fishing communities; several species are commonly traded in urban regions in different countries (Alves and Rosa 2007c, 2010; Ashwell and Walston 2008).

The commercialization for animal parts and their medicinal derivatives is contributing to the loss of some species (Lee 1999; Still 2003). As many factors affect animal populations in the world, the medicinal use of animals must be considered together with other anthropogenic pressures, such as habitat loss (Alves and Alves 2011). Habitat destruction is particularly pervasive in tropical areas where mangroves, coral reefs, and wetlands are being destroyed at alarming rates, and in temperate areas there are severe threats to wetland regions and estuaries; conflicts between industrial and tourist development and conservation are universal; furthermore, the threats from commercial fishing on biodiversity of coastal areas have been neglected (Gray 1997). Over-fishing, pollution, and other anthropogenic activities have resulted in the destruction of marine bioresources, especially acute degradation of biodiversity of marine medicinal animals, and as remarked by Fu et al. (2009), the feasible strategies for conservation of biodiversity and bioresources of marine medicinal species should be to improve the construction and management of marine natural reserves system as well as marine special reserves.

Medicinal species whose conservation status is a cause of concern should receive urgent attention, and aspects such as habitat loss/alteration should be discussed in connection with the present and future use of these species in folk medicine (Alves et al. 2007) and bioprospecting.

Furthermore, the importance of marine invertebrates in traditional medicine has extended to the field of bioprospecting, as some studies have demonstrated that animal by-products may be promising resources to the development of new pharmaceutical ingredients (Haefner 2003; Muller et al. 2004; Thomas et al. 2010). During the past two decades, marine invertebrates such as sponges, jellyfishes, sea anemones, gorgonians, corals, bryozoans, mollusks, echinoderms, tunicates, and crustaceans have furnished a large variety of metabolites with unusual structures. Several of these metabolites exhibit high order of biological activities, and some have biomedical potential (Pandey 2009). In fact, many important medical treatments and biotechnologies are based on those chemicals derived from marine organisms, such as sponges, soft corals, mollusks, bacteria, and algae (Letson 2008).

Marine species are increasingly sought in this research field (Pain 1996) and related papers tend to emphasize promising chemical compositions or pharmacological actions (Carté 1996; Scheuer 1988), generally without discussing the breadth of flora and fauna which could potentially be threatened by overharvesting. This decoupling of bioprospection and conservation is a cause for concern, especially when considering that, in addition to traditional systems of medicine, current practices can involve species which were not historically used (Vincent 1997), particularly as technologies such as deep-sea trawling make available marine resources which were previously unattainable or unknown (Cunningham and Zondi 1991).

Hunt and Vincent (2006) highlighted some conservation concerns about the exploitation of marine organisms in bioprospecting, including (1) nonselective or destructive collection methods (McMichael and Beaglehole 2000), (2) possible introduction of pathogens or exotic species by collectors (McMichael and Beaglehole 2000), and most often, (3) possible overcollection of target organisms (Alves et al. 2010c; Benkendorff 2002; Chivian et al. 2003). Furthermore, as remarked by Bruckner (2002), because most of the desired species for biotechnology have little value as a food fishery, management strategies for sustainable harvest have been lacking, and much of the information needed about the population dynamics or the life history of the organisms is unknown.

For high value, internationally traded species such as abalone (*Haliotis*) species are the focus of illegal harvest to fulfill the demand from Asia, a co-ordinated approach to CITES listing from all source countries is necessary. The example of the South African abalone (*Haliotis midae*) and the recent withdrawal of the species from CITES listing shows how politicized this issue can be (Burgener 2010).

Marques (1997) pointed out that considerations of negative impacts on biodiversity should not be limited to the traditional use of animal or animal-derived products, but must also extend to their exploitation by the pharmaceutical industry. In this sense, conservation measures applied to medicinal species should consider both their exploitation in traditional medicines and in the pharmaceutical industry (Alves et al. 2007). More recently, various studies have highlighted the necessity of the rational use and conservation of the species most frequently used for that purpose, and discussed the implications of the medicinal use of species under legal protection (Almeida and Albuquerque 2002; Alves and Rosa 2005).

As a first step in promoting continued biomedical research for marine natural products, countries must involve the development of management plans (through joint efforts involving scientists, resource managers in the source country, and industry) for sustainable harvest of potentially valuable invertebrates–a step that must occur before large-scale extraction takes place (Bruckner 2002).

In several cases, when medicinal species are also used as food, it becomes more difficult to assess the magnitude of the impact of traditional medicines, since the consumption of the species may be related to its therapeutic value. Nature-based traditional food and medicine are generally viewed as interchangeable, diet being highly regarded as the primary basis for sustaining and/or restoring health and well-being. Consequently, foods are considered and oftentimes chosen for their distinctive medicinal or healing values (Alves and Rosa 2007b; Alves et al. 2007; Begossi 1992; Begossi and Braga 1992).

For instance, the presence of foods in a Chinese medicine shop reflects the belief that food and medicine are part of a continuous spectrum (Wong and Dahlen 1994), and this can be illustrated by the case of holothurids. Traditionally, Chinese ate dried sea cucumbers primarily by their tonic value, since they have been considered a tonic food for many years. Known as "ginseng of the sea", those animals are recognized both as a tonic and as traditional medicine in the ancient literature (Chen 2003).

According to Traditional Chinese Medicine, sea cucumbers nourish the blood and vital essence (*jing*), tonifies kidney (*qi*) (treats disorders of the kidney system, including reproductive organs), and reduces dryness, especially of the intestines (Chen 2004; Conand 2006).

Another interface between zootherapeutic animals and food is derived from the use of by-products in the preparation of animal-based remedies. As an example, Moura and Marques (2008) pointed that one characteristic in common among zootherapeutic animals in Brazil is the use of by-products that would otherwise be discarded. Indeed, many of the animal-based medicinal products commercialized in Brazil are by-products from specimens used for other purposes (e.g., food, magical-religious, etc.). These multiple uses of fauna and their impact on animal populations must be properly assessed and taken into consideration when implementing recovery plans, especially for species that are highly exploited (Alves and Rosa 2006, 2007b, c; Alves et al. 2007, 2010b).

Connections between traditional medicine, biodiversity, and human health have recently been addressed by different authors (Alves and Rosa 2007a, 2008; Anyinam 1995; Chivian 2002; McMichael and Beaglehole 2000), who have drawn attention to the fact that biodiversity loss can have indirect and direct effects on human well-being. The reliance on traditional uses of animals as food and as medicine by communities around the world highlights the need for further interdisciplinary research in ethnozoology which can be used in strategies to conserve biodiversity (Alves and Souto 2010; Alves et al. 2010c). Furthermore, loss of wildlife resources, apart from threatening people's health and well-being, affects their cultural integrity. In this sense, understanding the contexts of traditional therapeutic uses of animals is central for elucidating their potential impact on public health and biodiversity conservation.

## 12.5 Conclusions

The marine environment represents an extraordinary resources supplier for human medicinal use. Our review shows that at least 266 marine invertebrates are used for medicinal purposes in the world, underlining the importance of this animals group as alternative therapeutic sources.

The highest levels of marine biodiversity are found in tropical countries which are developing. In these countries, local people use traditional medicine as one of the main therapeutic alternatives and marine invertebrates are a source of a number of popular remedies. Additionally, studies on bioprospecting have revealed the huge potential of those animals for the pharmaceutical industry. In light of the above, it is evident that this use of marine biodiversity must be considered under a conservationist context, since it may represent an additional pressure to threatened species.

Animals provide the raw materials for remedies used to treat physical and/or spiritual diseases. Besides being influenced by cultural aspects, the relations between humans and biodiversity in the form of zootherapeutic practices are conditioned by the social and economic relations between humans themselves. Sustainability of harvesting of medicinal animals is challenged by many factors, from both social and ecological perspectives. It is important to respect differing views of the value of wildlife, while, at the same time, conserving biodiversity. Strategies should take place partly for protection of biodiversity, but also to ensure sustainable use of coastal habitat resources. Sustainable use of these resources will require that all stakeholders are involved in the assessments and the decision-making processes that follow.

Using animal products as components of bioprospecting has implications for medicine, the environment, economy, public health, and culture. Although widely diffused, zootherapeutic practices remain virtually unstudied, and so far there has been neither a demonstration of the clinical efficacy of the popularly used remedies nor an evaluation of the sanitary implications of the prescription of animal products for the treatment of diseases. New studies of medicinal fauna, which seek a better understanding of this form of therapy—including ecological, cultural, and pharmacological aspects—are necessary.

Lastly, the depletion of medicinal resources not only poses a challenge for conservation but represents a serious threat to the health of many human communities, and efforts to stabilize the status of these species are important not only to conservationists but to millions of people who make use of the use of traditional remedies (Alves et al. 2010a, b; Still 2003). In that context, Nations (1997) emphasizes the utilitarian value of species' protection, and the perspective of long-term use of biological diversity, while Soejarto (1996) remarks that conservation permits the continuing use of the resources in ways that are non-destructive and sustainable; from the pharmaceutical point of view, it provides time to eventually demonstrate fully the potentially medicinal value of the resources.

## References

Adeola MO (1992) Importance of wild animals and their parts in the culture, religious festivals, and traditional medicine, of Nigeria. Environ Conserv 19(2):125–134

Alino PM, Cajipe GJB, Ganzon-Fortes ET, Licuanan WRY, Montano NE, Tupas LM (1990) The use of marine organisms in folk medicine and horticulture: a preliminary study. Seaweed Information Center (SICEN). Supplement of SICEN Newsletter

Almeida AV (2007) Zooterapia indígena brasileira do século XVIII nas obras de Guilherme Piso, Georg Marcgrave e Joannes de Laet. Sitientibus Série Ciências Biológicas 7(3):261–272

Almeida AV, Alves AGC, Lucena RFP, Albuquerque UP (2005) Prescrições zooterápicas indígenas brasileiras nas obras de Guilherme Piso (1611–1678). In: Atualidades em Etnobiologia e Etnoecologia. Nupeea; Sociedade Brasileira de Etnobiologia e Etnoecologia, Recife

Almeida CFCBR, Albuquerque UP (2002) Uso e conservação de plantas e animais medicinais no Estado de Pernambuco (Nordeste do Brasil): Um estudo de caso. Interciencia 27(6):276–285

Alves RRN (2008) Commercialization of *Uranoscodon superciliosus* Linnaeus, 1758 (Tropiduridae) for magical-religious purposes in North and Northeastern of Brazil. Sitientibus Série Ciências Biológicas 8(2):257–258

Alves RRN (2009) Fauna used in popular medicine in Northeast Brazil. J Ethnobiol Ethnomed 5(1):1–30

Alves RRN, Alves HN (2011) The faunal drugstore: animal-based remedies used in traditional medicines in Latin America. J Ethnobiol Ethnomed 7(9)

Alves RRN, Barbosa JAA, Santos SLDX, Souto WMS, Barboza RRD (2009a) Animal-based remedies as complementary medicines in the semi-arid region of Northeastern Brazil. Evidence-based complementary and alternative medicine 2011

Alves RRN, Barboza RRD, Souto WMS (2010a) A global overview of canids used in traditional medicines. Biodivers Conserv 19(6):1513–1522

Alves RRN, Dias TLP (2010) Usos de invertebrados na medicina popular no Brasil e suas implicações para conservação. Trop Conserv Sci 3(2):159–174

Alves RRN, Léo Neto NA, Santana GG, Vieira WLS, Almeida WO (2009b) Reptiles used for medicinal and magic religious purposes in Brazil. Appl Herpetol 6(3):257–274
Alves RRN, Lima HN, Tavares MC, Souto WMS, Barboza RRD, Vasconcellos A (2008a) Animal-based remedies as complementary medicines in Santa Cruz do Capibaribe, Brazil. BMC Complement Altern Med 8:44
Alves RRN, Neto NAL, Brooks SE, Albuquerque UP (2009c) Commercialization of animal-derived remedies as complementary medicine in the semi-arid region of Northeastern Brazil. J Ethnopharmacol 124(3):600–608
Alves RRN, Rosa IL (2005) Why study the use of animal products in traditional medicines? J Ethnobiol Ethnomed 1(5):1–5
Alves RRN, Rosa IL (2006) From cnidarians to mammals: the use of animals as remedies in fishing communities in NE Brazil. J Ethnopharmacol 107:259–276
Alves RRN, Rosa IL (2007a) Biodiversity, traditional medicine and public health: Where do they meet? J Ethnobiol Ethnomed 3(14):9
Alves RRN, Rosa IL (2007b) Zootherapeutic practices among fishing communities in North and Northeast Brazil: a comparison. J Ethnopharmacol 111:82–103
Alves RRN, Rosa IL (2007c) Zootherapy goes to town: the use of animal-based remedies in urban areas of NE and N Brazil. J Ethnopharmacol 113:541–555
Alves RRN, Rosa IL (2008) Medicinal animals for the treatment of asthma in Brazil. BMC Complement Altern Med 14:350–351
Alves RRN, Rosa IL (2010) Trade of animals used in Brazilian traditional medicine: trends and implications for conservation. Hum Ecol 38(5):691–704
Alves RRN, Rosa IL, Santana GG (2007) The role of animal-derived remedies as complementary medicine in Brazil. Bioscience 57(11):949–955
Alves RRN, Souto WMS (2010) Etnozoologia: conceitos, considerações históricas e importância. In: Alves RRN, Souto WMS, Mourão JS (eds) A Etnozoologia no Brasil: importância, itatus atual e perspectivas, vol 7, 1st edn. NUPEEA, Recife
Alves RRN, Souto WMS, Barboza RRD (2010b) Primates in traditional folk medicine: a world overview. Mamm Rev 40(2):155–180
Alves RRN, Souto WMS, Mourão JS (2010c) A etnozoologia no Brasil: Importância, Status atual e Perspectivas, vol 7. Estudos & Avanços, 1 edn. NUPEEA, Recife
Alves RRN, Vieira WLS, Santana GG (2008b) Reptiles used in traditional folk medicine: conservation implications. Biodivers Conserv 17:2037–2049
Anyinam C (1995) Ecology and ethnomedicine: exploring links between current environmental crisis and indigenous medical practices. Soc Sci Med 40(3):321–329
Ashwell D, Walston N (2008) An overview of the use and trade of plants and animals in traditional medicine systems in Cambodia, 1st edn. Greater Mekong Programme, Ha Noi, TRAFFIC Southeast Asia
Barajas E (1961) Los animales usados en la medicina popular mexicana. Imprenta Universitaria, Ciudad de México
Begossi A (1992) Food taboos at Búzios Island (SE Brazil): their significance and relation to folk medicine. J Ethnobiol 12(1):117–139
Begossi A, Braga F (1992) Food taboos and folk medicine among fishermen from the Tocantins river. Amazoniana 12:341–352
Benkendorff K (2002) Potential conservation benefits and problems associated with bioprospecting in the marine environment. In: Lumley D, Dickman C (eds) A zoological revolution: using native fauna to assist in its own survival. Royal Zoological Society of New South Wales and Australian Museum, Sydney
Benoist R (1951) Contribution a l'Histoire de l'utilisation en pharmacie et en thérapeutique des animaux aquatiques (autres que les poisons) et de leur produits. Université de Strasbourg, Strasbourg
Bensky D, Clavey S, Stöger E (2004) Chinese Herbal Medicine: Materia medica. Eastland Press, Seatle

Bruckner AW (2002) Life-saving products from coral reefs. Issues Sci Technol. http://www.nap.edu/issues/18.3/p_bruckner.html. Accessed 16 July 2011

Burgener M (2010) Evaluation of the CITES appendix III listing and delisting of South African abalone *Haliotis midae*. TRAFFIC Bull 23(1):42–48

Carté BK (1996) Biomedical potential of marine natural products. Bioscience 46(4):271–286

Chen J (2003) Overview of sea cucumber farming and sea ranching practices in China. SPC beche-de-mer Inf Bull 18:18–23

Chen J (2004) Present status and prospects of sea cucumber industry in China. FAO Fisheries technical paper, pp 25–38

Chen J (2005) Present status and prospects of sea cucumber industry in China. In: Lovatelli A (ed) Advances in sea cucumber aquaculture and management. Food Agric Organiz, Rome

Chivian E (2002) Biodiversity: its importance to human health. Center for Health and the Global Environment, Harvard Medical School, Cambridge

Chivian E, Roberts CM, Bernstein AS (2003) The threat to cone snails. Science 302(5644):391

Choo PS (2004) Fisheries, trade and utilization of sea cucumbers in Malaysia. In: Lovatelli A, Conand C, Purcell S, Uthicke S, Hamel J-F, Mercier A (eds) Advances in sea cucumber aquaculture and management. FAO fisheries technical paper 463, FAO, Rome

Choo PS (2008) Population status, fisheries and trade of sea cucumbers in Asia. FAO fisheries and aquaculture technical paper 516, FAO, Rome

Chopra RN, Chopra IC, Handa KL, Kapur LD (2006) Indigenous drugs of India, 2nd edn. Academic Publishers, Calcuta

CITES (2001) List of animal species used in traditional medicine. In: Seventeenth meeting of the animals committee, CITES, Hanoi. http://www.cites.org/eng/com/ac/18/E18-13-1.doc. Acessed 21 Aug 2011

Conand C (1990) The fishery resources of Pacific Island countries, part 2: holothurians. FAO fisheries technical paper 272.2, FAO, Rome

Conand C (2006) Harvest and trade: utilization of sea cucumbers; sea cucumber fisheries; current international trade; illegal, unreported and unregulated trade; bycatch; socio-economic characteristics of the trade in sea cucumbers. In: Bruckner AW (ed) Proceedings of the CITES workshop on the conservation of sea cucumbers in the families Holothuriidae and Stichopodidae. NOAA Technical Memorandum NMFS-OPR-34, Silver Spring, pp 51–73

Costa-Neto EM (1999a) Cultura pesqueira, desenvolvimento e sustentabilidade no litoral norte do estado da Bahia: um estudo de caso. TecBahia 14(1):131–139

Costa-Neto EM (1999b) Healing with animals in Feira de Santana City, Bahia, Brazil. J Ethnopharmacol 65:225–230

Costa-Neto EM (1999c) Traditional use and sale of animals as medicines in Feira de Santana City, Bahia, Brazil. Indig Knowl Dev Monit 7(2):6–9

Costa-Neto EM (2000a) Sustainable development and traditional knowledge: a case study in a Brazilian artisanal fishermen's community. Sustainable Dev 8(2):89–95

Costa-Neto EM (2000b) Zootherapy based medicinal traditions in brazil. Honey Bee 11(2)

Costa-Neto EM (2001) A cultura pesqueira do litoral Norte da Bahia. EDUFBA/EDUFAL, Salvador/Macéio, Salvador/Macéio, Brazil

Costa-Neto EM, Marques JGW (2000) Conhecimento ictiológico tradicional e a distribuição temporal e espacial de recursos pesqueiros pelos pescadores de Conde, Estado da Bahia, Brasil. Etnoecológica 4(6):56–68

Cunningham AB, Zondi AS (1991) Use of animal parts for the commercial trade in traditional medicines. Institute of Natural Resources, Pietermaritzburg

Descola P (1996) The spears of twilight: life and death in the Amazon jungle. The New Press, NY

Dias TLP, Leo Neto NA, Alves RRN (2011) Molluscs in the marine curio and souvenir trade in NE Brazil: species composition and implications for their conservation and management. Biodivers Conserv, pp 1–13

Ferreira FS, Brito S, Ribeiro S, Saraiva A, Almeida W, Alves RRN (2009) Animal-based folk remedies sold in public markets in Crato and Juazeiro do Norte, Ceara, Brazil. BMC Complement Altern Med 9(1):17

Figueiredo N (1994) Os 'bichos' que curam: os animais e a medicina 'folk' em Belém do Pará. Boletim do Museu Paraense Emílio Göeldi 10(1):75–91

Fredalina BD, Ridzwan BH, Abidin AA, Kaswandi MA, Zaiton H, Zali I, Kittakoop P, Jais AM (1999) Fatty acid compositions in local sea cucumber: stichopus chloronotus, for wound healing. Gen Pharmacol 33(4):337–340

Fu X, Wang CY, Shao CL, Han L, Wu YF, Wu HL, Qian SB, Guan HS (2009) Endangered and rare species of marine medicinal organisms and Their protection in China. Period Ocean Univ Chin 39(4):719–734

Gopal R, Vijayakumaran M, Venkatesan R, Kathiroli S (2008) Marine organisms in Indian medicine and their future prospects. Nat Prod Radiance 7(2):139–145

Gössling S, Kunkel T, Schumacher K, Zilger M (2004) Use of molluscs, fish, and other marine taxa by tourism in Zanzibar, Tanzania. Biodivers Conserv 13(14):2623–2639

Gray JS (1997) Marine biodiversity: patterns, threats and conservation needs. Biodivers Conserv 6(1):153–175

Groark KP (1997) To warm the blood, to warm the flesh: the role of the steambath in highland maya (Tzeltal-Tzotzil) ethnomedicine. J Lat Am Lore 20(1):3–96

Haefner B (2003) Drugs from the deep: marine natural products as drug candidates. Drug Discov Today 8(12):536–544

Hamidah A, Rustam ZA, Tamil AM, Zarina LA, Zulkifli ZS, Jamal R (2009) Prevalence and parental perceptions of complementary and alternative medicine use by children with cancer in a multi ethnic Southeast Asian population. Pediatr Blood Cancer 52(1):70–74

Herbert DG, Hamer ML, Mander M, Mkhize N, Prins F (2003) Invertebrate animals as a component of the traditional medicine trade in KwaZulu-Natal, South Africa. Afr Invertebr 44(2):327–344

Hu SY (1980) An enumeration of Chinese materia medica. The Chinese University Press, Hong Kong

Hunt B, Vincent ACJ (2006) Scale and sustainability of marine bioprospecting for pharmaceuticals. AMBIO. J Hum Environ 35(2):57–64

Kang S, Phipps M (2003) A question of attitude: South Korea's traditional medicine practitioners and wildlife conservation, 1st edn. TRAFFIC East Asia, Hong Kong

Lee SKH (1999) Trade in traditional medicine using endangered species: an international context. In: Paper presented at the 2nd Australian symposium on traditional medicine and wildlife conservation, Melbourne

Léo Neto NA, Brooks SE, Alves RRN (2009) From Eshu to Obatala: animals used in sacrificial rituals at Candomble "terreiros" in Brazil. J Ethnobiol Ethnomed 5(1):1–23

Letson D (2008) Oceans and human health: human dimensions. In: Walsh PJ, Smith SL, Fleming LE, Solo-Gabriele HM, Gerwick WH (eds) Oceans and human health: risks and remedies from the seas. Academic Press is an imprint of Elsevier, pp 91–98

Lev E (2003) Traditional healing with animals (zootherapy): medieval to present-day Levantine practice. J Ethnopharmacol 85(1):107–118

Lev E (2007) Drugs held and sold by pharmacists of the Jewish community of medieval (11–14th centuries) cairo according to lists of materia medica found at the Taylor-Schechter Genizah collection, Cambridge. J Ethnopharmacol 110(2):275–293

Lev E, Amar Z (2002) Ethnopharmacological survey of traditional drugs sold in the Kingdom of Jordan. J Ethnopharmacol 82:131–145

Liu C, Tseng A, Yang S (2005) Chinese herbal medicine: modern applications of traditional formulas. Alternative medicine. CRC Press, Boca Raton

Lovatelli A, Conand C, Purcell S, Uthicke S, Hamel JF (2004) Advances in sea cucumber aquaculture and management. Can J Fish Aquat Sci 61:519–528

Machado ABM, Drummond GM, Paglia AP (2008) Livro Vermelho da fauna brasileira ameaçada de extinção, vol 2 MMA. Fundação Biodiversitas, Brasília

Mahawar MM, Jaroli DP (2008) Traditional zootherapeutic studies in India: a review. J Ethnobiol Ethnomed 4(1):17

Mallmann MLW (1996) A farmacopéia do mar: invertebrados marinhos de interesse médico e a etnomedicina alagoana. Universidade Federal de Alagoas, Maceió
Marean CW, Bar-Matthews M, Bernatchez J, Fisher E, Goldberg P, Herries AIR, Jacobs Z, Jerardino A, Karkanas P, Minichillo T (2007) Early human use of marine resources and pigment in South Africa during the Middle Pleistocene. Nature 449(7164):905–908
Marques JGW (1997) Fauna medicinal: recurso do ambiente ou ameaça à biodiversidade? Mutum 1(4)
Marshall NT, Milledge SAH, Afonso PS (2001) Stormy seas for marine invertebrates: trade in sea cucumbers, sea shells and lobsters in Kenya. Tanzania and Mozambique, TRAFFIC East/ Southern Africa
McMichael AJ, Beaglehole R (2000) The changing global context of public health. The Lancet 356(9228):495–499
Morwood MJ, O'Sullivan PB, Aziz F, Raza A (1998) Fission-track ages of stone tools and fossils on the east Indonesian island of Flores. Nature 392(6672):173–176
Moura FBP, Marques JGW (2008) Zooterapia popular na Chapada Diamantina: uma Medicina incidental? Ciência Saúde Coletiva 13:2179–2188
Muller WEG, Batel R, Schroder HC, Muller IM (2004) Traditional and modern biomedical prospecting: Part I–the history: sustainable exploitation of biodiversity (sponges and invertebrates) in the Adriatic Sea in Rovinj (Croatia). Evidence-based Complement Altern Med 1(1):71–82
Nadkarni KM (1994) Indian materia medica, vol 2, 3rd edn. Popular Prakashan, Mumbai
Nations JD (1997) A ecologia profunda encontra o mundo em desenvolvimento. In: Wilson EO (ed) Biodiversidade. Nova Fronteira, Rio de Janeiro
Niir Board of Consultants and Engineers (2003) Hand book on unani medicines with formulae, processes, uses and analysis. Asia Paific Business Press Inc., Delhi
Nishida AK, Nordi N, Alves RRN (2006) Mollusc gathering in Northeast Brazil: an ethnoecological approach. Hum Ecol 34(1):133–145
Padmanabhan P, Sujana KA (2008) Animal products in traditional medicine from Attappady hills of Western Ghats. Indian J Tradit Knowl 7(2):326–329
Pain S (1996) Hostages of the deep. New Sci 151(2047):38–42
Pajaro MG, Meeuwig JJ, Giles BG, Vincent ACJ (2004) Biology, fishery and trade of sea moths (Pisces: Pegasidae) in the central Philippines. Oryx 38(04):432–438
Pandey P (2009) Endangered medicinal species of the Indian Ocean: radical need for conservation. Chem Biodivers 6(7):990–1001
Pereira J (1842) The elements of materia medica and therapeutica, vol 2, 2nd edn. Longman, Brown, Green and Longmans, London
Perezrul DH (2006) National report–Mexico. In: Bruckner AW (ed) Proceedings of the CITES workshop on the conservation of sea cucumbers in the families holothuriidae and stichopodidae. NOAA technical memorandum NMFSOPR 34, Silver Spring, pp 166–168
Perry A (2000) Global survey of marine medicinals. In: Moreau M-A, Hall HJ, Vincent ACJ (eds) Proceedings of the first international workshop on the management and culture of marine species used in traditional medicines. Project Seahorse, Montreal, pp 35–43
Prabhakar AK, Roy SP (2009) Ethno-medical uses of some shell fishes by people of Kosi River basin of North-Bihar, India. Ethnomedicine 3:1–4
Purcell SW, Lovatelli A, Vasconcellos M, Yimin Y (2010) Managing sea cucumber fisheries with and ecosystem approach. FAO fisheries and aquaculture technical paper no. 520, FAO, Rome
Quave CL, Lohani U, Verde A, Fajardo J, Rivera D, Obón C, Valdes A, Pieroni A (2010) A comparative assessment of zootherapeutic remedies from selected areas in Albania, Italy, Spain and Nepal. J Ethnobiol 30(1):92–125
Radbill SX (1976) Child hygiene among the Indians. Texas Rep Biol Med 3(4):419–512
Read BE, Li S, Pak C (1937) Chinese materia medica: turtle & shellfish drugs, avian drugs, a compendium of minerals and stones. Peking Natural History Bulletin Reprinted by Southern Materials Center, Inc., Taipei

Rosa IL, Oliveira TPR, Osório FM, Moraes LE, Castro ALC, Barros GML, Alves RRN (2011) Fisheries and trade of seahorses in Brazil: historical perspective, current trends, and future directions. Biodivers Conserv 20(9):1951–1971
Ruppert EE, Barnes RD, Fox RS (1994) Invertebrate zoology. Saunders College Publishing, Orlando
Salamanca AM, Pajaro MG (1996) The utilization of seashells in the Philippines. TRAFFIC Bull Wildl Trade Monit Unit 16:61–72
Scheuer PJ (1988) Ethno-natural historical leads. In: Fautin DG (ed) Biomedical importance of marine organisms, vol 13. California Academy of Sciences, San Francisco, pp 37–40
Schröder T (2010) World ocean review 2010. Maribus, Hamburg
Seraj S, Rahmatullah M, Monjur-E-Khudha M, Aporna SA, Khan MSH, Jahan R (2011) Amulets and other uncommon treatments prescribed by traditional medicinal practitioners of the bede community residing in Porabari village of Dhaka district, Bangladesh. J Altern Complem Med 17(11):987–993
Silva MLV, Alves ÂGC, Almeida AV (2004) A zooterapia no Recife (Pernambuco): uma articulação entre as práticas e a história. Biotemas 17(1):95–116
Sodeinde OA, Soewu DA (1999) Pilot study of the traditional medicine trade in Nigeria. TRAFFIC Bull 18(1):35–40
Soejarto DD (1996) Biodiversity prospecting and benefit-sharing: perspectives from the field. J Ethnopharmacol 51:1–15
Sousa RS (2010) Etnobotânica e etnozoologia de comunidades pesqueiras da Área de Proteção Ambiental (APA) do Delta do Parnaíba, Nordeste do Brasil. Universidade Federal do Piauí, Teresina
Still J (2003) Use of animal products in traditional Chinese medicine: environmental impact and health hazards. Complem Ther Med 11:118–122
Stringer CB, Finlayson JC, Barton RNE, Fernández-Jalvo Y, Cáceres I, Sabin RC, Rhodes EJ, Currant AP, Rodríguez-Vidal J, Giles-Pacheco F (2008) Neanderthal exploitation of marine mammals in Gibraltar. Proc Natl Acad Sci 105(38):14319–14324
Tang WC (1987) Chinese medicinal materials from the sea. Chin Med 1(4):571–600
Thomas TRA, Kavlekar DP, LokaBharathi PA (2010) Marine drugs from sponge-microbe association–a review. Mar Drugs 8(4):1417
Titcomb M, Fellows DB, Pukui MK, Devaney DM (1978) Native use of marine invertebrates in old Hawaii. Pac Sci 32(4):325–386
Traffic Europe-Russia (2006) Harvest and trade of trepang (far east sea cucumber) from the Russian far east. In: Bruckner AW (ed) Proceedings of the CITES workshop on the conservation of sea cucumbers in the families Holothuriidae and Stichopodidae. NOAA technical memorandum NMFSOPR 34, Silver Spring, p 217
Van NDN, Tap N (2008) An overview of the use of plants and animals in traditional medicine systems in Viet Nam, 1st edn. TRAFFIC Southeast Asia, Greater Mekong Programme, Ha Noi
Vincent ACJ (1997) Trade in pegasid fishes (sea moths), primarily for traditional Chinese medicine. Oryx 31(3):199–208
Vincent ACJ (2006) Live food and non-food fisheries on coral reefs, and their potential management. In: Côté IM, Reynolds JD (eds) Coral reef conservation. Cambridge University Press, Cambridge, pp 183–236
Vohora SB, Khan M (1978) Animal origin drugs used in unani medicine. Indian J Pharmacol 10(3):255
Voultsiadou E (2010) Therapeutic properties and uses of marine invertebrates in the ancient Greek world and early Byzantium. J Ethnopharmacol 130(2):237–247
Walker P, Wood E (2005) Life in the sea: people and the sea. Facts on File, Inc., NY
Whiting MJ, Williams VL, Hibbitts TJ (2011) Animals traded for traditional medicine at the Faraday market in South Africa: species diversity and conservation implications. J Zool 284:84–96
WHO (2011) International statistical classification of diseases and related health problems (ICD-10). http://apps.who.int/classifications/icd10/browse/2010/en. Accessed 20 Oct 2011

Wong K, Dahlen M (1994) Streetwise guide: Chinese herbal medicine. Red Mansions Publishing, San Francisco

Xiyin L, Guanghui Z, Qiang Z, Liang W, Benxue G (2004) Studies on hatchery techniques of the sea cucumber, Apostichopus japonicus. In: Lovatelli A, Conand C, Purcell S, Uthicke S, Hamel J-F, Mercier A (ed) Advances in sea cucumber aquaculture and management. FAO fisheries technical paper 463, FAO, Rome

Yinfeng G, Xueying Z, Yan C, Di W, Sung W (1997) Sustainability of wildlife use in traditional Chinese medicine. Conserving China biodiversity: reports of the biodiversity working group (BWG), China council for international cooperation on environment and development

Zaidnuddin I, Ibrahim K (2006) Natinal report—Malaysia. In: Bruckner AW (ed) Proceedings of the CITES workshop on the conservation of sea cucumbers in the families Holothuriidae and Stichopodidae. NOAA technical memorandum NMFSOPR 34, Silver Spring, pp 169–180

# Chapter 13
# The Role of Animal-Derived Remedies as Complementary Medicine in Brazil

**Rômulo Romeu Nóbrega Alves, Gindomar Gomes Santana and Ierecê Lucena Rosa**

**Abstract** Although animal-derived remedies constitute an integral part of folk medicine in many parts of the world, particularly for people with limited or no access to mainstream medical services, their role in health care has generally been overlooked in discussions about public health, conservation, and management of faunistic resources and ecosystem protection. Brazil's high biological and socio-cultural diversity translates into a wealth of traditional knowledge and practices, including the use of animals for medicinal purposes. In this chapter, we report on the use of 354 medicinal animal species in Brazil, 96% of which are wild caught and 21% of which are on one or more lists of endangered species. Further population declines may limit users' access to these bioresources and diminish the knowledge base upon which traditional medicine is built.

## 13.1 Introduction

The relationship between humans and nature, especially between humans and other animals, is one of the oldest cultural assets in history. There have been extremely close connections of dependence and co-dependence between humans

R. R. N. Alves (✉)
Ethnozoology, Conservation and Biodiversity Research Group, Departamento de Biologia, Universidade Estadual da Paraíba, Campina Grande-PB 58109-753, Brazil
e-mail: romulo_nobrega@yahoo.com.br

G. G. Santana
Laboratório de Ecofisiologia Animal, Departamento de Sistemática e Ecologia, Universidade Federal da Paraíba, João Pessoa-PB 58051-900, Brazil

I. L. Rosa
Departamento de Sistemática e Ecologia, Universidade Federal da Paraíba, João Pessoa 58059-970, Brazil

R. R. N. Alves and I. L. Rosa (eds.), *Animals in Traditional Folk Medicine*, 289
DOI: 10.1007/978-3-642-29026-8_13, © Springer-Verlag Berlin Heidelberg 2013

and animals throughout history (Alvard et al. 1997; Alves et al. 2009a; Baker 1930; Emery 2007; Foster and James 2002; Frazier 2007; Silvius et al. 2004). As O'Hara-May (1971) noted, animals and their products are a primary resource used by ancient peoples for food or for treating their illnesses. Archives, papyruses, and other early written historical sources attest to the medical use of animals, their parts, and their products (Lev 2003). The classical medical literature also indicates the use of animals as remedies, including Hippocrates' record of cattle's milk, chickens' eggs, mammals' horns, and sea sponges being used as medicine (Riddle 1987), and Dioscorides' Materia medica, in which about 10% of the substances mentioned were body parts or products of animals (Gunther 1959).

Ethnic folk medicine still makes use of animals and products derived from animal organs (Lev 2003); examples of current uses of animal-derived remedies can be found in many urban, semiurban, and more remote localities in different parts of the world (Almeida and Albuquerque 2002; Alves and Alves 2011; Alves et al. 2007; Apaza et al. 2003; Quave et al. 2010; Sodeinde and Soewu 1999; Vázquez et al. 2006). In recent years, the awareness has grown that the unsustainable use of medicinal animals contributes to the risk of extinction of certain species (Alves et al. 2010b; Lee 1999), yet the links between that body of knowledge and concerns about public health, harvesting impacts, and stakeholders' involvement remain understudied.

Brazil's high biological and sociocultural diversity (Alves and Rosa 2007a; Elisabetsky and Wannmacher 1993) translates into a wealth of traditional knowledge and practices, including the use of animals for medicinal purposes. It is estimated that Brazil hosts between 15 and 20% of the world's biological diversity, and the greatest number of endemic species. Brazil has more than 200 indigenous tribes and a large number of traditional communities that all possess considerable knowledge about the local fauna and flora and exhibit an array of natural resource use strategies and each ethnic culture possesses a broad knowledge regarding the medicinal properties of wildlife species (Alves 2008; Alves and Rosa 2007b). The extensive medicinal use of animal parts and products has been documented both in rural and urban areas (Almeida and Albuquerque 2002; Alves 2009; Alves et al. 2006, 2009b, 2010a; Alves and Pereira Filho 2007; Alves and Rosa 2006, 2007a, b, 2008; Branch and Silva 1983; Confessor et al. 2009; Costa-Neto 1999; Ferreira et al. 2009; Figueiredo 1994; Oliveira et al. 2010; Silva 2008), and is sustained by a thriving trade in medicinal animals conducted by herbalists in markets throughout Brazil (Alves et al. 2010a; Alves and Rosa 2007a, 2010; Oliveira et al. 2010).

Animals have been used as sources of medicine in Brazil since ancient times, and have played a significant role in local healing practices. Many people still use animal medicines as an alternative (or supplement) to visiting modern healthcare practitioners (Alves and Rosa 2006). However, little attention has been paid to the cultural, medical, economic, or ecological significance of zootherapeutic practices, even though the federal government's National Policy of Pharmaceuticals (Política Nacional de Medicamentos, Portaria no. 3916/98) specifies that "the support to research aiming to use the therapeutic potential of the national flora and fauna, with emphasis on certification of their medical properties, should be continued and expanded".

In that direction, we published a review article on medicinal animals in Brazil (Alves et al. 2007), exploring following the publication of this text in the journal Bioscience (Alves et al. 2007), focusing on the historical context in which zootherapy developed in Brazil, gaps in knowledge, and animal-based remedies and treatments currently used in the country. The two main goals of the aforementioned article were to assist in the construction of a national data bank of animal-derived remedies that could be used in conservation and management initiatives, and to document the traditional medicinal knowledge of communities that are rapidly losing certain of their socioeconomic and cultural characteristics.

This chapter updates the information provided in Alves et al. (2007), therefore expanding the Brazilian database of medicinal animals.

## 13.2 History of Medicinal Animal and Plant Use

Europeans who reached Brazil in the sixteenth century found a diversified fauna quite different from the European and African faunas; some of the explorers described the most common or conspicuous species, indicating their main useful or dangerous properties (Almaça 2002). The exuberant flora and the colorful animals fascinated Europeans who visited the country and their compatriots in Europe, who received descriptions, illustrations, and specimens during the first two centuries following the arrival of Portuguese settlers (Peixoto and Escudeiro 2002).

As colonization went on, however, amazement turned to exploitation based on slavery, initially of the indigenous peoples and later of Africans from various ethnic groups. From a medical perspective, colonization brought together elements of indigenous, African and European systems of diagnosis and cure. The European system came to dominate, relegating the indigenous and African systems to a folk category outside the mainstream medicine practiced in the country (Alves et al. 2007).

The present-day thriving trade in medicinal plants, however, testifies to the resilience of the merged indigenous African system of diagnosis and treatment, which gained official recognition in 2006 through the establishment of the National Policy for Integrative and Complementary Practices (Política Nacional de Práticas Integrativas e Complementares) in the federal public health care system. Animal-derived remedies, although less well known than plant-derived ones, also coexist with conventional medicine in contemporary Brazil. In a vacuum of studies or regulations, various animals have been used medicinally since colonial times, including *Iguana iguana* (Linnaeus, 1758), *Caiman latirostris* (Daudin, 1801), *Dasyatis guttata* (Bloch and Schneider, 1801), *Goniopsis cruentata* (Latreille, 1803), *Crotalus durissus* (Wagler, 1824), and *Micrurus ibiboboca* (Merrem, 1820) (Almeida et al. 2005; Silva et al. 2004).

## 13.3 Collecting Data on Medicinal Animals

We assembled information from scientific articles and gathered data through field surveys as follows: from January to May 2002, and from February to March 2003, we visited markets and shops in the cities of Belém, São Luís, Teresina, Goiânia, Natal, João Pessoa, Campina Grande, Recife, Maceió, Sergipe, Salvador, Vitória, Niterói, Florianópolis, and Porto Alegre, where we documented the medicinal species traded and obtained information about trade routes. Between 2003 and 2005, we visited outdoor markets in Teresina, Belém, João Pessoa, and Campina Grande, where we interviewed 79 merchants (45 men and 34 women) about the use of and trade in medicinal animals (23 interviewees in Belém, 21 in São Luís, 21 in Teresina, 10 in João Pessoa, and 4 in Campina Grande).

In structured interviews of the 79 merchants, respondents were asked to supply, for each animal traded, the vernacular name, folk use, parts used, and preparation and administration of remedy. Information obtained from these structured interviews was complemented by unstructured interviews (Huntington 2000), during which respondents addressed issues such as the clandestine nature of the trade and alternative uses of a particular species. We conducted an additional 137 interviews with traditional users of animal-derived remedies (67 men and 70 women) in the following fishing communities: the municipality of Cajueiro da Praia ($n = 36$); Pesqueiro Beach, municipality of Soure ($n = 41$); Environmental Protected Area Barra do Rio Mamanguape, municipality of Rio Tinto ($n = 30$); and the municipality of Raposa ($n = 30$), as described in Alves and Rosa (2007a).

Zoological material was identified with the aid of taxonomists familiar with the study areas' faunas through examination of voucher specimens donated by the interviewees; photographs of the animals or their parts, taken during interviews; and vernacular names. Voucher specimens and photographs were deposited at the Departamento de Sistemática e Ecologia, Universidade Federal da Paraíba. Only taxa that could be identified to the species level were included in the database.

Scientific names cited in publications were updated in accordance with the Integrated Taxonomic Information System's "Catalogue of Life: 2011 Annual Checklist" (www.catalogueoflife.org/search.php) and the American Museum of Natural History's "Amphibian Species of the World" (http://research.amnh.org/herpetology/amphibia/index.php). The conservation status of animal species follows IUCN (World Conservation Union; www.iucnredlist.org/); CITES (Convention on International Trade in Endangered Species of Wild Fauna and Flora; www.cites.org/eng/resources/species.html); and Brazil's official list of endangered species (Machado et al. 2008) and national list of species of aquatic invertebrates and fishes endangered, overexploited, or threatened by exploitation. Diseases treated with zootherapeutical products were grouped on the basis of the classification used by the Centro Brasileiro de Classificação de Doenças (Brazilian Center for the Classification of Diseases), as described by Alves and Rosa (2007b).

## 13.4 Medicinal Species and Conservation Concerns

Of the 354 animal species whose medical use has been recorded in Brazil, fishes (93 species), mammals (66), reptiles (57), and birds (47) were the most used vertebrates; insects (39), molluscs (18) and crustaceans (17) were the most used invertebrates. Other groups quoted by interviewees were Echinodermata (7), Chelicerata (1), Annelida (1), Cnidaria (2), and Amphibia (6) (Table 13.1). Most of the species used (n = 339; 96%) are wild caught and 75 of them (21%) are on some list of endangered species.

Of the 354 medicinal species recorded, 218 (62%; mainly vertebrates) are also used as food and are sold for that purpose in some parts of Brazil. Examples are *Balistes vetula* (Linnaeus, 1758), *Crassostrea rhizophorae* (Guilding, 1828), and *Ucides cordatus* (Linnaeus, 1763); additionally, some medicinal animals are sold as souvenirs or for magical or religious purposes. Examples are the products derived from the dolphin species *Sotalia fluviatilis* (Gervais and Deville, 1853), *Sotalia guianensis* (P. J. Van Bénéden, 1864), and *Inia geoffrensis* (Blainville, 1817); the seahorse *Hippocampus reidi* (Ginsburg, 1933); and the snakes *Boa constrictor* (Linnaeus, 1758) and *Caudisona durissa* (Linnaeus, 1758).

Fewer than 7% ($n = 26$) of the medicinal animals were reported as being used medicinally in more than four states, suggesting that local faunistic composition is a determinant of regional zootherapeutic practices. For instance, the medicinal use of the Amazonian species *Podocnemis expansa* (Schweiger, 1812), *Melanosuchus niger* (Spix, 1825), and *Arapaima gigas* (Schinz, 1822) was recorded only in the northern region. Nevertheless, in many cities, medicinal animals are widely traded at stalls in outdoor markets or, occasionally, in small markets specifically dedicated to this activity, which indicates that zootherapeutic practices have come to coexist with allopathic (conventional) medicine in Brazil. Trade routes traverse not only municipalities but also different states, with 236 species being used in rural areas, 43 in urban areas, and 75 in both areas.

## 13.5 Animal-Derived Remedies and Illnesses Treated

Animal-derived remedies were used for treating 194 diseases and conditions; asthma, rheumatism, wounds, thrombosis, and bronchitis were the most common ones. A single illness could be treated with various animal species, and most animals were prescribed for treating multiple illnesses; for example, products obtained from the trahira (*Hoplias malabaricus* [Bloch,1794]) were used to treat 22 conditions, and 16 conditions were treated with Amazonian dolphin (*S. fluviatilis*) products. Ingredients reportedly used in the medicines were fat (the most cited one), flesh, bone, bone marrow, cartilage, skin, tail, feather, liver, bile ("fel"), milk, rattle (from rattlesnakes), spine, shell, honey, wax, scale, rostral expansion, otolith, penis, carapace, blood, gizzard, beak, cocoon, teeth, tongue, egg, eggshells, tibia,

**Table 13.1** Animal taxa recorded as having medicinal properties in Brazilian traditional medicine

**CNIDARIANS**

Mussidae: *Mussismilia* (1), Physaliidae: *Physalia* (1)

**MOLLUSCS**

Ampullariidae: *Pomacea* (1), Cassidae: *Cassis* (1), Cardiidae: *Dallocardia* (1), Achatinidae: *Achatina* (1), Donacidae: Iphigenia (1), Littorinidae: *Littorina* (1), Lucinidae: *Lucina* (1), Megalobulimidae: *Megalobulimus* (1), Melongenidae: *Pugilina* (1), Mytilidae: *Mytella* (2), Ostreidae: *Crassostrea* (1), Octopodidae: *Octopus* (1), Strombidae: *Strombus* (1), Teredinidae: *Neoteredo* (1), *Teredo* (1), Vasidae: *Turbinella* (1), Veneridae: *Anomalocardia* (1)

**ANNELIDA**

Lumbricidae: *Lumbricus* (1)

**CRUSTACEANS**

Armadillidiidae: *Armadillidium* (1), Calappidae: *Calappa* (1), Gecarcinidae: *Cardisoma* (1), Grapsidae: *Goniopsis* (1), *Plagusia* (1), Hippidae: *Emerita* (1), Majidae: *Mithrax* (1), Ocypodidae: *Ocypode* (1), *Ucides* (1), *Uca* (1), Palaemonidae: *Macrobranchium* (3), Penaeidae: *Xiphopenaeus* (2), Pseudosquillidae: *Cloridopsis* (1), Sesarmidae: *Aratus* (1), Trychodactylidae: *Trychodactylus* (1)

**CHELICERATA**

Buthidae: *Rhopalurus* (1)

**INSECTS**

Apidae: *Apis* (1), *Cephalotrigona* (1), *Frieseomelitta* (1), *Melipona* (6), *Partamona* (1), *Plebeia* (1), *Tetragonisca* (1), *Trigona* (2), Blattidae: *Periplaneta* (1), *Eurycotis* (1), Chrysomelidae: *Coraliomela* (1), *Pachymerus* (1), Curculionidae: *Rhynchophorus* (1), *Rhinostomus* (1), Formicidae: *Atta* (2), *Dinoponera* (1), *Solenopsis* (1), Gryllidae: *Acheta* (1), Paragryllus (1), Meloidae: *Palembus* (1), Muscidae: *Musca* (1), Psychidae: *Eurycotis* (1), *Oiketicus* (1), Pediculidae: *Pediculus* (1), Termitidae: *Microcerotermes* (1), *Nasutitermes* (2), Vespidae: *Apoica* (1), *Brachygastra* (1), *Polistes* (1), *Polybia* (1), *Protopolybia* (1), *Synoeca* (1)

**ECHINODERMS**

Echinasteridae: *Echinaster* (2), Echinometridae: *Echinometra* (1), Luidiidae: *Luidia* (1), Mellitidae: *Mellita* (1), Oreasteridae: *Oreaster* (1), Toxopneustidae: *Lytechinus* (1)

**FISHES**

Auchenipteridae: *Trachelyopterus* (1), Anostomidae: *Leporinus* (2), *Schizodon* (1), Ariidae: *Bagre* (1), *Genidens* (2), *Netuma* (1), *Sciadeichthys* (1), Aspredinidae: *Aspredo* (1), *Aspredinichthys* (1), Balistidae: *Balistes* (2), Batrachoididae: *Thalassophryne* (1), Callichthyidae: *Callichthys* (1), Carcharhinidae: *Carcharhinus* (3), *Galeocerdo* (1), *Rhizoprionodon* (2), *Sphyrna* (1), Centropomidae: *Centropomus* (2), Characidae: *Astyanax* (1), *Brycon* (1), *Colossoma* (1), *Hydrolycus* (1), Clupeidae: *Opisthonema* (1), Dasyatidae: *Dasyatis guttata* (2), Doradidae: *Franciscodoras* (1), *Lithodoras* (1), *Megalodoras* (1), *Platydoras* (1), *Pterodoras* (1), *Oxydoras* (1), Echeneidae: *Echeneis* (1), Electrophoridae: *Electrophorus* (1), Erythrinidae: *Erythrinus* (1), *Hoplias* (2), Gadidae: *Gadus* (1), Ginglymostomatidae: *Ginglymostoma* (1), Heptapteridae: *Pimelodella* (1), Holocentridae: *Holocentrus* (1), Megalopidae: *Megalops* (1), Muraenidae: *Gymnothorax* (3), Myliobatidae: *Aetobatus* (1), Narcinidae: *Narcine* (1), Odontaspididae: *Eugomphodus* (1), Ogcocephalidae: *Ogcocephalus* (1), Osteoglossidae: *Arapaima* (1), *Osteoglossum* (1), Pimelodidae: *Phractocephalus* (1), *Pseudoplatystoma* (2), *Sorubimichthys* (1), *Rhamdia* (1), *Zungaro* (1), Potamotrygonidae: *Paratrygon* (1), *Potamotrygon* (4), *Plesiotrygon* (1), Pristidae: *Pristis* (2), Prochilodontidae: *Prochilodus* (4), Rajidae: *Atlantoraja* (1), Serrasalmidae: *Mylossoma* (1), *Serrasalmus* (1), Sciaenidae: *Cynoscion* (2), *Micropogonias* (1), *Pachyurus* (1), *Plagioscion* (2), Sparidae: *Calamus* (1), Synbranchidae: *Synbranchus* (1), Syngnathidae: *Hippocampus* (2), Tetraodontidae: *Colomesus* (1), *Sphoeroides* (1), Trichiuridae: *Trichiurus* (1), Urolophidae: *Urotrygon* (1)

(continued)

**Table 13.1** (continued)

| |
|---|
| **AMPHIBIANS** |
| Bufonidae: *Rhinella* (2), Hylidae: *Hyla* (1), Leptodactylidae: *Leptodactylus* (3) |
| **REPTILES** |
| Gekkonidae: *Hemidactylus* (1), Polychrotidae: *Polychrus* (2), *Anolis* (1), Iguanidae: *Iguana* (1), Teiidae: *Ameiva* (1), *Cnemidophorus* (1), *Tupinambis* (3), Tropiduridae: *Tropidurus* (3), *Uranoscodon* (1), Boidae: *Boa* (1), *Corallus* (2), *Eunectes* (2), *Epicrates* (2), Colubridae: *Leptophis* (1), *Mastigodryas* (1), *Oxyrhopus* (3), *Spilotes* (1), Crotalidae: *Caudisona* (1), Elapidae: *Micrurus* (1), Viperidae: *Bothrops* (1), *Lachesis* (1), Chelidae: *Chelus* (1), *Phrynops* (2), *Mesoclemmys* (1), *Kinosternon* (1), Cheloniidae: *Caretta* (1), *Chelonia* (1), *Eretmochelys* (1), *Lepidochelys* (1), Dermochelyidae: *Dermochelys* (1), Geoemydidae: *Rhinoclemmys* (1), Podocnemididae: *Podocnemis* (4), *Peltocephalus* (1), Testudinidae: *Chelonoidis* (2), Alligatoridae: *Caiman* (3), *Melanosuchus* (1), *Paleosuchus* (2) |
| **BIRDS** |
| Anatidae: *Anser* (1), *Cairina* (1), Anhimidae: *Anhima* (1), Ardeidae: *Ardea* (1), *Casmerodius* (1), *Tigrisoma* (1), Caprimulgidae: *Nyctidromus* (1), Cathartidae: *Coragyps* (1), *Cathartes* (1), Cracidae: *Penelope* (2), Ciconidae: *Ciconia* (1), Columbidae: *Leptotila* (1), *Columba* (1), Corvidae: *Cyanocorax* (1), Cotingidae: *Procnias* (1), Cuculidae: *Crotophaga* (1), *Guira* (1), Charadriidae: *Vanellus* (1), Emberezidae: *Coereba* (1), Falconidae: *Herpetotheres* (1), *Caracara* (1), *Falco* (1), Furnaridae: *Furnarius* (1), Meleagrididae: *Meleagris* (1), Odontophoridae: *Odontophorus* (1), Phasianidae: *Gallus* (1), *Numida* (1), *Pavo* (1), Picidae: *Dryocopus* (1), Rallidae: *Aramides* (1), Psittacidae: *Amazona* (1), Ramphastidae: *Ramphastos* (2), *Pteroglossus* (2), Rheidae: *Rhea* (1), Tinamidae: *Crypturellus* (2), *Nothura* (2), *Rhynchotus* (1), *Tinamus* (1), Trochilidae: *Eupetomena* (1), Tyrannidae: *Fluvicola* (1), *Pitangus* (1), Trochilidae: *Eupetomena* (1), Turdidae: *Troglodytes* (1) |
| **MAMMALS** |
| Phyllostomidae: *Desmodus* (1), Agoutidae: *Cuniculus* (1), Balaenopteridae: *Balaenoptera* (1), *Megaptera* (1), Bovidae: *Bos* (1), *Bubalus* (1), *Ovis* (1), *Capra* (1), Bradypodidae: *Bradypus* (2), Canidae: *Canis* (1), *Cerdocyon* (1), *Chrysocyon* (1), *Dusicyon* (1), *Speothos* (1), Caviidae: *Cavia* (1), *Kerodon* (1), Cebidae: *Alouatta* (4), *Cebus* (1), Cervidae: *Blastocerus* (1), *Mazama* (3), *Ozotocerus* (1), Dasypodidae: *Cabassous* (1), *Dasypus* (1), *Euphractus* (1), *Tolypeutes* (1), Dasyproctidae: *Dasyprocta* (1), Delphinidae: *Sotalia* (2), Didelphidae: *Didelphis* (3), Erethizontidae: *Coendou* (3), *Chaetomys* (1), *Sphiggurus* (1), Equidae: *Equus* (2), Felidae: *Leopardus* (1), *Felis* (1), *Puma* (1), *Panthera* (2), Hydrochaeridae: *Hydrochaeris* (1), Iniidae: *Inia* (1), Leporidae: *Sylvilagus* (1), *Oryctolagus* (1), Mustelidae: *Conepatus* (1), *Lontra* (1), Myrmecophagidae: *Myrmecophaga* (2), Procyonidae: *Nasua* (1), *Procyon* (1), Physeteridae: *Physeter* (1), Suidae: *Sus* (1), Tapiridae: *Tapirus* (1), Tayassuidae: *Pecari* (1), *Tayassu* (1), Trichechidae: *Trichechus* (2) |

secretions, head, heart, urine, foot, leg, nest, guts, pollen, ear, spawn, nail, horn, sucking dish, eye, and, more rarely, whole animals. Hard parts, such as teeth, nails, shells, rattles from snakes, fish scales, bone, and cartilage, generally are dried in the sun, grated and crushed to powder, and then administered as tea or taken during meals. Fat, body secretions, and oil are either ingested or used as an ointment.

Traditional Brazilian medicine is also associated with a belief system locally termed simpatias. One frequently mentioned aspect of simpatias is the secretive nature of the treatment—for a remedy to be effective, people receiving it must not know what they are taking. Popular beliefs may have implications for the way species are used: some can be used live or dead, depending on the beliefs of the community. One form of spiritual treatment uses amulets containing animal parts, supposedly to protect from the evil eye and, often, to prevent diseases; amulets may be hung around the neck, glued on a piece of cloth, kept in one's pocket or wallet, or kept in homes for protection from evil energies. For example, the teeth of caimans (*C. latirostris* [Daudin, 1801], *M. niger*, and *Paleosuchus palpebrosus* [Cuvier, 1807]) and the skin of the crab-eating raccoon (*Procyon cancrivorus* [G. Cuvier, 1798]) are used as protection against snakebite, and the skin of the coati *Nasua nasua* (Linnaeus, 1766) is attached to a person's belt to prevent backache.

## 13.6 Discussion

The extensive use of foods as medicinal remedies reported in our study is in line with recent field investigations around the world, which show that more than 40 animal foods are used as remedies (Pieroni and Grazzini 1999). In addition to their use as food, some medicinal species are also sold as souvenirs or for magical or religious purposes. Such multiple uses should be considered when implementing recovery plans, especially for the highly exploited species.

Our results reveal that most medicinal animals are prescribed to treat multiple diseases, which seems to be the common practice in other traditional medicine systems. For example, in Mexico, the carapace and tail of the *Dasypus novemcinctus* (Linnaeus, 1758) are used to treat diarrhea, tuberculosis, and whooping cough, and to accelerate parturition (Vázquez et al. 2006). In India, the fat, skin, and bile duct of the land monitor *Varanus bengalensis* (Daudin, 1758) are used for treating piles, rheumatism, burns, and bites from spiders and snakes (Kakati et al. 2006). In Bolivia, products derived from the *Agouti paca* (Linnaeus, 1766) have been documented as remedies for general body pain, leishmaniasis, snakebite, rheumatism, heart pain, pain in bones, liver pain, fever, and pain during childbirth (Apaza et al. 2003).

The reported multiple therapeutic actions and the use of various animals for the same condition presumes different properties either of parts used or of modes of preparation, corroborating the remark by Iwu (1993) that different chemical constituents are enhanced by different factors, such as preparation, dosage, or part used. Additionally, the possibility of using various remedies for the same ailment is popular because it permits adapting to the availability of the animals (Alves and Rosa 2006).That particular species have similar uses in different places indicates that there is a solid information base of zootherapeutic practices within Brazil and elsewhere (Alves and Alves 2011; Apaza et al. 2003; Branch and Silva 1983; El-Kamali 2000; Lev 2003; Mahawar and Jaroli 2006, 2008; Sodeinde and Soewu 1999; Vázquez et al. 2006).

The influence of magical beliefs integrated into traditional medicine in Brazil is apparent, and zootherapeutic practices are closely connected with Afro-Brazilian religions (which have also incorporated elements from indigenous peoples).

Various animals are used for the magical-religious practices of the Afro-Brazilian cults, mainly in urban areas (Alves et al. 2009c; Alves and Pereira Filho 2007; Leo Neto and Alves 2010; Léo Neto et al. 2009), a ritualistic use that emphasizes the holistic nature of traditional medicine and addresses problems with the spiritual, physical, and social-psychological aspects of people's daily lives. Animals are employed within a magical-prophylactic perspective, with the purpose of warding off diseases of "unnatural" origin—a practice that encompasses perceptions related to the belief that supernatural forces are involved in causing diseases, as well as in their treatment.

Because Brazil is highly heterogeneous socially and profoundly unequal in distribution of income, socioeconomic aspects play a role in the perseverance of zootherapeutic practices. For the majority of the population, access to hospital care is available within the public sector, but the organization of the health care system reflects the schisms within Brazilian society: high-technology private care is available to the rich, but only inadequate public care is available to the poor (Barros and Porto 2002; Haines 1993), which makes the use of available, affordable animal and plant remedies an important alternative. Finally, traditional home remedies may be used, in part, to resist urban modern medicine (Boltanski 1989) and to support the traditional culture (Ngokwey 1995). All these aspects need to be addressed in conjunction with policies that may have impacts on the resource base used by the local populations; otherwise, clandestine, unreported, unmonitored extraction and commercialization of medicinal animals will persist.

Implementation of a top-down model of management that is detrimental to local populations, or tightening of regulations decoupled from attempts to empower communities, may limit dialog among stakeholders rather than help to conserve and manage zootherapeutic resources. Costa-Neto (2001) has shown that on the coast of Bahia, new regulatory measures have turned formerly free collection of animals for medicine into an exogenous taboo, generating cultural disconnections between the animal source and fishers, a situation that highlights the importance of Seixas and Begossi's (2001) suggestion that new regulations should be in tune with local population needs, thereby increasing compliance in management. Furthermore, as Alves and Rosa (2005) remarked, there is a need to shift the focus from how to obtain the greatest amount of zootherapeutical resources to how to ensure future uses, as well as a need for a transdisciplinary approach to evaluating the ecological and social components of zootherapy and how they fit together.

Traditional medicine is widely accepted, and the animal materials sometimes are preferred to the pure pharmacologically active constituents for cultural or financial reasons, as observed in this study. Furthermore, the economic activity of traders of medicinal animals and plants may become lucrative in a context of high unemployment rates and a low level of formal education (Huntley et al. 1989)—a situation not uncommon in developing countries such as Brazil.

Lastly, from the perspective both of the faunistic resources and of traditional knowledge holders, the growing number of habitats being completely lost or severely altered in Brazil and elsewhere (Myers et al. 2000) suggests growing threats to the survival of many potential valuable medicinal animals and the cultural aspects associated with them.

## 13.7 Conclusion and Future Research

Our findings suggest that although socioeconomic factors determine the perseverance of zootherapeutic practices, interest in animal-derived remedies seems to extend beyond people lacking access to modern medicinal services. Even in cities where such services are more accessible, many people continue to go to traditional healers, showing the cultural acceptability of such practices. In that context, we consider the following goals as central to understanding zootherapeutic practices long established in Brazil: (1) identifying target populations and estimating the number of individuals being removed from the wild for medicinal purposes; (2) assessing the conservation status, ecology, life-history traits, and population parameters of heavily exploited species; (3) developing some form of collaborative research program and monitoring of the trade; and (4) addressing clinical and sanitary concerns. Efforts should initially focus on taxa believed to be most at risk of overcollection, especially those already on one or more lists of endangered species.

Zootherapy is intertwined with sociocultural and religious beliefs that must be understood by those engaged in modern conservation and protection of Brazil's biodiversity; effective ways to include socioeconomic information and expertise in conservation are needed. It is vital to know precisely the species concerned, their correct names, and their critical habitats—this aspect alone justifies further studies on zootherapeutic practices in Brazil.

From a biological perspective, there is a need to increase our understanding of the biology and ecology of species commonly used as remedies to better assess the impacts of harvesting them (for medicinal or other purposes) on their wild populations. Medicinal species whose conservation status is in question should receive urgent attention, and aspects such as habitat loss and alteration should be discussed in connection with present and future medicinal uses. As Anyinam (1995) remarked, environmental degradation affects users of traditional medicine both by limiting their access to the resources traditionally used and by diminishing the knowledge base in their community upon which traditional medicine is constructed.

In this context, research on Brazil's long tradition of using animals in healing needs to be prioritized and expanded. Equally important, the available information about medicinal uses, trade of animals for medicinal purposes, and users should become an integral part of the country's conservation strategies.

**Acknowledgments** We are grateful for financial support from PROBIO/MMA/IBRD/GEF/CNPq and PADI Foundation. Thanks are also due to CNPq (Conselho Nacional de Desenvolvimento Científico e Tecnológico), for providing us with research productivity scholarships which helped the development of our studies on the medicinal uses of animals in Brazil.

# References

Almaça C (2002) Reino animal. Episteme 15:97–106
Almeida AV, Alves AGC, Lucena RFP, Albuquerque UP (2005) Prescrições zooterápicas indígenas brasileiras nas obras de Guilherme Piso (1611–1678). In: Atualidades em Etnobiologia e Etnoecologia. Nupeea; Sociedade Brasileira de Etnobiologia e Etnoecologia, Recife, pp 45–60
Almeida CFCBR, Albuquerque UP (2002) Uso e conservação de plantas e animais medicinais no Estado de Pernambuco (Nordeste do Brasil): Um estudo de caso. Interciencia 27(6):276–285
Alvard MS, Robinson JG, Redford KH, Kaplan H (1997) The sustainability of subsistence hunting in the neotropics. Conserv Biol 11(4):977–982
Alves RRN (2008) Animal-based remedies as complementary medicine in Brazil. Forsch Komplementmed 15:226–227. doi:10.1159/000144178
Alves RRN (2009) Fauna used in popular medicine in Northeast Brazil. J Ethnobiol Ethnomed 5(1):1–30. doi:10.1186/1746-4269-5-1
Alves RRN, Alves HN (2011) The faunal drugstore: animal-based remedies used in traditional medicines in Latin America. J Ethnobiol Ethnomed 7(9):1–43
Alves RRN, Filho GAP, Lima YCC (2006) Snakes used in ethnomedicine in Northeast Brazil. Environ Dev Sustain 9(4):455–464
Alves RRN, Mendonça LET, Confessor MVA, Vieira WLS, Lopez LCS (2009a) Hunting strategies used in the semi-arid region of northeastern Brazil. J Ethnobiol Ethnomed 5(12): 1–50. doi:10.1186/1746-4269-5-12
Alves RRN, Léo Neto NA, Santana GG, Vieira WLS, Almeida WO (2009b) Reptiles used for medicinal and magic religious purposes in Brazil. Appl Herpetology 6(3):257–274
Alves RRN, Neto NAL, Brooks SE, Albuquerque UP (2009c) Commercialization of animal-derived remedies as complementary medicine in the semi-arid region of Northeastern Brazil. J Ethnopharmacol 124(3):600–608
Alves RRN, Oliveira MGG, Barboza RRD, Lopez LCS (2010a) An ethnozoological survey of medicinal animals commercialized in the markets of Campina Grande, NE Brazil. Hum Ecol Rev 17(1):11–17
Alves RRN, Souto WMS, Barboza RRD (2010b) Primates in traditional folk medicine: a world overview. Mammal Rev 40(2):155–180. doi:10.1111/j.1365-2907.2010.00158.x
Alves RRN, Pereira Filho GA (2007) Commercialization and use of snakes in North and Northeastern Brazil: implications for conservation and management. Biodivers Conserv 16:969–985. doi:10.1007/s10531-006-9036-7
Alves RRN, Rosa IL (2005) Why study the use of animal products in traditional medicines? J Ethnobiol Ethnomed 1(5):1–5. doi:10.1186/1746-4269-1
Alves RRN, Rosa IL (2006) From cnidarians to mammals: the use of animals as remedies in fishing communities in NE Brazil. J Ethnopharmacol 107:259–276
Alves RRN, Rosa IL (2007a) Zootherapy goes to town: the use of animal-based remedies in urban areas of NE and N Brazil. J Ethnopharmacol 113:541–555. doi:10.1016/j.jep2007.07.015
Alves RRN, Rosa IL (2007b) Zootherapeutic practices among fishing communities in North and Northeast Brazil: a comparison. J Ethnopharmacol 111:82–103. doi:10.1016/j.jep2006.10.033
Alves RRN, Rosa IL (2008) Use of tucuxi dolphin sotalia fluviatilis for medicinal and magic/ religious purposes in North of Brazil. Hum Ecol 36:443–447. doi:10.1007/s10745-008-9174-5
Alves RRN, Rosa IL (2010) Trade of animals used in Brazilian traditional medicine: trends and implications for conservation. Hum Ecol 38(5):691–704
Alves RRN, Rosa IL, Santana GG (2007) The role of animal-derived remedies as complementary medicine in Brazil. Bioscience 57(11):949–955
Anyinam C (1995) Ecology and ethnomedicine: exploring links between current environmental crisis and indigenous medical practices. Soci Sci Med 40(3):321–329
Apaza L, Godoy R, Wilkie D, Byron E, Huanca T, Leonard WR, Peréz E, Reyes-García V, Vadez V (2003) Markets and the use of wild animals for traditional medicine: a case study among the Tsimané ameridians of the Bolivian rain forest. J Ethnobiol 23(1):47–64

Baker FC (1930) The use of animal life by the mound-building Indians of Illinois. Trans Ill State Acad Sci 22:41–64

Barros E, Porto S (2002) Health care in Brazil: equity as challenge. Global Development Network, Washington

Boltanski L (1989) As classes sociais eo corpo. Graal, Rio de Janeiro

Branch L, Silva MF (1983) Folk medicine in Alter do Chão, Pará, Brasil. Acta Amazônica 13:737–797

Confessor M, Mendonca L, Mourao J, Alves R (2009) Animals to heal animals: ethnoveterinary practices in semi-arid region, Northeastern Brazil. J Ethnobiol Ethnomed 5(1):37

Costa-Neto EM (1999) Healing with animals in Feira de Santana City, Bahia, Brazil. J Ethnopharmacol 65:225–230

Costa-Neto EM (2001) A cultura pesqueira do litoral Norte da Bahia. EDUFBA/EDUFAL, Salvador/Macéio, Salvador

El-Kamali HH (2000) Folk medicinal use of some animal products in Central Sudan. J Ethnopharmacol 72:279–282

Elisabetsky E, Wannmacher L (1993) The status of ethnopharmacology in Brazil. J Ethnopharmacol 38(2–3):129–135

Emery KF (2007) Assessing the impact of ancient maya animal use. J Nat Conserv 15(3): 184–195

Ferreira FS, Brito S, Ribeiro S, Almeida W, Alves RRN (2009) Zootherapeutics utilized by residents of the community Poco Dantas, Crato-CE, Brazil. J Ethnobiol Ethnomed 5(1):1–21

Figueiredo N (1994) Os 'bichos' que curam: os animais e a medicina 'folk' em Belém do Pará. Boletim do Museu Paraense Emílio Göeldi 10(1):75–91

Foster MS, James SR (2002) Dogs, Deer, or Guanacos: zoomorphic figurines from pueblo grande, Central Arizona. J Field Archaeol 29(1):165–176

Frazier J (2007) Sustainable use of wildlife: the view from archaeozoology. J Nat Conserv 15(3): 163–173

Gunther RT (1959) The Greek herbal of dioscorides. Hafner, New York

Haines A (1993) Health care in Brazil. BMJ 306:303–306

Huntington HP (2000) Using traditional ecological knowledge in science: methods and applications. Ecol Appl 10(5):1270–1274

Huntley B, Siegfried R, Sunter C (1989) South African environments into the 21st century. Human & Rousseau, Cape Town

Iwu MM (1993) Handbook of African medicinal plants. CRC, Boca Raton (FL)

Kakati LN, Ao B, Doulo V (2006) Indigenous knowledge of zootherapeutic use of vertebrate origin by the Ao Tribe of Nagaland. Hum Ecol 19(3):163–167

Lee SKH (1999) Trade in traditional medicine using endangered species: an international context. Paper presented at the 2nd Australian symposium on traditional medicine and wildlife conservation, Melbourne, Australia

Leo Neto NA, Alves RRN (2010) A Natureza sagrada do Candomblé: análise da construção mística acerca da Natureza em terreiros de candomblé no Nordeste de Brasil. Interciencia 35(8):568–574

Léo Neto NA, Brooks SE, Alves RRN (2009) From eshu to obatala: animals used in sacrificial rituals at candomble "terreiros" in Brazil. J Ethnobiol Ethnomed 5(1):1–23

Lev E (2003) Traditional healing with animals (zootherapy): medieval to present-day Levantine practice. J Ethnopharmacol 85:107–118. doi:10.1016/S0378-8741(02)00377-X

Machado ABM, Drummond GM, Paglia AP (2008) Livro Vermelho da fauna brasileira ameaçada de extinção, vol 2. MMA, Fundação Biodiversitas, Brasília, DF

Mahawar MM, Jaroli DP (2006) Animals and their products utilized as medicines by the inhabitants surrounding the ranthambhore national park, India. J Ethnobiol Ethnomed 2(46): 1–5. doi:10.1186/1746-4269-2-46

Mahawar MM, Jaroli DP (2008) Traditional zootherapeutic studies in India: a review. J Ethnobiol Ethnomed 4(1):1–17

Myers N, Mittermeier RA, Mittermeier CG, Fonseca GAB, Kent J (2000) Biodiversity hotspots for conservation priorities. Nature 403(6772):853–858

Ngokwey N (1995) Home remedies and doctors' remedies in Feira (Brazil). Soc Sci Med 40(8):1141–1153

O'Hara-May J (1971) Foods or medicines? A study in the relationship between foodstuffs and materia medica from the sixteenth to the nineteenth century. Trans Br Soc Hist Pharm 1(2): 61–97

Oliveira ES, Torres DF, Brooks SE, Alves RRN (2010) The medicinal animal markets in the metropolitan region of Natal city, Northeastern Brazil. J Ethnopharmacol 130(1):54–60

Peixoto AL, Escudeiro A (2002) Pachira aquatica (Bombacaceae) na obra "História dos Animais e Árvores do Maranhão" de Frei Cristóvão de Lisboa. Rodriguésia 53(82):123–130

Pieroni A, Grazzini A (1999) Alimenti-medicina di origine animale. In: Pieroni A (ed) Herbs, humans and animals/Erbe, uomini e bestie, 1st edn. VDM Verlag, England, pp 155–171

Quave CL, Lohani U, Verde A, Fajardo J, Rivera D, Obón C, Valdes A, Pieroni A (2010) A comparative assessment of zootherapeutic remedies from selected areas in Albania, Italy, Spain and Nepal. J Ethnobiol 30(1):92–125

Riddle JM (1987) Folk tradition and folk medicine: recognition of drugs in classical antiquity. In: Scarboroough J (ed) Folklore and folk medicine. American Institute of the History of Pharmacy, Madison

Seixas C, Begossi A (2001) Ethnozoology of caiçaras from Aventureiro, Ilha grande. J Ethnobiol 21(1):107–135

Silva AL (2008) Animais medicinais: conhecimento e uso entre as populações ribeirinhas do rio Negro, Amazonas, Brasil. Boletim do Museu Paraense Emílio Göeldi 3(3):343–357

Silva MLV, Alves ÂGC, Almeida AV (2004) A zooterapia no Recife (Pernambuco): uma articulação entre as práticas e a história. Biotemas 17(1):95–116

Silvius KM, Bodmer RE, Fragoso JMV (2004) People in nature: wildlife conservation in South and Central America. Columbia University Press, New York

Sodeinde OA, Soewu DA (1999) Pilot study of the traditional medicine trade in Nigeria. Traffic Bull 18(1):35–40

Vázquez PE, Méndez RM, Guiascón ÓGR, Piñera EJN (2006) Uso medicinal de la fauna silvestre en los Altos de Chiapas, México. Interciencia 31(7):491–499

# Chapter 14
# Mediterranean Zootherapy: A Historical to Modern Perspective

**Cassandra L. Quave and Andrea Pieroni**

**Abstract** Animals have been used as a source of human medicine for millennia. In the Mediterranean, these ancient practices were documented in historic texts such as Dioscoride's *De Materia Medica* (40–90 A.D.) and continue to be documented even in current day ethnobiological surveys. Here, we summarize a few recent ethnobiological literature on Southern European zootherapy and compare these "modern" traditional medical applications of animals and their byproducts with those ancient practices documented >2,000 years ago. In doing so, we reflect on the continuity between ancient and modern medicine and examine the implications that such practices hold for both animal conservation and drug discovery.

## 14.1 Introduction

Animals and animal products have constituted an important portion of the Mediterranean pharmacopeia for millennia. Collectively recognized as zootherapeutic remedies today, many of these ancient therapies have persisted in current day traditional medical practices and even become integrated into modern pharmaceuticals. Their use has been documented in both ancient medical texts and recent ethnobiological field studies. In this chapter, we review the results of our field studies on the zootherapeutic practices of the Mediterranean. In doing this,

C. L. Quave (✉)
Center for the Study of Human Health, Emory University, Candler Library 107E,
550 Asbury Circle, Atlanta, GA 30322, USA
e-mail: cquave@emory.edu

A. Pieroni
University of Gastronomic Sciences, Piazza Vittorio Emanuele 9,
I-12060, Pollenzo/Bra, Italy

R. R. N. Alves and I. L. Rosa (eds.), *Animals in Traditional Folk Medicine*,
DOI: 10.1007/978-3-642-29026-8_14, 

we employ the use of animals as medicine to better understand the ethnomedical practices of this geographic region. We use the seminal work (*De Materia Medica*) of the Greek physician Pedanius Dioscorides as our main resource for comparison of the state of zootherapy during the time of the Emperor Nero in the Roman Empire (circa first century A.D.) to present day. Here, we discuss the following questions:

- How have ancient zootherapeutic practices become integrated into current day health practices, encompassing both ethnomedical and allopathic care?
- Have the types of animals used in Mediterranean zootherapy changed over the past two millennia? Or, rather, are the same animals used but for different applications?

## 14.2 Historic Zootherapeutic Practices

Records of zootherapeutic practices have been identified in the surviving texts of ancient cultures, beginning with the earliest written records. For example, historical documents of ancient Egypt such as the *Ebers Paypyrus* (1550 B.C.) include medicinal descriptions of animal substances such as honey, lizard blood, sperm whale ambergis, and musk deer glands, among others (Lev 2006; Nunn 1996; Bryan 1930). Animals have clearly played a central role in the medical pharmacopeias of mankind for at least the past 3,000–4,000 years.

Today, animal-based medicines continue to play an important role in diverse systems of traditional medicine worldwide, as well as in modern pharmaceuticals. Two good examples of this include ACE (angiotensin converting enzyme) inhibitors from pit viper snake [*Bothrops jararaca* (Wied 1824)] venom (Bisset 1991) and dietary supplements of Omega-3 PUFAs (polyunsaturated fatty acids) from certain fish oils (Costa-Neto 2005). The continued use of zootherapeutics from ancient times to present day suggests that their use may be associated with some medical efficacy.

There are several excellent studies that describe the state of medical pharmacopeias in different eras ranging from the tenth to nineteenth centuries in the Mediterranean basin. This includes studies on Medieval zootherapy in the Levant (Lev 2003), Egypt (Lev 2007; Lev and Amar 2006), and Serbia (Jarić et al. 2011), among others. We are interested in gaining a better understanding of how these zootherapeutic practices have evolved over a longer temporal period. Thus, specific to our focus on Mediterranean ethnomedicine, here we assess the state of current-day zootherapy with that documented in this region 2,000 years ago.

### *14.2.1 Pedanius Dioscorides*

Pedanius Dioscorides (40–90 A.D.) was a Greek physician who traveled throughout the Roman Empire (including Asia Minor, Greece, Italy, and France) as an army doctor during the time of the Emperor Nero. It was during this period of travel and practicing medicine that he amassed a wealth of knowledge therapeutic remedies of that era. In particular, he collected medicinal plants, studied disease processes, and collected information on other healing materials. Dioscorides recorded the knowledge that he accumulated over this period in his *De Materia Medica*—a text that has arguably been known as the ultimate authority on plants and medicine for the past 2,000 years. Later medical texts and herbals, including the *Canon of Medicine* (1025 A.D.) by the great Persian polymath Ibn Sīnā (or Avicenna), drew heavily from Dioscorides' work.

Although *De Materia Medica* is best known as an extensively descriptive herbal, details regarding the use of animals or their byproducts in medicine are also abundant. Specifically, Chap. 2 on "Living Creatures" provides a detailed account of more than 100 zootherapeutic remedies. For the purposes of this chapter, we have chosen to use both Goodyer's old English translation (1655), reprinted and edited (Gunther 1959), as well as a more recent modern English translation (Osbaldeston 2000) as key references in comparing the state of zootherapy in the Mediterranean as documented in our various contemporary field studies to that of two millennia ago.

## 14.3 Results and Discussion

Over the past decade, we have been actively involved in the collection and documentation of traditional knowledge of medical practices in the Mediterranean. Specifically, our research efforts to date have been focused in the areas of modern day Italy (Pieroni and Quave 2005; Pieroni et al. 2002, 2004a, b; Quave and Pieroni 2005, 2007; Quave et al. 2008), Albania (Quave et al. 2010; Pieroni et al. 2005), Serbia (Pieroni et al. 2011), Croatia (Pieroni et al. 2003; Pieroni and Giusti 2008), and Romania (unpublished data). The results of these studies are reported in Tables 14.1–14.5. In these studies, we have endeavored to collect and record data on not only the use of botanical remedies in traditional medicine, but also the use of animals and their byproducts for medical purposes.

### *14.3.1 Livestock as Medicine*

In our field studies, we have noted the prominent incorporation of domestic animals and livestock into the local folk pharmacopeia. In particular, the use of animals like swine, cattle, poultry, goats, and dogs, are most common.

**Table 14.1** Animal remedies recorded (2004–2005) in the folk medicine of the Northern Albanian Alps (Pieroni et al. 2005)

| Remedy | Preparation and administration | Quoted medical use |
|---|---|---|
| Bees wax | Liquefied and arranged on the top of a small piece of cloth which is inserted (still warm) in the ear | To treat earache |
| Blood | The blood gathered after having cut the ear of the animal is put in the mouth of the same animal | To treat fever in animals |
| Cheese (fresh) | Applied externally | To heal wounds |
| | Eaten raw or cooked with flour and eaten hot | "To strengthen the stomach"; anti-diarrheic |
| Cobweb | Applied externally | Hemostatic |
| Cow feces | Used fresh and applied externally | To heal burns |
| Cow milk | Drunk | To treat intestinal pains and poisonings (especially in children, and also in animals) |
| | Drunk very hot | To treat mumps |
| Dog hair | Applied externally | To treat wounds from dog bites |
| Donkey milk | Drunk fresh | To heal coughs in the elderly |
| Egg | Eaten raw | Antidote against poisoning. Also used for cattle and sheep |
| | The raw egg is put on a piece of raw wool which is adhered to the body with the help of oil (ritual) | To treat pains (the egg moves on the wool and where the yolk stops, the yolk "takes the pain away" and comes out of the membrane; the whole treatment last at least 1 h!) |
| | Cooked | Anti-diarrheic |
| | Eaten | To treat stomach ache |
| Fish | Left alive in a small amount of water | Diagnostic means to establish the length of hepatitis. While the fish is still alive, the affected person will remain ill (ritual) |
| Goat fat | Heated and drunk (one spoonful) | To treat asthma |
| Hen muscular stomach | The membrane of the muscular stomach is extracted and dried, then ground and made into a decoction | To treat kidney stones |
| Honey | Applied externally under the ears | Mumps |
| | Applied on the mucosa | To treat mouth inflammations |

(continued)

**Table 14.1** (continued)

| Remedy | Preparation and administration | Quoted medical use |
|---|---|---|
| | Dissolved in hot water and milk and drunk | Given as a reconstituent after women have given birth |
| | Eaten | Given as a reconstituent after women have given birth |
| Human feces | Used fresh, applied externally | To treat snake bites |
| *Jardun* (dense yogurt-like dairy product obtained boiling fresh sheep milk with salt) | Drunk | Reconstituent; used to prevent many illnesses |
| Medicinal leech[a] | Applied externally | To relieve muscular pains ("they suck the bad blood") |
| Milk cream | Applied externally | To treat chapped lips |
| Petroleum[a] | Applied externally on the legs for 24 h | To treat rheumatism[b] |
| Pork fat | Heated and applied externally | Used as veterinary preparation to heal wolf bites (cattle, sheep, goats), and as a symptomatic for treating skin inflammations due to erysipelas (*Erysipelothrix rhusiopathiae*) in pigs |
| Rennet (from the calf abomasus) | Dried, added in food | To treat severe digestive troubles in animals[b] |
| Snail | Fresh meat ground up and mixed with sugar, in compress | To treat eye inflammations |
| Turtle meat | Eaten cooked | To heal coughs in the elderly[b] |
| Urine | Applied topically | Toothache, earache, symptomatic in relieving the pains of measles |
| Yogurt | Eaten | To treat food poisoning (also used for animals) |
| | Eaten | To treat stomach ache |
| Wool | Special singlet (*krahol*) and socks made with raw wool, and to worn only when affected by high fever | Diaphoretic |
| Whey | Drunk | To treat kidney stones; nutraceutical |
| | Applied externally in washes | To treat sunburn |

[a] Product bought in local market
[b] Disappeared use in the last decades

This propensity to use these animals for common ailments is likely related to the ease of access to these products as compared to the rarer, wild terrestrial, or marine organisms.

**Table 14.2** Animal remedies recorded (2011) in the folk medicine of a North-Eastern Italian diaspora in Dobruja, Eastern Romania

| Remedy | Preparation and administration | Quoted medical use |
|---|---|---|
| Cat's tie | Rubbed externally on the affected part | Herpes |
| Dog's lick | Direct external application | Wounds |
| Donkey's milk | Drunk | Cough (especially in children) |
| Egg albumen | Externally applied on the eye | Eye inflammations |
| | Mixed with hemp fibres, topically wrapped | Sprains |
| Eggs, mixed with sugar (sbatudin) | Eaten | Gracility (strengthening food, especially for children) |
| Hare fat (also aged) | Applied externally | Ingrown nails; suppurative |
| Hedgehog spines | Burned, and the resulting ashes inserted into the vagina | Vaginal complaints |
| Hen | Soup (*sopa da galina*) | Post-partum reconstituent |
| Hen's muscular stomach membrane | Dried and powdered, ingested in spoons | Diarrhea |
| Honey | Externally applied | Burns |
| Pork lard | Old pork fat is externally applied | Hemorrhoids; sores |
| Turtle meat | Boiled, then eaten | Tumors |
| Woman's milk | Inserted in the eye | Eye inflammations |

Dioscorides reported some remedies that we have never documented in our studies on contemporary zootherapy. This includes the use of the lungs of swine, lamb or bear for the treatment of foot blisters, and sores; ground horse or ass hooves for epilepsy; goat hooves for baldness; goat liver for blindness; boar liver for snake bite; poultry parts for snake bite; pig knucklebone for colic; whey for epilepsy and skin diseases; and blood as a poison antidote.

Some domestic zootherapeutic remedies, however, have survived the passage of time relatively intact. This may be also due to the fact that a remarkable part of these home-medicines are in fact "food-medicines" (Etkin 2006; Pieroni and Price 2006), i.e. food items, which are consumed to obtain specific therapeutic effects. They are still popular in Southern Europe, especially in rural, and agro-pastoral contexts.

Current zootherapeutical practices involving domestic animal-based food products include for example the use of eggs to treat diarrhea and inflammation; milk (cow, goat, ass, and mare) for treating cough or as a laxative; whey and yogurt for digestive troubles; cheese as an intestinal astringent and topical anti-inflammatory; butter for topical applications (skin infections, inflammations, and burns); and the fats of numerous domestic (and, sometimes, wild) animals are still quoted for various topical applications.

Regarding the treatment of dog bites, there are some similarities in past and current remedies. Today, this is treated using a part of the dog that bit the patient (such as the dog hair), and other wounds may be treated with dog saliva or feces.

Dioscorides, on the other hand, reported treatment of dog bite by eating the roasted liver of the infected dog and also wearing the tooth of this dog as an amulet. Similarly, in modern-day Serbia, wolf teeth are worn as Evil-eye amulets. In both cases, a key component of the treatment is to use some part of the rabid dog that bit the patient in the formulation of the curative therapy. This tradition of treating an animal bite with the culpable animal (or another of same species) is also evident in the case of scorpion stings, which are treated by applying ground up scorpion topically to the affected area.

## *14.3.2 Wild Terrestrial Animals*

Based on the number of remedies reported in *De Materia Medica*, the medical use of wild animals in the Mediterranean appears to has been more prominent in the past. Today, some wild terrestrial animals are employed in the pharmacopeia, and this includes various snakes, insects (bees, scorpions), turtles, rabbits, hedgehog, badger, fox, wolf, birds (storks, pigeons. common cuckoo), leeches, snails, and slugs. Relatively few of these, however, were reported in more than one of our field sites. For example, reports of the use of snails for eye inflammations was only recorded in the north Albanian Alps, despite the availability of this and other similar species in other study sites. Likewise, the use of hedgehog spines for treating vaginal complaints was reported only in eastern Romania, despite the presence of wild populations in other areas of the Mediterranean. Interestingly, this use of the hedgehog is quite different from that reported by Dioscorides. In the past, the hedgehog was used to treat baldness, dropsy, elephantitis, and diarrhea—but nothing related to gynecological issues.

Some of the remedies of the past that were notably absent from reports in our study sites included the use of snake skins for otitis; hares for sterility and as a poison antidote; hippopotamus or beaver testicles for snakebite; burnt weasels for snakebite, gout, scrofulous tumors, and epilepsy; frogs for snakebite, toothache, and baldness; bed bugs for quartian fever (malaria); cockroaches for otitis pain; woodlouse for tonsillitis, otitis pain, painful urination and jaundice; seagull liver for placental expulsion following birth; grasshoppers for bladder problems; locusts for difficult urination; osprey for kidney stones; skylark for colic, swallows for epilepsy and tonsillitis; elephant tooth for abscesses (whitlows) of the finger or toenails; skink as an aphrodisiac; and earthworms for toothache, among some others. Bees wax, honey and propolis, however, are used medicinally in much the same way as 2,000 years ago in both topical and oral (medicinal food) applications.

Although certain exotic animals, such as the hippopotamus, elephant and skink were included in the ancient pharamacopeia, they are notably absent today. In the past, these animal products would have likely been included in the trade network responsible for moving plant materials throughout the Mediterranean. Today, however, such trade is highly restricted due to international laws that govern the movement of biological materials. Trade restrictions undoubtedly influence the use

**Table 14.3** Animal remedies recorded (2010) in the folk medicine of Serbians and (bosniakicized) Albanians in the Pešter plateau, South-Western Serbia (Pieroni et al. 2011)

| Remedy | Preparation and administration | Quoted medical use |
|---|---|---|
| Badger (*jazavac*)'s internal organs | Topical applications of the fresh internal organs, immediately after the animal has been killed | Hemorrhoids |
| Bee's wax | Externally applied | Earache |
| | | Bruises |
| Butter and clarified butter | Consumed | Panacea |
| | External applications | Warts, chilblains, and wounds |
| Cheese | Consumed | Galactagogue |
| Clotted cream (*kajmak*) and cream | Consumed | Reconstituent |
| | External applications | Emollient and chilblains |
| Cow/buffalo/sheep fat | Mixed with bee's wax and honey, in a cream and externally applied | Wounds |
| Dairy products (all) | Consumed | Prevention of bone fractures, panacea |
| Donkey's milk | Drunk | Pertussis |
| Donkey's urine | Instilled in the nose (urine has to come from young animals only) | Sinusitis |
| Dog's saliva (lick) | External lick given by young dogs | Warts |
| Ewe's milk | A piece of cloth imbibed with ewe's milk and put on top of a child's abdomen | Anthelmintic |
| Ewe's cheese (fresh) | Consumed | "Good for the heart," diabetes, reconstituent |
| Fat-based foods | Consumed | Galactagogue |
| Fox's veins | Dried veins of a killed fox, put inside the ear | Earache |
| Goat cheese | Topically applied | Wounds |
| Goat milk | Drunk | Cough |
| Goat or sheep skin | Topically applied (warm) on the chest, (with a piece of paper to divide the human and goat skins) | Bronchitis |
| Honey | Consumed | Cough, sore throat, galactagogue, heart tonic, "good for the circulation," panacea |
| | Topically applied | Burns |
| Horse's hair | Tied to the wart for 2 days | Warts |
| Human urine | Topically applied | Skin burns, furuncles |

(continued)

**Table 14.3** (continued)

| Remedy | Preparation and administration | Quoted medical use |
|---|---|---|
| *Jardum* (dairy product obtained by gently heating fresh ewe's milk—milked in July and August only—with salt) | Consumed | Panacea |
| Milk (generally cow's milk) | Boiled, and then drunk | Sore throats, fever, headache, hypertension, constipation, "healthy food" |
| | | Galactagogue |
| Mare milk (milked after the mare has given the first birth) | Drunk | Sore throats, cough, pertussis |
| Mother's lick | Mother licking in the central part of the front of the child, then simulating spitting three times on the right and three times on the left | Evil Eye |
| Pork lard | External massages with lard, at the end with *rakija* | Wounds, chilblains, fever (children) |
| Snake | Snake dried in the shadow of a juniper shrub, then the fat extracted and stored; snake fat, mixed with lemon balm tea and flour, to make a poultice (*mehlem*) | Every skin disease |
| Stork (*Ciconia ciconia, leileku/ roda*)'s beak or bone | A dried piece of stork—generally the beak or a bone—in a necklace, or sewn in the internal part of a cloth and dressed, as an amulet; alternatively, a stork's feather is boiled and the resulting water used in external washes | Evil Eye amulet |
| Yogurt (*kos, kiselo mlijeko*) | Drunk | Stomachache, hypertension, "good for the circulation," "healthy food," panacea |
| Whey (*hirra, surutka*) | Drunk | Digestive troubles, diabetes, obesity; Cold, bronchitis |
| Wolf' tooth | Used in a necklace, as an amulet | Evil Eye amulet |
| Woman's milk | External application | Earache, eye inflammations |
| Wool | External application | Chilblains |
| | Dress in warm wool clothes | Rheumatisms, fever |

of various animal products as people tend to use those resources which are most readily available to them—either by way of their environment, agricultural practices, or local markets.

**Table 14.4** Animal remedies recorded (2001–2006) in Arbëreshë (ethnic Albanian) villages, southern Italy (Quave et al. 2008; Pieroni et al. 2002)

| Remedy | Preparation and administration | Quoted medical use |
|---|---|---|
| Dog saliva | External application | Anti-furuncles, antiseptic[a] |
| Donkey hair (braided rope) | Ritual object (tool used for the topical application of red wine) | Ritual healing of *mal vjint* |
| Egg albumen | Scrambled, local application with salt and cotton or wheat bran | Anti-bruises[a] |
| Hair | External amulet | Amulet against *malocchio* (evil-eye)[a] |
| Hen meat | Cooked in a soup as food | Reconstituent after giving birth |
| Honey | Consumed | Against sore throat |
| Horse blood | Consumed raw | Anti-anemia[a] |
| Pig gall bladder | Left outside for one night and then applied topically | Anti-chilblain[a] |
| Ricotta cheese | Consumed | Light anti-diarrheal |
| Scorpion | Oleolite (cold decoction with olive oil to be instilled in the ears) | Anti-otitis[a] |
| Whey (liquid precipitate from the cheese making process) | Drunk | Mild laxative |
| Wood affected by woodworms (*Anobium punctatum*) | External application | Hemostatic; anti-mastitis[a] |

[a] No longer used

### *14.3.3 Marine Life*

Roughly 24% of the animal-based remedies reported by Dioscorides relate to marine animals. The difference in knowledge of marine animals for medicine between then and now is quite remarkable as only one example of a marine animal used for human medicine (fish used as a diagnostic tool for hepatitis) was reported in our studies. This could be due to the fact that our study sites were not located in close proximity to the sea—but were rather in land-locked or even in mountainous zones. Dioscorides, on the other hand, would have spent much time in coastal communities during his maritime travels throughout the Mediterranean with the army.

The diversity of sea life used in ancient medicine reported by Dioscorides is astonishing—and ranged from various types of fish (scorpion fish, spiny-finned fish, cuttlefish, red mullet, sheath fish, maena, smelt, and tunny), to sea urchins, shellfish (purpura), mussels, seahorse, whelks, bivalves, crabs, sea centipedes, electric rays, sea hares, sting rays, sea gudgeons, and jellyfish. Examples of their medicinal uses included the sea urchin being good for the stomach and intestines, and a diuretic; the sea horse for treating baldness; mussels for eye remedies and to treat dog bite; crabs for snakebite and other insect bites and to treat consumption; sea centipedes and sea hares for depilation (hair removal); sea scorpion fish gall for

**Table 14.5** Animal derived remedies recorded in the folk medicine of the Dolomiti Lucane (Castelmezzano and Pietrapertosa), inland Southern Italy (Pieroni et al. 2004a; Quave et al. 2008, 2010)

| Remedy | Preparation and administration | Quoted medical use |
|---|---|---|
| Two-headed salamander | Head cut and stored in alcohol | Good omen |
| Cerumen | Topical application | Heal purulent skin abscesses (caused by thorns) |
| Common coockoo | To hear the bird singing | Good omen for long life |
| Cow feces | Topical application | To heal skin burns |
| | Smelling it early in the morning | Against pertussis |
| Dog feaces | Topical application | To heal skin burns |
| Donkey milk | Drunk | Reconstituent for children |
| Dried ricotta | Mixed with hot water, which remains after having boiled noodles, and then the mixture used as sauce for the same noodles; mixed with boiled bread | Anti-diarrhea; galactagogue |
| Egg albumen | Scrambled and topically applied with a cloth | Anti-bruises |
| Egg | Boiled eggs | Anti-diarrhea |
| Fermented cream (from cow milk) | Topical application | Emollient for healing skin inflammations of babies |
| Four-lined snake | Fat (*a sunzē*) extracted when the snake is still alive, used as an ointment | Anti-rheumatism |
| Goat milk | Drunk hot with honey | Anti-tussive; reconstituent (children) |
| Human hair | Cut on 1st Friday of March | Good omen for preventing headache |
| Hen meat | Soup | Post-partum depurative (even given as gift to a woman who has just given birth); reconstituent during various illnesses |
| Human milk | Topical application | To heal eye inflammations |
| Human sweat | Topical application of the sweat soaked inner brim of a hat | Anti-wounds |
| Insect (non identified) | The crystal of the insect placed on necklaces to be dressed | Amulet against the evil-eye (*affascēnē*) |
| Leather (extracted from a black dog) | To be worn as an amulet | Against the evil-eye |
| Leech | Topical application | To heal a not clearly identified disease related with skin troubles, as an apotropoaic: if the animals survive after the application, it is seen as good omen |
| Mouse | Eaten boiled or cooked | Anti-enuresis |

(continued)

**Table 14.5** (continued)

| Remedy | Preparation and administration | Quoted medical use |
|---|---|---|
| Pig lard | Soup | Laxative |
| Pigeon meat | Soup | Postpartum depurative (even given as gift to the women who have had a birth); reconstituent during an illness |
| | Soup | Galactagogue |
| Sheep milk | Drunk | Laxative |
| Silk ribbon | Bound around the wart | Anti-warts |
| Slug | Topical application | To heal warts: after the treatment, which has to be carried out when full moon is decreasing, the slug is hung on a *Rubus ulmifolius* thorn; when the animal has dried up, the wart will have disappeared |
| Urine | Topical application | Hemostatic |
| Whey | Drunk | Laxative; digestive troubles |
| Wood affected by | woodworms | Topical application |
| Antiseptic (babies) | | |

treating white spots on the cornea; sting rays for tooth extraction; red mullet to treat pains of spider bites and sea dragon and scorpion stings; sheath fish for sciatica and dysentery; sea gudgeon as a laxative; and smelt to treat warts, corns, and gangrenous ulcerations. Further ethnobiological studies conducted in coastal communities in this region would be of great utility in gaining a better understanding of the continuity of traditional knowledge of marine zootherapeutics in comparison with this and other ancient texts of the region.

## 14.4 Conclusions

Zootherapy is still a thriving practice in the Mediterranean, though based on our field studies in interior and mountainous regions; it is relatively restricted to domestic and some wild terrestrial animals. This differs greatly from Dioscorides reports which included a great diversity of marine life and also several valuable exotic species that would have likely been traded throughout the region by land and sea during his lifetime. Most of the animals included in the ancient and contemporary pharmacopeias of this region shared the characteristic of being readily available to the people using them. Today, domestic animals and some wild species that are common in agricultural communities and the surrounding areas are used most often.

It is interesting to note that many of the same types of livestock used 2,000 years ago for these medicinal purposes are still used today, though the preparation and application of the products may vary by region. Thus, while many of these medicinal products continue to play an important role in local ethnomedical practices; very few have made the transition to modern pharmaceutical applications. Further pharmacological investigation of those remedies that have withstood the test of time is necessary and could lead to both the validation of these traditional medical practices and the development of new pharmaceuticals for the global market.

## References


Bisset NG (1991) One man's poison, another man's medicine? J Ethnopharmacol 32:71–81

Bryan CP (1930) Ancient egyptian medicine. The Papyrus Ebers Ares, Chicago

Costa-Neto EM (2005) Animal-based medicines: biological prospection and the susatinable use of zootherapeutic resources. Anais da Academia Brasileira de Ciências 77(1):33–43

Etkin N (2006) Edible medicines, An ethnopharmacology of food. University of Arizina Press, Tucson

Gunther RT (1959) The greek herbal of dioscorides. Hafner Publishing Co, New York

Jarić S, Mitrović M, Djurdjević L, Kostić O, Gajić G, Pavlović D, Pavlović P (2011) Phytotherapy in medieval Serbian medicine according to the pharmacological manuscripts of the Chilandar Medical Codez (15–16th centuries). J Ethnopharmacol 137:601–619. doi:10.1016/j.jep.2011.06.016

Lev E (2003) Traditional healing with animals (zootherapy): medieval to present-day Levantine practice. J Ethnopharmacol 85:107–118. doi:10.1016/S0378-8741(02)003377-X

Lev E (2006) Healing with animals in the Levant from the 10th to the 18th century. J Ethnobiol Ethnomed 2:11

Lev E (2007) Drugs held and sold by pharmacists of the Jewish community of medieval (11th-14th centuries) Cairo according to lists of *materia medica* found at the Taylo-Schechter Genizah collection, Cambridge. J Ethnopharmacol 110:275–293. doi:10.1016/j.jep.2006.09.044

Lev E, Amar Z (2006) Reconstruction of the inventory of *materia medica* used by members of the Jewish community of medieval Cairo according to prescriptions found in the Taylor-Schechter Genizah collection, Cambridge. J Ethnopharmacol 108:428–444. doi:10.1016/j.jep.2006.06.005

Nunn JF (1996) Ancient egyptian medicine. University of Oklahoma Press, Norman

Osbaldeston TA (2000) Dioscorides De Materia Medica: Being an Herbal with Many other Medicinal Materials Written in Greek in the First Century of the Common Era; A New Indexed Version in Modern English. IBIDIS Press, Johannesburg

Pieroni A, Dibra B, Grishaj G, Grishaj I, Gjon Maçai S (2005) Traditional phytotherapy of the Albanians of Lepushe Northern Albanian Alps. Fitoterapia 76(3–4):379–399. doi:10.1016/j.fitote.2005.03.015

Pieroni A, Giusti ME, Münz H, Lenzarini C, Turković G, Turković A (2003) Ethnobotanical knowledge of the Istro-Romanians of Žejane in Croatia. Fitoterapia 74(7–8):710–719. doi:10.1016/j.fitote.2003.06.002

Pieroni A, Giusti ME (2008) The remedies of the folk medicine of the croatians living in Cidarija Northern Istria. Coll Antropol 32(2):623–627

Pieroni A, Giusti ME, Quave CL (2011) Cross-cultural ethnobiology in the Western Balkans: medical ethnobotany and ethnozoology among Albanians and Serbs in the Pešter Plateau, Sandžak South-Western Serbia. Anglais 39(3):17. doi:10.1007/s10745-011-9401-3

Pieroni A, Price LL (2006) Eating and healing. Traditional food as medicine. Haworth Press, Binghamton

Pieroni A, Quave CL (2005) Traditional pharmacopoeias and medicines among Albanians and Italians in southern Italy: a comparison. J Ethnopharmacol 101(1–3):258–270

Pieroni A, Quave CL, Nebel S, Heinrich M (2002) Ethnopharmacy of the ethnic Albanians (Arbëreshë) of northern Basilicata Italy. Fitoterapia 73(3):217–241. doi:10.1016/S0367-326X(02)00063-1

Pieroni A, Quave CL, Santoro RF (2004a) Folk pharmaceutical knowledge in the territory of the Dolomiti Lucane, inland southern Italy. J Ethnopharmacol 95(2–3):373–384. doi:10.1016/j.jep.2004.08.012

Pieroni A, Quave CL, Villanelli ML, Mangino P, Sabbatini G, Santini L, Boccetti T, Profili M, Ciccioli T, Rampa LG, Antonimi G, Girolamini C, Cecchi M, Tomasi M (2004b) Ethnopharmacognistic survey on the natural ingredients used in folk cosmetics, cosmeceuticals and remedies for healing skin diseases in the inland Marches, Central-Eastern Italy. J Ethnopharmacol 91:331–344. doi:10.1016/j.jep.2004.01.015

Quave CL, Lohani U, Verde A, Fajardo J, Rivera D, Obón C, Valdes A, Pieroni A (2010) A comparative assessment of zootherapeutic remedies from selected areas in Albania, Italy Spain and Nepal. J Ethnobiol 30(1):92–125

Quave CL, Pieroni A (2005) Ritual healing in Arbëreshë Albanian and Italian communities of Lucania, southern Italy. J Folklore Res 42(1):57–97

Quave CL, Pieroni A (2007) Traditional health care and food and medicinal plant use among historic Albanian migrants and Italians in Northern Lucania, southern Italy. In: Pieroni A, Vandebroek I (eds) Travelling cultures, plants and medicines. The ethnobiology and ethnopharmacy of migrations. Berghahn Press, Oxford pp 204–226

Quave CL, Pieroni A, Bennett BC (2008) Dermatological remedies in the traditional pharmacopoeia of Vulture-Alto Bradano, inland southern Italy. J Ethnobiol Ethnomed 4:5. doi:10.1186/1746-4269-4-5

# Chapter 15
# A Review of Fauna Used in Zootherapeutic Remedies in Portugal: Historical Origins, Current Uses, and Implications for Conservation

**Luis Miguel Pires Ceríaco**

**Abstract** Portugal has a very rich ethnozoological heritage due to its rich biodiversity, history, and culture. However, there are very few works dedicated to this, and little is known about the zootherapeutic remedies used in Portugal. Due to its location in a biodiversity hotspot, its agro-pastoralist traditions and also its role during the maritime discoveries of the Renaissance, Portugal has a long list of animals and animal parts that have been used in its folk medicine. These uses can still be found in many historical documents and pharmacopeias. Although these uses started to decline in the nineteenth century, many of them survived, being passed from generation to generation by oral tradition, and are still used by local populations. We can still find 225 remedies, using 54 animal species, mainly domestic animals and common species, in use in modern day Portugal. The use of some animals in traditional medicine, such as Lataste's viper, the Iberian wolf, and the European pond turtle, can be an additional pressure on these already endangered species. Studies are much needed to better understand zootherapeutic uses in Portuguese folk medicine and their implications for conservation. This chapter intends to be a starting point to these future investigations.

## 15.1 Introduction

The use of animals for medical purposes is part of a body of traditional knowledge that is becoming increasingly more relevant to discussions on conservation biology, public health policies, the sustainable management of natural resources,

L. M. P. Ceríaco (✉)
Centro de Estudos de História e Filosofia da Ciência (CEHFCi), Universidade de Évora, Palácio do Vimioso, Largo Marquês de Marialva n°8, 7000-554 Évora, Portugal
e-mail: luisceriaco@netcabo.pt

R. R. N. Alves and I. L. Rosa (eds.), *Animals in Traditional Folk Medicine*,
DOI: 10.1007/978-3-642-29026-8_15, © Springer-Verlag Berlin Heidelberg 2013

biological prospection, and patents (Alves and Rosa 2005), partly due to the large amount of historical, ethnographic, and anthropological information that these practices carry (Lev 2002). Portugal is the westernmost country in continental Europe; its territory is located in the western part of the Iberian Peninsula. The Portuguese territory is bordered on the north and east by Spain and on the south and west by the Atlantic Ocean. Like people in all countries of the world, the Portuguese have used and still use rocks, minerals, plants, fungi, animals, and animal body parts as sources of medicines in popular pharmacopeia. As a Mediterranean/Atlantic country, Portugal is inserted in a region that is considered to be a biodiversity hotspot (Myers et al. 2000), and its agro-pastoralist traditions and long rural history gave birth to a very strong and complex relationship between its people and nature. Game such as deer, wild boar, and foxes are quite common in the more natural areas in Portugal, as well as rabbits, hares, partridges, pigeons, badgers, lizards, snakes, and invertebrates. Top predators like the Iberian wolf (*Canis lupus signatus*) and the Eurasian eagle owl (*Bubo bubo*) also exist in the country. Apart from the larger cities such as Lisbon and Oporto, and their metropolitan areas, human settlements in Portugal are well inserted in, or at least in close contact with, natural areas such as cork oak forests (the Portuguese *Montado*) and other *Quercus*-related ecosystems, pine woods, and pastures. Livestock and domestic animals are similar to those in Europe, with sheep, goats, pigs, cows, horses, donkeys, and chickens being the most common animals present. Also, Portugal's long historical background and geographic position, with direct and indirect influences from many cultures such as the Celts, Romans, Arabs, and Europeans, and even its direct contact with cultures from the four continents during the Portuguese maritime expansion, gathered a large amount of knowledge regarding nature and the use of natural resources such as sources of food, raw materials, and medicines, for example. Many references to these different uses of natural resources are found in historical documents from different ages, and even from different natures. We can find references to these uses in thirteenth century medicinal information, fifteenth century witchcraft accusations (Bethencourt 1987), sixteenth and seventeenth century travel reports, eighteenth century pharmacopeias, and also in eighteenth century naturalist documents, such as from natural history museums and cabinet investigators. For example, in the *Pharmacopea Lusitana*, D. Caetano de Santo António (1725) cited about 39 animal species that were used in medicine, some of them exotic, like lions and elephants. Other eighteenth century pharmacopeias also cited similar numbers of species, such as in the *Medicina Lusitana* (Henriques 1731) and *Pharmacopea Tubalense* (Coelho 1760), where 37 and 35 animal species, respectively, were cited as being used in dozens of types of medicine. Another interesting aspect is the importance of ethnozoological products, mostly for medicine and mainly from Brazil, Africa, and Asia, in the Portuguese natural history museums and Cabinets. Loureiro (1710–1791), a Jesuit missionary and member of the Royal Academy of Sciences of Lisbon, well known for his major work in ethnobotany and pharmacopeia *Flora Conchichinensis* (Loureiro 1790), focusing on the botany and ethnobotanical uses of Conchichina (currently Laos, Cambodia and Vietnam), left several parts of animals used in the medicine of that colony in the Royal Academy of Sciences Museum, the most famous of which (and still present in the Lisbon Academy of Sciences Museum) are the "petrified crabs". Also, in the instructions for naturalists exploring the vast tropical empire that Portugal controlled at that time, there were always references to the importance of reporting animal uses in the economy, food industry, and medicine, which could be of importance to the kingdom, and also references to the importance of sending specimens of these to the national museums. The different uses of several types of exotic animals, reported in the pharmacopeias cited above, such as lions, camels, and elephants (their tusks), suggest that museums and naturalist collections also served to provide the kingdom with ethnozoological medicines, in the same way that the same collections, and also the botanical gardens, served as repositories of ethnobotanical products (Brigola 2002).

Although the use of animal parts as sources of medicines, food, and raw materials is as old and as common as the use of botanical parts, when compared to ethnobotanical studies, ethnozoological studies, especially those dedicated to the use of animals and their body parts in traditional medicine, have been neglected (Alves and Rosa 2005). Ethnobotanical studies are common in Portugal (Novais et al. 2004; Frazão-Moreira and Fernandes 2006; Salgueiro 2010) and knowledge about the different uses of plants or parts of plants as remedies is much larger than that of zootherapeutics. Ethnozoological studies in Portugal are still very few and recent (Brito et al. 2001; Ceríaco 2010a; Ceríaco et al. 2011). Zootherapy can be defined as the use of animals or animal parts as remedies for the treatment of some diseases or health conditions (Alves and Rosa 2005). The lack of these types of studies in Portugal, with a prevalence of ethnobotanical studies, can be explained by the number of botanical species used and by the number of medicines made from plants, which is much higher than the number made from animal parts. Also, plants are regarded as being clean, healthy, and harmless, whereas animals are mostly regarded as being impure and dangerous, and as disease vectors (Fontes and Sanches 2000). Despite this, zootherapeutic practices also exist in Portugal. This is not unusual since all countries and cultures of the world have a large proportion of zootherapeutic medicines in their traditional pharmacopeias (WHO/IUCN/WWF 1993). The most famous and well-known zootherapeutic uses in traditional medicines are those of traditional Chinese medicines, or even traditional Brazilian and Indian medicines, for which Alves et al. (2007) and Mahawar and Jaroli (2008), respectively, documented the use of 283 and 109 animal species. If countries with the highest biodiversity indices, such as Brazil and India, or very ancient cultures like China, have rather large numbers of animals in pharmacopeias, other countries will also use these practices. Quave et al. (2010) reported 34 species for Spain, 21 for Italy, 11 for Albania, and 37 for Nepal. Although little is known about zootherapeutic practices in Portugal, its pharmacopaeia should not be very different from other Mediterranean countries such as Spain or Italy. This lack of knowledge about the Portuguese situation constitutes a real problem, given the rapid loss of traditional knowledge (TK) due to the abandonment of traditional practices, increasing urbanization, and the adverse effects of globalization—aspects which have been regarded by many as a crisis similar to the loss of biodiversity (Singh 2002; Sutherland 2003; Ramstad et al. 2007). Studies on TK, and also on the way people relate to their surrounding biodiversity, are therefore very important (Alves and Rosa 2005). Also, since we know very little about those traditions, we may be overlooking some pressing issues, particularly with respect to the impact of certain practices on the conservation of different animal species.

This chapter comprises the first study entirely devoted to a review of all uses of zootherapeutic remedies in Portugal, using historical documents, already published sources, and original data from field surveys. We will also highlight some possible conservation problems that those uses might cause for some species of fauna.

## 15.2 Fauna used in Zootherapeutic Remedies

Although in the second half of the nineteenth century, due to advances in science and society, traditional medicine practices and the use of ancient pharmacopeias gradually disappeared from "official" medical and pharmaceutical practices (at this time pharmacy and medicine were already based more on industrial chemistry and the more commonly studied plants than on "old" practices that were more naturalistic and empiric), knowledge about the popular uses of plants and animals was not lost or abandoned. In fact, this change in practices only really happened in the large urban centers and in hospitals, universities, and academies. The ancient uses of fauna and flora as medicines were maintained by most of the urban and, in particular, rural people who had relied on them for many generations, and who were not informed about what was taking place in the intellectual establishments. As Quave et al. (2010) reported, close interactions between rural populations and the environment, often associated with an economic dependence on local natural resources, fostered the accumulation of a wealth of zootherapeutic knowledge. Also, these natural and traditional medicines were the only type of health care that rural people had access to without having to pay large amounts of money or travel to a city to see a doctor. These practices were not abandoned and can still be found in many regions of Portugal today.

This review of the popular uses of animals as zootherapeutic remedies that are still being used in Portugal today was based on bibliographic investigations, namely on Felgueiras (1956), Lobo (1995), Mantero (1998), Fontes and Sanches (2000), and de Almeida et al. (2009), and on field research, from January 2009 to March 2011 in most regions of Portugal. The review is presented in Table 15.1, and follows the organization of similar previous studies (Quave et al. 2010). This is the first specific study on this topic in Portugal, so there are probably many more zootherapeutic remedies still being used in the country than those reported here; further investigations are needed to gain a better knowledge of the topic.

The use of approximately 54 species of animals in medicine has been reported in different parts of Portugal. Mammals comprise the highest number of animals

**Table 15.1** Popular uses of animals as zootherapeutic remedies in Portugal

| Scientific name | English common name | Vernacular name | Popular use | Preparation and application[a] |
|---|---|---|---|---|
| **Phylum**: Annelida **Class**: Clitellata **Order**: Hirudinea | | | | |
| *Hirudo medicinalis* (Linnaeus, 1758) | European medicinal leech | *Sanguessugas* | Heals contusions | A whole live animal is placed directly over the wound to suck up blood (M) |
| | | | Anti-toothache | An animal is placed over the throat to suck blood (M) |
| | | | Anti-cancer | A whole live animal is placed directly over the cancer to suck the "bad" blood (M) |
| **Phylum**: Annelida **Class**: Oligochaeta **Order**: Haplotaxida | | | | |
| *Lumbricus terrestris* (Linnaeus, 1758) | European earthworm | *Minhoca* | Anti-jaundice | A whole animal is baked and eaten by the patient (M) |
| | | | Anti-rheumatism | A soup of pork meat and lard, butter, a pint of oil, veal or lamb lard, alcohol, earthworms, and a black cat, minus head and tail |
| | | | Anti-poison | A poultice of worms, meruges and a lily's (*Amaryllis sp.)* onion |
| **Phylum**: Arthropoda **Class**: Insecta **Order**: Hymenoptera | | | | |
| *Apis mellifera* (Linnaeus, 1758) | Common bee | *Abelha* | Heals contusions | A chicken egg yolk is mixed with honey, sugar, and olive oil and applied to the wound (M) |
| | | | Heals hernias | Panels covered in honey are placed on the afflicted area (M) |
| | | | Against mouth sores | Rinse the mouth with vinegar, rose water, and honey (M) |
| | | | Against erysipelas | Honey is applied to the afflicted area (M) |
| | | | Against furuncles | A poultice of yellow soap, lime, rye, and honey is placed over the furuncle (M) |
| | | | Against herpes | A poultice of olive oil, vinegar, and honey is placed over the afflicted area (M) |
| | | | Heals burns | Honey is applied directly to the burn (M) |

(continued)

Table 15.1 (continued)

| Scientific name | English common name | Vernacular name | Popular use | Preparation and application[a] |
|---|---|---|---|---|
| | | | Against dandruff | A poultice of milk and honey is applied to the hair during a bath (M) |
| | | | Against anorexia | A syrup of honey, eggs, white wine, grapes and figs (M) |
| | | | Against jaundice | Human excrement covered in honey is applied over the afflicted area (M) |
| | | | Promotes abortion | Hot honey with white wine and cinnamon is drunk by the woman (M) |
| | | | Against anemia | A soup of bread with honey (MF) |
| | | | Heals coughs | Hot honey with milk (MF) |
| | | | | Onion with honey (MF) |
| | | | | Lemon tea with honey (MF) |
| | | | Heals sore eyes; anti-conjunctivitic | Honey is applied to the afflicted eye with a chicken feather (M) |
| | | | Postpartum reconstituent | Honey is dissolved in hot water and milk and drunk (MF) |
| | | | Against bronchitis | Syrup of honey, sugar, and rosemary (MF) |
| | | | Calmative | Honey is eaten when a person wants to sleep (MF) |
| **Phylum**: Arthropoda **Class**: Insecta **Order**: Diptera | | | | |
| *Musca domestica* (Linnaeus, 1758) | House fly | *Mosca* | Heals sore eyes; anticonjunctivitic | Rub a live animal in the affected eye (M) |
| | | | Anti-herpes | A dead animal is applied directly over the herpes (M) |
| **Phylum**: Arthropoda **Class**: Insecta **Order**: Phthiraptera | | | | |
| Unknown | Human lice | *Piolhos* | Anti-jaundice | Nine live lice are eaten, with or without flour, honey, and sugar |
| **Phylum**: Arthropoda **Class**: Insecta **Order**: Coleoptera | | | | |
| *Berberomeloe majalis* (Linnaeus, 1758) | Red-stripped oil beetle | *Vaca-loura, Vaca-loira, Vaca-loirão, Vaca de Deus* | Anti-warts | Beetle oil is applied over the wart (M) |

(continued)

**Table 15.1** (continued)

| Scientific name | English common name | Vernacular name | Popular use | Preparation and application[a] |
|---|---|---|---|---|
| | | | Anti-furuncles | Live animal is fried in olive oil and placed over the furuncle (M) |
| **Phylum**: Arthropoda **Class**: Arachnida **Order**: Araneae | | | | |
| Unknown | Spider | *Aranhas* | Hemostatic | A spider's web is placed on the laceration (M) |
| | | | Anti-colitis | Spiders' webs are used to make an oleolite that is placed over the belly (M) |
| | | | Anti-fever | A live animal is placed inside the hollow stem of the giant reed *Arundo donax* and the reed is hung from the sick person's neck as a necklace |
| **Phylum**: Arthropoda **Class**: Arachnida **Order**: Scorpiones | | | | |
| *Buthus occitanus* (Amoreux, 1789) | Common European scorpion | *Alacrau, Lacrau, Escorpião* | Heals scorpion bites | The animal is fried in olive oil and placed over the bite (M) |
| | | | | The dead animal is placed over the bite (M) |
| | | | Anti-cancer | Scorpion stings the place of the cancer. If the person does not die within 24 h, the cancer is cured |
| **Phylum**: Mollusca **Class**: Gastropoda **Order**: Pulmonata | | | | |
| Unknown | Snail | *Caracol* | Anti-furuncle | Snail porridge is placed directly over the furuncle (M) |
| *Arion ater* (Linnaeus, 1758) | Slug | *Lesma* | Anti-wart | A live animal is placed directly over the wart (M) |
| | | | Anti-tuberculosis/Anti-asthma | A slug soup is served to the patient without the patient knowing what it is (M) |
| | | | Anti-cancer | A live animal is placed directly over the cancer (M) |
| | | | Anti-anemic | A squashed animal is placed over the patient's belly (M) |
| | | | Painkiller | Eat cooked animals (M) |
| | | | Stops hands sweating | Rub a live animal into the hands (M) |

(continued)

**Table 15.1** (continued)

| Scientific name | English common name | Vernacular name | Popular use | Preparation and application[a] |
|---|---|---|---|---|
| **Phylum**: Chordata **Class**: Actinopterygii **Order**: Gadiformes | | | | |
| *Gadus morhua* (Linnaeus, 1758) | Atlantic cod | *Bacalhau* | Anti-anorexia | Eat dried cod (M) |
| | | | Anti-abdominal pains | A poultice of bread and a cod's tail is applied to the belly (M) |
| | | | Anti-anemic | Drink cod liver oil (M) |
| | | | Anti-parasitic | Eat dried cod (M) |
| | | | Anti-fever | Eat fried cod (M) |
| | | | Heals bone fractures | Cod skin is placed over the affected zone, with two splints (M) |
| | | | Anti-madness | Eat three cod tails (M) |
| **Phylum**: Chordata **Class**: Actinopterygii **Order**: Clupeiformes | | | | |
| *Sardina sp.* and *Sardinella sp.* | Sardine | *Sardinha* | Anti-alcoholism | A sardine dried for more than one year is turned into a powder and dissolved in wine (M) |
| **Phylum**: Chordata **Class**: Actinopterygii **Order**: Siluriformes | | | | |
| *Ictalurus melas* (Rafinesque, 1820) | Black bullhead | *Peixe-gato* | Anti-rheumatic | Fish liver oil is applied to the joints that hurt (M) |
| **Phylum**: Chordata **Class**: Actinopterygii **Order**: Anguilliformes | | | | |
| *Conger conger* (Linnaeus, 1758) | European conger eel | *Congro, Safio* | Anti-rheumatic | Fish liver oil is applied to the joints that hurt (M) |
| **Phylum**: Chordata **Class**: Amphibian **Order**: Anura | | | | |
| *Bufo bufo* (Linnaeus, 1758) | Common toad | *Sapo* | Heals wounds | An animal is boiled and the resulting oil is placed directly over the wound (M) |
| *Bufo calamita* (Laurenti, 1768) | Natterjack toad | *Sapo* | Against anthrax | A small part of the animal's skin is soaked in vinegar for 2 weeks and then placed directly over the anthrax wound (M) |
| *Pelobates cultripes* (Cuvier, 1829) | Western spadefoot | *Sapo* | Against herpes | Toad slime is applied to the herpes (M) |

(continued)

**Table 15.1** (continued)

| Scientific name | English common name | Vernacular name | Popular use | Preparation and application[a] |
|---|---|---|---|---|
| | | | Against mumps | A burning cigar is given to a toad, which starts to smoke it. When the toad explodes, it is placed over the patient's throat (M) |
| | | | Amulet | A toad is impaled on a stick and left in the middle of the harvest (MR) |
| | | | Taboo | If a person says the word "*sapo*" (toad), they have to spit three times to avoid the appearance of mouth sores in their mouth (MR) |
| *Pelophylax perezi* (Seoane 1885) | Perez's frog | *Rã* | Heals groin wounds. | An open animal is placed over the wound (M) |
| | | | Heals mouth asthma | A live animal is placed inside the mouth and slowly sucked (M) |
| | | | Anti-rheumatism | A frog is fried in olive oil and the oleolite is applied to the affected area (M) |
| | | | Stops hands sweating | Rub a live animal into the hands (M) |
| **Phylum**: Chordata **Class**: Amphibian **Order**: Caudata | | | | |
| *Salamandra salamandra* (Linnaeus, 1758) | Fire salamander | *Salamandra, Salamantiga* | Heals burns | Powder from a dried salamander is applied directly over the burn (M) |
| | | | Taboo | If a person shows their teeth to a salamander, the teeth will rot (MR) |
| *Pleurodeles waltl* (Michahelles 1830) | Spanish ribbed newt | *Salamantiga, Galipato, Salamandra* | Taboo | There is a belief that the newt cleans the water from wells and springs, making it potable (MR) |
| **Phylum**: Chordata **Class**: Reptilia **Order**: Squamata | | | | |
| *Timon lepidus* (Daudin, 1802) | Ocellated lizard | *Lagarto, Sardão* | Heals sore eyes; anti-conjunctivitic | Lizard feces are applied to the eyes of humans and animals (M) (EV) |
| | | | Anti-toothache | A small amount of lizard blood is collected from the head of a live animal and placed on the tooth that hurts (M) |

(continued)

Table 15.1 (continued)

| Scientific name | English common name | Vernacular name | Popular use | Preparation and application[a] |
|---|---|---|---|---|
| | | | | Powder obtained from toasted lizard skin is placed over the tooth that hurts (M) |
| | | | Anti-mumps | The patient chews cheese and then places it in front of a lizard burrow. If the lizard eats it, the mumps will transfer from the patient to the lizard (MR) |
| *Psammodromus algirus* (Linnaeus, 1758) | Large | psammodromus | *Lagartixa* | Anti-fever |
| A live animal is placed inside the hollow stem of the giant reed *Arundo donax* and the reed is hung from the sick person's neck as a necklace (M) | | | | |
| *Podarcis hispanica* (Steindachner, 1870) | Iberian wall lizard | *Lagartixa* | Anti-fever | A live animal is placed inside the hollow stem of the giant reed *Arundo donax* and the reed is hung from the sick person's neck as a necklace (M) |
| *Elaphe scalaris* (Schinz, 1822) | Ladder snake | *Cobra* | Heals wounds | Snake's lard is directly applied to the wound (M) |
| *Malpolon monspessulanus* (Hermann, 1804) | Montpellier snake | *Cobra* | Heals wounds | Snake skin is mixed with olive oil and applied over the wound (M) |
| *Coluber hippocrepis* (Linnaeus, 1758) | Horseshoe whip snake | *Cobra* | Against tuberculosis | A snake's head is used as a necklace, touching the skin (MR) |
| | | | | Snake skin is toasted and milled and then eaten (M) |

(continued)

**Table 15.1** (continued)

| Scientific name | English common name | Vernacular name | Popular use | Preparation and application[a] |
|---|---|---|---|---|
| | | | Against furuncles | Snake soup is eaten by the patient (M) |
| | | | Anti-herpes | Snake lard is applied to the herpes (M) |
| | | | Against scabies | Snake lard is applied to the herpes (M) |
| | | | | Take a dead snake, hold it by the head and hit it with a stick so that all the poison runs to the tail. Then, the tail containing the all of the poison is sliced and fried in olive oil and the ointment applied to the scabies (M) |
| | | | Heals burns | Milled snake skin is applied to the burn (M) |
| | | | Against toothache | Snake skin is mixed with olive oil and placed on the afflicted tooth (M) |
| | | | Against mumps | A snake's spine is used as a necklace (MR) |
| | | | Against *espinhela-caída* (stomach aches and general illness) | Toasted and milled snake's skin is mixed with alcohol and drunk by the patient (M) |
| | | | Anti-cough | Snake skin tea (M) |
| | | | Anti-fever | Snake lard with olive oil (M) |
| | | | Anti-rheumatism | Eat snake lard soup (M) |
| | | | | Snake lard is applied to the joints (M) |
| | | | Painkiller | Snake lard is applied to the afflicted area (M) |
| | | | Pig's cough | Snake skin wrapped in cabbage is given to the animal (EV) |
| *Natrix maura* (Linnaeus, 1758) | Viperine snake | *Cobra, Cobra de água, Víbora* | | A snake's head is carried in the pocket as a good luck amulet (MR) |
| *Vipera latastei* (Boscá, 1878) | Lataste's viper | *Víbora, Bíbera* | Amulet | A viper's head is carried in the pocket as a good luck amulet (MR) |

(continued)

**Table 15.1** (continued)

| Scientific name | English common name | Vernacular name | Popular use | Preparation and application[a] |
|---|---|---|---|---|
| **Phylum**: Chordata **Class**: Reptilia **Order**: Testudines | | | | |
| *Emys orbicularis* (Linnaeus, 1758) | European pond turtle | *Cágado* | Anti-jaundice | Blood from a turtle's right paw is applied to the face(M) |
| | | | Anti-fever | A live animal is placed inside the hollow stem of the giant reed *Arundo donax*. and the reed is hung on the sick's person neck as a necklace (M) |
| *Mauremys* | | | | *leprosa*(Schwieegger, 1812) |
| Spanish pond turtle | *Cágado* | Anti-jaundice | Blood from a turtle right paw is applied in the face (M) | |
| | | | Anti-fever | A live animal is placed inside the hollow stem pf the giant reed (*Arundo donax*) and the reed is hung on the sick's person neck as a necklace (M) |
| **Phylum**: Chordata **Class**: Aves **Order**: Galliformes | | | | |
| *Gallus gallus domesticus* (Linnaeus, 1758) | Chicken | *Galinha, Galo, Pinto* | Against erysipelas | Chicken's blood is applied over the afflicted area (M) |
| | | | | Olive oil is applied over the afflicted area with a chicken feather (M) |
| | | | Heals burns | Blood from the chicken's crest (M) |
| | | | | Chicken's egg yolk is applied to the burn (M) |
| | | | Against toothache | Chicken excrement is applied to the afflicted tooth (M) |
| | | | | Egg yolk is applied to the afflicted tooth (M) |
| | | | Against belly ache/ cramps | Chicken's lard is placed over the belly (M) |
| | | | | Eat many eggs (MF) |
| | | | | Chicken excrement tea (M) |

(continued)

**Table 15.1** (continued)

| Scientific name | English common name | Vernacular name | Popular use | Preparation and application[a] |
|---|---|---|---|---|
| | | | Against jaundice | Two raw chicken eggs are eaten in the morning (MF) |
| | | | Promotes abortion | Eat soup of a black chicken (MF) |
| | | | Postpartum reconstituent | Eat scrambled eggs (MF) |
| | | | Against mumps | Chicken's lard is applied to the afflicted person's chin and throat (M) |
| | | | Against fever | Dried chicken excrement tea (M) |
| | | | | Two live animals are opened up and the patient's feet are placed inside both animals' bodies (M) |
| | | | Heals sore eyes | Wash the afflicted eyes with water and chicken egg yolk (M) |
| | | | Against tumors | Chicken excrement is placed over the tumor (M) |
| | | | Against bronchitis | Hot wine with chicken lard (M) |
| | | | | Orange tea with two eggs (MF) |
| | | | | Eggs with sugar (MF) |
| | | | Against asthma | Powdered eggshell with tea or coffee (MF) |
| | | | Against rheumatism | Chicken's lard is placed over the afflicted joints (M) |
| | | | Against obesity | Chicken eggs with vinegar (MF) |
| | | | Against rabid dog bite | Three egg yolks mixed with olive oil (MF) |
| | | | Against anorexia | A syrup of honey, eggs, white wine, grapes and figs (M) |
| | | | Against rachitis | Chicken eggs eaten every morning (MF) |
| *Alectoris rufa* (Linnaeus, 1758) | Red-legged partridge | *Perdiz* | Promotes expulsion of placenta | Partridge soup is given to the woman (M) |
| | | | Prevents birth pains and induces child birth | Partridge feathers are burned over the woman's legs (M) |
| | | | | Drink water in which partridge feathers were cooked (M) |
| | | | Anti-headache | Burn partridge feathers and inhale the smoke (M) |
| | | | Against sleeping disorders | Burn partridge feathers and inhale the smoke (M) |

(continued)

**Table 15.1** (continued)

| Scientific name | English common name | Vernacular name | Popular use | Preparation and application[a] |
|---|---|---|---|---|
| | | | Heals cow diseases | Burn partridge feathers and inhale the smoke (EV) |
| **Phylum**: Chordata **Class**: Aves **Order**: Columbiformes | | | | |
| *Columba livia* (Gmelin, 1789) | Rock pigeon | *Pombo, Pomba* | Heals flu/influenza | The patient inserts their feet inside the body of a dead pigeon (M) |
| | | | Anti-intestinal infections | A live pigeon is opened up and placed over the belly. If the pigeon turns black the next day, the disease is cured (M) |
| | | | Anti-asthma | The patient should eat pigeon feces (M) |
| | | | Painkiller | A live pigeon is opened up and placed over the afflicted region (M) |
| **Phylum**: Chordata **Class**: Aves **Order**: Ciconiiformes | | | | |
| *Ardea purpurea* (Linnaeus, 1766) | Purple heron | *Garça* | Anti-arthritis | Heron's fat is applied to the place that hurts (M) |
| | | | Heals sore eyes; anti-conjunctivitic | Heron's fat is placed directly in the eyes (M) |
| **Phylum**: Chordata **Class**: Aves **Order**: Strigiformes | | | | |
| *Bubo bubo* (Linnaeus, 1758) | Eurasian eagle owl | *Mocho, Bufo* | Anti-anorexia | A soup with owl's meat is eaten (M) |
| **Phylum**: Chordata **Class**: Aves **Order**: Cuculiformes | | | | |
| *Cuculus canorus* (Linnaeus, 1758) | Common cuckoo | *Cuco* | Augury (good omen) | When a person hears a cuckoo singing, they should roll onto the floor to have good luck (MR) |
| **Phylum**: Chordata **Class**: Mammalia **Order**: Rodentia | | | | |
| *Mus musculus* (Linnaeus, 1758) | House mouse | *Rato, Ratazana* | Anti-otitic | Several live newborn mice are placed inside a jar of olive oil and fried. The oil is then filtered and drops of it are instilled in the affected ear (M) |
| | | | Anti-colitis | A tea is made from mouse feces (M) |
| **Phylum**: Chordata **Class**: Mammalia **Order**: Soricomorpha | | | | |

(continued)

**Table 15.1** (continued)

| Scientific name | English common name | Vernacular name | Popular use | Preparation and application[a] |
|---|---|---|---|---|
| *Talpa europaea* (Linnaeus, 1758) | European mole | *Toupeira* | Against erysipelas/ Dermatological | Carry around the neck, as a necklace, a bag with the hand of a mole (MR) |
| | | | | Kill a mole and do not tell anyone for a year (MR) |
| | | | Against all pains | Kill a mole and do not tell anyone for a year (MR) |
| | | | Anti-toothache | Place a mole's tooth near the tooth that hurts (MR) |
| **Phylum**: Chordata **Class**: Mammalia **Order**: Lagomorpha | | | | |
| *Lepus europaeus* (Pallas, 1778) | European hare | *Lebre* | Increases a woman's fecundity | Hare's cheese is eaten (M) |
| | | | Heals sore eyes; anticonjunctivitic | Hare's blood is placed in the injured eyes (M) |
| | | | Calmative | Drink hare's bile (M) |
| *Oryctolagus cuniculus* (Linnaeus, 1758) | European rabbit | *Coelho* | Anti-blindness | A rabbit's head is cooked and the smoke is wafted to the eyes (M) |
| | | | Amulet | Rabbit's paws are carried in a pocket (MR) |
| **Phylum**: Chordata **Class**: Mammalia **Order**: Artiodactyla | | | | |
| *Bos taurus* (Linnaeus, 1758) | Cattle | *Vaca, Boi* | Heals wounds | Cow's milk cream is applied to the wound (M) |
| | | | Against dandruff | The afflicted person's head must be licked by a cow (M) |
| | | | Heals burns | The animal's tongue is passed over the burn (M) |
| | | | | Cow's butter is applied to the burn (M) |
| | | | Against warts | A cow's hair is stuck to the wart (M) |
| | | | Against cramping pains | Drink cow's urine (M) |
| | | | Heals sore eyes | Cold cow's milk and tepid water are placed in the afflicted eye (M) |
| | | | Stops nosebleeds | Cow excrement is inserted in the bleeding nose (M) |
| | | | Against rheumatism | Cow's butter is placed on the afflicted joints (M) |

(continued)

**Table 15.1** (continued)

| Scientific name | English common name | Vernacular name | Popular use | Preparation and application[a] |
|---|---|---|---|---|
| | | | | Cow's lard is placed on the afflicted joints (M) |
| | | | Against toothache | Cow excrement is placed on the afflicted tooth (M) |
| | | | Against anemia | Male cow's blood is drunk whilst still hot (M) |
| | | | Heals mumps | Cow's oxbow is placed over the afflicted person's neck (M) |
| | | | Against tuberculosis | Male cow's blood is drunk whilst still hot (M) |
| *Capra aegagrus hircus* (Linnaeus, 1758) | Goat | *Cabra, Cabrito, Cabrão, Bode* | Against roundworm | A mixture of goat's feces is placed in the afflicted person's eyes (M) |
| | | | Against bronchitis | Goat's butter with spirit (M) |
| | | | Against epilepsy | Milk of two goats, one must be the mother of the other goat (M) |
| *Cervus elaphus* (Linnaeus, 1758) | Red deer | *Cervo, Veado* | Anti-herpes | A deer's horn is placed in water. Then, this water is applied to the herpes (M) |
| | | | Anti-toothache | A deer's horn is turned into powder and then smoked (M) |
| | | | Heals abdominal pains | A deer's horn is turned into powder and drunk with water (M) |
| | | | Heals animal bites | A deer's penis is placed in water. Then, this water is applied to the bite (M) |
| | | | | A deer's horn is burned and then applied to the bite (M) |
| | | | Taboo | Carry deer horn powder in the pocket (MR) |
| *Ovis aries* (Linnaeus, 1758) | Domestic sheep | *Ovelha, Carneiro, Cordeiro* | Hemostatic | Sheep's fat is directly applied to the wound (M) |
| | | | Against erypiselas | Sheep's wool soaked in olive oil is directly applied over the affected region (M) |
| | | | Anti-mumps | Sheep's wool is placed over the patient's chin (M) |
| | | | Anti-epistaxis | Sheep excrement is inserted into the nose (M) |
| | | | | Wind one strand of wool around the thumb, and lift the arm on the same side (right or left) as the nostril that bleeds (MR) |

(continued)

**Table 15.1** (continued)

| Scientific name | English common name | Vernacular name | Popular use | Preparation and application[a] |
|---|---|---|---|---|
| | | | Anti-fever | Sheep's milk should be drunk for nine days (M) |
| *Sus scrofa domestica* (Linnaeus, 1758) | Domestic pig | *Porco, Javardo, Marrano* | Heals wounds | Pig's gal is applied to the wound (M) |
| | | | | Pig's belly button is placed over the wound (M) |
| | | | Heals eczema/ dermatological problems | Pig's lard is applied to the eczema (M) |
| | | | | The marrow of the bone from the chin of a pig is placed over the afflicted area (M) |
| | | | Heals burns | Pig's lard is applied to the burn (M) |
| | | | Against cramps | Pig's lard is applied to the belly (M) |
| | | | Promotes menstruation | Hot wine, with pig's lard and sugar (MF) |
| | | | Anti-anemia | Uncooked pig's liver (MF) |
| | | | Against mumps | Pig's lard is rubbed over a lettuce leaf and placed over the afflicted person's chin (M) |
| | | | Against scabies | Pig's lard and sulfur is placed over the afflicted area (M) |
| | | | Promotes intelligence in children | Cooked pig's brains (MF) |
| | | | Against asthma | Cooked pig's heart (MF) |
| | | | | A syrup of wine and pig's lard and sugar (M) |
| | | | Against rheumatism | Pig's lard is placed on the afflicted joints (M) |
| | | | Against ulcers | Pig's lard (M) |
| **Phylum**: Chordata **Class**: Mammalia **Order**: Perissodactyla | | | | |
| *Equus caballus* (Linnaeus, 1758) | Horse | *Cavalo* | Anti-anemia | Horse's blood is drunk (M) |
| | | | Anti-headache | Horse's lard is applied externally to the head (M) |

(continued)

**Table 15.1** (continued)

| Scientific name | English common name | Vernacular name | Popular use | Preparation and application[a] |
|---|---|---|---|---|
| | | | Taboo | Horseshoes are used as good luck amulets (MR) |
| *Equus asinus* (Linnaeus, 1758) | Donkey | *Burro* | Anti-warts | A donkey's hair is placed near the wart (M) |
| | | | Taboo | If a child drools a lot, the child must kiss a donkey to stop it (MR) |
| | | | Anti-epistaxis | Donkey feces are smelled to stop the blood (M) |
| | | | Anti-alcoholism | Nine drops of donkey's blood are mixed with the alcoholic person's wine (M) |
| **Phylum**: Chordata **Class**: Mammalia **Order**: Carnivora | | | | |
| *Felis silvestris catu* (Linnaeus, 1758) | Cat | *Gato* | Heals snake bites | Warm cat's meat is placed over the snake bite (M) |
| *Meles meles* (Linnaeus, 1758) | European badger | *Texugo* | Anti-rheumatism | Badger's lard is applied to the joints (M) |
| | | | Human milk production | A badger's paw is placed on the woman's chest to facilitate milk production (MR) |
| *Mustela nivalis* (Linnaeus, 1766) | Least weasel | *Doninha* | Heals weasel bites | Weasel lard is applied to the bite (M) |
| *Vulpes vulpes* (Linnaeus, 1758) | Red fox | *Raposa, Raposo* | | The fried liver of a fox is eaten by children to invigorate their strength (M) |
| | | | Extracts splinters | |
| *Canis lupus signatus* (Cabrera, 1907) | Iberian wolf | *Lobo* | Anti-rheumatism | |
| | | | | Rye that has crossed a wolf's "*gola*" (part of a wolf's trachea) is given to the sick animal (EV) |
| | | | | Water that has crossed a wolf's "*gola*" (part of a wolf's trachea) is given to the sick animal (EV) |

(continued)

**Table 15.1** (continued)

| Scientific name | English common name | Vernacular name | Popular use | Preparation and application[a] |
|---|---|---|---|---|
| *Canis lupus familiaris* (Linnaeus, 1758) | Dog | *Cão* | Hemostatic | The dog licks the wound and heals it (M) |
| | | | | Toasted and powdered dog feces are applied to the wound (M) |
| | | | | Dog's urine is applied to the wound (M) |
| | | | Anti-toothache | Carry a dog's tooth in a pocket (MR) |
| | | | Heals rabid dog bites | Eat the head or the heart of the dog (M) |
| | | | | Rub the wound with hair from the dog that bit the patient (M) |
| | | | Anti-fever | The patient's nails are placed in bread and given to a dog to eat (MR) |
| | | | Anti-otitic | Several live newborn mice are placed in a jar of olive oil and fried. The oil is then filtered and drops of it are instilled in the affected ear (M) |
| | | | Anti-madness | A dog or a puppy is opened up and placed on the mad person's head so the dog's blood covers his face (MR) |
| **Phylum**: Chordata **Class**: Mammalia **Order**: Primates | | | | |
| *Homo sapiens sapiens* (Linnaeus, 1758) | Human | *Homem* | Against jaundice | Human excrement is eaten with honey (M) |
| | | | | Drink the urine from a little boy (M) |
| | | | Against fever | Powdered human bones with wine (M) |
| | | | | Drink the urine from a little girl (M) |
| | | | Heals tumors | Human excrement is applied to the afflicted area (M) |
| | | | Heals warts | Drink menstrual blood from a virgin (M) |
| | | | Heals wounds | Human urine is applied over the wound (M) |
| | | | Against dermatophytosis | Children's urine mixed with mud is applied to the afflicted area (M) |
| | | | Against toothache | A pulp of urine and flour is placed on the afflicted tooth (M) |

(continued)

**Table 15.1** (continued)

| Scientific name | English common name | Vernacular name | Popular use | Preparation and application[a] |
|---|---|---|---|---|
| | | | Against cramps | Drink the urine from a little boy (M) |
| | | | Heals intestinal infections | Drink the urine from a little boy (M) |
| | | | Heals wasp stings | Children's urine mixed with mud is applied to the afflicted area (M) |

[a] Type of medical application: *M* medicine; *MF* medicinal food; *MR* magic-religious; *C* cosmetic; *EV* ethnoveterinary remedy

**Table 15.2** Distribution of fauna and corresponding remedies in Portugal

| Phylum | Class | Order | Animal species ($N = 54$) | | Remedies ($N =$) | |
|---|---|---|---|---|---|---|
| | | | No. | % | No. | % |
| Annelida | Clitellata | Hirudinea | 1 | 1.8 | 3 | 1.3 |
| | Oligoqueta | Haplotaxida | 1 | 1.8 | 3 | 1.3 |
| Arthropoda | Insecta | Hymenoptera | 1 | 1.8 | 19 | 8.4 |
| | | Díptera | 1 | 1.8 | 2 | 0.8 |
| | | Phtiraptera | 1 | 1.8 | 1 | 0.4 |
| | | Coleoptera | 1 | 1.8 | 2 | 0.8 |
| | Arachnida | Araneae | 1 | 1.8 | 3 | 1.3 |
| | | Scorpiones | 1 | 1.8 | 3 | 1.3 |
| Mollusca | Gastropoda | Pulmonata | 2 | 3.6 | 7 | 3 |
| Chordata | Actinopterygii | Gadiformes | 1 | 1.8 | 7 | 3 |
| | | Clupeiformes | 1 | 1.8 | 1 | 0.4 |
| | | Siluriformes | 1 | 1.8 | 1 | 0.4 |
| | | Anguliformes | 1 | 1.8 | 1 | 0.4 |
| | Amphibia | Anura | 4 | 7.2 | 10 | 4.4 |
| | | Caudata | 2 | 3.6 | 3 | 1.3 |
| | Reptilia | Squamata | 8 | 14.8 | 26 | 12 |
| | | Testudines | 2 | 3.6 | 2 | 0.8 |
| | Aves | Galliformes | 2 | 3.6 | 31 | 14 |
| | | Columbiformes | 1 | 1.8 | 4 | 1.7 |
| | | Ciconiiformes | 1 | 1.8 | 2 | 0.8 |
| | | Strigiformes | 1 | 1.8 | 1 | 0.4 |
| | | Cuculiformes | 1 | 1.8 | 1 | 0.4 |
| | Mammalia | Rodentia | 1 | 1.8 | 2 | 0.8 |
| | | Soricomorpha | 1 | 1.8 | 4 | 1.7 |
| | | Lagomorpha | 2 | 3.6 | 5 | 2 |
| | | Artiodactyla | 5 | 9 | 44 | 20 |
| | | Perissodactyla | 2 | 3.6 | 7 | 3 |
| | | Carnivora | 6 | 10.8 | 18 | 8 |
| | | Primates | 1 | 1.8 | 12 | 5 |
| Total | | | 54 | 100 | 225 | 100 |

(*Note* Some species are used in the same remedies, so, if are from the same family, the remedy is counted only as one. For example, the use of Toad slime against herpes is a remedy extracted from three different species (*Bufo bufo*, *Bufo calamita* and *Pelobates cultripes*), thus it is only counted as one remedy)

used for medicinal purposes: 18 (33%) mammals, 10 (19%) reptiles, 10 (19%) invertebrates, 6 (11%) amphibians, 6 (11%) birds, and 4 (7%) fishes have been reported for medical purposes (Table 15.1). Approximately 225 remedies that use these species have been reported, with the majority from domestic animals such as pigs, cows, and goats; sheep (Order Artiodactyla) and chickens (Order Galliformes) with 44 (20%) and 25 (11%) remedies, respectively; and reptiles, such as snakes and lizards (Order Squamata), with 26 (12%) remedies; and insects, namely bees (Order Hymenoptera), with 19 (8.4%) remedies (Table 15.2). Furthermore, some

**Fig. 15.1** Viper heads as zootherapeutics and amulets. **a** A local woman from Gralheira village, Serra de Montemuro, showing a recently removed viper head. Photo adapted from a report in the *Tal & Qual* journal, on 21st May 1993 (Campos 1993). **b** A Lataste's viper head in a jar with alcohol, in Sobral de São Miguel, 2008, courtesy of Mrs. Susete Ferreira

other groups such as carnivores, including dogs, badgers, and wolves (Order Carnivora), and even primates, as humans (Order Primates) have been used in 18 (8%) and 12 (5%) remedies, respectively (Table 15.2).

Similar to other Mediterranean case studies, a large portion of the animal-based remedies cited in this review were prepared from domestic animals, and this, as reported by Quave et al. (2010) is most likely to be "due to their role in local subsistence practices and the consequential ease of access and availability of these species to humans".

Many of the species and remedies recorded to Portugal are similar to those used in Spain and Italy. At least 33 remedies using the same species were reported in Spain and Italy (see Quave et al. 2010), but the majority are exclusive to Portugal (or at least they have never been reported elsewhere).

## 15.3 Implications for Conservation

The use of endangered species in all forms of traditional medicine is a cause for growing concern (Alves and Rosa 2005). Although the animals used in Portuguese zootherapeutic remedies are mostly domestic livestock or represent very common wild animals with no conservation issues, there are some others that are not, and of these species there are some with conservation problems. One example is Lataste's viper, *Vipera latastei*, which is classified as Vulnerable (VU) in the Portuguese Vertebrate Red List (Cabral et al. 2006) and is used in the northern parts of Portugal as an amulet and also in zootherapeutics (Fig. 15.1). According to Brito et al. (2001), that animal is collected and traded for superstitious and medical purposes, and it is even sold in cities such as Lisbon and Oporto, and also on some internet websites dedicated to witchcraft. The collection of this animal has roots in very ancient times (according to some pharmacopeias and historical documents it was collected in the Medieval ages), as reported by authors such as Nobre (1928) and Ferreira (1935). Although collection was quite harmless to viper populations in the past because of their abundance and lack of conservation threats, nowadays this pressure may aggravate the already delicate situation that viper populations are suffering due to the loss, fragmentation, and degradation of their habitat by anthropogenic pressures such as forest fires, intensive forestry, intensive land reclamation for agriculture, urban development, and the implementation of various infrastructures (Cabral et al. 2006; Brito et al. 2001). Also, direct human persecution due to fear and misunderstanding can constitute a real threat to the viper (Cox et al. 2006; Brito et al. 2001; Ceríaco 2010a, b). Brito et al. (2001) estimated that between 50 and 100 vipers are captured per year in the vicinities of Gerês village alone, and this situation could be similar in other localities in that region. Vipers are sold in some villages as an "underground" business, since their capture and trade are considered illegal. This illegal activity, the trade of wildlife body parts for traditional medicine, occurs worldwide and the annual global trade in animal-based medicines, according to Marques (1997), accounts for billions of dollars per year. Most of the poachers state that it is now more difficult to find a viper than before, suggesting that viper populations are declining at an accelerating pace (Brito et al. 2001). Some of these poachers have started to collect other snakes, like the viperine snake, *Natrix maura*, which is very similar to Lataste's viper (Brito et al. 2001), and they have even been reported to cross the border to collect vipers in Spain. This situation may become even worse if the poachers begin to capture other viper species that occur in the region, such as the Iberian adder, *Vipera seoanei*, which has an Endangered (EN) conservation status (Cabral et al. 2006). Further investigations into the full extent and the dimensions of this practice are needed to understand the real implications that it can have on viper populations.

Another interesting case is the use of Iberian wolf (*Canis lupus signatus*) parts, also in northern regions of Portugal. The relationship between wolves and human populations is not easy or peaceful, with the wolf being seen as a threat to livestock, domestic animals, and even humans (Álvares 2004). Also, the large number of stories and the folklore and beliefs about this animal give it the reputation of being a diabolic, supernatural, and evil animal, adding to the unpopularity that it already suffers for being a threat to the *modus vivendi* of pastoral communities economically dependent on their flocks. According to many authors, in zootherapy, body parts that belong to animals perceived as being strong, venomous, or evil are more effective and thus the remedy will be more powerful (Fontes and Sanches 2000; Quave et al. 2010). This is the case for the wolf in Portugal, with a long history of its uses in medieval and renaissance pharmacopeias, and also with a particular use that still occurs in a few localities in northern Portugal. The use of the wolf's "*gola*" still occurs in Portugal to cure respiratory problems in humans and animals. The "*gola*" is a dried part of the wolf's trachea that is used to "filter" the water given to animals or people (Fig. 15.2). It is believed that the presence or proximity of a wolf in a house where domestical animals, as pigs and cows, sleep transmits a respiratory disease named "*lobagueira*" or "*mau ar do lobo*" ("wolf bad air"),

**Fig. 15.2** Wolf "golas" from northern Portugal. **a** Woman using a "gola" to filter water. **b** An old dried "gola". Courtesy of Grupo Lobo and Prof. Francisco Fonseca

which particularly affects pigs (Álvares 2004). There are even "*golas*" in use today that are more than 150 years old, which were passed through generations of the same family (Álvares 2004). Today this use is restricted to a very few localities in northern Portugal, and its use is gradually disappearing as the TK also disappears; most of the reports refer to the use of old "*golas*" instead of new ones. This use, however, should not be excluded as a potential conservation issue because many wolves are still being killed and illegally persecuted in Portugal, often for no clear reason. Most of the killings are probably related to the protection of flocks or to illegal hunting, but killing wolves to collect parts for zootherapy can also still occur. Wolf populations in Portugal are currently stable but, because of their limited geographical distribution and loss of habitat (Cabral et al. 2006), they are classified as Endangered (EN) in the Portuguese Vertebrate Red List (Cabral et al. 2006). Further investigations are needed to understand the real impact of the medicinal use on wolf populations, and also to preserve the knowledge of this use and even some of the older "*golas*" as anthropological artifacts.

Both the wolf and the viper, as well as other snakes are considered to be powerful and evil animals, so the "power" and "efficacy" of remedies based on these animals are thought to be very strong. Also, the difficulty of obtaining products from these animals due to their rarity and scarcity reinforces the belief that remedies based on them are strong and potent. As reported by Quave et al. (2010), the sociocultural value assigned to threatened species considered "powerful" due to their rarity or because of the peril associated with their collection should be taken into consideration when designing biodiversity conservation measures.

Turtle populations, particularly European pond turtle (*Emys orbicularis*) populations, may also be affected by capture for zootherapeutic uses. *Emys orbicularis* is classified as Endangered (EN) in the Portuguese Vertebrate Red List (Cabral et al. 2006) due to the loss and degradation of its habitat, pollution, and competition with invasive species such as the Pond slider (*Trachemys scripta*). If we add the phenomenon of direct poaching by fishermen who believe that turtles are a menace to the fish population, and the capture of many animals by the same fishermen to illegally keep as pets, or even the use of turtles for local gastronomy (Ceríaco 2010b), it can be seen that the level of pressure that these populations suffer is very high. The capture of turtles for zootherapeutics is indeed an extra threat to this species.

Specifically in the case of Portugal, the capture of species for zootherapeutic uses is not one of the main conservation issues, but rather an additional pressure on some already vulnerable populations. Although we can identify some species where capture for zootherapeutic uses constitutes another pressure on their survival, further studies focused on Portuguese ethnozoology are needed to better understand the diversity, extent, and real implications of such practices to science and conservation, as well as to preserve the folklore and cultural heritage related to animals.

## 15.4 Conclusions

The increased interest in the knowledge that traditional populations possess about the use of animals for medical purposes is partly because the empirical basis that developed over centuries may have, in many cases, scientific corroboration, but above all the historical, economical, sociological, anthropological, and environmental aspects of such practices are of great interest for science and conservation (Lev 2002). In a historical context, the use of animals and animal parts in medical practices in Portugal has very old origins and constitutes a very rich historical, ethnographical, anthropological, and biological heritage. This can be explained by the strong cultural and historical relationship between people and nature, but it is also due to the exchange and trade of knowledge between Portugal and other countries, during the Portuguese maritime expansion. The current use of animals and their body parts in traditional medicine did not occur by chance, being rather the result of centuries of use by Portuguese people. Zootherapeutic medicines were very common in popular and official pharmacopeias, although this popularity was always lower than for botanical-based medicines.

Many similarities in uses and even the same uses for the same species can be found in, for example, eighteenth century Portuguese pharmacopeias. This suggests that what we call folk medicine today, or, in a more vernacular language, what we call witchcraft, are remnants of the uses of "old science", which were used and applied in ancient apothecaries, pharmacies, and royal hospitals all over the country. Since the nineteenth century revolution in chemistry, pharmacy, and medicine was mainly focused on big cities, especially in academies, universities, and hospitals, the rest of the country had to rely on the old, better known, and more widely available ethnozoological (and ethnobotanical) medicines. The important cultural and medical heritage that is zootherapy can be an excellent starting point for understanding the social, cultural, and even scientific evolution of pharmacy and medicine. Again, further studies are needed to fully document Portuguese ethno-pharmacopeia, and to compare the present data with historical documents and ancient pharmacopeias. Furthermore, comparing Portuguese pharmacopeias with pharmacopeias from countries where Portugal had colonies, such as Brazil, Angola, Mozambique, Cape Verde, and others, would be an interesting investigation into the exchange of knowledge and practices during the first era of the globalization of Portuguese discoveries and the Portuguese maritime expansion. The presence of exotic animals, such as elephants and lions, in eighteenth century pharmacopeias, and the presence of zootherapeutic specimens in natural history museums, is a striking evidence of this exchange, and it would be interesting to discover whether these practices still remain today.

This traditional knowledge is fast eroding in Portugal, as in the rest of the world, as a side effect of modernization and globalization, leaving to ethnobiologists the great responsibility of not only inventorying traditional biological resources but also conserving and revitalizing traditional beliefs so that age-old cultures are not lost (Mahawar and Jaroli 2008). This review has shown that the Portuguese zootherapeutic heritage is very rich and diverse. Portuguese people still use at least 54 different animal species as sources of 225 remedies. This considerable number is interesting because it is larger than reported in ancient pharmacopeias, and even larger than reported in other Mediterranean countries such as Spain and Italy (Quave et al. 2010). Due to its unique geographical position, the biodiversity index in Portugal is considerably high, providing human communities with a large number of animal and plant species that can be used for food, raw materials, and medicines. In the traditional Portuguese pharmacopeia, the majority of medicines used were made from domestic livestock such as cows, pigs, goats, bees, dogs, and chickens, since they were the most common animals that people had access to, and they are also very typical of agro-pastoral economies. Bovine, ovine, and caprine animals are very common for milk and meat production, and pigs are also commonly used for meat production. Additionally, apiculture constitutes an important income for the local economy, and chickens are kept in almost every rural house for eggs and food; the presence of dogs, as pets or working dogs, is also a very common scenario. This situation is similar to that found in other Mediterranean countries where communities with an agro-pastoral economy use more products from animals that they are in regular contact with for the management of human health (Quave et al. 2010).

These uses encompass various types of medical and food uses, including the use of animal excrement and blood. Despite the prevalent use of medicines based on parts of domestic animals, many use parts from wild animals, as exemplified by reptiles, which play an important role in Portuguese zootherapy, with eight species being used to make 26 medicines. Reptiles are still quite common in traditional agricultural landscapes, and also near human settlements. Also, many snakes and lizards use human constructions such as walls, fences, and houses as refuges or lairs, living side by side with humans. As in Spain and other Mediterranean countries, the time spent by Portuguese populations on forest activities, like hunting or working, provides the population with a greater level of access to wild fauna such as reptiles and other species that are used in popular zootherapeutic remedies (Quave et al. 2010). The use of reptiles in traditional medicine is a worldwide phenomenon that, in many cases, constitutes an extra pressure on already endangered and declining populations (Alves et al. 2008). The illegal trading of reptiles and the commercialization of their products has caused a significant reduction in the number of snake species in certain parts of the world (Fitzgerald et al. 2004). Data from Portugal were not available until this study was undertaken; thus, there are actually more reptile species being affected worldwide by their use in traditional folk medicine than was previously reported by Alves et al. (2008) in their world review. This reinforces the idea that further ethnobiological studies are required to get the full picture.

Documentation of the traditional uses of different animal species in zootherapy is not the only reason for studying those practices. It is also important to integrate the cultural and biological aspects of such practices into a broader discourse encompassing conservation, cooperative management, and sustainability (Alves and Rosa 2005), contributing toward better conservation programs for endangered species. As throughout the entire world, wild fauna in Portugal suffer from habitat loss, reduced habitat quality, pollution, and overhunting, amongst others. Thus poaching, which is driven by the use of certain species in traditional medicines, can be an extra source of pressure on already threatened populations. For Portugal, the most striking case is the use of Lataste's viper heads as amulets, leading to the illegal poaching of hundreds of individuals of this endangered species per year, being a cause of growing concern (Brito et al. 2001). This situation is an important conservation issue and needs to be better studied and understood by biologists, conservationists, and public managers.

The illegal nature of the captures, however, may render studies difficult in Portugal. Moreover, it must be remembered that reporting the knowledge of these practices is not easy: while still occurring in the country, those practices generally are not discovered by direct means. There is a social stigma that whoever uses such practices must be a wizard or witch, and it is only through certain types of discourses that some of these practices can be detected, since nobody likes to admit having such knowledge (Lobo 1995).

As Lev (2002) pointed out, scholarly investigations and studies of the medicinal uses of animals and their products provide us with a better understanding of the historical, ethnic, and traditional aspects of medicine and *Materia Medica,* and, in future studies and management decisions based on a comprehensive body of knowledge such as this, we will have a better understanding of human relationships with nature that will surely lead to better judgments and decisions concerning environmental studies, biodiversity conservation, and sustainable management.

The main conclusion of this chapter is that further studies on this topic are required in order to fully understand the whole picture of the use of animals and animal parts in Portuguese traditional medicines and the possible implications for conservation. Questions such as (1) is the distribution of these uses equal in all regions of the country; (2) are these uses disappearing or growing; (3) how many animals are actually captured for these practices; (4) are there any public health implications and (5) how old are these uses and where do they originate from, among others, must be answered in future studies. This chapter is a first step toward that goal and I personally hope that it will be a helpful basis for further studies and works.

**Acknowledgments** The author first thanks Dr Rômulo Alves and Dr Ierecê Rosa, the editors, for their kind invitation to contribute a chapter to this book. He also thanks Dr Francisco Álvares and all the members of Grupo Lobo for their help and kind contributions of photographs and documents. He thanks Mrs. Susete Ferreira, from Sobral de São Miguel, who kindly gave him the photograph of the viper's head. He owes special thanks to all of his colleagues who helped with the data collection, during the field surveys and also with the bibliography collection in libraries and archives, especially Mariana Marques. Furthermore, he appreciates all the people they interviewed, who gave their time and precious information about zootherapeutic uses in Portugal.

## References

Álvares F (2004) O lobo o imaginário popular. In: Nunes M (ed) Serra da Aboboreira—a terra, o homem e os lobos. Câmara Municipal de Amarante, Amarante

Alves RRN, Rosa IL (2005) Why study the use of animal products in traditional medicines. J Ethnobiol Ethnomed doi: 10.1186/1746-4269-1-5

Alves RRN, Rosa IL, Santana GG (2007) The role of animal-derived remedies as complementary medicine in Brazil. Bioscience 57(11):949–955

Alves RRN, Vieira WLS, Santana GG (2008) Reptiles used in traditional folk medicine: conservation implications. Biodivers Conserv 17(8):2037–2049. doi:10.1007/s10531-007-9305-0

Bethencourt F (1987) O Imaginário da Magia—feiticeiras, saludadores e nigromantes no séc, vol XVI. Universidade Aberta, Lisbon

Brigola JCP (2002) Colecções, Gabinetes e Museus em Portugal no Século XVIII. Fundação Calouste Gulbenkian/Fundação para a Ciência e Tecnologia, Lisbon

Brito JC, Rebelo A, Crespo EG (2001) Viper killings for superstitious reasons in Portugal. Boletin de la Asociacion Herpetologica Espñola 12:101–104

Cabral MJ et al (2006) Livro vermelho dos vertebrados de Portugal. Assírio e Alvim, Lisbon

Campos A (1993) Víboras em saldo. Tal e Qual 21(05/93):17

Ceríaco LMP (2010a) Gecko's folklore in portuguese oral tradition. In: Conde SP (ed) Proceedings of the international conference on oral tradition, Concello de Ourense, Ourense

Ceríaco LMP (2010b) Human attitudes towards herpetofauna. Dissertation, University of Évora, Portugal

Ceríaco LMP, Marques MM, Madeira NC, Vila-Viçosa CMM, Mendes P (2011) Folklore and traditional ecological knowledge of Geckos in Southern Portugal: implications for conservation and science. J Ethnobiol Ethnomed 7(26):1–10. doi:10.1186/1746-4269-7-26

Coelho MR (1760) Pharmacopea tubalense. Officina Balio Geredini, Rome
Cox N, Chanson J, Suart S (2006) The status and distribution of reptiles and amphibians of the Mediterranean Basin, 1st edn. IUCN, Gland
de Almeida AG, Magalhães M, Guimarães AP (2009) Artes de cura e espanta-males—Espólio de medicina popular recolhido por Michel Giacometti, 1st edn. Gradiva, Lisbon
Felgueiras G (1956) Os batráquios no conceito popular e na superstição. Acta do Primeiro Congresso de Etnografia e Folclore de Braga 2:65–92
Ferreira JB (1935) O ofidismo no seu aspecto histórico e actual, 1st edn. Academia de Ciências de Lisboa, Lisboa
Fitzgerald LA, Painter CW, Reuter A, Hoover C (2004) Collection, trade, and regulation of reptiles and amphibians of the Chihuahuan Desert ecoregion. TRAFFIC North America World Wildlife Fundo, Washington D.C
Fontes A, Sanches JG (2000) Medicina Popular—Ensaio de Antropologia Médica. Ñncora Editora, Lisbon
Frazão-Moreira A, Fernandes MM (2006) Plantas e saberes. No limiar da etnobotânica em Portugal. Edições Colibri, Lisbon
Henriques FF (1731) Medicina Lusitana. Caza Miguel Diaz, Amsterdam
Lev E (2002) Traditional healing with animals (zootherapy): medieval to present-day Levantine practice. J Ethnopharmacol 85:107–118
Lobo E (1995) A doença e a cura: recorrência à bruxaria na procura de saúde, 1st edn. Estratégias Criativas Vila Nova de Gaia, Portugal
Loureiro J (1790) Flora cochinchinensis. 1 edn. Typis, et Expensis Academicis, Lisbon
Mahawar M, Jaroli DP (2008) Traditional zootherapeutic studies in India: a review. J Ethnobiol Ethnomed 4(17):1–12. doi:10.1186/1746-4269-4-17
Mantero I (1998) Portugal Misterioso. Reader's Digest, Lisbon
Marques JGW (1997) Fauna medicinal: Recurso do ambiente ou ameaça à biodiversidade. Mutum 1:1–4
Myers N, Mittermeier RA, Mittermeier CG, da Fonseca GAB, Kent J (2000) Biodiversity hotspots for conservation priorities. Nature 403:858–858
Nobre AF (1928) Animais venenosos de Portugal. Instituto de zoologia da Universidade do Porto, Porto
Novais MH, Santos I, Mendes S, Pinto-Gomes C (2004) Studies on pharmaceutical ethnobotany in Arrabida Natural Park (Portugal). J Ethnopharmacol 93:183–195
Quave CL, Lohani U, Verde A, Fajardo J, Rivera D, Obón C, Valdes A, Pieroni A (2010) A comparative assessment of zootherapeutic remedies from selected areas in Albania, Italy, Spain and Nepal. J Ethnobiol 30:92–125
Ramstad KM, Nelson NJ, Paine G, Beech D, Paul A, Paul P, Allendorf FW, Daugherty CH (2007) Species and cultural conservation in New Zealand: Maori traditional ecological knowledge of Tuatara. Conserv Biol 21:455–464
Salgueiro J (2010) Ervas, usos e saberes—Plantas medicinais no Alentejo e outros produtos naturais. Edições Colibri, Lisboa
Santo António C (1725) Pharmacopea Lusitana. Real Mosteiro de São Vicente de Fora, Lisbon
Singh JS (2002) The biodiversity crisis: a multifaceted review. Curr Sci 82:638–647
Sutherland WJ (2003) Parallel extinction risk and global distribution of languages and people. Nature 423:276–279
WHO/IUCN/WWF (1993) Guidelines on conservation of medicinal plants. WHO/IUCN/WWF, Gland

# Chapter 16
# Zootherapy and Biodiversity Conservation in Nigeria

**Durojaye A. Soewu**

**Abstract** Zootherapy, the use of animal products in health care delivery is widely accepted among the various constituent nationalities in Nigeria. This practice involves domesticated and wild fauna resources, cutting across all taxa. In addition to health challenges treated by conventional medicine, zootherapy in Nigeria also accommodates situations that are psychological, spiritual or even mystical. The choice of species/parts employed is guided by several factors which include the perceived bioactive constituents, some morpho-physiological characteristics and behavioural ecology, as well as some mythological conceptions associated with the animal. Utilisation of wild animals however, has no consideration for either the sustainability of such uses or the conservation status of the species. Rare, threatened and even endangered species of animals are being used extensively in zootherapeutic practices. This indiscriminate utilisation underscores the need to strike a delicate balance between peoples' custom, health care needs and biodiversity conservation interests.

## 16.1 Introduction

Africa has a long, valued tradition of using natural resources, especially animals and their products for medicinal purposes. Traditional African medicine (TAM), in its broad diversity and deep richness had been in existence long before the advent of more orthodox modern medicine and the people depended largely on traditional

D. A. Soewu (✉)
Department of Biological Sciences, Covenant University,
P.M.B 1023 Ota, Ogun, Nigeria
e-mail: durosoewu@hotmail.com

R. R. N. Alves and I. L. Rosa (eds.), *Animals in Traditional Folk Medicine*,
DOI: 10.1007/978-3-642-29026-8_16, 

medicine (TM) as their only source of health care (Adeola 1992; Soewu and Adekanola 2011). The utilisation of animals and animal parts in the preparation of several products employed in diverse ways for health care delivery via traditional medical practices enjoys very wide acceptance across Nigeria. TM plays such a significant role in meeting the health care needs of the majority of Nigerians that 75–80% of the Nigerian population use the services of traditional healers (Gamaniel et al. 2005). The figure quoted for Nigeria in that report agrees totally with the submission of World Health Organisation (WHO) that 80% of the world population relies on TM prepared mainly by the use of natural products (animals and plants) to meet their daily health requirements (Soewu and Ayodele 2009). This utilisation involves domestic as well as wild animals. However, the use of animals taken directly from the wild naturally calls for more concern as it has all the potential to influence the functional dynamics of the ecosystem and ultimately the quality of life and even health care delivery for the people.

## 16.2 Traditional Medicine

The WHO stated that TM refers to health practices, approaches, knowledge and beliefs incorporating animal and mineral-based medicines, spiritual therapies, manual techniques and exercises, applied singularly or in combination to treat, diagnose and prevent illnesses or maintain well-being (WHO 2000). TM was further defined by WHO as the sum total of all knowledge and practices, whether explicable or not, used in diagnosis, prevention and elimination of physical, mental or social imbalance and relying exclusively on practical experiences and observations handed from generation to generation, whether verbally or in writing (Soewu and Adekanola 2011; Adesina 2007). These descriptions incorporate several elements of TM and encompass the various forms of medicines and therapies such as herbal medicine, massage, homoeopathy, mud bath, music therapy, wax bath, reflexology, dance therapy, hydrotherapy, mind and spirit therapies, self-exercise therapies, radiation and vibration, osteopathy, chiropractice, aromatherapy, preventive medicine, radiant heat therapy, therapeutic fasting and dieting spinal manipulation, psychotherapy, etc. (Adesina 2007). This definition could be further extended for Africa by including a phrase such as "while bearing in mind the original concept of nature which include the material world, the sociological environment, whether living or dead and the metaphysical forces of the universe" (Sofowora 1993). TM is also a comprehensive term used to refer both to systems such as traditional Chinese medicine, Indian ayurveda and Arabic unani medicine, and to various forms of indigenous medicine (Alves and Rosa 2007).

It is therefore expected that a large country of the size of Nigeria, with diverse cultures and traditions, should be rich in TM and should have eminent and respected traditional healers to take care of the teeming population (Adesina 2007). In countries like Nigeria where the dominant health care system is based on allopathic medicine, or where TM has not been incorporated into the national health care system, TM is often termed "complementary", "alternative" or "non-conventional" medicine (Alves and Rosa 2007).

The traditional healer, as defined by WHO (1978), is a person who is recognised by the community in which he lives as competent to provide health care by using vegetable, animal and mineral substances and certain other methods based on the social, cultural and religious background, as well as on the knowledge, attributes and beliefs that are prevalent in the community, regarding physical, mental and social well-being and the causation of disease and disability (Adesina 2007).

## 16.3 The Federal Republic of Nigeria

Nigeria is located in western Africa on the Gulf of Guinea and has a total area of 923,768 $km^2$ (356,669 sq mi) making it the world's 32nd largest country. It shares a 4,047 km (2,515 mi) border with Benin (773 km), Niger (1,497 km), Chad (87 km), Cameroon (1,690 km), and has a coastline of at least 853 km. Nigeria lies between latitudes 4°N and 14°N, and longitudes 2°E and 15°E. The highest point in Nigeria is Chappal Waddi at 2,419 m (7,936 ft). The main rivers are the Niger and the Benue River which converge and empty into the Niger Delta, one of the world's largest river deltas and the location of a large area of Central African Mangroves. It has a population size of 155,215,573; a population density of 164.8/$km^2$ (2010 estimate); total GDP $377.949 {$216.803 billion} billion and per capita income $2,422 {$1,389} (2009 estimate). The official currency is Naira (₦) (NGN). English is the official language of Nigeria, but it is the second most spoken language in all states of the federation after regional and local indigenous languages (Anon 2010a)

Nigeria has a varied landscape. By virtue of its geographical extent, it spans different climatic and ecological zones. The mean annual rainfall ranges from about 450 mm in the northeast to about 3,500 mm in the coastal southeast, with rains falling within 90–290 days, respectively. The mean annual temperature ranges from 21°C in the south to 30°C in the north with extremes of 14 and 45°C and a latitude range of 0–1,000 m above sea level. The far south is defined by its tropical rainforest climate, where annual rainfall is around 1,524–2,032 mm a year. In the southeast stands the Obudu Plateau. Coastal plains are found in both the southwest and the southeast. This forest zone's most southerly portion is a saltwater swamp, also known as a mangrove swamp because of the large number of mangroves in the area. North of this is a freshwater swamp, containing different vegetation from the saltwater swamp, and north of that is a rain forest. The highly varied climatic conditions and physical features have consequently endowed Nigeria with a very rich biodiversity, making the country an important centre for biodiversity. It is widely believed that the areas surrounding Calabar, Cross River State, contain the world's largest diversity of butterflies. The drill monkey is only found in the wild in southeast Nigeria and neighbouring Cameroon (Anon 2010a, b, 2011).

## 16.4 Biodiversity in Nigeria

Nigeria is rich in biodiversity. The country is endowed with a variety of plant and animal species. There are about 7,895 plant species identified in 338 families and 2,215 genera. There are over 22,000 vertebrate and invertebrate species. These species include about 20,000 insects, 1,000 birds, 1,000 fishes, 247 mammals and 123 reptiles. Of these animals about 0.14% is threatened while 0.22% is endangered. One of the avian species, which is forest-based, *Mahimbus ibadanensis*, is reported to be endemic to the country. There are as many as 56 forest snakes out of which one, *Mehelya egbensis*, is reported to be endemic. There are also a large number of amphibian species, 18 of which are rare, being found only in Nigeria and the Cameroons. One of the 18, *Bufo perreft*, is reported to be endemic to Nigeria (Anon 2010b, 2011). Table 16.1 gives a list of wild animals under various categories of threat in Nigeria with their conservation status. With the exception of those species that are already extinct, all the species listed in the table are still being used indiscriminately in TM. The medicinal values of each of these wild fauna species vary greatly from one region of the country to the other. Even within a particular region, each animal has numerous applications regarding their involvement in zootherapeutic practices.

## 16.5 Traditional Versus Conventional Medicine in Nigeria

Reliance on TM by the majority of the populace in most parts of Africa is due to several factors. Traditional medical practices are often deemed a more appropriate method of treatment, notwithstanding the array of options within reach. Non-availability of facilities and personnel, and the cost implications of orthodox medicine are other factors that tend to promote the use of traditional medicinal preparations (Soewu and Adekanola 2011). Also the number of traditional medical healers practicing in most regions of the continent is far greater than the number of western medical practitioners; hence, availability of TM invariably outweighs that of orthodox medicine (Soewu and Ayodele 2009). Although there are no accurate records of the total number of traditional healing providers in Nigeria, it has been estimated that there are well over 200, 000 healers spread all over the country. In almost every village within the country, there is at least one traditional healer (Gamaniel et al. 2005). As at 1989, there were approximately 40,000 traditional medical practitioners in the old Bendel state of Nigeria (Taylor and Fox 1992). (Bendel was just one out of a nineteen-state structure of the country then). This implies that a very significant number of Nigerians, especially those living in the countryside, still receive their medical care through the traditional system (Gamaniel et al. 2005).

The perception of TM as an integral part of the people's culture does arouse positive sentiments towards it and encourages its sustained usage. From a cultural

**Table 16.1** Wildlife species presently classified as rare, threatened or endangered in Nigeria

| S/no | Order | Family | Common name | Scientific name | Status[a] |
|---|---|---|---|---|---|
| 1 | Chelonia | Pelomedusidae | African keeled mud turtle | *Pelusios carinus* | Endangered |
| 2 | | | African keeled mud turtle | *Pelusios castaneus* Schweigger 1812 | Endangered |
| 3 | | | African keeled mud turtle | *Pelusios nanus* | Endangered |
| 4 | | | William's African mud turtle | *Pelusios williamsi* | Endangered |
| 5 | | Trionychidae | Abry's flapshell turtle | *Cycloderma aubryii* | Endangered |
| 6 | | | Namibian flapshell turtle | *Cyclonorbis elegans* | Endangered |
| 7 | | | Senegal flapshhell turtle | *Cyclonorbis senegalensis* | Endangered |
| 8 | | Dermochelidae | Leatherback turtle | *Dermochelys coriacea* | Endangered |
| 9 | | Chelonidae | Green turtle | *Chelonia mydas* | Endangered |
| 10 | | | Olive ridley | *Lepidochelys olivacea* | Endangered |
| 11 | | | Hoaksbill turtle | *Eretmochelys imbricata* | Endangered |
| 12 | Crocodylia | | Nile crocodile | *Crocodylus niloticus* | Endangered |
| 13 | | | Slender-snouted crocodile | *Crocodylus catapractus* | Endangered |
| 14 | | | African dwarf crocodile | *Osteolamus tetrapis* | Endangered |
| 15 | Squamata | Veranidae | Nile monitor lizard | *Varanus niloticus* | Endangered |
| 16 | | | Monitor lizard | *Varanus exanthematicus* | Endangered |
| 17 | | Pythonidae | Royal python | *Python regius* | Endangered |
| 18 | | | Rock python | *Python sebae* | Endangered |
| 19 | Struthioniformes | Struthionidae | Ostrich | *Struthio camelus* | Endangered |
| 20 | Pelecaniformes | Pelethronodae | Pinkbacked pelecan | *Pelecanus rufescens* | Endangered |
| 21 | Coconiforms | Ardeidae | Grey heron | *Ardea cinerea* | Endangered |
| 22 | | | Goliath heron | *Ardea goliath* | Endangered |
| 23 | | | Breen heron | *Bruorides virescens* | Endangered |
| 24 | | | Purple heron | *Ardea purpurea* | Endangered |
| 25 | | | Great egret | *Egretta alba* | Endangered |
| 26 | | | Little egret | *Egretta garzetta* | Endangered |
| 27 | | | Cattle egret | *Ardeola ibis* | Endangered |
| 28 | | | Squocco heron | *Ardeola rolloides* | Endangered |

(continued)

**Table 16.1** (continued)

| S/no | Order | Family | Common name | Scientific name | Status[a] |
|---|---|---|---|---|---|
| 29 | | | Black-crowned night heron | *Nycticorax nycticorax* | Endangered |
| 30 | | Scopidae | Hammercop | *Scopus unbretta* | Terminated |
| 31 | | Ciconidae | White stork | *Ciconia ciconia* | Endangered |
| 32 | | | Abdims stork | *Ciconia abdimii* | Endangered |
| 33 | | | Saddle-billed stork | *Ephippiorhnchus senegalensis* | Endangered |
| 34 | | | Marabou stork | *Leptotilus crumeniferus* | Endangered |
| 35 | | | Wood ibis | *Ibis ibis* | Endangered |
| 36 | | Threskiornithidae | African spoonbill | *Platelea alba* | Endangered |
| 37 | | | Sacred ibis | *Threskiomis aethiopica* | Endangered |
| 38 | | | Glossy ibis | *Plegadis falcinelus* | Endangered |
| 39 | | | Hadada ibis | *Bostrychia hagedash* | Endangered |
| 40 | Falconiforms | Accipitridae | Nubian vulture | *Aegypius tracheliotus* | Endangered |
| 41 | | | Rappel's griffon vulture | *Gyps rupellii* | Endangered |
| 42 | | | White-backed vulture | *Gyps bengalensis* | Endangered |
| 43 | | | Palm-nut vulture | *Gypohierax angolensis* | Endangered |
| 44 | | | Hooded vulture | *Neophron monachus* | Endangered |
| 45 | | | West African River eagle | *Haliaetus vocifer* | Endangered |
| 46 | | | Short-toed eagle | *Cricaetus gallicus* | Endangered |
| 47 | | | Marital eagle | *Polemaetus bellicosus* | Endangered |
| 48 | | | Bateleur eagle | *Terathopius ecaudatus* | Endangered |
| 49 | | | Common buzzard | *Butoe butoe* | Threatened |
| 50 | | | Montaguas harrier | *Cyrcus pygargus* | Threatened |
| 51 | | | Goshawk | *Accipitar genitilis* | Threatened |
| 52 | | | Sparrow hawk | *Accipitar nisus* | Threatened |
| 53 | | Falconidae | Hobby | *Falco subbuteo* | Threatened |
| 54 | | | Kestrel | *Falco innunculus* | Threatened |
| 55 | | Sagittariidae | Secretary bird | *Sagittarius serpentarious* | Endangered |
| 56 | | Phasianidae | Helmet guinea-fowl | *Numida meleagris* | Threatened |
| 57 | | | Crested guinea-fowl | *Guttera edourdi* | Endangered |

(continued)

**Table 16.1** (continued)

| S/no | Order | Family | Common name | Scientific name | Status[a] |
|---|---|---|---|---|---|
| 58 | | | Blue-breasted kingfisher | *Halcyon malimbica* | Threatened |
| 59 | | | Malachite kingfisher | *Alcedo cristata* | Threatened |
| 60 | | | Pied kingfisher | *Ceryle rudis* | Threatened |
| 61 | | | Pigmy kingfisher | *Ceryx picta* | Threatened |
| 62 | | | Senegal kingfisher | *Halcyon senegalensis* | Threatened |
| 63 | | Upupidae | Hoopoe | *Upupa epops* | Endangered |
| 64 | | Bucerotidae | Abyssianian Ground Hornbill | *Bucorvus abyssinicus* | Endangered |
| 65 | | Ploeceidae | Ibadan malimbus | *Malimbus ibadanensis* | Endangered |
| 66 | | | Black mountain weaver | *Ploceus melanogaster* | Endangered |
| 67 | Primates | Cercopithecidae | Colobus monkey (Guereza) | *Colobus polykomos* | Endangered |
| 68 | | | Olive colobus | *Procolobus verus* | Endangered |
| 69 | | | Red-eared Guenon | *Cercopithecus erythrotis* | Endangered |
| 70 | | | Moustached monkey | *Cercopithecus cephus* | Endangered |
| 71 | | | Mona monkey | *Cercopithecus mona* | Threatened |
| 72 | | | White throated monkey | *Cercopithecus erythrogaster* | Endangered |
| 73 | | | Patas monkey | *Erythrocebus patas* | Threatened |
| 74 | | | Olive baboon | *Papio anubis* | Threatened |
| 75 | | Cercopithecidae | White-hosed monkey | *Cercopithecus nicititans* | Extinct |
| 76 | | | Green (tantelus) monkey | *C. aethiops* | Extinct |
| 77 | | | Rensiss monkey | *C. preussi* | Extinct |
| 78 | | | Ground monkey | *C. poganis* | Extinct |
| 79 | | | Grey-checked mangabey | *C. albigenia* | Extinct |
| 80 | | | Red-capped mangabey | *C.torguatus* | Extinct |
| 81 | | | Drill baboon | *Mandrillus leucocphaeus* | Endangered |
| 82 | | | Chimpanzee | *Pan troglodytes* | Endangered |
| 83 | | Pongidae | Western lowland gorilla | *Gorilla gorila* | Endangered |
| 84 | Pholiodota | Manidae | Giant pangolin | *Manis gigantea* | Threatened |
| 85 | | | Tree pangolin | *Manis tricuspis* | Threatened |

(continued)

**Table 16.1** (continued)

| S/no | Order | Family | Common name | Scientific name | Status[a] |
|---|---|---|---|---|---|
| 86 | Hystricomorpha | Hystricidae | Crested porcupine | *Hystrix cristata* | Threatened |
| 87 | | | Brush-tailed porcupine | *Atherurus african* | Threatened |
| 88 | Carnivora | Canidae | Hunting dog | *Lycaon pictus* | Endangered |
| 89 | | | Side-striped jackal | *Canis adustus* | Rare |
| 90 | | | Pale fox | *Vulpes pallida* | Rare |
| 91 | | | Honey badger | *Mellivora capensis* | Rare |
| 92 | | Mustelidae | Cape clawless otter | *Aonys capensis* | Rare |
| 93 | | Viverridae | African civet cat | *Civettictis civetta* | Endangered |
| 94 | | | Cusimanse | *Crossarchus crossarchs* | Rare |
| 96 | | Hyaenidae | Spotted hyaena | *Crocuta crocuta* | Rare |
| 97 | | | Stripped hyaena | *Hyaena hyaena* | Endangered |
| 98 | | Felidae | Serval cat | *Leptailurus serval* | Rare |
| 99 | | | Caracal or desert hynx | *Caracal caracal* | Rare |
| 100 | | | Leopard | *Panthera pardus* | Endangered |
| 101 | | | Lion | *Panthera leo* | Endangered |
| 102 | | | Cheetah | *Acinonyx jubatus* | Endangered |
| 103 | Tubulidentata | Oryeteropidae | Aardvark | *Orycteropus afer* | Extinct |
| 104 | Proboscidea | Elephanitidae | African bush elephant | *Loxodonta africana africana* | Endangered |
| 105 | | | African forest elephant | *Loxodonta africana cyclotis* | Endangered |
| 106 | Hyracoidea | Procaviidae | Rock hyrax | *Procavia capensis* | Rare |
| 107 | | | Three hyrax | *Dendrohyrax arboreus* | Rare |
| 108 | Sirenia | Trichecchidae | Manatee | *Trichechus senegalensis* | Endangered |
| 109 | Artiodactyla | Suidae | Red river hog | *Potamochoerus aethipicus* | Rare |
| 110 | | | Wart hog | *Phocochoerus* aethiopicus | Threatened |
| 111 | | | Giant forest hog | *Hylochoerus meinertzhagani* | Endangered |
| 112 | | Hippopotamidae | African hippoptamus | *Hippoptamus amphibious* | Endangered |
| 113 | | | Pigmy hippoptamus | *Hexaprotodon liberensis helsopi* | Endangered |
| 114 | | Tragulidae | Water chevrotain | *Hymoschus acquaticus* | Endangered |
| 115 | | Giraffidae | Giraffe | *Giraffa camelopardalis* | Endangered |

(continued)

**Table 16.1** (continued)

| S/no | Order | Family | Common name | Scientific name | Status[a] |
|---|---|---|---|---|---|
| 116 | | Bovidae | African buffalo | *Cyncerus cafer cafer* | Threatened |
| 117 | | | Dwarf buffalo | *Cyncerus cafer nanus* | Threatened |
| 118 | | | Mountain reedbuck | *Redunca fulvirufula* | Endangered |
| 119 | | | Bohor reedbuck | *Redunca redunca* | Endangered |
| 120 | | | Giant eland | *Taurotragus derbianus* | Endangered |
| 121 | | | Western hartebeest | *Alcelaphus b. major* | Endangered |
| 122 | | | Roan antelope | *Hippotragus equines* | Endangered |
| 123 | | | Korrigum (topi) | *Damaliscus l korrigum* | Endangered |
| 124 | | | Western kob | *Kobus kob kob* | Endangered |
| 125 | | | Bush buck | *Tragelaphus scriptus* | Endangered |
| 126 | | | Sitatunga | *Tragelaphus spekii* | Endangered |
| 127 | | | Red-fronted gazelle | *Gazella rufifrons* | Threatened |
| 128 | | | Dorcas gazelle | *Gazela dorcas* | Endangered |
| 129 | | | Dama gazelle | *Gazella dama* | Endangered |
| 130 | | | Yellow backed duiker | *Cephalophus sylvicultor* | Endangered |
| 131 | | | Red flanked duiker | *Cephalophus rufilatus* | Endangered |
| 132 | | | Maxwell's duiker | *Cephalophus maxwellii* | Endangered |
| 133 | | | Black duiker | *Cephalophus niger* | Endangered |
| 134 | | | Blue duiker | *Cephalophus monticlla* | Endangered |
| 135 | | | Bay duiker | *Cephalophus dorsali* | Endangered |
| 136 | | | Klipspringer | *Oreotragus oreotragus* | Endangered |
| 137 | | | Royal antelope | *Neotragus pygmaeus* | Endangered |

*Source*: Anon 2010b; [a] IUCN 2010a, b

perspective, TM has become a part of the people's culture despite the fact that this form of medicine is not well organised (Sofowora 1993). People are just more comfortable with TM and are satisfied with its results; hence, many choose TM regardless of the existence and availability of orthodox medicine. These situations both reflect and perpetuate the importance of TM, stressing that notwithstanding any reason adduced, people tend to simply prefer TM to all other options available (Marshall 1998).

Most synthetic drugs are no longer as effective as they used to be; hence the trend in most parts of the world is the resurgence of interest in traditional medicines or natural remedies with little or no side effects (Olopade 2002). This anticipated side effects of synthetic drugs administered in orthodox medicine is another factor that promotes patronage of traditional medicines. Non-synthetic drugs are believed to be either totally free from residual effects or produce such at a very minimal, almost insignificant level. A side effect is usually regarded as an undesirable secondary effect which occurs in addition to the desired therapeutic effect of a drug or medication. Side effects may vary for each individual depending on the person's disease state, age, weight, gender, ethnicity and general health (Soewu 2006).

TM as practiced today continues to live side by side with orthodox medicine in Nigeria where the traditional medical practitioners continuously make new discoveries which have cured major ailments in the society. Such discoveries stemmed from the consistent efforts of traditional healers to eradicate dangerous diseases that have plagued the society in recent times and that are apparently incurable through orthodox medicine. Major diseases such as epilepsy, cancer, convulsion, paralysis, snake bites, mental illness and even other ailments having hereditary origins are now being cured by TM (Odu 1987; Soewu 2008).

TM in Nigeria believes that illness is due not only to organic causes but also arises from the supernatural, indicating the displeasure of ancestral gods and evil spirits, or is the effect of black magic, spirit possession or the intrusion of an object into the body. The psychological aspect of traditional medical practice looms very large and is perhaps the strength of this form of medicine and its practitioners, as even in strictly somatic cures, the traditional practitioner often depends heavily on manipulating the psychology of his patient before prescribing a treatment (Sofowora 1993).

## 16.6 Zootherapy in Nigeria

Ethnozoology refers to the study of the relationship between human societies and the animal resources around them (Jaroli et al. 2010). Zootherapy is an important component of ethnozoology. It deals with the healing of human ailments by using therapeutics based on medicine obtained from animals or ultimately derived from them (Costa-Neto 1999). Zootherapeutic resources constitute the essential ingredients in different traditional systems (Mahawar and Jaroli 2007).

Zootherapeutic practices in Nigeria employ domestic as well as wild species of animals to treat various conditions and situations. For every animal species being utilised in zootherapy in Nigeria, virtually all the parts find a use in treating one situation or the other. On some occasions, even waste products like urine and faecal droppings of an animal are employed in preparing zootherapeutic recipes used to treat some conditions in humans.

Appeasing gods, witches, deities and ancestral spirits through the use of sacrifices also constitutes an integral part of the zootherapeutic healing systems in

Nigeria. There are various kinds of sacrifices depending on the situation at hand. The choice of animal to be used is influenced by the same factors applicable to other facets of the zootherapeutic practices as discussed in Sect. 16.6.3. It has been observed that the sacrifices have over the years proved to be efficacious as specific healing requests presented through the sacrifices had been fully satisfied. On some occasions, the response from the 'astral realm' could be visible and instantaneous. In such an instance, the materials presented as sacrifice would either be "consumed" by unseen forces, sometimes with specific leftovers in a weird but predictable pattern, within a stipulated period of time, signifying acceptance. Otherwise, it could be 'ignored' to mean total rejection. Rejection of a sacrifice is a bad omen and would require further moves to avert an impending situation or remedy a prevailing unpleasant condition.

Another important component of zootherapy in Nigeria is the use of incantations and mystical word chants. Disease is considered as possession by evil spirits and should be treated using some natural products along with incantations (Soewu 2008; Sawandi 2006). These incantations are often reflective of the basic components of the preparation; the perceived behavioural and ecological features as well as mythological conceptions that informed the choice of that particular animal in the preparation (Soewu 2008).

Non-conventional conditions accommodated in zootherapy across Nigeria is as diverse as the variety of life situations that can ever be imagined and vary among the constituent nationalities. Sales boosters, fortune rousers, hypnotism for several purposes including sexual abuse, and money making charms are some of those preparations that have no application in orthodox medicine. Others are preparations for good luck, bad dreams, making or preventing rains, to confer invisibility, to prevent accidents, as well as amulets used as machetes and bullet proofs (Soewu and Adekanola 2011). Inflicting, curing and preparing antidotes for sexual poisons and the production of spells and curses are some other situations in zootherapy that have no equivalence in conventional medicine (Soewu 2006, 2008).

The use of these resources for healing purposes comes in various forms of preparation and administration in Nigeria depending on the situation being treated, (Sodeinde and Soewu 1999; Soewu and Ayodele 2009). Preparation of fauna-based traditional medicines could be processed in different forms: a single, specific part of an animal; various combinations of different parts of the same animal species; and combinations of different parts of several animals, plants, minerals and other natural resources (Soewu 2006). The mode of practice of TM in Nigeria varies from one community to another. Almost every practitioner prepares his or her own remedies without reference to any universal protocol or formulated standards and essentially all of them operate outside the conventional health system. Most traditional healers employ a low level of technology in the preparation of their products. The medicaments being produced are largely non-standardised in quantity and uncontrolled in quality (Gamaniel et al. 2005). The practices vary from one locality to another. Medicinal preparations in traditional systems are prescribed and administered in various ways, depending on the identified causes, availability of required natural resources, expertise of the traditional medical practitioner as well as socio-economic considerations (Sofowora 1993). The perceived causative factor(s) are expected to play a major role in determining the method of treatment to employ. The causes of disease have been classified in TM into five categories and according to that classification human beings are affected in five different ways. It is therefore believed that traditional practitioners tend to have greater appeal to patients because only one of these causes is known to orthodox medical practitioners. According to Lambo (1979) cited in Sofowora (1993), the five causes of disease according to TM are:

1. Physical ailments—These are disease caused by injurious elements entering the human system through food, drink, skin, etc. These causes of disease are well known to orthodox medical practitioners.
2. Psychological causes—Grouped under this title are diseases caused to a person when his/her will is not in harmony with the laws of nature. The diseased body is said to be sometimes affected by a diseased state of the mind. Some people tend to believe that they are sick when in actual fact nothing is wrong with their system (hypochondria).
3. Astral influences—'It is known in occult science that the radiations from cosmic agents e.g. sun, moon and planets, have an influence on human beings either for good or evil'. The moon is said to influence the brain, and it is said that for this reason lunatics become wild and act abnormally when the new moon appears. It is indeed generally believed, not only in Nigeria but in the entire West African region, that lunatics become more violent at the birth of the new moon.
4. Spiritual causes—Evil thoughts, evil desires and machinations by enemies (i.e. by remote influences, for instance, hypnotism) including soul projection or evil telepathic messages are all grouped together under spiritually caused diseases. Diseases caused by witchcraft would probably also come into this class.
5. Esoteric causes—Diseases originating from the soul, or those caused by the deeds of an individual in his former life (before reincarnation), are said to have esoteric causes.

From these five causes of disease it would appear that the traditional practitioner's concept of disease is on a wider plane than that of the modern doctor. As a fallout of such emphasis placed by TM on supernatural forces, the TM practitioner is contacted not only in times of sickness but also when an evil omen is noticed or when actual misfortune, such as an accident, economic losses in commercial endeavours, quarrels in the family, etc. has occurred or is envisaged.

### *16.6.1 Zootherapy and Biodiversity Conservation in Nigeria*

A common dilemma facing all fauna species is the soaring demand for their body parts for use in medicinal practices (Soewu and Ayodele 2009). There is little doubt then that the use of endangered species in zootherapeutic preparations presents some of the most challenging issues in wildlife conservation today and by

**Fig. 16.1** *Gorilla gorilla* skulls on display in a market in Nigeria

touting the medicinal properties of these species, traditional medicines fuel the continuing demand (Soewu 2008; Gaski and Johnson 1994). The continued depletion of medicinal wildlife resources not only embodies a challenge for conservation, but more importantly represents a serious threat to the health status of the human population (Soewu 2006; Marshall 1998). There is a risk that a growing herbal market and its great commercial benefit might pose a threat to biodiversity through the over-harvesting of the raw materials for herbal medicines and other natural health care products. Zootherapeutic practices, especially the aspects that concern wild animal species, if not controlled can lead to the extinction of endangered species and the destruction of natural habitats and resources (WHO 2003).

Presently, Nigeria is facing potentially massive loss of wildlife due to over-hunting. Out of the 137 species listed in Table 16.1, one is ecologically extinct, 7 are already extinct, 12 are rare, 23 are threatened while 94 are endangered. The practice of TM in Nigeria is so broad that virtually all known species have been associated with one medicinal value or the other. Regrettably, the demand created by TM is one major cause for the over-exploitation of the wild population of numerous animal species (Alves and Rosa 2005). As indicated in the Table 16.1, many animal species in Nigeria are presently under varying degrees of threat due to indiscriminate killing while some species have actually slipped into extinction. There is no indication that the level of use of medicinal wildlife resources for zootherapy will diminish. On the contrary, there is every reason to believe that the quantities of animals (and plants) required for TM will increase substantially in the years to come as the human population grows and acceptance of TM and natural products increases in the market (Soewu and Adekanola 2011; Marshall 1998).

Of the various uses of wild fauna and their products in Nigeria, use for traditional medicinal preparation stands out as the broadest and perhaps the largest route of consumptive utilisation of wild animals. This is a result of many factors. While the use for other purposes is selective based on traditional taboos, prejudices or formal religious dicta, use in TM transcends all barriers of institutions placed on other uses. People who would ordinarily not have utilised a particular species for other uses still find it permissible to consume or apply medicinal preparations concocted with such 'forbidden' species and some other ingredients (Soewu 2006; Fayenuwo 1999).

For instance Moslems are forbidden by religious dicta to consume Warthog, monkeys, pigs and scavengers such as crows and raven. However, an Islamic faithful, holding firmly to these religious inhibitions would readily consume or apply traditional drugs prepared with these animals because it is no longer the raw animal but a medicinal preparation and this is permissible to the same religious dicta forbidding its direct consumption. Also, indigenes of an area where consumption of certain animal species is forbidden by traditional beliefs—for example ancestral linkage—have been found to consume, without reservations traditional decoctions made from such species of animals. Consumption via zootherapeutic preparations of otherwise forbidden species is permissible and acceptable because such a drug is an integral part of the people's culture (Soewu 2006; WHO 2000).

It is also worthwhile to note that zootherapy in Nigeria utilises a number of species that are usually not considered fit for human consumption. These include animals like Chameleon, skink, shrew, small-sized species of birds, smaller rodents and some insects to mention a few (Fayenuwo 1999; Banjo et al. 2002). Some other species which are considered sacred and "untouchable" in one part of the country also do readily find acceptance in zootherapeutic practices of another region in the country. All these constitute additional pressure on populations of animals in the wild stemming solely from the medicinal values of these animals. In the long run, this extra pressure on some species may create a significant impact on the ecosystem and the overall conservation of biodiversity.

### *16.6.2 Economic Importance of Zootherapy*

Another factor that actively encourages the continued depletion of these resources is their market value or the viability of trade in them and their products as a source of income. Animal-derived compounds have been documented to bring considerable benefits to quite a number of people—hunters, vendors, zootherapeutic practitioners and the end users in terms of monetary value and human welfare. Trade in wild animal products used as medicines represents a source of income for many people as the animal often changes hands several times before reaching the end user, thus allowing various levels of traders to profit from the commercialisation. Hunters sell to intermediaries (taxi drivers, wholesale vendors) who in turn supply to retailers in town. These retailers eventually sell either in parts or whole to the professional medical practitioners, or sometimes for domestic utilisation in trado-medical practices. Due to their economic and social background, many of these people—vendors, zootherapeutic practitioners, both men and women—make a living selling medicinal herbs and animal parts, raw or processed (Soewu 2006).

Wild animals in zootherapy are an important exploitable natural resource that has a prominent role to play in the economy of this nation (Ayodele and Ihongbe 1995). For most zootherapeutic practitioners and vendors, the art/trade is commonly the main, oftentimes the only source of livelihood. As long as there is sufficient money to be made in trade and processing of wildlife resources, not only will individual species suffer but conservation at regional or even world level may be threatened (Simmonds 1998).

### *16.6.3 Guiding Factors*

The choice of animal for a particular purpose is probably made on the basis of a unique co-evolution between social and ecological systems (Soewu 2008; IUCN 1997). The choice of animal species utilised in trado-medicinal preparations in Nigeria is guided by several factors. Some of the leading factors include:

1. Bioactive constituents of such animals or the specific parts concerned. Identification of and correlation of such bioactive compounds with certain disease conditions often come through trial and error undertaken over a long period of time. Once the efficacy of preparations of such animals/parts has been proven over time, it becomes accepted and established;
2. Some behavioural and ecological tendencies naturally associated with the animal or part concerned is another guiding factor;
3. Certain mythological conceptions associated with the animal(s) also influence the choice of situations the animal, in whole or parts, is employed to treat;
4. Another factor is the array of complementary ingredients. These ingredients sometimes possess some behavioural and ecological traits believed to complement that of the main ingredient in a particular pattern as far as the situation to be treated is concerned.

### *16.6.4 Substitute Animals*

Some substitute animal species may be used in place of the regular species if it is not readily available. However, there are certain limitations to this use of substitute animals. The factors guiding the choice of the array of ingredients combined to treat a particular situation is often not flexible enough to make such a substitution feasible. Consequently, substituting the main animal species may require some changes in the complementary ingredients. Also, the use of a different species may not have the level of efficacy ascribed to the preparation. Acceptance of a new recipe involving substitute species by the end-users is also a major factor. End-users are reluctant to accept changes and are sceptical about such preparations.

Such scepticism may even influence the consumption and administration of the new recipe, thereby rubbing off on the results obtained from the use of such preparation. This scepticism about the use of substitute animals is one major cause of the desperation to get the regular animal at all costs which results in deliberate searches and contract-hunting for preferred wild animals.

Another point which is of conservation interest as regards the use of substitute animals is that most of the animals presently being utilised as regular species actually started off as substitute species for some other species that became difficult to obtain then. As the conservation of the hitherto preferred regular species got worse off, at least locally over the years, some of them slipping into ecological extinction, the use of the present day regular species became continuously pronounced until it got established as the preferred species (Soewu and Ayodele 2009; Soewu 2008). Advocating and promoting the use of substitute animal species requires a good degree of caution. Although the species are still thriving, it is a prerequisite to undertake a thorough investigation into the population dynamics of such animals so as to establish the sustainability of such uses.

### 16.6.5 ExSitu Animals in Zootherapy

Not all wild animals reared under ex-situ conservation schemes like domestication and captive breeding will readily be applicable in zootherapeutic preparations. In a situation where the main attribute informing the use of an animal is attached to some behavioural traits exhibited by the animal in relation to ecological factors, ex-situ animals may be considered unfit for the features desired (Soewu 2008). Simulation of the natural habitat can provide some, but not all the ecological characteristics capable of provoking or supporting the behavioural traits of choice in the animal.

Essentially, it means zootherapy in Nigeria would continue to rely heavily on animals cropped from the wild. This obviously has the tendency to worsen the challenges for conservation of the concerned species as well as overall biodiversity conservation. The trend may eventually lead to unavailability of essential medicinal resources, thereby impacting negatively on health care delivery systems.

### 16.6.6 Sustainability

Use of wild animals in TM has no consideration for the sustainability of such utilisation as regards populations of animals in the wild. This is attributable, though not necessarily limited to the following reasons:

1. The majority of the residents in the country are not conscious of the conservation status of any material, fauna or flora, which are employed in the various

preparations. The perception is that they are God's/nature's given gifts that must be exploited as long as they are available;

2. These legal machineries that are supposed to confer some degree of protection on these wild medicinal resources by regulating their trade and utilisation are hardly applied. Therefore, in situations where the people even have the required awareness to certain restrictions regarding a particular species, they usually take solace in these lapses in the implementation of the purported protection accorded to such animals.

It is therefore a common practice to openly trade in and utilise endangered and threatened species for their medicinal properties. Figure 16.1 shows skulls of *Gorilla gorilla* on display in a market. Virtually all the endangered species are openly displayed in various markets stalls across the country and many of these species are even sometimes displayed for sale along roadsides throughout the nation.

Recent reports of ethnozoological studies in Nigeria are already pointing in a direction of utilisation of wild fauna species beyond the sustainable levels for the various species in the ecosystems. Continuous decline in size, reduction in availability and abundance are some variables suggesting this unsustainable trend in utilisation of wild animals (Soewu and Ayodele 2009; Soewu 2008; Sodeinde and Soewu 1999).

In addition to exerting pressure directly on populations of wild animals, zoo-therapeutic practices in Nigeria also have some age and gender specificity that may drastically alter the natural equilibrium of the dynamics in the populations of these animals. It has also been reported that some preparations required newborn, juvenile or even pregnant animals. Such a practice may reduce the number of individuals that will have the chance to participate in reproductive activities.

## 16.7 Conclusion

While the impacts of non–consumptive values of wildlife (scientific research, park viewing and bird watching) may be overlooked, those of the consumptive values, which often cause depletion of resources in the wild, are of critical conservation interest. The use for food and TM appears to exert more pressure on wild populations of animals than all other uses. Hence, a critical examination of these uses as an index of cropping pressure on resources in the wild is a prerequisite for any conservation programme to be meaningful and effective.

Conservation of biodiversity, most especially those species having medicinal properties in Nigeria require some basic steps:

1. A more detailed record of indigenous knowledge relating to zootherapeutic practices among the constituent nationalities in the country is urgently required as a first step towards devising strategies for sustainable exploitation of these natural resources;

2. Holistic improvement on the availability of orthodox medical facilities so as to place health care delivery within the reach of the teeming rural population at a drastically reduced price. This may reduce the need for traditional medicinal preparations requiring wild animal species;
3. Capacity building for the entire population to generally improve the standard of living and enjoin the support of the people for conservation programmes;
4. Inclusion of conservation education in school curriculum from pre-nursery to the tertiary levels of education;
5. A sincere, serious commitment, backed with the necessary will power by the government to apply the terms of all treaties and conventions on biological diversity to which the country is a signatory.

## References

Adeola MO (1992) Importance of wild animals and their parts in the culture, religious festivals, and traditional medicine, of Nigeria. Environ Conserv 19(2):125–134

Adesina SK (2007) Traditional medical care in Nigeria. http://www.onlinenigeria.com/links/LinksReadPrint.asp?blurb=574. Accessed July 2011

Alves RRN, Rosa IL (2005) Why study the use of animal products in traditional medicines? J Ethnobiol Ethnomed 1(5):1–5

Alves RRN, Rosa IL (2007) Biodiversity, traditional medicine and public health: where do they meet? J Ethnobiol Ethnomed 3(14):1–9

Anon (2010a) Nigeria. http://en.wikipedia.org/wiki/Nigeria. Accessed 12 June 2011

Anon (2010b) Federal Republic of Nigeria (2010) Fourth National Biodiversity Report. Abuja

Anon (2011) Biodiversity in Nigeria. www.onlinenigeria.com/links/bioticadv.asp?blurb=525. Accessed 17 July 2011

Ayodele IA, Ihongbe AU (1995) Utilisation of wild animals in traditional medicine in the sub-humid region of Nigeria. The Conservator (FACS). University of Ibadan, Nigeria, pp 11–17

Banjo AD, Lawal OA, Owolana OA, Olubanjo OA, Ashidi JS, Dedeke GA, Soewu DA, Owa SO, Sobowale OA (2002) An Ethno-zoological survey of insect and their allies among the Remos (Ogun State). Afr J Indigenous Knowl Syst 2(1):61–67

Costa-Neto EM (1999) Traditional use and sale of animals as medicines in Feira de Santana City, Bahia, Brazil. Indigenous Knowl Dev Monit 7(2):6–9

Fayenuwo JO (1999) Status of rodent conservation and damage problems in Osun state, Nigeria, Ph.D thesis. Department of Wildlife and Fisheries Management University of Ibadan

Gamaniel KS, Fakeye T, Sofowora A (2005) Federal republic of Nigeria. In: Bodeker G, Ong CK, Grundy C, Burford G, Shein K (eds) WHO global atlas of traditional, complimentary and alternative medicine, 1st edn. World Health Organization Centre for Health Development, Kobe, Japan, p 30

Gaski LA, Johnson KA (1994) Prescription for extinction: endangered species and patented oriental medicines in trade, 1st edn. Traffic USA, Washington

IUCN (1997) Inter-commission task force on indigenous people indigenous peoples and sustainability: cases and actions, 1st edn. IUCN, Gland, Switzerland

IUCN (International Union for Conservation of Nature) (2010) IUCN Red List of Threatened Species. Version 2010.4. http://www.iucnredlist.org

IUCN (2010) Red list of threatened species. The World Conservation Union. http://www.iucn.org/redlist

Jaroli DP, Mahawar MM, Vyas N (2010) An ethnozoological study in the adjoining areas of Mount Abu wildlife sanctuary India. J Ethnobiol Ethnomed 6(6):1–8
Lambo JO (1979) The healing powers of herbs, with special reference to obstetrics and gyneacology in African Medicinal Plants (ed Sofowora, EA). University of Ife Press, Ife, Nigeria, pp 22–31
Lambo TA (1979) Communication/World Health Organization meeting on research and training in traditional systems of medicine in developing countries, with particular reference to medicinal plants and herbs. Report, World Health Organization
Mahawar MM, Jaroli DP (2007) Traditional knowledge on zootherapeutic uses by the Sahari tribe of Rajasthan, India. J Ethnobiol Ethnomed 3(25):1–6. doi:10.1186/1746-4269-3-25
Marshall NT (1998) Searching for a cure: conservation of medicinal wildlife resources in East and Southern Africa, 1st edn. Traffic International, Cambridge
Odu MO (1987) The art of traditional healing in Nigeria. National Concord, p 5
Olopade EO (2002) The herbs for good health: the 50th anniversary lecture of the University of Ibadan, 1st edn. NARL Specialist Clinic, Ibadan, Nigeria
Sawandi T (2006) Yorubic medicine: the art of divine herbology. Online articles. http://www.planetherbs.com. Accessed July 2011
Simmonds M (1998) Changing conservation aims: who will represent wildlife? Oryx 32(2):82
Sodeinde OA, Soewu DA (1999) Pilot study of the traditional medicine trade in Nigeria. Traffic Bull 18(1):35–40
Soewu DA (2006) Utilisation of wild animals in traditional medicine in Ogun state, Nigeria, Ph.D thesis, Department of Wildlife and Fisheries Management, University of Ibadan, Ibadan, Nigeria
Soewu DA (2008) Wild animals in ethnozoological practices among the Yorubas of soutwestern Nigeria and the implications for biodiversity conservation. Afr J Agric Res 3(6):421–427
Soewu DA, Adekanola TA (2011) Traditional-medical knowledge and perception of pangolins (*Manis sps*) among the Awori people, Southwestern Nigeria. J Ethnobiol Ethnomed 7(25):1–11
Soewu DA, Ayodele IA (2009) Utilisation of Pangolin (*Manis sps*) in traditional Yorubic medicine in Ijebu pronvince, Ogun State, Nigeria. J Ethnobiol Ethnomed 5(39):1–11
Sofowora A (1993) Medicinal plants and traditional medicine in Africa, 1st edn. Spectrum Books, Ibadan, Nigeria
Taylor ME, Fox J (1992) The fetish market, Lome, Togo. Nigerian Field 57:119–125
World Health Organisation (WHO) (1978) The Promotion and development of Traditional Medicine. Technical Report series 622
World Health Organisation (WHO) (2000) Traditional Medicine. http://www.wpro.who.int/health_topics/traditional_medicine. Accessed July 2011
World Health Organisation (2003) Traditional Medicine. http://www.worldhealthorganisation.com. Accessed July 2011
World Health Organisation (WHO) (2005) Ecosystems and human well-being: health synthesis: a report of the Millennium Ecosystem Assessment. 1 edn. Geneva, Switzerland

# Chapter 17
# Wild Mammals Trade for Zootherapeutic and Mythic Purposes in Benin (West Africa): Capitalizing Species Involved, Provision Sources, and Implications for Conservation

**Chabi A. M. S. Djagoun, Hugues A. Akpona, Guy. A. Mensah, Clive Nuttman and Brice Sinsin**

**Abstract** A common problem affecting many animal species is the soaring demand for their body parts for use in medicinal products. In Benin, in spite of intense commercial exploitation of wildlife for medicinal purposes, no official statistics on the use of animals for medicinal and magic/religious purposes are available and consequently, there is little consideration of the issue in laws, decision-making processes, and conservation strategies. The aim of this study was to list the mammal species sold on the medicinal market and the conservation implications of the use of mammal species in traditional folk medicines. Among the 87 mammal species traded on the traditional medicine market in Benin, 46 were sold by at least half of those traders surveyed; the conservation status of these animals included rare, vulnerable, and threatened species. Moreover, it was noticed that the source of animals is not limited to Benin since some species available at markets are not listed in the Benin's fauna. This study also found that rarer species were more costly and this constitutes an economic motivation for sellers to develop strategies for the availability of threatened species on their displays. Urgent conservation actions are needed to reduce the pressure that this activity sector might contribute to biodiversity loss.

C. A. M. S. Djagoun (✉) · H. A. Akpona · Guy. A. Mensah · B. Sinsin
Laboratoire d'Ecologie Appliquée, Faculté des Sciences Agronomiques,
Université d'Abomey-Calavi, 01 BP 526 Cotonou, Benin
e-mail: sylvestrechabi@gmail.com

H. A. Akpona
Ministry of Environment and Protection of Nature,
Direction Générale des Forêts et des Ressources Naturelles,
BP 393 Cotonou, Benin

C. Nuttman
Department of Zoology, Tropical Biology Association,
Downing Street, Cambridge, CB2 3EJ, UK

R. R. N. Alves and I. L. Rosa (eds.), *Animals in Traditional Folk Medicine*,
DOI: 10.1007/978-3-642-29026-8_17, 

**Keywords** Mammal · Traditional medicine market · Trade · Conservation

## 17.1 Introduction

Most Africans believe that there are some magical powers which are attached to special healing acts when wild animals' by-products are used as directed by a traditional healer (Adeola 1992; Adjakpa and Ogouvide 1998). Animals are 'therapeutic arsenals' that have been playing significant roles in the healing processes, magic rituals, and religious practices of peoples from the five continents; human societies that have a developed medical system will utilize animals as medicines (Costa-Neto and Marques 2000). The trade of animal-based medicine is becoming much more common in the majority of markets, thus increased understanding of the use of animals' body parts as folk medicines is relevant because such use exerts additional pressure on wild populations (Lee 1999; Alves and Pereira-Filho 2007; Alves et al. 2009).

In Benin, the exploitation of wildlife as zootherapeutic resources is one of the economic diversification strategies developed by local populations. This exploitation is underpinned by the reduced capacity of populations to access modern medicine and to the socioeconomic and cultural importance of this activity. The trade of animal-based medicine is found in the majority of Benin markets but little is known about the markets of animal-based medicines. Similarly, knowledge is limited about the impact of this activity on wildlife declines and the sustainable use of zootherapeutic resources. This lack of attention to market traders is surprising given their preponderance and economic importance. Since people have been using animals for a long time, suppression of exploitation is unlikely to be a viable strategy to stave off the threat of extinction for some species. As noted by Kunin and Lawton (1996), those species directly involved in traditional medicines should be among the highest priorities for conservation.

Some authors have investigated the medicinal importance of Benin wildlife (Coubéou 1995; Adjakpa and Ogouvide 1998; Assogbadjo 2000; Akpona 2004; Djagoun 2005). A study conducted by Adjakpa and Ogouvide (1998) highlighted the diversity of birds used in animal-based medicine in Benin; however, there are no data available on the other wildlife such as mammals, which represent the most commonly exploited wildlife due to their importance for family income through hunting, artisan handicrafts, eco-tourism, and game appeal.

Within that context, the present work focused on an inventory of mammal species sold in the Benin traditional medicine market, evaluated the diversity, abundance, source of provision, and socio-economical context of the use of these animals for therapeutic purposes, and discusses the implications for sustainable biodiversity conservation.

## 17.2 Methods and Survey Design

### *17.2.1 Study Area*

The Republic of Benin is situated in West Africa between latitudes 6°10′N and 12°25′N and longitudes 0°45′E and 3°55′E, covering a land area of 112,622 km$^2$. It is bordered by the republics of Togo in the west, Nigeria in the east, Burkina Faso and Niger in the north, and the Atlantic Ocean to the south (Fig. 17.1). The population has been estimated at 6,752,569 inhabitants with an average density of 57 inhabitants per km$^2$ (INSAE-RGPH 2006). The mean annual rainfall varies from 900 to 1,300 mm while the mean annual temperatures range from 26 to 28°C and may exceptionally reach 35–40°C in northern localities. About 24% of the land is covered by forest, while only 22.7% of the total land area is legally protected (CENATEL 1992; FAO 2001); the vegetation in several of the protected sites has almost entirely vanished and has even been invaded by human settlements.

### *17.2.2 Data Collection*

Benin has more than 40 cities in which we were aware of at least one important market (and several sub-markets) that contained animal-based medicine activities.

*Computing existing data* All available references or reports on folk remedies based on animal parts in Benin and West African countries were examined. Only taxa that could be identified to species level were included in the database. The conservation status of the animal species follows the IUCN Red List (http://www.iucnredlist.org/); Benin's official list of endangered species (Neuenshwander et al. 2011) was also used.

*Survey of animal-based medicine market* Information on the use and commercialization of animals for medicinal purposes was collected through a semi-structured survey among selected animal-based medicine traders throughout the country, using a questionnaire. We surveyed 22 markets throughout the country with a total of 110 sellers (Table 17.1); 18 districts were selected for the study, from a total of 77 districts in the country (Fig. 17.1). The choice of animal-based medicine markets has a double advantage because the merchants concerned combine their profession of traditional healers with the trade of animals. Visits were also made to outdoor markets, temporary markets, and religious articles stores where products derived from wildlife are commonly sold. Markets were selected by taking into account their proximity to a protected area, their nature (local, national or international), and the ethnic groups. The animals recorded were identified by direct inspection or from photographs on the stand (Fig. 17.2) and others parameters, such as richness and abundance, frequency, provision sources were recorded through interviews with the sellers.

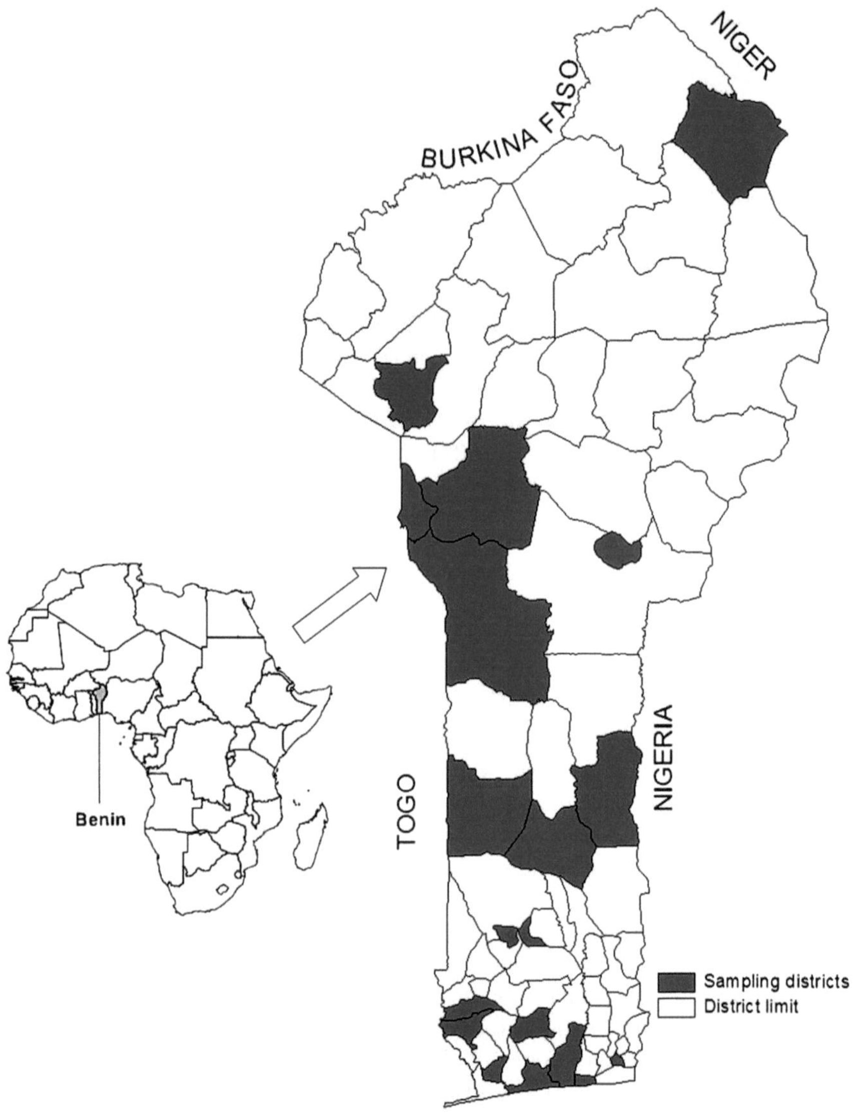


Fig. 17.1 Location of Benin and sampling districts

**Table 17.1** Number of markets surveyed with the number of interviewees per zone

| Zone | Number of markets | Sample size |
|---|---|---|
| South | 10 | 60 |
| Centre | 7 | 30 |
| North | 5 | 20 |
| Total | 22 | 110 |

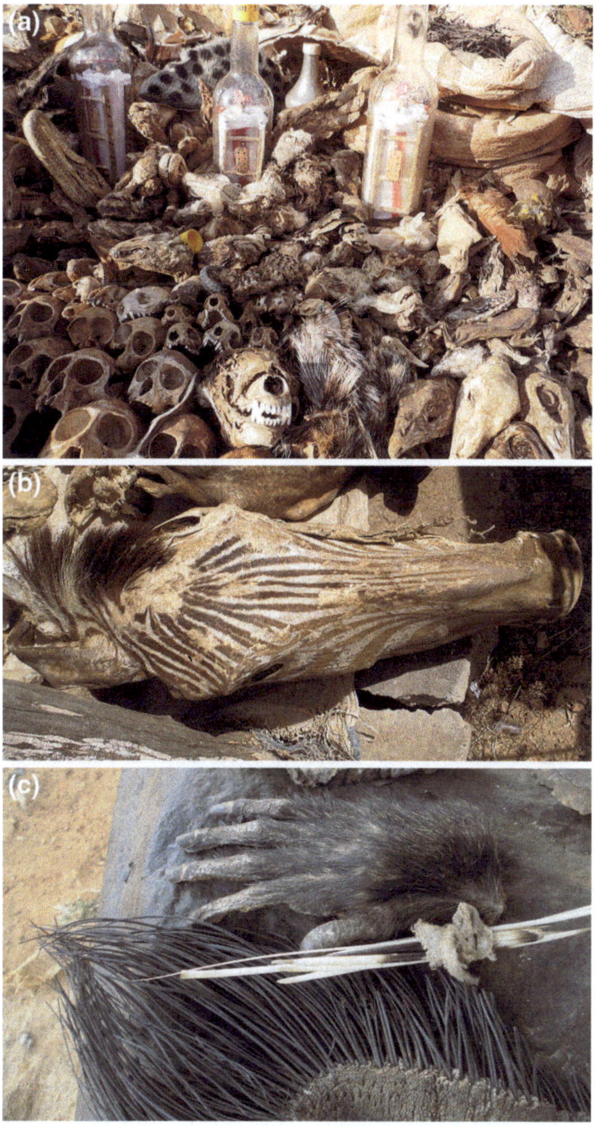


**Fig. 17.2** Pictures taken in the traditional medicine markets surveyed: **a** One of the displays of the Bohicon fetish market; **b** The head of a zebra (*Equus zebra* Linnaeus, 1758), a species not present in Benin, found in a local market; **c** The gorilla (*Gorilla gorilla* Savage and Wyman, 1847) hand found in the northern Benin market whereas the distribution of this species does not extend to Benin

**Table 17.2** Number of mammal species per order recorded in Benin traditional medicine market

| Mammal order | Number of species observed[a] | Number of possible species[b] |
|---|---|---|
| Carnivora | 20 | 28 |
| Chiroptera | 9 | 50 |
| Primates | 10 | 12 |
| Erinaceomorpha | 1 | 1 |
| Soricomorpha | 2 | 8 |
| Lagomorpha | 1 | 2 |
| Artiodactyla | 12 | 17 |
| Perissodactyla | 1 | 0 |
| Pholidota | 2 | 2 |
| Proboscidea | 1 | 1 |
| Rodentia | 25 | 56 |
| Tubulidentata | 1 | 1 |
| Hyracoidea | 2 | 2 |
| Sirenia | 0 | 1 |
| TOTAL | 87 | 180 |

[a] The number of species observed is the total number of species found at the medicinal markets

[b] The total possible species number refers to the number of species per order existing in Benin, according to the literature (Lamarque 2004; De Visser et al. 2001; Sinsin et al. 2008; Neuenshwander et al. 2011)

## 17.3 Results and Discussion

### *17.3.1 Mammals Traded at the Animal-Based Medicine Market in Benin*

Excluding domestic animals, we identified 87 species of mammal traded on the animal-based medicine markets in Benin, representing 13 mammal orders (Table 17.2). From the total of 87 wild mammal species recorded, the species from the Rodentia (28.7%) were the most common in the surveyed markets. This was followed by the Carnivora (23.0%), Artiodactyla (13.8%), Primates (11.5%), Chiroptera (10.3%) and Soricomorpha, Hyracoidea, and Pholidota (2.3% each). The orders Lagomorpha, Perissodactyla, Proboscidea, Tubulidentata were all represented by a single species. Given the known occurrence of individual species within Benin, all members of the orders Tubulidentata, Hyracoidea, Erinaceomorpha, and Pholidota were found on the traditional medicine market. No species of the order Perissodactyla was expected to be found at the market, as there are no representatives recorded in Benin. Nonetheless, we recorded one zebra species, indicating that some species are imported from outside the country by the traders. A list of all medicinal wild mammal species identified is given in Appendix A.

### *17.3.2 The Common Mammal Species Traded on the Traditional Medicine Market in Benin*

Table 17.3 lists the common species traded at the traditional medicine market in Benin. We defined the 'common' species as the species which were sold by more than five traders interviewed during the traditional medicine markets survey. Hence, the species sold by less than five traders from our total sample size (110 traders) were considered as 'occasional' and were not included in the list.

Overall, 46 mammal species were categorized as common; from these, 20 species were found to be sold by more that 50% of the traders interviewed. This group included ungulates (*Sylvicapra grimmia* (Linnaeus, 1758); *Syncerus caffer* (Sparrman, 1779); *Ourebia ourebi* (Zimmermann, 1783); *Kobus kob* (Erxleben, 1777); *Cephalophus silvicultor* (Afzelius, 1815)); rodents (*Xerus erythropus* (Desmarest, 1817); *Cricetomys gambianus* Waterhouse, 1840; *Cricetomys emini* Wroughton, 1910; *Atherurus africanus* Gray, 1842; *Arvicanthis niloticus* (Desmarest, 1822); *Thryonomys swinderianus* (Temminck, 1827)) and primates (*Papio Anubis* Lesson, 1827; *Cercopithecus mona* (Schreber, 1774); *Chlorocebus aethiops* (Linnaeus, 1758)). Other notable species such as *Equus zebra* Linnaeus, 1758; *Gorilla gorilla* (Savage and Wyman, 1847); *Lycaon pictus* (Temminck, 1820); *Pan troglodytes* (Blumenbach, 1775); *Orycteropus afer* (Pallas, 1766) were sold by less than 10% of the total traders interviewed. Generally, the abundance of species sold in the traditional medicine market followed the same trend of species abundance in their natural habitat; i.e., species listed as abundant in the traditional medicine market were also commonly recorded in the literature as abundant in their habitat (e.g., Mensah et al. 2006; Sinsin et al. 2008; Djagoun and Gaubert 2009). On the other hand, some species considered as highly threatened at national or international level were found in abundance at the market [see: *Atherurus africanus* Gray, 1842; *Loxodonta africana* (Blumenbach, 1797)]. Furthermore, several of the less represented mammal species at the traditional medicine market are also of conservation concern. In fact, most of the recorded species (4 out of 6) are on the IUCN Red List of Threatened Species.

### *17.3.3 Economical Importance of Rare Versus Common Traded Mammal Species*

Figure 17.3 shows the relationship between the percentages of traders recorded selling each mammal species (see Table 17.3) and the mean income generated by selling the head of a given species. It should be noted that species such as *Hippopotamus amphibius* Linnaeus, 1758 and *Loxodonta africana* (Blumenbach, 1797) were considered as outliers because the high mean income attributable to these species was due to the large size of their heads. Also, exotic species such as *Equus zebra* Linnaeus, 1758, *Pan troglodytes* (Blumenbach, 1775), and *Gorilla gorilla*

**Table 17.3** Percentage of traders ($N = 110$) recorded selling mammal species in Benin traditional medicine market, with indication of the conservation status of the traded species

| Scientific names | Frequency of traders selling the species (%) | IUCN Red list[a] | Benin Red list[b] |
|---|---|---|---|
| *Alcelaphus buselaphus* (Pallas, 1766) | 30.9 | LC | VU |
| *Acinonyx jubatus* (Schreber, 1775) | 8.2 | VU | EN |
| *Atelerix albiventris* (Wagner, 1841) | 80.9 | LC | Not listed |
| *Atherurus africanus* Gray, 1842 | 69.1 | NT | NT |
| *Atilax paludinosus paludinosus* (G.[Baron] Cuvier, 1829) | 56.4 | LC | VU |
| *Arvicanthis niloticus* (Desmarest, 1822) | 85.5 | LC | Not listed |
| *Cephalophus silvicultor* (Afzelius, 1815) | 74.5 | LC | DD |
| *Cercopithecus mona* (Schreber, 1774) | 67.3 | LC | VU |
| *Chlorocebus aethiops* (Linnaeus, 1758) | 54.5 | LC | LC |
| *Civettictis civetta* (Schreber, 1776) | 31.8 | LC | VU |
| *Cricetomys gambianus* Waterhouse, 1840 | 94.5 | LC | Not listed |
| *Cricetomys emini* Wroughton, 1910 | 72.7 | LC | Not listed |
| *Crocuta crocuta* (Erxleben, 1777) | 17.3 | LC | NT |
| *Crossarchus obscurus* F. G. Cuvier, 1825 | 78.2 | LC | LC |
| *Dendrohyrax arboreus* (A. Smith, 1827) | 23.6 | DD | EN |
| *Equus zebra* Linnaeus, 1758 | 3.6 | VU | Not listed |
| *Erythrocebus patas* (Schreber, 1775) | 38.2 | LC | LC |
| *Felis silvestris* Schreber, 1777 | 20.9 | LC | VU |
| *Genetta genetta* (Linnaeus, 1758) | 65.5 | LC | LC |
| *Gorilla gorilla* (Savage and Wyman, 1847) | 10.0 | EN | Not listed |
| *Herpestes ichneumon* (Linnaeus, 1758) | 35.5 | LC | LC |
| *Hippopotamus amphibius* Linnaeus, 1758 | 35.5 | VU | VU |
| *Hippotragus equinus* (Desmarest, 1804) | 37.3 | LC | VU |
| *Kobus kob* (Erxleben, 1777) | 52.7 | LC | NT |
| *Loxodonta africana* (Blumenbach, 1797) | 30.6 | VU | VU |
| *Lutra maculicollis* Lichtenstein, 1835 | 12.7 | LC | VU |
| *Lycaon pictus* (Temminck, 1820) | 7.3 | EN | CR |
| *Manis gigantea* Illiger, 1815 | 14.5 | VU | CR |
| *Manis tricuspis* Rafinesque, 1821 | 26.4 | LC | VU |
| *Orycteropus afer* (Pallas, 1766) | 4.5 | LC | EN |
| *Ourebia ourebi* (Zimmermann, 1783) | 68.2 | LC | VU |
| *Pan troglodytes* (Blumenbach, 1775) | 7.3 | EN | Not listed |
| *Panthera leo* (Linnaeus, 1758) | 13.6 | VU | VU |
| *Panthera pardus* (Linnaeus, 1758) | 11.8 | NT | VU |
| *Papio Anubis* Lesson, 1827 | 80.9 | LC | LC |
| *Perodicticus potto* (Müller, 1766) | 44.5 | LC | LC |
| *Phacochoerus aethiopicus* (Pallas, 1766) | 67.3 | LC | NT |
| *Potamochoerus porcus* (Linnaeus, 1758) | 47.3 | LC | VU |
| *Rousettus aegyptiacus* (E. Geoffroy, 1810) | 49.1 | LC | VU |
| *Sylvicapra grimmia* (Linnaeus, 1758) | 99.1 | LC | LC |
| *Syncerus caffer* (Sparrman, 1779) | 82.7 | LC | NT |
| *Thryonomys swinderianus* (Temminck, 1827) | 77.3 | LC | LC |
| *Tragelaphus scriptus* (Pallas, 1766) | 37.3 | LC | NT |

(continued)

**Table 17.3** (continued)

| Scientific names | Frequency of traders selling the species (%) | IUCN Red list[a] | Benin Red list[b] |
|---|---|---|---|
| *Tragelaphus spekii* P. L. Sclater, 1863 | 25.5 | LC | EN |
| *Vulpes pallida* (Cretzschmar, 1826) | 19.1 | DD | VU |
| *Xerus erythropus* (Desmarest, 1817) | 91.8 | LC | Not listed |

*EN* Endangered, *VU* Vulnerable, *NT* Near threatened, *DD* Data deficient, *LC* Least concern
[a] http://www.iucnredlist.org/
[b] See: Neuenshwander et al. (2011)

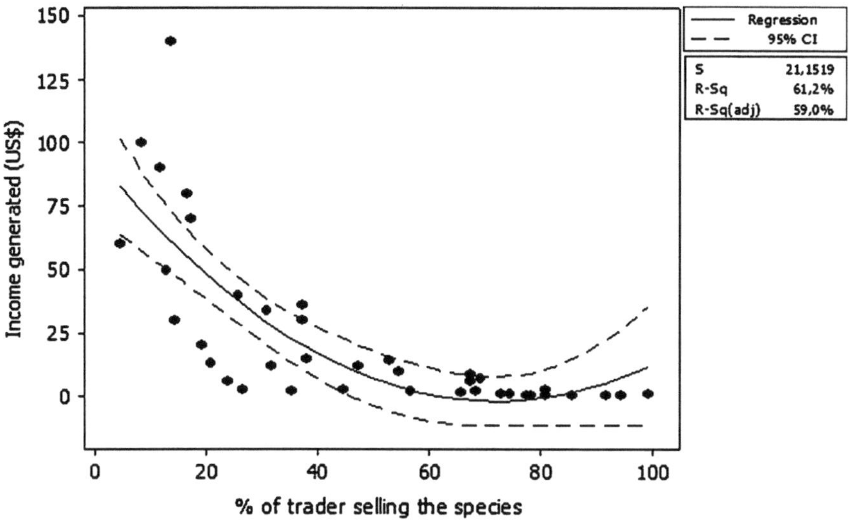


**Fig. 17.3** Relationship between incomes generated and the percentage of the traders selling the given species on the market

(Savage and Wyman, 1847) were excluded from the analysis. The relationship between the percentage of traders recorded selling a mammal species and the mean income generated in selling the head of the given species fitted well with a quadratic model and was statistically significant ($p = 0.001$) with the explained variance of 61.2%. From this model we concluded that the rare species traded at the traditional medicine market attained higher economic value than the common species.

### *17.3.4 Source of Provision of Mammal Species Traded at the Traditional Medicine Market in Benin*

According to our survey, most of the animals sold at the Benin markets come from National parks or Classified forest; we found that 68% of the mammal species sold

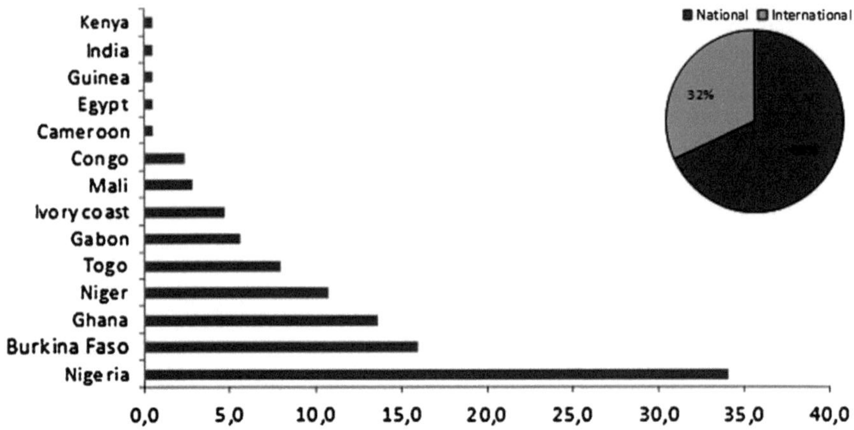


**Fig. 17.4** The source of the animals provisioning the traditional medicine market according to the traders

in the traditional medicine markets were provided locally by the hunters, while 32% came from elsewhere. Some specimens were also imported by the traders, the main source of provision being Nigeria (34.1%), followed by Burkina Faso (15.9%), Ghana (13.6%), Niger (10.7%), Togo (7.9%), Gabon (5.6%), Ivory Coast (4.7%), Mali (2.8%), and Congo (2.3%). A smaller percentage (0.5%) of specimens came from Cameroon, Egypt, Guinea, India, and Kenya (Fig. 17.4). Nigeria was quoted by most respondents (65%) as the main country supplying the traditional medicine market in Benin, and this may be related to the geographic proximity of the two countries. Some of the supplier countries are known to harbour exotic species traded in Benin, such as *Equus zebra* Linnaeus, 1758 (Fig. 17.2b), *Pan troglodytes* (Blumenbach, 1775) and *Gorilla gorilla* (Savage and Wyman, 1847) (Fig. 17.2).

### *17.3.5 Conservation Implications of Mammal Species Traded in the Benin Traditional Medicine Market*

The unsustainable use of medicinal animals has been highlighted as a potential threat to many species populations (Lee et al. 1998). Although human activities such as slash and burn agriculture, goat and cattle raising, and extensive subsistence hunting are thought to be causing severe environmental impoverishment and a loss of biodiversity in Benin (Djagoun and Gaubert 2009), the medicinal use of animals creates an additional threat and must be considered in conjunction with other anthropogenic pressures. In Benin, we found that the rarest species are more costly than the commonest species, and this may be an important factor in declines of wild populations of certain species. Additionally, the supply of the rarest mammal species is inferior to the existing demand. Thus, the high price of the rare species and the continued demand for this species makes hunting pressure greater, as traders will prefer to commercialize on their stand high value, rare species. The economic value of zootherapy, as expressed by those who trade animal-based medicines, should be taken into account whenever policies and environmental measures are designed against the trade (Costa-Neo 2005). Instead of sending the practitioners of the zootherapy to prisons, or creating policies aiming to force traders to abandon such a practice, decision makers should attempt to contextualise this form of human/nature connection within its cultural dimension.

Traditional medicine based on animals and their products is of high importance to urban livelihood in Benin; in particular, the traditional medicine markets are more developed in the southern Benin, the sparsely populated part of the country. All the main cities in southern Benin have a traditional medicine market with up to 15 traders per market. Conversely, the local wildlife habitats are very patchily distributed into small forested islands and have been continuously logged for agricultural development, while sizable forest habitats can now only be found in northern Benin, where few of the traditional medicine markets exist, with an estimated number of three traders per market. Nonetheless, the paucity of the traditional medicine market and the higher number of remaining natural areas in northern Benin do not guarantee wildlife conservation in that area because the majority of the large mammals traded in the southern markets are collected in the north.

The dilemma facing all fauna species is the soaring demand for their body parts for use in medicinal products (Soewu 2008). From a biological perspective, there is a need to increase our understanding of the biology and ecology of species commonly used as remedies to better assess the impacts of harvesting them (for medicinal or other purposes) on their wild populations (Alves et al. 2007). Moreover, it is important to promote research which can integrate all factors (including traditional medicine) affecting the species listed in this study, in order to develop a model to assess the sustainability of the current exploitation strategies, and to propose feasible conservation measures. Medicinal species that are threatened should receive urgent attention, and efforts to tackle their habitat loss or alteration could be further supported by highlighting their present and future medicinal uses.

**Acknowledgments** This work was fully supported by Rufford Small Grants Foundation (RSG REFERENCE NUMBER: 8291–1). We are grateful to the traders who participated in this research. Our acknowledgments also go to the colleagues Tewogbade Jean-Didier Akpona, Chrystelle Dakpogan, and Gboja Mariano Houngbedji for their assistance during the field data collection phase.

## 17.4 Appendix A: List of the medicinal wild mammal species inventoried on the markets

**Carnivora (20 species)**

- *Atilax paludinosus* (G.[Baron] Cuvier, 1829)
- *Ichneumia albicauda* (G.[Baron] Cuvier, 1829)
- *Herpestes ichneumon* (Linnaeus, 1758)
- *Galerella sanguinea* (Rüppell, 1835)
- *Crossarchus obscurus* (F. G. Cuvier, 1825)
- *Lutra maculicollis* (Lichtenstein, 1835)
- *Mellivora capensis* (Schreber, 1776)
- *Ictonyx striatus* (Perry, 1810)
- *Genetta genetta* (Linnaeus, 1758)
- *Genetta pardina* (I. Geoffroy Saint-Hilaire, 1832)
- *Civettictis civetta* (Schreber, 1776)
- *Nandinia binotata* (Gray, 1830)
- *Caracal caracal* (Schreber, 1776)
- *Felis silvestris* (Schreber, 1777)
- *Canis adustus* (Sundevall, 1847)
- *Panthera leo* (Linnaeus, 1758)
- *Panthera pardus* (Linnaeus, 1758)
- *Acinonyx jubatus* (Schreber, 1775)
- *Crocuta crocuta* (Erxleben, 1777)
- *Lycaon pictus* (Temminck, 1820)

**Chiroptera (9 species)**

- *Epomophorus gambianus* (Ogilby, 1835)
- *Epomops franqueti* (Tomes, 1860)
- *Hypsignathus monstrosus* (H. Allen, 1861)
- *Megaloglossus woermanni* (Pagenstecher, 1885)
- *Eidolon helvum* (Kerr, 1792)
- *Hipposideros cyclops* (Temminck, 1853)
- *Micropteropus pusillus* (Peters, 1867)
- *Nanonycteris veldkampi* (Jentink, 1888)
- *Epomophorus gambianus* (Ogilby, 1835)

**Primates (10 species)**

- *Chlorocebus aethiops* (Linnaeus, 1758)
- *Cercopithecus mona* (Schreber, 1774)
- *Erythrocebus patas* (Schreber, 1775)
- *Papio Anubis* (Lesson, 1827)
- *Colobus vellerosus* (I. Geoffroy, 1834)
- *Procolobus verus* (Van Beneden, 1838)

- *Galago senegalensis* (É. Geoffroy Saint-Hilaire, 1796)
- *Gorilla gorilla* (Savage and Wyman, 1847)
- *Pan troglodytes* (Blumenbach, 1775)
- *Perodicticus potto* (Müller, 1766)

**Soricomorpha (2 species)**

- *Crocidura olivieri* (Lesson, 1827)
- *Crocidura poensis* (Fraser, 1843)

**Erinaceomorpha (1 species)**

- *Atelerix albiventris* (Wagner, 1841)

**Lagomorpha (1 species)**

- *Lepus capensis* (Linnaeus, 1758)

**Artiodactyla (12 species)**

- *Syncerus caffer* (Sparrman, 1779)
- *Hippotragus equinus* (Desmarest, 1804)
- *Kobus ellipsiprymnus defassa*
- *Kobus kob* (Erxleben, 1777)
- *Ourebia ourebi* (Zimmermann, 1783)
- *Alcelaphus buselaphus* (Pallas, 1766)
- *Redunca redunca* (Pallas, 1767)
- *Tragelaphus scriptus* (Pallas, 1766)
- *Cephalophus rifulatus;*
- *Sylvicapra grimmia* (Linnaeus, 1758)
- *Cephalophus silvicultor* (Afzelius, 1815)
- *Tragelaphus spekii* (P. L. Sclater, 1863)

**Perissodactyla (1 species)**

- *Equus zebra* (Linnaeus, 1758)

**Pholidota (2 species)**

- *Manis tricuspis* (Rafinesque, 1821)
- *Manis gigantea* (Illiger, 1815)

**Proboscidea (1 species)**

- *Loxodonta africana* (Blumenbach, 1797)

**Rodentia (25 species)**

- *Uranomys ruddi* (Dollman, 1909)
- *Tatera guineae* (Thomas, 1910)
- *Lemniscomys zebra* (Heuglin, 1864)
- *Funisciurus leucogenys* (Waterhouse, 1842)
- *Xerus erythropus* (Desmarest, 1817)

- *Anomalurus derbianus* (Gray, 1842)
- *Tatera guineae* (Thomas, 1910)
- *Steatomys jacksoni* (Hayman, 1936)
- *Protoxerus stangeri* (Waterhouse, 1842)
- *Myomys derooi* (Van der Straeten and Verheyen, 1978)
- *Mus haussa* (Thomas and Hinton, 1920)
- *Malacomys longipes* (Milne-Edwards, 1877)
- *Mastomys natalensis* (Smith, 1834)
- *Lophuromys sikapusi* (Temminck, 1853)
- *Lemniscomys striatus* (Linnaeus, 1758)
- *Hylomyscus alleni* (Waterhouse, 1838)
- *Graphiurus lorraineus* (Dollman, 1910)
- *Heliosciurus gambianus* (Ogilby, 1835)
- *Funisciurus substriatus* (de Winton, 1899)
- *Arvicanthis niloticus* (Desmarest, 1822)
- *Atherurus africanus* (Gray, 1842)
- *Cricetomys emini* (Wroughton, 1910)
- *Cricetomys gambianus* (Waterhouse, 1840)
- *Thryonomys swinderianus* (Temminck, 1827)
- *Acomys cineraceus* (Heuglin, 1877)

**Tubulidentata (1 species)**

- *Orycteropus afer* (Pallas, 1766)

**Hyracoidea (2 species)**

- *Procavia capensis kerstingi* (Matschie, 1899)
- *Dendrohyrax dorsalis sylvestris* (Temminck, 1855)

## References

Adeola MO (1992) Importance of wild animals and their parts in the culture, religious festivals, and traditional medicine, of Nigeria. Environ Conserv 19:125–134

Adjakpa JB, Ogouvide FT (1998) Contribution à l'étude économique et socioculturelle des oiseaux sauvages utilisés en pharmacopée béninois. CEROE, Cotonou

Akpona H (2004) Facteurs de conservation des loutres au Sud du Bénin : cas de forêt classée de la Lama et des corridors avec les zones humides de la Vallée de l'Ouémé. Thèse d'ingénieur agronome. FSA/UAC. Abomey-Calavi, Benin

Alves RRN, Pereira-Filho GA (2007) Commercialization and use of snakes on North and Northeastern Brazil: implications for conservation and management. Biodiver Conserv 16: 969–985

Alves RRN, Rosa IL, Santana GG (2007) The role of animal-derived remedies as complementary medicine in Brazil. BioScience, 57(11):949–955

Alves RRN, Leo Neto NA, Brooks SE, Albuquerque UP (2009) Commercialization of animal-derived remedies as complementary medicine in the semi-arid region of northeastern Brazil. J Ethnopharmacol 124:600–608

Assogbadjo AE (2000) Etude de la biodiversité des ressources forestières alimentaires et évaluation de leur contribution à l'alimentation des populations locales de la forêt classée de la Lama. Thèse d'ingénieur agronome. FSA/UNB. Abomey-Calavi, Benin

CENATEL (1992) Carte des aires protégées du Bénin extraite de la carte générale du Bénin au 1/600.000, IGN 1992, Cotonou, Bénin

Costa-Neto EM (2005) Animal-based medicines: biological prospection and the sustainable use of zootherapeutic resources. An Acad Bras Cienc 77(1):33–43

Costa-Neto EM, Marques JGW (2000) Faunistic resources used as medicines by artisanal fishermen from Siribinha Beach, State of Bahia, Brazil. J Ethnobiol 20:93–109

Coubéou PT (1995) Diversité faunistique des différents biotopes de la forêt classée de la Lama. Thèse d'ingénieur agronome. FSA/UNB. Abomey-Calavi, Bénin

De Visser J, Mensah GA, Codjia JTC, Bokonon-Ganta AH (2001) Guide préliminaire de reconnaissance des rongeurs du Bénin. C.B.D.D./Ecooperation/ReRE/VZZ - République du Bénin/Royaume des Pays-Bas

Djagoun CAMS (2005) Abondance et répartition des espèces de mangouste (Crossarchus obscurus Cuvier, 1825...) dans la forêt de Niaouli. Mémoire d'obtention du DIT. APE/EPAC/UAC, Benin

Djagoun CAMS, Gaubert P (2009) Small carnivorans from southern Benin: a preliminary assessment of diversity and hunting pressure. Small Carnivore Conserv 40:1–10

FAO (2001) La situation des forêts et de la faune sauvage en Afrique. Commission régionale de la FAO pour l'Afrique, Rome

INSAE-RGPH (2006) Recensement général de la population et de l'habitat. Résultats provisoires. Cotonou, Bénin

Kunin WE, Lawton JH (1996) Does biodiversity matter? Evaluating the case for conserving species. In: Gaston KJ (ed) Biodiversity: a biology of numbers and differences. Blackwell Science, Oxford, pp 283–308

Lamarque F (2004) Les grands mammifères du Complexe WAP. Union Européenne/CIRAD/ECOPAS

Lee S, Hoover C, Gaski A, Mills J (1998) A world apart? Attitudes toward traditional Chinese medicine and endangered species in Hong Kong and the United States. TRAFFIC EastAsia, TRAFFIC North America, and World Wildlife Fund—US, Washington, DC

Lee S (1999) Trade in traditional medicine using endangered species: an international context. Proceedings of the second australian symposium on traditional medicine and wildlife conservation, Melbourne, Australia

Mensah A, Pomalegni B, Anagonou G, Anani C, Gnanhoui David S (2006) Inventaire des mammifères rongeurs et des reptiles dans la Réserve de Biosphère de la Pendjari. Rapport technique final. GTZ, Bénin

Neuenshwander P, Sinsin B, Goergen G (2011) Protection de la Nature en Afrique de l'Ouest: Une Liste Rouge pour le Bénin. Nature conservation in West Africa: red list for benin. International institute of tropical agriculture, Ibadan

Sinsin B, Sogbohossou EA, Nobime G, Mama A (2008) Dénombrement aérien de la faune dans la Réserve de Biosphère de la Pendjari: Rapport technique. CENAGREF/Projet Pendjari—CTZ-GFA Consulting, Benin

Soewu DA (2008) Wild animals in ethnozoological practices among the Yorubas of southwestern Nigeria and the implications for biodiversity conservation. Afr J Agric Res 3:421–427

# Chapter 18
# Birds of a Feather: Quantitative Assessments of the Diversity and Levels of Threat to Birds Used in African Traditional Medicine

**Vivienne L. Williams, Anthony B. Cunningham, Robin K. Bruyns and Alan C. Kemp**

**Abstract** This chapter reviews the richness of bird use for traditional medicines across Africa. At least 354 species from 205 genera, 70 families, and 25 orders are used for traditional medicine in 25 African countries. Most birds are in the order Passeriformes (108 species used and 82 species traded), with the starlings (Sturnidae) the most commonly used family of passerines (nine species). Of all the bird families in trade, the Accipitridae had the most number of recorded genera (26 genera; 37 species; including kites, hawks, eagles, vultures), followed by the Ardeidae (11 genera; 15 species; including herons and egrets). The Bucerotidae (hornbills), Cuculidae (cuckoos) and Strigidae (owls) were the next most specious families in trade. The Ostrich was the most widely used bird species (11 African countries), although it was only recorded as sold in markets of four countries. Barn owls were the most widely sold. Using a widely accepted method for grouping species according to commonness or rarity, we show that 24% of traded bird species are very common and locally abundant in several habitats over a large geographic area. Ten percent of traded species are, however, rare, occurring in low numbers in specific habitats over a small geographic area. The order with the highest proportion

V. L. Williams (✉) · R. K. Bruyns
School of Animal, Plant and Environmental Sciences,
University of the Witwatersrand, Private Bag 3, Wits,
Johannesburg 2050, South Africa
e-mail: vivwill@netdial.co.za

A. B. Cunningham
School of Plant Biology, University of Western Australia,
Nedlands 6009, Australia
e-mail: tonyc05@bigpond.net.au

A. C. Kemp
Research Associate, FitzPatrick Institute of African Ornithology,
University of Cape Town, Rondebosch, 7700 South Africa
e-mail: leadbeateri@gmail.com

R. R. N. Alves and I. L. Rosa (eds.), *Animals in Traditional Folk Medicine*,
DOI: 10.1007/978-3-642-29026-8_18, © Springer-Verlag Berlin Heidelberg 2013

of rare species was the Bucerotiformes (hornbills). Excluding non-breeding Palearctic (PAL) migrants, based on the 2011 IUCN Red List, 12 traded bird species are Threatened (three Endangered; nine Vulnerable) and eight are Near Threatened.

## 18.1 Introduction

At a global scale, at least 45.7% of extant bird species (4,561 species) are used by people, mainly as pets (37.0%) or are hunted for food (14.2%), with fewer bird species used in sport hunting, ornamentation, or traditional medicine (Butchart 2008). In contrast to several quantitative studies on the bird trade for pets both internationally (Thomsen et al. 1992) or nationally (Jepson and Ladle 2005), relatively few studies have been done documenting the trade in birds for traditional medicine in southern Africa (Whiting et al. 2011), Asia (Mahawar and Jaroli 2008), or Latin America (Alves and Rosa 2010). Although Mahawar and Jaroli's (2008) review for India covers a large part of South Asia, this is a first attempt at a continent-wide review of the richness of bird use for traditional medicine. Our inventory of birds used and traded for traditional medicine in Africa was compiled from published accounts (for example by Adjakpa et al. 2002; Marshall 1998; Nikolaus 2000; Simelane and Kerley 1998) supplemented by our own research (Cunningham and Zondi 1991; Whiting et al. 2011) and personal observations in various markets across Africa. We had two overall aims in this study. First, to determine the prevalence and diversity of bird species in African traditional medicine markets and second, to assess the levels of vulnerability posed by the trade in traditional medicine on bird species.

## 18.2 Methods

### *18.2.1 Data Sources*

Our inventory of birds used and traded for traditional medicine in Africa (Fig. 18.1) was compiled from published accounts, personal observations, and photographs taken in traditional medicine markets over the past 22 years (1989–2011) (Fig. 18.2). The published accounts include: Adeola (1992); Adjakpa et al. (2002); Baskind and Birbeck (2005); Beninguisse and De Brouwere (2004); BirdLife International (2011); Buckley (1985); Cocker (2000); Cocker and Mikkola (2001); Crump (2003); Cunningham (1990); Derwent and Mander (1997); El-Kamali (2000); Kizungu et al.(1998); Lacaille (1951); Loeb et al. (1956); Macdonald (1985); Mander et al. (2007); Marshall (1998); Nevadomsky (1988); Ngwenya (2001); Ntiamoa-Baidu (1987a, b); Ogada (2007); Serge (2010); Serle and Morel (1979); Simelane and Kerley (1998); Sodeinde and Soewu (1999); Soewu (2008); Taylor and

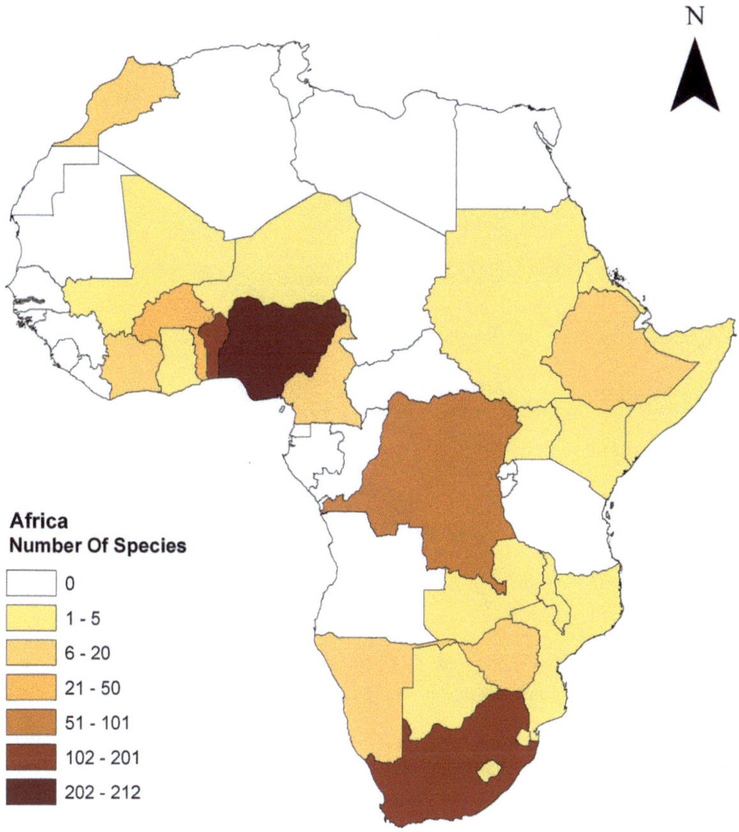


**Fig. 18.1** The richness of bird species recorded used for traditional medicine in African countries

Fox (1992); White et al. (2004); Whiting et al. (2011). The most comprehensive published information available was for birds sold in traditional medicine markets in Benin (BJ), Nigeria (NG), and South Africa (ZA) (e.g. Adjakpa et al. 2002; Nikolaus 2000; Whiting et al. 2011). In addition, market data for BJ (Dantokpa market), Burkina Faso (BF) (Ouagadougou), Côte D'Ivoire (CI) (Abidjan and Bouake), ZA (Johannesburg), Togo (TG) (Lome market), and Zimbabwe (ZW) (Bulawayo) were supplemented with identifications made from photographs taken during fieldwork (Fig. 18.3). However, the availability of information for other African countries was patchy. Hence, we do not consider our inventory to be conclusive and complete for Africa since there are countries for which we did not source information. Therefore, some of the results of the analyses must be viewed in light of the paucity and/or absence of published accounts for many countries. All references used to compile the species list are cited with the inventory in Table 18.1.

**Fig. 18.2** A single stall at the traditional medicine "fetish" market in Lome, Togo, West Africa showing the diversity of birds, mammals, and reptiles on sale (Photo credit: Peter Howard)

### *18.2.2 Ornithological Classification and Enumeration*

We followed the classification and nomenclature of BirdLife International (www.birdlife.org) because the taxonomic list is kept current and it is the basis for the avian IUCN Red List assessments. Furthermore, it was important that quantitative information obtained for the majority of species was of a consistent quality and derived primarily from the same source; hence, by using the BirdLife website, we could standardize the values of the variables used in the analyses. While the BirdLife ornithological classification differs slightly from other taxonomic lists available online, it was necessary to adhere to one system—however, this meant that five taxa that have been considered separate species in the ethno-avian literature were 'lumped' with other species for this paper, namely: (1) Burchell's Coucal (*Centropus burchellii*) with White-browed Coucal (*Centropus superciliosus*) (ZA, BJ); (2) Dark-capped Bulbul (*Pycnonotus tricolor*) with Common Bulbul (*Pycnonotus barbatus*) (BJ, NG, ZA); (3) Sahel Paradise-Whydah (*Vidua orientalis*) with Eastern Paradise-Whydah (*Vidua paradisaea*) (NG); (4) Cape White-eye (*Zosterops virens*) with Pale White-eye (*Zosterops pallidus*) (ZA); and (5) African Hoopoe (*Upupa africana*) with Eurasian Hoopoe (*Upupa epops*, although the BirdLife Taxonomic Working Group

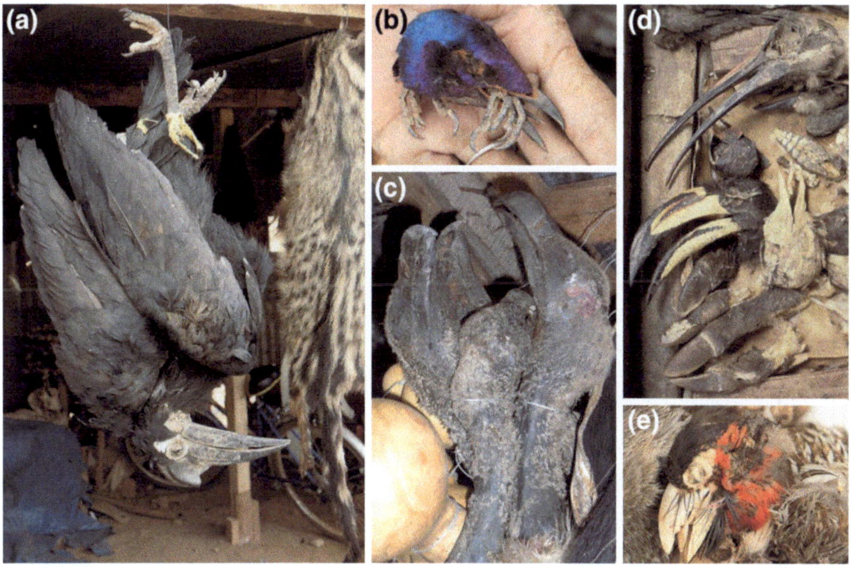

**Fig. 18.3** Africa has a high diversity of bird species in the traditional medicine trade, including: **a** Large-bodied, long-lived species such as the Abyssinian Ground-hornbill (Ouagadougou market, Burkina Faso (BF). **b** Starlings (Sturnidae), the most commonly used family amongst passerine birds (9 species), characteristically sold in BF with the feet inserted in the beak. **c** Vultures, another high conservation priority group, sold here in Xipamanine market, Mozambique. **d** Heads of hornbills next to aardvark claws (Abidjan, Cote d'Ivoire). **e** A Double-toothed Barbet head (Ouagadougou market, BF) (Photos: A. B. Cunningham)

is reviewing this treatment) (BF, MA, NG, SD, ZA). Setting species limits always has an element of subjectivity, so we have opted for a single taxonomic list rather than judge the merits of individual subspecific units 'lumped' and/or 'split' that were used in previous analyses but contradict our list.

It is important to note that we were conservative in our enumeration of the total number of avian taxa, and that we counted only those birds that could be identified to species level. If a bird could only be recognized to genus (e.g. Glossy-starling, *Lamprotornis* sp.), then it was not included in the total bird count nor included for analysis. Also excluded from most of the results (except where specified) are:

- Two exotic species recorded in the markets, namely Indian Peafowl (*Pavo cristatus*) and the Common Myna (*Acridotheres tristis*) since, while these species might be used now, they would not have formed part of traditional inventories;
- 15 species of Palearctic (PAL) non-breeding migrants in Africa (Common Cuckoo, Spotted Flycatcher, Montagu's Harrier, Pallid Harrier, Northern House-martin, Red-necked Nightjar, Northern Pintail, Kentish Plover, European Roller, Great Snipe, Jack Snipe, White Stork, Barn Swallow, Yellow Wagtail, Eurasian Wryneck). The PAL were omitted because much of the data used to assess threats to African species would not apply (even for the few species with a small

**Table 18.1** Species inventory checklist of avian species found to be used and/or traded for traditional medicine in 25 African countries

| Classification (**ORDER**, family) | Species | Common name | IUCN category[a] | Rarity classification | Population trend | Countries traded in (max $n = 7$) |
|---|---|---|---|---|---|---|
| **ANSERIFORMES** | | | | | | |
| Anatidae | *Anas acuta*[PAL] | Pintail, Northern | LC | B | D | NG |
| | *Anas undulata* | Duck, Yellow-billed | LC | B | S | NG |
| | *Dendrocygna bicolor* | Whistling-duck, Fulvous | LC | B | D | NG |
| | *Dendrocygna viduata* | Whistling-duck, White-faced | LC | B | I | BJ, NG, ZA |
| | *Plectropterus gambensis* | Goose, Spur-winged | LC | A | I | BJ, NG |
| | *Pteronetta hartlaubii* | Duck, Hartlaub's | LC | F | D | NG |
| | *Thalassornis leuconotus* | Duck, White-backed | LC | D | D | ZA |
| – | – | Duck, unidentified | – | – | – | TG, ZA |
| **APODIFORMES** | | | | | | |
| Apodidae | *Apus affinis* | Swift, Little | LC | A | I | NG |
| | *Apus caffer* | Swift, White-rumped | LC | B | I | NG |
| **BUCEROTIFORMES** | | | | | | |
| Bucerotidae | *Bycanistes bucinator* | Hornbill, Trumpeter | LC | H | S | ZA |
| | *Bycanistes cylindricus* | Hornbill, Brown-cheeked | NT[b] | H | D | BJ, NG |
| | *Bycanistes fistulator* | Hornbill, Piping | LC | F | S | CI, NG, ZA |
| | *Bycanistes subcylindricus* | Hornbill, Black-and-white-casqued | LC | F | ? | BJ, NG |
| | *Ceratogymna atrata* | Hornbill, Black-casqued | LC | F | D | BJ, NG |
| | *Ceratogymna elata* | Hornbill, Yellow-casqued | NT[c] | H | D | NG |
| | *Tockus camurus* | Hornbill, Red-billed Dwarf | LC | F | S | NG |
| | *Tockus erythrorhynchus* | Hornbill, Red-billed | LC | B | S | BF, BJ, NG, ZA |
| | *Tockus fasciatus* | Hornbill, African Pied | LC | F | ? | BF, BJ, CI, NG, TG |
| | *Tockus hartlaubi* | Hornbill, Black Dwarf | LC | F | S | NG |
| | *Tockus nasutus* | Hornbill, African Grey | LC | A | S | BF, BJ, NG |
| | *Tockus* sp. | Hornbill, unidentified | – | – | – | TG, ZA |
| | *Tropicranus albocristatus* | Hornbill, White-crested | LC | F | S | NG |
| Bucorvidae | *Bucorvus abyssinicus* | Ground-hornbill, Abyssinian | LC | H | S | BF, BJ, NG |

(continued)

**Table 18.1** (continued)

| Classification (**ORDER**, family) | Species | Common name | IUCN category[a] | Rarity classification | Population trend | Countries traded in (max $n = 7$) |
|---|---|---|---|---|---|---|
| | *Bucorvus cafer* | Ground-hornbill, Southern | VU[d] | D | D | ZA, ZW |
| Musophagidae | *Corythaeola cristata* | Turaco, Great Blue | LC | F | S | BJ, NG |
| | *Crinifer piscator* | Plantain-eater, Western Grey | LC | F | S | BF, BJ, NG, TG |
| | *Musophaga rossae* | Turaco, Ross's | LC | F | S | – |
| | *Musophaga violacea* | Turaco, Violet | LC | F | S | BJ, NG |
| | *Tauraco corythaix* | Turaco, Knysna | LC | H | S | ZA |
| | *Tauraco macrorhynchus* | Turaco, Yellow-billed | LC | F | S | NG |
| | *Tauraco persa* | Turaco, Guinea | LC | F | S | BJ, NG |
| | *Tauraco porphyreolophus* | Turaco, Purple-crested | LC | F | D | ZA |
| | – | Turaco, unidentified | – | – | – | – |
| **CAPRIMULGIFORMES** | | | | | | |
| Caprimulgidae | *Caprimulgus climacurus* | Nightjar, Long-tailed | LC | C | S | BJ, NG |
| | *Caprimulgus nigriscapularis* | Nightjar, Black-shouldered | LC | F | S | NG |
| | *Caprimulgus ruficollis*[PAL] | Nightjar, Red-necked | LC | D | D | – |
| | *Caprimulgus* sp. | Nightjar, unidentified | – | – | – | BJ, ZA |
| | *Macrodipteryx longipennis* | Nightjar, Standard-winged | LC | E | S | BJ, BG |
| | *Macrodipteryx vexillarius* | Nightjar, Pennant-winged | LC | G | S | – |
| **CHARADIIFORMES** | | | | | | |
| Burhinidae | *Burhinus capensis* | Thick-knee, Spotted | LC | A | S | NG, ZA |
| | *Burhinus senegalensis* | Thick-knee, Senegal | LC | C | ? | BF, NG |
| | *Burhinus* sp. | Thick-knee, unidentified | – | – | – | ZA |
| Charadriidae | *Charadrius alexandrinus*[PAL] | Plover, Kentish | LC | B | D | BJ |
| | *Vanellus albiceps* | Lapwing, White-headed | LC | B | S | BJ |
| | *Vanellus armatus* | Lapwing, Blacksmith | LC | F | I | ZA |
| | *Vanellus coronatus* | Lapwing, Crowned | LC | B | I | ZA |

(continued)

**Table 18.1** (continued)

| Classification (**ORDER**, family) | Species | Common name | IUCN category[a] | Rarity classification | Population trend | Countries traded in (max $n = 7$) |
|---|---|---|---|---|---|---|
| | *Vanellus crassirostris* | Lapwing, Long-toed | LC | F | ? | NG |
| | *Vanellus lugubris* | Lapwing, Senegal | LC | H | ? | BJ, NG, ZA |
| | *Vanellus senegallus* | Lapwing, Wattled | LC | B | S | NG |
| | *Vanellus tectus* | Lapwing, Black-headed | LC | F | ? | BF, NG |
| Glareolidae | *Glareola pratincola* | Pratincole, Collared | LC | B | D | BJ |
| | *Pluvianus aegyptius* | Plover, Egyptian | LC | D | D | BJ |
| | *Rhinoptilus chalcopterus* | Courser, Bronze-winged | LC | A | S | NG |
| Jacanidae | *Actophilornis africanus* | Jacana, African | LC | B | S | BJ |
| Laridae | *Larus cirrocephalus* | Gull, Grey-headed | LC | B | S | ZA |
| | *Larus* sp. | Gull, unidentified | – | – | – | ZA |
| | *Sterna caspia* | Tern, Caspian | LC | B | I | ZA |
| Recurvirostridae | *Recurvirostra avosetta* | Avocet, Pied | LC | B | S | ZA |
| Scolopacidae | *Gallinago media*[PAL] | Snipe, Great | NT[e] | C | D | BJ |
| | *Lymnocryptes minimus*[PAL] | Snipe, Jack | LC | C | S | BJ |
| **CICONIIFORMES** | | | | | | |
| Ardeidae | *Ardea cinerea* | Heron, Grey | LC | A | S | NG, TG, ZA |
| | *Ardea goliath* | Heron, Goliath | LC | D | S | BJ, CI |
| | *Ardea melanocephala* | Heron, Black-headed | LC | A | I | NG |
| | *Ardea purpurea* | Heron, Purple | LC | B | D | BJ |
| | *Ardea* sp. | Heron, unidentified | – | – | – | BF, BJ |
| | *Ardeola ralloides* | Heron, Squacco | LC | B | D | NG |
| | *Bubulcus ibis* | Egret, Cattle | LC | A | I | BF, BJ, NG, ZA |
| | *Butorides striata* | Heron, Striated | LC | A | D | BJ, NG |
| | *Casmerodius albus* | Egret, Great | LC | A | ? | ZA |
| | *Egretta garzetta* | Egret, Little | LC | A | I | BJ |
| | *Egretta gularis* | Reef-egret, Western | LC | H | S | TG |
| | *Egretta* sp. | Egret, unidentified | – | – | – | BF |
| | *Gorsachius leuconotus* | Night-heron, White-backed | LC | D | S | NG |

(continued)

**Table 18.1** (continued)

| Classification (**ORDER**, family) | Species | Common name | IUCN category[a] | Rarity classification | Population trend | Countries traded in (max $n = 7$) |
|---|---|---|---|---|---|---|
| | *Ixobrychus sturmii* | Bittern, Dwarf | LC | D | ? | NG |
| | *Mesophoyx intermedia* | Egret, Intermediate | LC | A | D | NG |
| | *Nycticorax nycticorax* | Night-heron, Black-crowned | LC | B | D | BJ, NG, TG, ZA |
| | *Tigriornis leucolopha* | Tiger-heron, White-crested | LC | H | ? | NG |
| | – | Egret, unidentified | – | – | – | BF, TG |
| | – | Night-Heron, unidentified | – | – | – | TG |
| Ciconiidae | *Anastomus lamelligerus* | Openbill, African | LC | B | D | NG |
| | *Ciconia abdimii* | Stork, Abdim's | LC | B | D | NG |
| | *Ciconia ciconia*[PAL] | Stork, White | LC | B | I | ZA |
| | *Leptoptilos crumeniferus* | Stork, Marabou | LC | A | I | BJ, NG |
| | *Mycteria ibis* | Stork, Yellow-billed | LC | B | D | BF |
| | – | Stork, unidentified | – | – | – | ZA |
| Scopidae | *Scopus umbretta* | Hamerkop | LC | B | D | BF |
| **COLIIFORMES** | | | | | | |
| Collidae | *Colius striatus* | Mousebird, Speckled | LC | A | ? | ZA |
| | – | Mousebird, unidentified | – | – | – | ZA |
| **COLUMBIIFORMES** | | | | | | |
| Columbidae | *Columba arquatrix* | Olive-pigeon, African | LC | B | D | – |
| | *Columba guinea* | Pigeon, Speckled | LC | A | S | NG, ZA |
| | *Columba livia* | Pigeon, Rock | LC | A | D | ZA |
| | *Columba uncinata* | Pigeon, Afep | LC | F | S | BJ |
| | *Columba* sp. | Pigeon, unidentified | – | – | – | BJ, BG |
| | *Stigmatopelia senegalensis* | Dove, Laughing | LC | A | S | ZA |
| | *Streptopelia capicola* | Dove, Ring-necked | LC | A | I | NG, ZA |
| | *Streptopelia semitorquata* | Dove, Red-eyed | LC | A | I | BJ, NG, ZA |
| | *Streptopelia* sp. | Dove, unidentified | – | – | – | ZW |
| | *Treron calvus* | Green-pigeon, African | LC | A | D | BJ |
| | *Treron waalia* | Green-pigeon, Bruce's | LC | E | D | NG |

(continued)

**Table 18.1** (continued)

| Classification (**ORDER**, family) | Species | Common name | IUCN category[a] | Rarity classification | Population trend | Countries traded in (max $n = 7$) |
|---|---|---|---|---|---|---|
| | *Turtur afer* | Wood-dove, Blue-spotted | LC | A | S | NG |
| | – | Dove, unidentified | – | – | – | TG, ZA, ZW |
| **CORACIIFORMES** | | | | | | |
| Alcedinidae | *Alcedo cristata* | Kingfisher, Malachite | LC | B | S | BJ, NG, ZA |
| | *Ceryle rudis* | Kingfisher, Pied | LC | B | ? | BJ, ZA |
| | *Ceyx pictus* | Pygmy-kingfisher, African | LC | A | S | BJ, NG |
| | *Halcyon albiventris* | Kingfisher, Brown-hooded | LC | E | S | ZA |
| | *Halcyon chelicuti* | Kingfisher, Striped | LC | A | S | NG |
| | *Halcyon leucocephala* | Kingfisher, Grey-headed | LC | A | S | BJ, NG |
| | *Halcyon malimbica* | Kingfisher, Blue-breasted | LC | E | D | NG |
| | *Halcyon senegalensis* | Kingfisher, Woodland | LC | A | S | BJ, CI, NG |
| | *Megaceryle maxima* | Kingfisher, Giant | LC | B | D | NG, ZA |
| | – | Kingfisher, unidentified | – | – | – | BF, ZA, ZW |
| Coraciidae | *Coracias abyssinicus* | Roller, Abyssinian | LC | E | I | BF, BJ, NG, TG |
| | *Coracias cyanogaster* | Roller, Blue-bellied | LC | H | D | BJ, NG |
| | *Coracias garrulus*[PAL] | Roller, European | NT[f] | A | D | BF, BJ, NG |
| | *Coracias gularis* | Roller, Blue-throated | LC | F | D | – |
| | *Coracias naevia* | Roller, Purple | LC | C | D | BF, NG |
| | *Eurystomis glaucurus* | Roller, Broad-billed | LC | A | S | BJ |
| | *Eurystomis* sp. | Roller, unidentified | – | – | – | BJ, NG |
| Meropidae | *Merops albicollis* | Bee-eater, White-throated | LC | B | S | BJ, NG |
| | *Merops malimbicus* | Bee-eater, Rosy | LC | F | ? | NG |
| | *Merops nubicus* | Bee-eater, Northern Carmine | LC | B | D | BJ, NG |
| | *Merops oreobates* | Bee-eater, Cinnamon-chested | LC | F | ? | – |
| | *Merops pusillus* | Bee-eater, Little | LC | A | D | NG |
| **CUCULIFORMES** | | | | | | |
| Cuculidae | *Centropus grillii* | Coucal, Black | LC | D | S | BJ, NG |

(continued)

**Table 18.1** (continued)

| Classification (**ORDER**, family) | Species | Common name | IUCN category[a] | Rarity classification | Population trend | Countries traded in (max $n = 7$) |
|---|---|---|---|---|---|---|
| | *Centropus leucogaster* | Coucal, Black-throated | LC | F | S | BJ, NG |
| | *Centropus monachus* | Coucal, Blue-headed | LC | E | S | – |
| | *Centropus senegalensis* | Coucal, Senegal | LC | C | S | BF, BJ, NG |
| | *Centropus superciliosus* | Coucal, White-browed | LC | B | S | BJ, ZA |
| | *Ceuthmochares aereus* | Yellowbill | LC | H | S | BJ, NG |
| | *Chrysococcyx caprius* | Cuckoo, Didric | LC | A | S | BJ, NG |
| | *Clamator glandarius* | Cuckoo, Great Spotted | LC | A | S | NG |
| | *Clamator jacobinus* | Cuckoo, Pied | LC | A | S | NG |
| | *Clamator levaillantii* | Cuckoo, Levaillant's | LC | A | S | BJ, NG |
| | *Cuculus canorus*[PAL] | Cuckoo, Common | LC | A | D | – |
| | *Cuculus clamosus* | Cuckoo, Black | LC | A | S | NG |
| | *Cuculus gularis* | Cuckoo, African | LC | A | S | BJ, NG |
| **FALCONIFORMES** | | | | | | |
| Acciptridae | *Accipiter badius* | Shikra | LC | A | S | BJ, NG |
| | *Accipiter erythropus* | Sparrowhawk, Red-thighed | LC | H | D | NG |
| | *Accipiter tachiro* | Goshawk, African | LC | B | D | NG |
| | *Aquila rapax* | Eagle, Tawny | LC | A | S | NG, ZA |
| | *Aquila wahlbergi* | Eagle, Wahlberg's | LC | A | S | NG |
| | *Aviceda cuculoides* | African Baza | LC | D | S | NG |
| | *Butastur rufipennis* | Buzzard, Grasshopper | LC | A | D | NG |
| | *Buteo auguralis* | Buzzard, Red-necked | LC | E | I | BJ, NG |
| | *Buteo rufofuscus* | Buzzard, Jackal | LC | E | S | ZA |
| | *Chelictinia riocourii* | Kite, African Swallow-tailed | LC | F | D | NG |
| | *Circaetus beaudouini* | Snake-eagle, Beaudouin's | VU[g] | H | D | NG |
| | *Circaetus cinerascens* | Snake-eagle, Banded | LC | G | D | NG |
| | *Circaetus cinereus* | Snake-eagle, Brown | LC | C | S | NG |
| | *Circus aeruginosus* | Marsh-harrier, Western | LC | B | I | BJ |
| | *Circus macrourus*[PAL] | Harrier, Pallid | NT[h] | D | D | BJ |
| | *Circus pygargus*[PAL] | Harrier, Montagu's | LC | C | D | BJ |

(continued)

**Table 18.1** (continued)

| Classification (**ORDER**, family) | Species | Common name | IUCN category[a] | Rarity classification | Population trend | Countries traded in (max $n = 7$) |
|---|---|---|---|---|---|---|
| | *Dryotriorchis spectabilis* | Serpent-eagle, Congo | LC | H | D | NG |
| | *Elanus caeruleus* | Kite, Black-winged | LC | A | S | BJ, NG, ZA |
| | *Gypaetus barbatus* | Lammergeier | LC | D | D | ZA |
| | *Gypohierax angolensis* | Vulture, Palm-nut | LC | B | S | NG |
| | *Gyps africanus* | Vulture, White-backed | NT[i] | C | D | BJ, NG, ZA |
| | *Gyps coprotheres* | Vulture, Cape | VU[j] | H | D | ZA |
| | *Gyps fulvus* | Vulture, Griffon | LC | C | I | – |
| | *Gyps rueppellii* | Vulture, Rueppell's | NT[k] | C | D | BJ, CI, NG |
| | *Haliaeetus vocifer* | Fish-eagle, African | LC | B | S | BJ, NG, ZA, ZW |
| | *Kaupifalco monogrammicus* | Buzzard, Lizard | LC | A | S | BJ, NG |
| | *Lophaetus occipitalis* | Eagle, Long-crested | LC | B | I | BJ, NG |
| | *Melierax gabar* | Goshawk, Gabar | LC | B | S | NG |
| | *Melierax metabates* | Chanting-Goshawk, Dark | LC | B | S | NG |
| | *Milvus migrans* | Kite, Black | LC | A | D | BJ, NG |
| | *Milvus* sp. | Kite, unidentified | – | – | – | BF |
| | *Necrosyrtes monachus* | Vulture, Hooded | EN[l] | A | S | BJ, CI, NG, TG, ZA |
| | *Neophron percnopterus* | Vulture, Egyptian | EN[m] | C | D | BJ, NG |
| | *Polemaetus bellicosus* | Eagle, Martial | NT[n] | C | D | – |
| | *Polyboroides typus* | Harrier-hawk, African | LC | A | S | NG, TG, ZA |
| | *Sagittarius serpentarius* | Secretarybird | LC | D | D | ZA |
| | *Stephanoaetus coronatus* | Hawk-eagle, Crowned | LC | D | D | BJ, ZA |
| | *Terathopius ecaudatus* | Bateleur | NT[o] | B | D | NG, ZA |
| | *Torgos tracheliotos* | Vulture, Lappet-faced | VU[p] | D | D | BJ, ZA |
| | *Trigonoceps occipitalis* | Vulture, White-headed | VU[q] | D | D | NG, ZA |
| | – | Hawk, unidentified | – | – | – | BF, BJ, NG |
| | – | Eagle, unidentified | – | – | – | BF, TG, ZA, ZW |
| | – | Vulture, unidentified. | – | – | – | TG, ZA, ZW |
| Falconidae | *Falco alopex* | Kestrel, Fox | LC | H | S | BJ |

(continued)

**Table 18.1** (continued)

| Classification (**ORDER**, family) | Species | Common name | IUCN category[a] | Rarity classification | Population trend | Countries traded in (max $n = 7$) |
|---|---|---|---|---|---|---|
| | *Falco ardosiaceus* | Kestrel, Grey | LC | B | S | BJ, NG |
| | *Falco biarmicus* | Falcon, Lanner | LC | A | D | NG |
| | *Falco chicquera* | Falcon, Red-necked | LC | D | S | NG |
| | *Falco cuvierii* | Hobby, African | LC | H | D | NG |
| | *Falco tinnunculus* | Kestrel, Common | LC | A | D | BJ, NG |
| | *Falco* sp. | Kestrel, unidentified | – | – | – | ZA |
| **GALLIFORMES** | | | | | | |
| Numididae | *Guttera plumifera* | Guineafowl, Plumed | LC | F | D | – |
| | *Guttera pucherani* | Guineafowl, Crested | LC | F | S | BJ, NG, ZA |
| | *Numida meleagris* | Guineafowl, Helmeted | LC | A | S | BF, BJ, NG, ZA, ZW |
| | – | Guineafowl, unidentified | – | – | – | NG |
| Phasianidae | *Afropavo congensis* | Peafowl, Congo | VU[r] | H | D | – |
| | *Coturnix chinensis* | Quail, Blue | LC | D | S | BJ |
| | *Coturnix coturnix* | Quail, Common | LC | B | D | NG, ZA |
| | *Coturnix delegorguei* | Quail, Harlequin | LC | B | S | BJ, NG |
| | *Coturnix* sp. | Quail, unidentified | – | – | – | – |
| | *Francolinus afer* | Spurfowl, Red-necked | LC | E | D | – |
| | *Francolinus ahantensis* | Francolin, Ahanta | LC | F | D | BJ, NG |
| | *Francolinus albogularis* | Francolin, White-throated | LC | H | S | BF |
| | *Francolinus bicalcaratus* | Fancolin, Double-spurred | LC | F | S | BF, NG |
| | *Francolinus* sp. | Francolin, unidentified | – | – | – | BF, BJ, CI, NG, TG, ZA |
| | *P. cristatus* | Peafowl, Indian[s] | LC | H | S | NG, ZA |
| | *Ptilopachus petrosus* | Partridge, Stone | LC | F | S | NG |
| **GRUIFORMES** | | | | | | |
| Gruidae | *Balearica pavonina* | Crowned-crane, Black | VU[t] | H | D | BJ, CI, NG |
| Heliornithidae | *Podica senegalensis* | Finfoot, African | LC | D | D | NG |
| Otididae | *Ardeotis arabs* | Bustard, Arabian | LC | D | D | CI, NG |
| | *Ardeotis kori* | Bustard, Kori | LC | H | D | – |
| | *Euphodotis melanogaster* | Bustard, Black-bellied | LC | D | D | NG |

(continued)

Table 18.1 (continued)

| Classification (**ORDER**, family) | Species | Common name | IUCN category[a] | Rarity classification | Population trend | Countries traded in (max $n = 7$) |
|---|---|---|---|---|---|---|
| | *Euphodotis senegalensis* | Bustard, White-bellied | LC | C | D | NG |
| | *Euphodotis* sp. | Bustard, unidentified | – | – | – | BF |
| | *Euphodotis savilei* | Bustard, Savile's | LC | G | S | CI |
| | *Neotis denhami* | Bustard, Denham's | NT[u] | D | D | NG |
| | – | Bustard, unidentified | – | – | – | – |
| Rallidae | *Amaurornis flavirostra* | Crake, Black | LC | B | S | BJ, NG, ZA |
| | *Canirallus oculeus* | Rail, Grey-throated | LC | H | D | BJ |
| | *Crecopsis egregia* | Crake, African | LC | C | S | NG |
| | *Gallinula angulata* | Moorhen, Lesser | LC | B | D | BJ, NG |
| | *Gallinula chloropus* | Moorhen, Common | LC | B | S | NG, ZA |
| | *Gallinula* sp. | Moorhen, unidentified | – | – | – | ZA |
| | *Himantornis haematopus* | Rail, Nkulengu | LC | F | D | NG |
| | *Porphyrio alleni* | Gallinule, Allen's | LC | B | D | BJ |
| | *Porphyrio porphyrio* | Swamphen, Purple | LC | B | ? | BJ, NG, ZA |
| | *Sarothrura pulchra* | Flufftail, White-spotted | LC | F | D | BJ |
| **PASSERIFORMES** | | | | | | |
| Campephagidae | *Coracina pectoralis* | Cuckooshrike, White-breasted | LC | F | D | NG |
| Cisticolidae | *Apalis porphyrolaema* | Apalis, Chestnut-throated | LC | F | S | – |
| | *Cisticola erythrops* | Cisticola, Red-faced | LC | B | S | – |
| | *Cisticola* sp | Cisticola, unidentified | – | – | – | ZA |
| | *Hypergerus atriceps* | Warbler, Oriole | LC | F | S | NG |
| | *Prinia bairdii* | Prinia, Banded | LC | E | S | – |
| | *Prinia subflava* | Prinia, Tawny-flanked | LC | A | ? | ZA |
| Corvidae | *Corvus albicollis* | Raven, White-necked | LC | E | D | ZA |
| | *Corvus albus* | Crow, Pied | LC | A | S | BF, BH, NG, TG, ZA, ZW |
| | *Corvus capensis* | Crow, Cape | LC | E | I | ZA |
| | *Corvus rhipidurus* | Raven, Fan-tailed | LC | G | D | – |
| | *Corvus* sp. | Crow, unidentified | – | – | – | ZA |
| | *Ptilostomus afer* | Piapiac | LC | E | S | NG |

(continued)

**Table 18.1** (continued)

| Classification (**ORDER**, family) | Species | Common name | IUCN category[a] | Rarity classification | Population trend | Countries traded in (max $n = 7$) |
|---|---|---|---|---|---|---|
| Dicruridae | *Dicrurus adsimilis* | Drongo, Fork-tailed | LC | A | S | BJ, NG |
| Estrildidae | *Cryptospiza salvadorii* | Crimson-wing, Abyssinian | LC | F | S | – |
| | *Estrilda astrild* | Waxbill, Common | LC | A | S | – |
| | *Estrilda atricapilla* | Waxbill, Black-headed | LC | E | S | – |
| | *Estrilda trogolodytes* | Waxbill, Black-rumped | LC | F | S | NG |
| | *Lagonostica senegala* | Firefinch, Red-billed | LC | A | S | BJ, NG |
| | *Lonchura bicolor* | Munia, Black-and-white | LC | E | S | – |
| | *Lonchura cucullata* | Munia, Bronze | LC | E | S | – |
| | *Nesocharis ansorgei* | Oliveback, White-collared | LC | F | S | – |
| | *Nigrita canicapillus* | Negrofinch, Grey-headed | LC | F | S | – |
| | *Spermophaga haematina* | Bluebill, Western | LC | F | S | NG |
| | *Spermophaga ruficapilla* | Bluebill, Red-headed | LC | E | S | – |
| Fringillidae | *Serinus sulphuratus* | Canary, Brimstone | LC | E | S | – |
| Hirundinidae | *Delichon urbicum*[PAL] | House-martin, Northern | LC | A | D | NG |
| | *Hirundo abyssinica* | Striped-swallow, Lesser | LC | A | I | BJ |
| | *Hirundo aethiopica* | Swallow, Ethiopian | LC | E | I | BJ, NG |
| | *Hirundo angolensis* | Swallow, Angola | LC | H | I | – |
| | *Hirundo rustica*[PAL] | Swallow, Barn | LC | A | D | BJ, NG |
| | *Hirundo smithii* | Swallow, Wire-tailed | LC | B | I | BJ |
| | *Hirundo* sp. | Swallow, unidentified | – | – | – | ZA |
| | *Psalidoprocne nitens* | Saw-wing, Square-tailed | LC | E | D | – |
| | – | Swallow, unidentified | – | – | – | NG |
| Laniidae | *Corvinella corvina* | Shrike, Yellow-billed | LC | F | S | BJ, NG |
| | *Lanius collaris* | Fiscal, Common | LC | A | I | BJ, ZA |
| Malaconotidae | *Dryoscopus gambensis* | Puffback, Northern | LC | F | S | BJ |
| | *Laniarius aethiopicus* | Boubou, Ethiopian | LC | E | S | NG |
| | *Laniarius barbarus* | Gonolek, Common | LC | E | S | NG |
| | *Laniarius ferrugineus* | Boubou, Southern | LC | E | S | ZA |

(continued)

**Table 18.1** (continued)

| Classification (**ORDER**, family) | Species | Common name | IUCN category[a] | Rarity classification | Population trend | Countries traded in (max $n = 7$) |
|---|---|---|---|---|---|---|
| | *Malaconotus blanchoti* | Bush-shrike, Grey-headed | LC | B | I | BJ, NG |
| | *Prionops plumatus* | Helmet-Shrike, White | LC | A | S | BJ, NG, ZA |
| | *Tchagra senegalus* | Tchagra, Black-crowned | LC | A | S | BF, BJ, NG |
| | *Telephorus multicolor* | Bush-shrike, Many-colored | LC | F | S | BJ |
| Monarchidae | *Terpsiphone rufiventer* | Paradise-Flycatcher, Black-headed | LC | F | D | BJ |
| | *Terpsiphone viridis* | Paradise-Flycatcher, African | LC | B | S | BJ |
| Motacillidae | *Macronyx* sp. | Longclaw, unidentified | – | – | – | ZA |
| | *Motacilla aguimp* | Wagtail, African Pied | LC | B | S | NG |
| | *Motacilla capensis* | Wagtail, Cape | LC | F | S | – |
| | *Motacilla flava*[PAL] | Wagtail, Yellow | LC | B | D | BJ |
| Muscicapidae | *Melaenornis edolioides* | Flycatcher, Northern Black | LC | E | S | NG |
| | *Melaenornis pammelaina* | Flycatcher, Southern Black | LC | E | S | ZA |
| | *Muscicapa striata*[PAL] | Flycatcher, Spotted | LC | A | D | BJ |
| | *Cossypha albicapilla* | Robin-chat, White-crowned | LC | H | S | NJ, NG |
| | *Cossypha natalensis* | Robin-chat, Red-capped | LC | F | S | ZA |
| | *Cossypha niveicapilla* | Robin-chat, Snowy-crowned | LC | F | S | NG |
| | *Cossypha* sp. | Robin-chat, unidentified | – | – | – | BJ, CI |
| Nectariniidae | *Nectarinia chloropygia* | Sunbird, Olive-bellied | LC | E | S | NG |
| | *Nectarinia coccinigaster* | Sunbird, Splendid | LC | E | S | BJ, NG |
| | *Nectarinia cuprea* | Sunbird, Copper | LC | B | S | NG |
| | *Nectarinia kilimensis* | Sunbird, Bronze | LC | E | S | – |
| | *Nectarinia senegalensis* | Sunbird, Scarlet-chested | LC | A | S | BJ, NG |
| | *Nectarinia superba* | Sunbird, Superb | LC | E | D | BJ |
| | *Nectarinia verticalis* | Sunbird, Green-headed | LC | E | S | NG |
| Oriolidae | *Oriolus auratus* | Oriole, African Golden | LC | B | S | BJ, NG |
| | *Oriolus brachyrhyncus* | Oriole, Western Black-headed | LC | F | D | BJ |
| | *Oriolus larvatus* | Oriole, African Black-headed | LC | F | I | – |
| Passeridae | *Passer domesticus* | Sparrow, House | LC | A | D | ZA |
| | *Passer griseus* | Sparrow, Northern Grey-headed | LC | A | S | NG |

(continued)

**Table 18.1** (continued)

| Classification (**ORDER**, family) | Species | Common name | IUCN category[a] | Rarity classification | Population trend | Countries traded in (max $n = 7$) |
|---|---|---|---|---|---|---|
| Platysteiridae | *Sporopipes frontalis* | Weaver, Speckle-fronted | LC | H | S | NG |
| | *Platysteira castanea* | Wattle-eye, Chestnut | LC | E | S | – |
| | *Platysteira cyanea* | Wattle-eye, Brown-throated | LC | E | S | NG |
| Plocidae | *Amblyospiza albifrons* | Weaver, Grosbeak | LC | E | S | – |
| | *Anaplectes rubriceps* | Weaver, Red-headed | LC | A | S | NG |
| | *Bubalornis albirostris* | Buffalo-weaver, White-billed | LC | E | S | BJ |
| | *Euplectes franciscanus* | Bishop, Orange | LC | F | S | BJ, NG |
| | *Euplectes* sp. | Bishop, unidentified | – | – | – | CI |
| | *Malimbus nitens* | Malimbe, Gray's | LC | F | S | BJ |
| | *Malimbus rubricollis* | Malimbe, Red-headed | LC | F | S | NG |
| | *Malimbus scutatus* | Malimbe, Red-vented | LC | F | S | BJ |
| | *Ploceus baglafecht* | Weaver, Baglafecht | LC | E | S | – |
| | *Ploceus cucullatus* | Weaver, Village | LC | A | S | BF, BJ, NG |
| | *Ploceus nigerrimus* | Weaver, Vieillot's Black | LC | F | S | BJ, NG |
| | *Ploceus nigricollis* | Weaver, Black-necked | LC | E | S | – |
| | *Ploceus xanthops* | Weaver, Holub's Golden | LC | F | S | – |
| | – | Weaver, unidentified | – | – | – | TG |
| Pycnonotidae | *Andropadus importunus* | Greenbul, Somber | LC | H | S | ZA |
| | *Andropadus latirostris* | Greenbul, Yellow-whiskered | LC | G | S | – |
| | *Andropadus tephrolaemus* | Greenbul, Western Mountain | LC | F | S | – |
| | *Bleda syndactylus* | Bristlebill, Common | LC | F | S | – |
| | *Nicator chloris* | Nicator, Yellow-spotted | LC | F | S | BJ |
| | *Phyllastrephus flavostriatus* | Greenbul, Yellow-streaked | LC | F | S | – |
| | *P. barbatus* | Bulbul, Common | LC | A | I | BJ, NG, ZA |
| | *Chlorocichla flavicollis* | Greenbul, Yellow-throated | LC | F | S | BJ |
| Sturnidae | *Acridotheres tristis* | Myna, Common[s] | LC | A | I | ZA |
| | *Cinnyricinclus leucogaster* | Starling, Violet-backed | LC | A | D | NG |

(continued)

**Table 18.1** (continued)

| Classification (**ORDER**, family) | Species | Common name | IUCN category[a] | Rarity classification | Population trend | Countries traded in (max $n = 7$) |
|---|---|---|---|---|---|---|
| | *Lamprotornis chalybaeus* | Glossy-starling, Greater Blue-eared | LC | B | S | NG |
| | *Lamprotornis nitens* | Glossy-starling, Red-shouldered | LC | E | S | ZA |
| | *Lamprotornis pulcher* | Starling, Chestnut-bellied | LC | H | S | NG |
| | *Lamprotornis purpureus* | Glossy-starling, Purple | LC | F | S | BF, BJ |
| | *Lamprotornis* sp. | Glossy-starling, unidentified | – | – | – | BJ, ZA |
| | *Lamprotornis splendidus* | Starling, Splendid Glossy | LC | F | S | BJ, NG |
| | *Onychognathus fulgidus* | Starling, Chestnut-winged | LC | F | D | BJ |
| | *Onychognathus morio* | Starling, Red-winged | LC | F | I | BJ |
| Sylviidae | *Melocichla mentalis* | Grass-warbler, Mustached | LC | B | S | NG |
| | *Sylvietta leucophrys* | Crombec, White-browed | LC | F | S | – |
| Timaliidae | *Turdoides plebejus* | Babbler, Brown | LC | G | S | NG |
| Turdidae | *Neocossyphus poensis* | Ant-thrush, White-tailed | LC | F | D | BJ, NG |
| | *Turdus olivaceus* | Thrush, Olive | LC | F | ? | ZA |
| | *Turdus pelios* | Thrush, African | LC | F | ? | BJ, NG |
| Viduidae | *Vidua macroura* | Whydah, Pin-tailed | LC | A | S | BJ, NG |
| | *V. paradisaea* | Paradise-Whydah, Eastern | LC | D | S | NG |
| | *Vidua* sp. | Paradise-Whydah, unidentified | – | – | – | BJ |
| Zosteropidae | *Z. pallidus* | White-eye, Pale | LC | E | ? | ZA |
| | *Zosterops senegalensis* | White-eye, African Yellow | LC | A | S | – |
| **PELICANIFORMES** | | | | | | |
| Pelicanidae | *Pelecanus onocrotalus* | Pelican, Great White | LC | B | S | BF, ZA |
| | *Pelecanus rufescens* | Pelican, Pink-backed | LC | B | S | BJ |
| | *Pelecanus* sp. | Pelican, unidentified | – | – | – | ZA |
| Sulidae | *Morus capensis* | Gannet, Cape | VU[v] | F | D | ZA |
| Threskiornithidae | *Bostrychia hagedash* | Ibis, Hadada | LC | A | I | CI, NG, ZA |
| | *Bostrychia rara* | Ibis, Spot-breasted | LC | H | S | NG |
| | *Geronticus calvus* | Ibis, Southern Bald | VU[w] | F | D | ZA |
| | *Threskiornis aethiopicus* | Ibis, African Sacred | LC | A | D | ZA |

(continued)

**Table 18.1** (continued)

| Classification (**ORDER**, family) | Species | Common name | IUCN category[a] | Rarity classification | Population trend | Countries traded in (max $n = 7$) |
|---|---|---|---|---|---|---|
| **PICIFORMES** | | | | | | |
| Indicatoridae | *Indicator* sp. | Honeyguide, unidentified | – | – | – | |
| Picidae | *Campethera abingoni* | Woodpecker, Golden-tailed | LC | E | S | ZA |
| | *Campethera cailliauti* | Woodpecker, Green-backed | LC | C | S | BJ |
| | *Campethera punctuligera* | Woodpecker, Fine-spotted | LC | F | S | BJ, NG |
| | *Dendropicos fuscescens* | Woodpecker, Cardinal | LC | A | S | – |
| | *Jynx torquilla*[PAL] | Wryneck, Eurasian | LC | C | D | NG |
| | *Mesopicos goertae* | Woodpecker, Grey | LC | A | S | BJ, NG |
| | – | Woodpecker, unidentified | – | – | – | TG, ZA |
| Ramphastidae | *Gymnobucco bonapartei* | Barbet, Grey-throated | LC | F | D | – |
| | *Lybius bidentatus* | Barbet, Doubled-toothed | LC | E | S | BF, BJ, NG |
| | *Lybius dubius* | Barbet, Bearded | LC | G | ? | BF, BJ, NG |
| | *Lybius torquatus* | Barbet, Black-collared | LC | A | D | – |
| | *Lybius vieilloti* | Barbet, Vieillot's | LC | H | ? | NG |
| | *Pogoniulus chrysoconus* | Tinkerbird, Yellow-fronted | LC | A | S | BJ, NG |
| | *Pogonolius bilineatus* | Tinkerbird, Yellow-rumped | LC | B | S | – |
| | *Trachyphonus vaillantii* | Barbet, Crested | LC | E | D | ZA |
| **PODICIPEDIFORMES** | | | | | | |
| Podicipedidae | *Tachybaptus ruficollis* | Grebe, Little | LC | B | D | NG |
| **PROCELLARIIFORMES** | | | | | | |
| Diomedeidae | *Thalassarche cauta* | Albatross, Shy | NT[x] | D | D | ZA |
| **PSITTACIFORMES** | | | | | | |
| Psittacidae | *Agapornis pullarius* | Lovebird, Red-headed | LC | E | D | BF, BJ |
| | *Poicephalus gulielmi* | Parrot, Red-fronted | LC | F | D | NG |
| | *Poicephalus senegalus* | Parrot, Senegal | LC | F | S | BF, BJ, NG |
| | *Poicephalus sp.* | Parrot, unidentified | – | – | – | NG,TG |
| | *Psittacula krameri* | Parakeet, Rose-ringed[s] | LC | E | I | BF, BJ, NG |
| | *Psittacula* sp. | Parakeet, unidentified | – | – | – | BF |
| | *Psittacus erithacus* | Parrot, Grey | NT[y] | F | D | BJ, NG, TG |
| **PTEROCLIDIFORMES** | | | | | | |

(continued)

Table 18.1 (continued)

| Classification (**ORDER**, family) | Species | Common name | IUCN category[a] | Rarity classification | Population trend | Countries traded in (max $n = 7$) |
|---|---|---|---|---|---|---|
| Pteroclididae | *Pterocles quadricinctus* | Sandgrouse, Four-banded | LC | F | S | NG |
| **SPHENISCIFORMES** | | | | | | |
| Spheniscidae | *Spheniscus demersus* | Penguin, African | EN[z] | F | D | ZA |
| **STRIGIFORMES** | | | | | | |
| Strigidae | *Asio capensis* | Owl, Marsh | LC | B | S | BJ, ZA |
| | *Athene noctua* | Owl, Little | LC | A | S | – |
| | *Bubo africanus* | Eagle-owl, Spotted | LC | A | S | BJ, NG, ZA |
| | *Bubo capensis* | Eagle-owl, Cape | LC | G | S | ZA |
| | *B. lacteus* | Eagle-owl, Giant | LC | C | S | NG |
| | *Bubo poensis* | Eagle-owl, Fraser's | LC | H | S | NG |
| | *Bubo* sp. | Eagle-owl, unidentified | – | – | – | CI, ZA |
| | *Glaucidium perlatum* | Owlet, Pearl-spotted | LC | B | S | BJ, NG |
| | *Glaucidium tephronotum* | Owlet, Red-chested | LC | H | S | BJ |
| | *Otus leucotis* | Scops-owl, White-faced | LC | C | S | BJ, NG |
| | *Otus scops* | Scops-owl, Common | LC | D | D | BJ |
| | *Otus senegalensis* | Scops-owl, African | LC | B | S | NG |
| | *Strix woodfordii* | Wood-owl, African | LC | B | S | BF, BJ, NG, ZA |
| | – | Owl, unidentified | – | – | – | TG, ZA, ZW |
| | – | Owl, unidentified | – | – | – | BF |
| Tytonidae | *Tyto alba* | Owl, Barn | LC | A | S | BF, BJ, CI, NG, TG, ZA, ZW |
| | *Tyto capensis* | Grass-owl, African | LC | H | D | ZA |
| **STRUTHIONIFORMES** | | | | | | |
| Struthionidae | *Struthio camelus* | Ostrich | LC | A | D | NG, ZA, ZW |
| **TROGONIFORMES** | | | | | | |
| Trogonidae | *Apaloderma narina* | Trogon, Narina | LC | D | S | NG |
| **UPUPIFORMES** | | | | | | |
| Phoeniculidae | *Phoeniculus casteneiceps* | Woodhoopoe, Forest | LC | H | D | NG |
| | *Phoeniculus purpureus* | Woodhoopoe, Green | LC | B | D | BJ, NG |
| | *Phoeniculus* sp. | Woodhoopoe, unidentified | – | – | – | ZA |

(continued)

**Table 18.1** (continued)

| Classification (**ORDER**, family) | Species | Common name | IUCN category[a] | Rarity classification | Population trend | Countries traded in (max $n = 7$) |
|---|---|---|---|---|---|---|
| | *Rhinopomastus aterrimus* | Scimitarbill, Black | LC | B | D | BJ |
| Upupidae | *Upupaepops* | Hoopoe, Eurasian | LC | A | D | BF, NG, ZA |

The list includes their IUCN conservation status (*LC* least concern, *NT* near threatened, *VU* vulnerable, *EN* endangered, *CR* critically endangered.), rarity classification (see Tables 18.2 and 18.3), population trend (*D* declining, *S* stable, *I* increasing, ? uncertain), and the countries in which the species were recorded in traditional markets (*BF* Burkina Faso, *BJ* Benin, *CI* Côte D'Ivoire, *NG* Nigeria, *TG* Togo, *ZA* South Africa, *ZW* Zimbabwe). The classification and nomenclature follows BirdLife International. *PAL* Palearctic migrant to Africa

[a] IUCN Red List Assessment criteria
[b] A2cd, A3cd, A4cd
[c] A2cd, A3cd, A4cd
[d] A4bcd
[e] A2cd, A3cd, A4cd
[f] A2bcd, A3bcd, A4bcd
[g] A2bcd, A3bcd, A4bcd, C1, C2a(ii)
[h] A2cde, A3cde, A4cde
[i] A2bcd, A3bcd, A4bcd
[j] C1 + 2a
[k] A2abcd, A3bcd, A4bcd
[l] A2acd, A3cd, A4acd
[m] A2bcde, A3bcde, A4bcde
[n] A2acde
[o] A2acde
[p] C2a
[q] C2a
[r] C2a(i)
[s] Exotic species
[t] A4bcd
[u] A2bcd, A3bcd, A4bcd
[v] A2acde, A3cde, A4acde, B2ab
[w] A3c, C2a
[x] D2
[y] A2bcd, A3bcd, A4bcd
[z] A2ace, A3ce, A4ace

breeding range extending into the PAL extremities of North Africa), further complicated by the extent of actual wintering (austral) range in Africa versus areas visited in transit to and from the summer (boreal) PAL breeding range;
- 45 species that are recorded as being used for TM but were not recorded in the markets (Table 18.1).

### *18.2.3 Patterns of Rarity and Commonness*

A way of examining avian vulnerability to utilization is to classify the species based on the probability of them becoming rare if they are over-exploited. Rabinowitz (1981) developed the 'seven forms of rarity' model that was originally applied to assessing the vulnerability of plants, with subsequent refinements (Rabinowitz et al. 1986). The model was based on three factors/variables that indicated the level of rarity, namely range size, habitat specificity, and local abundance (Table 18.2). When species are dichotomized for each of the variables, the result is an eight-cell model that Yu and Dobson (2000) adapted and used to create four ranks of rarity to assess the rarity and commonness of mammals (Table 18.3). For example, Category H contains species rare in all three factors and assigned a rank of 1, whereas Category A contains species that are common and assigned a rank of 4. The excessive utilization of species that are rare and targeted by hunters is therefore likely to result in more significant population declines than for species that are more common.

We applied Yu and Dobson's (2000) eight classes of rarity and placed the species in categories ranging from A to H. Thereafter, the categories were assigned ranks from 1 to 4 (most to least rare, respectively) (Table 18.3). The purpose was to examine patterns of rarity and commonness, and to compare these classifications with other variables in an attempt to detect species that may be threatened by the current levels of utilization. The data used to do the rarity assessments (and other quantitative analyses) were obtained from the BirdLife International website (www.birdlife.org/datazone/home, downloaded May 2011) and included: (1) population size estimates, (2) population trends (i.e. increasing, decreasing, stable), (3) Extent of Occurrence (EOO), (4) number of Level 1 habitats, and (5) number of African countries in which the species occurs.

Assigning species to the rarity categories requires dichotomizing the distribution (large or small), habitat (broad or narrow), and population abundance (high/dominant or small/non-dominant). For the distribution range, we determined the median EOO for all the inventoried species to be 4,630,000 km$^2$ (range: 82,800–99,700,000 km$^2$; $n = 297$) and therefore EOOs greater and smaller than the median were considered to be large and small ranges, respectively. Dichotomizing habitat and abundance were more subjective, since habitat types are essentially various graded combinations of geological, topographical, aquatic, and vegetational features ranging from desert to rainforest, and abundance ranges from local

**Table 18.2** Rabinowitz's seven forms of rarity based on three traits (adapted from Rabinowitz 1981; Cunningham 2001)

| Geographic range | | Large | | Small | |
|---|---|---|---|---|---|
| Local population size | | Large, dominant somewhere | Small, non-dominant | Large, dominant somewhere | Small, non-dominant |
| Habitat specificity | Wide | (A) Locally abundant in several habitats over a large geographic area | (C) Constantly sparse in several habitats over a large geographic area | (E) Locally abundant in several habitats over a small geographic area | (G) Constantly sparse in several habitats over a small geographic area |
| | Narrow | (B) Locally abundant in a specific habitat over a large geographic area | (D) Constantly sparse in a specific habitat over a large geographic area | (F) Locally abundant in a specific habitat over a small geographic area | (H) Constantly sparse in a specific habitat over a small geographic area |

Letters in brackets correspond with the rarity classes in Table 18.3

**Table 18.3** Classes of rarity after Rabinowitz (1981) and ranks (in parentheses) that are assigned to each class (derived from Yu and Dobson 2000)

| | | Distribution | | | |
|---|---|---|---|---|---|
| | | Large | | Small | |
| | Population | High | Low | High | Low |
| Habitat | Broad | A (4) | C (3) | E (3) | G (2) |
| | Narrow | B (3) | D (2) | F (2) | H (1) |

to widespread and rare to abundant. Habitat and status assessments in Sinclair and Ryan (2010) were used to assist in borderline judgments.

### *18.2.4 IUCN Red List Status and Index*

We further examined trends in the status and threats to avian diversity by calculating the IUCN Red List Index (RLI) of species survival. The RLI uses information from the Red List to monitor trends in the proportion of species expected to remain extant or become extinct in the near future (Butchart 2008). The index has been used to assess the state of the world's birds and the state of utilized bird species (BirdLife International 2008a, b, c). The RLI was calculated for all non-Data Deficient birds identified in this study for the years 2000, 2004, 2008, and 2011 from the Red List status available on the IUCN Red List of Threatened Species website (www.iucnredlist.org, downloaded July 2011) and BirdLife International (www.birdlife.org). The method used to calculate the index is described in Bubb et al. (2009) and Butchart (2008). RLI values close to one indicate that most species are categorized as Least Concern and are expected to remain extant in the near future, whereas values closer to zero indicate that a high proportion of species are either extremely threatened or extinct (Bubb et al. 2009). The index allows one to monitor temporal trends in the extinction risks to species.

### *18.2.5 Other Analyses*

Various other quantitative analyses were carried out. We calculated the richness and percentage similarity of species recorded in markets in seven countries using the Sørenson index (for incidence-based data). We also calculated the expected number of species in the rarity categories A–H, and hence whether there was a significant difference in the proportion of species observed in each rarity category using a Chi-squared test.

## 18.3 Results

### 18.3.1 Ornithological Richness

We recorded 354 species from 205 genera, 70 families, and 25 orders (Tables 18.1, 18.4) as being used and traded for traditional medicine in 25 African countries. When we excluded 45 species not recorded in markets, then the total ornithological richness of traded taxa at markets was 309 species from 189 genera, 69 families and 25 orders from seven African countries (BF, BJ, CI, NG, TG, ZA and ZW) (Table 18.4).

Perching birds (Passeriformes) had the highest number of recorded taxa in use (23 families, 60 genera, 108 species) and in trade (23 families, 50 genera, 82 species) (Table 18.4). Within the traded Passeriformes, the Sturnidae (starlings) was the most prevalent family of that order (nine species) (Fig. 18.3). Of all the families in trade, the Acciptridae (Order Falconiformes) were the family with the most recorded genera (26 genera; 37 species; including kites, hawks, eagles, vultures), followed by the Ardeidae (Order Ciconiiformes) (11 genera; 15 species; including herons and egrets). The Bucerotidae (hornbills), Cuculidae (cuckoos) and Strigidae (owls) were the next most specious families in trade (12, 11, and 11 species per family, respectively). The identified birds included two exotic species, namely the Indian Peafowl and the Common Myna. An additional 44 taxa were not identified to family, genus, and/or species.

NG had the most number of species recorded in the markets (200 species, plus one exotic, five PAL and six unidentified species), followed by BJ (134 species, plus nine PAL and nine unidentified), ZA (85 species, plus two exotic and 24 unidentified), BF (29 species, plus one PAL and 11 unidentified), CI (12 species, plus three unidentified), TG (11 species, plus 13 unidentified) and ZW (six species, plus six unidentified) (Fig. 18.1). The 15 PAL species were only recorded in NG, BJ, and BF. Five partial intra-African migrants were also identified (Wahlberg's Eagle, Bronze-winged Courser, Levaillant's Cuckoo, Violet-backed Starling, African Golden Oriole), but only in markets in NG and BJ, despite their total range in Africa extending to southern Africa.

### 18.3.2 Frequency and Similarity of Species Traded in Local Markets

While Ostriches were recorded as being used in more countries than other species (11 countries), this species was only recorded being sold in the markets of four countries (Morocco, NG, ZA, ZW) (Table 18.5; Fig. 18.4). The most prevalent species in the seven markets were Barn Owl (seven countries), Pied Crow (six countries), Hooded Vulture, Helmeted Guinea Fowl, and African Pied Hornbill (five countries each) (Table 18.5).

**Table 18.4** Summary of avian taxa used/sold for traditional medicine across 25 African countries

| Order (22) | All species used and traded | | | Traded species | |
|---|---|---|---|---|---|
| | No. families per order ($S = 70$)[a] | No. genera per order ($S = 205$) | No. species per order ($S = 354$)[b] | No. genera per order ($S = 189$) | No. species per order ($S = 309$)[c] |
| Anseriformes (Waterfowl) | 1 | 5 | 7 | 5 | 7 |
| Apodiformes (Swifts and relatives) | 1 | 1 | 2 | 1 | 2 |
| Bucerotiformes (Hornbills) | 3 | 9 | 22 | 9 | 21 |
| Caprimulgiformes (Nightjars and relatives) | 1 | 2 | 5 | 2 | 3 |
| Charadriiformes (Gulls and relatives) | 7 | 12 | 19 | 12 | 19 |
| Ciconiiformes (Storks) | 3 | 16 | 21 | 16 | 21 |
| Coliiformes (Mousebirds) | 1 | 1 | 1 | 1 | 1 |
| Columbiformes (Doves and pigeons) | 1 | 5 | 10 | 5 | 9 |
| Coraciiformes (Kingfishers and relatives) | 3 | 8 | 20 | 8 | 19 |
| Cuculiformes (Cuckoos and relatives) | 1 | 5 | 13 | 5 | 11 |
| Falconiformes (Diurnal birds of prey) | 2 | 28 | 45 | 27 | 43 |
| Galliformes (Gamebirds)[d] | 2 | 7 | 13 | 6 | 10 |
| Gruiiformes (Cranes and relatives) | 4 | 12 | 17 | 12 | 16 |
| Passeriformes (Perching birds)[d] | 23 | 60 | 108 | 50 | 82 |
| Pelecaniformes (Pelicans and relatives) | 3 | 5 | 7 | 5 | 7 |

(continued)

**Table 18.4** (continued)

| Order (22) | All species used and traded | | | Traded species | |
|---|---|---|---|---|---|
| | No. families per order ($S = 70$)[a] | No. genera per order ($S = 205$) | No. species per order ($S = 354$)[b] | No. genera per order ($S = 189$) | No. species per order ($S = 309$)[c] |
| Piciformes (Woodpeckers and relatives)[a] | 3 | 9 | 15 | 6 | 10 |
| Podicipediformes (Grebes) | 1 | 1 | 1 | 1 | 1 |
| Procellariiformes (Albatrosses and relatives) | 1 | 1 | 1 | 1 | 1 |
| Psittaciformes (Parrots) | 1 | 4 | 5 | 4 | 5 |
| Pteroclidiformes (Soundgrouses) | 1 | 1 | 1 | 1 | 1 |
| Sphenisciformes (Penguins) | 1 | 1 | 1 | 1 | 1 |
| Strigiformes (Owls) | 2 | 7 | 14 | 6 | 13 |
| Struthioniformes (Ratites) | 1 | 1 | 1 | 1 | 1 |
| Trogoniformes (Trogons and relatives) | 1 | 1 | 1 | 1 | 1 |
| Upupiformes (Hoopoes) | 2 | 3 | 4 | 3 | 4 |

The total excludes 44 taxa unidentified to family, order, and/or species

[a] The Family Indictoridae or Honeyguides are absent from the traded species list, hence $S = 69$ traded families

[b] Includes PAL bird species, but excludes 44 taxa that were unidentified to family, order, genus and/or species

[c] Includes PAL bird species, but excludes 38 taxa that were unidentified to family, order, genus and/or species

[d] Includes one exotic species

**Table 18.5** The most frequently recorded species in African markets and countries, excluding unidentified species identified only by a general category of bird (such as eagle, francolin, or vulture) without giving the species

| Common name | Species | No. countries traded in ($n = 7$) | No. countries use recorded in ($n = 25$) |
|---|---|---|---|
| Owl, Barn | *Tyto alba* | 7 | 7 |
| Crow, Pied | *Corvus albus* | 6 | 8 |
| Vulture, Hooded | *Necrosyrtes monachus* | 5 | 6 |
| Guineafowl, Helmeted | *Numida meleagris* | 5 | 6 |
| Hornbill, African Pied | *Tockus fasciatus* | 5 | 5 |
| Ostrich | *Struthio camelus* | 4 | 11 |
| Roller, Abyssinian | *Coracias abyssinicus* | 4 | 5 |
| Wood-owl, African | *Strix woodfordii* | 4 | 5 |
| Egret, Cattle | *Bubulcus ibis* | 4 | 4 |
| Plantain-eater, Western Grey | *Crinifer piscator* | 4 | 4 |
| Fish-eagle, African | *Haliaeetus vocifer* | 4 | 4 |
| Night-heron, Black-crowned | *Nycticorax nycticorax* | 4 | 4 |
| Hornbill, Red-billed | *Tockus erythrorhynchus* | 4 | 4 |
| Parrot, Grey | *Psittacus erithacus* | 3 | 6 |
| Eagle-owl, Spotted | *Bubo africanus* | 3 | 5 |
| Ground-hornbill, Abyssinian | *Bucorvus abyssinicus* | 3 | 5 |
| Hamerkop | *Scopus umbretta* | 3 | 5 |
| Hoopoe, Eurasian | *Upupa epops* | 3 | 5 |
| Ground-hornbill, Southern | *Bucorvus cafer* | 2 | 5 |

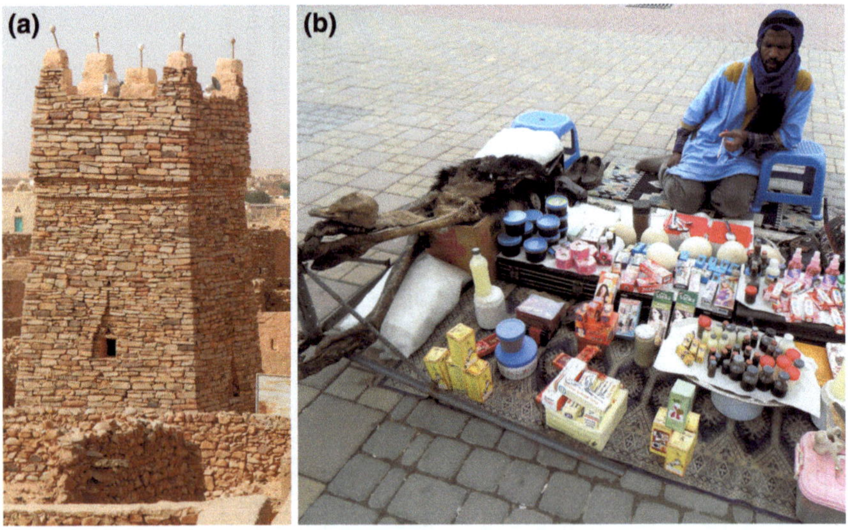


**Fig. 18.4** The most widely traded bird species is the ostrich, whose eggs and body parts are used for a variety of purposes. **a** Use of ostrich eggs at the tips of sticks on mosque minarets is common across West Africa. **b** A whole ostrich and ostrich eggs for sale by a Berber trader in Morocco (Photo credits: Peter Howard)

While there is a geographical difference in the occurrence of species and hence their presence in a market, selection of birds for medicine is often at a less specific 'morphospecies' level (i.e. a typological species that can only be identified as owl, vulture, kingfisher, etc.). When re-evaluating the frequencies of birds in the markets of seven countries, the most prevalent morphospecies were: (1) 'eagles' (including kites and buzzards) and owls (seven countries each); followed by (2) vultures, crows, hornbills, guineafowls, francolins, kingfishers, and herons (six countries); and (3) ground-hornbills (five countries). Parrots and turacos were also frequently represented (four and three countries each, respectively).

Based on Sørenson's index of similarity, the countries most similar in terms of the species present in the markets are BJ and NG (53% similar; 96 species common), followed by NG and ZA (26% similar, 41 species common), and BJ and ZA (23% similar, 30 species common). These three countries also have 23 species in common between them. Despite the limited survey work conducted in BF, the species sold in the market there are 25, 21, and 16% similar to the species sold in BJ, NG, and ZA, respectively.

### *18.3.3 Rabinowitz Classification and Patterns of Rarity and Commonness*

Results from the rarity classifications of species into categories from A to H, showed that 69 traded bird species (24%) are very common and that they are locally abundant in several habitats over a large geographic area (Category A, Fig. 18.5a, Table 18.6). Ten percent of traded species (28 species) are, however, rare and are considered to be constantly sparse in specific habitats over a small geographic area (Category H, Fig. 18.5a, Table 18.6). When comparing the observed number of traded species in the eight categories for ZA, NG, and BJ (Fig. 18.5b), the countries had similar proportions of species allocated to each category. The major differences seem to be, however, that ZA has a smaller proportion of species that are constantly sparse in several habitats over a large geographic area (category C) and species that are locally abundant in a specific habitat over a small geographic area (category F). NG and BJ have a larger proportion of species with narrower distributions.

When we examined the proportion of common and rare species amongst the various Orders, we found that the Strigiformes (Owls) had the highest proportion of species in Category H (23%) compared to other Orders, followed by Bucerotiformes (Hornbills) (19%) and Falconiformes (Diurnal birds of prey) (15%). When we calculated the mean rank per Order (based on Table 18.3) (excluding Orders with less than five species), it indicated that Bucerotiformes had the highest proportion of rare species (mean rank = 1.95; $S = 21$ species), followed by Gruiiformes (Cranes) (mean = 2.31; $S = 16$ species), Galliformes (Gamebirds) (mean rank = 2.33; $S = 9$ species) and Strigiformes (Owls) (mean = 2.54; $S = 13$) (Table 18.7). The Orders with the highest proportion of common species are Columbiformes (Doves and Pigeons)

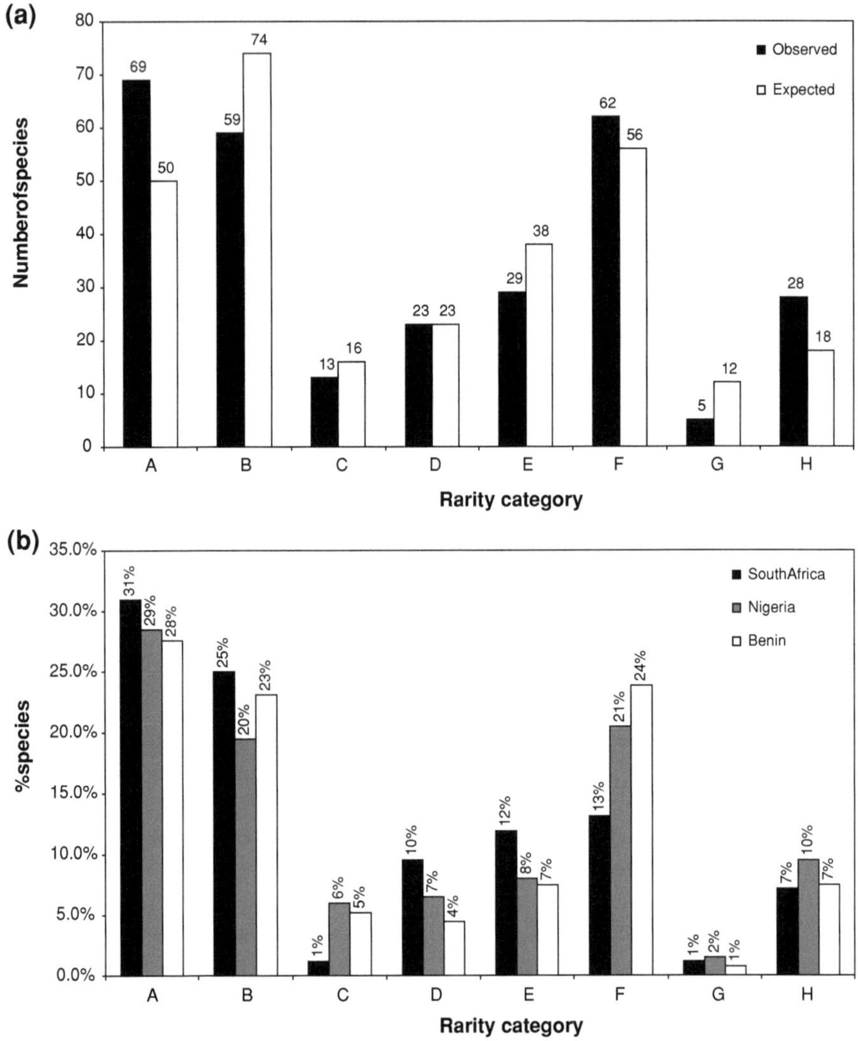


**Fig. 18.5** **a** The observed and expected number of traded species in each the eight categories of commonness or rarity. Differences were not significant at $p < 0.05$. **b** The observed number of traded species in each of the eight categories of commonness or rarity for two west African countries (BJ and NG) and for ZA

(mean = 3.67; $S = 9$), Coraciiformes (Kingfishers) (mean = 3.18; $S = 17$) and Cuculiformes (Cuckoos and relatives) (mean = 3.18; $S = 11$) (Table 18.7). The overall mean rank for all traded species was 2.73 ($S = 288$ species), indicating that there is a relatively high degree of 'uncommonness' among the traded species.

When examining the individual factors that contribute toward the rarity of avian species traded for traditional medicine (Table 18.8), we find that 43% of species have a 'small' geographic range (i.e. <5,980,000 $km^2$), 24% are non-dominant or

**Table 18.6** IUCN red list status, rarity categories, and population trends for traded species (excluding PALs)

| | IUCN red list status | | | | Total $N$ | Population trends | | | |
|---|---|---|---|---|---|---|---|---|---|
| Rarity | EN | VU | NT | LC | | Stable | Decreasing | Increasing | ? |
| A | 1 | | | 68 | 69 | 39 | 15 | 12 | 3 |
| B | | | 1 | 58 | 59 | 31 | 18 | 8 | 2 |
| C | 1 | | 2 | 10 | 13 | 7 | 5 | 0 | 1 |
| D | | 4 | 2 | 17 | 23 | 8 | 14 | 0 | 1 |
| E | | | | 29 | 29 | 17 | 6 | 5 | 1 |
| F | 1 | 2 | 1 | 58 | 62 | 36 | 17 | 2 | 7 |
| G | | | | 5 | 5 | 3 | 1 | 0 | 1 |
| H | | 3 | 2 | 23 | 28 | 13 | 12 | 0 | 3 |
| Total | 3 | 9 | 8 | 268 | 288 | 154 | 88 | 27 | 19 |
| % | 0.7 | 2.7 | 3.0 | 94 | – | 53.5 | 30.6 | 9.4 | 6.6 |

*EN* Endangered, *VU* Vulnerable, *NT* Near Threatened, *LC* Least Concern

constantly sparse within their range, and 60% are quite habitat specific. The mean EOO for species in Category A (Rank 4, the least rare species) is 22,040,851 km$^2$, compared to 2,369,222 km$^2$ for species in Category H (Rank 1, the most rare species) (Table 18.9). Species with intermediate rarity/commonness (Ranks 2 and 3, corresponding to categories B–G) range between 4,620,388 and 11,668,275 km$^2$. The mean EOO for all traded species for which data were available was 10,150,979 km$^2$ ($S = 239$), well above the ~4.6 million km$^2$ (range 2.4–22.0 million) median for 'all' (Sect. 2.3) species.

### 18.3.4 IUCN Red List Status and Index

Based on their 2011 IUCN Red List Status, 12 traded species are Threatened (three Endangered; six Vulnerable) and eight are Near Threatened (excluding PALs), and the majority of species (94%) are categorized as Least Concern (Tables 18.1, 18.10). Over the last 11 years, there has been a gradual decline in the RLI values of the traded species from 0.99 in 2000 and 2004, to 0.98 in 2008 and 0.976 in 2011, indicating that the species have become more threatened and are more at risk of extinction. For the most part, the populations of traded species are stable (54%) and/or increasing (9% Table 18.6). However, populations are declining for 31% of species. Major threats to birds vary with species (Owens and Bennett 2000; Julliard et al. 2003), so in devising conservation strategies, it is important not to focus unnecessarily on the traditional medicine trade. However, as we later discuss, threatened species that are specifically targeted by traditional medicine hunters (e.g. vultures, ground-hornbills, hornbills and various eagles) will experience more significant declines in population numbers if hunting is persistent.

**Table 18.7** Total number of traded species per order and mean rank scores of rarity for species per order

| Order | Total number traded species per order[a] | Mean rank[b] |
|---|---|---|
| Bucerotiformes (Hornbills) | 21 | 1.95 |
| Procellariiformes (Albatrosses and relatives) | 1 | 2.00 |
| Pteroclidiformes (Soundgrouses) | 1 | 2.00 |
| Sphenisciformes (Penguins) | 1 | 2.00 |
| Trogoniformes (Trogons and relatives) | 1 | 2.00 |
| Gruiiformes (Cranes and relatives) | 16 | 2.31 |
| Galliformes (Gamebirds) | 9 | 2.33 |
| Psittaciformes (Parrots) | 5 | 2.40 |
| Strigiformes (Owls) | 13 | 2.54 |
| Caprimulgiformes (Nightjars and relatives) | 3 | 2.67 |
| Pelecaniformes (Pelicans and relatives) | 7 | 2.71 |
| Passeriformes (Perching birds) | 72 | 2.74 |
| Charadriiformes (Gulls and relatives) | 16 | 2.75 |
| Upupiformes (Hoopoes) | 4 | 2.75 |
| Falconiformes (Diurnal birds of prey) | 43 | 2.76 |
| Piciformes (Woodpeckers and relatives) | 9 | 2.78 |
| Anseriformes (Waterfowl) | 6 | 2.83 |
| Podicipediformes (Grebes) | 1 | 3.00 |
| Ciconiiformes (Storks) | 20 | 3.05 |
| Cuculiformes (Cuckoos and relatives) | 11 | 3.18 |
| Coraciiformes (Kingfishers and relatives) | 17 | 3.18 |
| Apodiformes (Swifts and relatives) | 2 | 3.50 |
| Columbiformes (Doves and pigeons) | 9 | 3.67 |
| Coliiformes (Mousebirds) | 1 | 4.00 |
| Struthioniformes (Ostrich/Ratites) | 1 | 4.00 |
| Total | 288 | 2.73 |

Listed in order of most rare to least rare orders ([b] derived from rank scores for categories in Table 18.3)
[a] Excludes exotics and PALs

**Table 18.8** Number and percentage of traded species dichotomized according to distribution range, population size, and habitat specificity factors

| Factor | | No. of species ($S$ = 188) | % |
|---|---|---|---|
| *Distribution range* | | | |
| Large | Geographic area EOO >4,630,000 km$^2$ | 164 | 57 |
| Small | Geographic area EOO <4,630,000 km$^2$ | 124 | 43 |
| *Population* | | | |
| High | Dominant somewhere/locally dominant | 219 | 76 |
| Low | Non-dominant/constantly sparse | 69 | 24 |
| *Habitat* | | | |
| Broad | Several habitats | 116 | 40 |
| Narrow | Specific habitats | 172 | 60 |

**Table 18.9** Mean EOOs in the four rarity categories (corresponding with the ranks in Table 18.3)

| | Valid *N* | Mean (km$^2$) | S. D. | Minimum | Maximum |
|---|---|---|---|---|---|
| 1 | 27 | 2,369,222 | 1,178,383 | 400,000 | 3,880,000 |
| 2 | 85 | 4,620,388 | 4,092,192 | 130,000 | 23,900,000 |
| 3 | 80 | 11,668,275 | 10,243,678 | 125,000 | 52,400,000 |
| 4 | 47 | 22,040,851 | 19,764,934 | 7,180,000 | 99,700,000 |
| Total | 239 | 10,150,979 | 12,762,529 | 125,000 | 99,700,000 |

**Table 18.10** IUCN red list status, rarity categories, and population trends for traded species (excluding PALs)

| | IUCN red list status | | | | Total *N* | Population trends | | | |
|---|---|---|---|---|---|---|---|---|---|
| Rarity | EN | VU | NT | LC | | Stable | Decreasing | Increasing | ? |
| A | 1 | | | 68 | 69 | 39 | 15 | 12 | 3 |
| B | | | 1 | 58 | 59 | 31 | 18 | 8 | 2 |
| C | 1 | | 2 | 10 | 13 | 7 | 5 | 0 | 1 |
| D | | 4 | 2 | 17 | 23 | 8 | 14 | 0 | 1 |
| E | | | | 29 | 29 | 17 | 6 | 5 | 1 |
| F | 1 | 2 | 1 | 58 | 62 | 36 | 17 | 2 | 7 |
| G | | | | 5 | 5 | 3 | 1 | 0 | 1 |
| H | | 3 | 2 | 23 | 28 | 13 | 12 | 0 | 3 |
| Total | 3 | 9 | 8 | 268 | 288 | 154 | 88 | 27 | 19 |
| % | 0.7 | 2.7 | 3.0 | 94 | – | 53.5 | 30.6 | 9.4 | 6.6 |

*EN* Endangered, *VU* Vulnerable, *NT* Near Threatened, *LC* Least Concern

## 18.4 Discussion

Based on Butchart's (2008) global assessment, we know that at least 45.7% of extant bird species (4,561 species) are used by people, primarily as pets (37.0%) or for food (14.2%). Much less is known about bird use for traditional medicine. This study aimed to start filling that gap, showing that there is a greater prevalence and diversity of bird species in African traditional medicine markets than are used for traditional medicines on any other continent. In South America, for example, only six bird species (out of 97 animal species) were recorded in the comprehensive surveys of traditional medicine markets in Brazil (Alves and Rosa 2010). In India, only 12 bird species were used for traditional medicine out of 109 animal species (Mahawar and Jaroli 2008). In Brazil and India, medicinal use of domestic chickens was recorded. No birds are recorded as commonly traded in Chinese traditional medicine Zhao (2004), apart from one notable exception. This is the massive regional trade in swiftlet nests (from *Collocalia fuciphagus* and *C. maximus*) from southeast Asia to China. These are marketed by companies such as Eu Yan Sang (www.euyansang.com) as a "health food" and are sold in China as a luxury food (*yàn wō*) that we know as "bird's nest soup". Trade in swiftlet nests are in the grey area between food and medicine described by Etkin and Ross (1982), with this

Asian trade well studied and managed (Hobbs 2003; Sankaran 2001). In contrast, relatively few studies have been carried out on trade in birds for traditional medicine or to assess the levels of vulnerability of bird species to this trade.

In their seminal study of the main causes of extinction risk in bird species, Owens and Bennett (2000) made the important point that different bird lineages may follow different paths to extinction. The first of these paths was for large-bodied, slow-breeding bird species that become threatened when the fecundity-mortality balance is disrupted, for example by human use or by introduced predators. At a global scale, they pointed to the kiwis (Apterigidae), cassowaries (Casuariidae), megapodes (Megapodiidae), penguins (Spheniscidae), and albatrosses, petrels and allies (Procellariidae) as being prone to extinction in this way. The second path they suggested was the effects of habitat loss on ecologically specialized bird species, giving the examples of hummingbirds (Trochilidae), trogons (Trogonidae), scrub-birds (Menuridae), and logrunners (Orthonychidae), as families prone to this path. Owens and Bennett (2000) also suggested that a small number of bird families faced both sources of extinction risk, such as parrots (Psittacidae), rails (Rallidae), pheasants, partridges and francolins (Phasianidae), pigeons (Columbidae), cranes (Gruidae), and white-eyes (Zosteropidae).

Bird use in traditional medicine and the risks of extinction need to be considered in context against other threats. For example, habitat loss and the reduction in tree cover due to clearing by people, affects ground-hornbills (Vernon 1986; Msimanga 2004). Elephants also reduce tree cover, and vultures in Swaziland were noted not to be nesting in areas where elephants occurred (Monadjem and Garcelon 2005). Vultures are further affected by habitat loss, land-use change, collisions with aircraft and power lines (Herholdt and Anderson 2006, and poisoning. Large-scale mortalities have been caused by deliberate and inadvertent poisoning with pesticides such as the arbamate-based pesticide Furadan$^{TM}$ Otieno et al. 2010). In ZA, poisoning has been the main factor causing a decline in Bateleur Eagle populations (de Kock and Watson 1985). However, threatened species that are specifically targeted by traditional medicine hunters (e.g. vultures, ground hornbills, hornbills, and various eagles) will experience more significant declines in population numbers if hunting persists and no viable alternatives are available.

In the African traditional medicines trade, our data showed that 24% of traded bird species are very common and locally abundant in several habitats over a large geographic area, and so are unlikely to be of conservation concern. We also suggest that compared to the multiple factors influencing habitat loss, the traditional medicine trade plays a lesser role in the decline of species. Even in the case of birds in families prone to extinction (such as penguins (Spheniscidae), albatrosses and petrels (Procellariidae) (Owens and Bennett 2000) that are also used for traditional medicine, our personal observations are that these birds have probably died indirectly from other causes that may include stranding. Furthermore, the presence of species such as the Shy Albatross in a South African market (Table 18.1), may be the result of an opportunistic discovery rather than targeted harvesting events.

Nevertheless, the African traditional medicines trade can be a significant factor that affects the conservation status of many of the large-bodied, slow-breeding bird species

**Fig. 18.6** **a** A typical open daily market for traditional medicines in west Africa: Ouagadougou, BF. **b** A traditional medicines trader showing the diversity of products sold, from *Achatina* sp. land-snail shells and aardvark feet to reptiles and birds. **c** Heads from black Crowned-cranes and vultures on sale in bouake market, Cote d'Ivoire in 1989 (Photos: A. B. Cunningham)

that are already of conservation concern (Table 18.2). In particular, some Accipitridae (all vultures and some eagle species), large hornbills (Bucerotidae), particularly the Abyssinian and Southern Ground-hornbills, and in some countries, such as Cote d'Ivoire, trade in cranes (Gruidae), such as the Black Crowned Crane (Fig. 18.6).

Appropriate conservation strategies for utilized species should ideally be based on a thorough understanding of bird biology. The pilot project to harvest and rear redundant second-hatched chicks, that would otherwise starve and/or be killed by the first hatched elder sibling, as a sustainable source of Southern Ground-hornbill,

African Hawk-eagle (*Hieraaetus spilogaster*) and Milky Eagle Owl (Bubo lacteus) individuals (Kemp 2000) is one such example. Ostrich farming is another. Although ostriches are widely sold for traditional medicine and are the largest-bodied African bird, most ostrich parts (eggs, feet, feathers) sold in southern Africa are from farmed birds. In other parts of Africa, including Morocco (Fig. 18.4), ostrich parts are probably obtained from wild birds that have been hunted. Conversely, effective conservation strategies should also be based on a good understanding of why particular birds are selected for use across diverse African healing traditions. Such an understanding requires an appreciation of the cultural context that has long been appreciated by many anthropologists, but which is rare amongst many conservation biologists or ornithologists.

**Acknowledgments** We would like to thank the many traditional healers we have worked with over the years, most of whom preferred to stay anonymous. TC would like to thank Siyabonga Zondi for his good company and assistance in surveying traditional medicine markets in the 1980s. We also appreciate the permission from Peter Howard to use some of his photographs taken in Morocco and Togo. VW was provided funding for this research by the University of the Witwatersrand SPARC Prestigious and URC Postdoctoral Fellowships. RB acknowledges funding from the National Research Foundation.

## References

Adeola MO (1992) Importance of wild animals and their parts in the culture, religious festivals, and traditional medicine. Environ Conserv 19:125–134

Adjakpa JB, Tchabi A, Ogouvide FT (2002) Oiseaux utilisés en pharmacopée traditionnelle au Bénin. Malimbus 24:1–14

Alves RRN, Rosa IL (2010) Trade of animals used in Brazilian traditional medicine: trends and implications for conservation. Hum Ecol 38:691–704. doi:10.1007/s10745-010-9352-0

Baskind R, Birbeck G (2005) Epilepsy care in Zambia: a study of traditional healers. Epilepsia 46(7):1121–1126. doi:10.1111/j.1528-1167.2005.03505.x

Beninguisse G, De Brouwere V (2004) Tradition and modernity in Cameroon: the confrontation between social demand and biomedical logistics of health services. Afr J Reprod Health 8(3):152–175

BirdLife International (2008a) The status of the world's birds has deteriorated over the last 20 years. Presented as part of the birdlife state of the world's birds website. http://birdlife.org/datazone/sowb/casestudy/72. Accessed 21 Dec 2010

BirdLife International (2008b) Nearly half of the bird species are used directly by people. Presented as part of the birdlife state of the world's birds website. http://birdlife.org/datazone/sowb/casestudy/98. Accessed 20 Dec 2010

BirdLife International (2008c) The red list index for utilized bird species illustrates their deterioration in status. Presented as part of the birdlife state of the world's birds website. http://birdlife.org/datazone/sowb/casestudy/161. Accessed 20 Dec 2010

BirdLife International (2011) IUCN red list for birds. http://www.birdlife.org. Accessed 4 Feb 201

Bubb PJ, Butchart SHM, Collen B, Dublin H, Kapos V, Pollock C, Stuart SN, Vié J-C (2009) IUCN red list index—guidance for national and regional use. IUCN, Switzerland

Buckley AD (1985) Yoruba medicine. Oxford University Press, New York

Butchart SHM (2008) Red list indices to measure the sustainability of species use and impacts of invasive alien species. Bird Conserv Int 18(Suppl 1):S245–S262. doi:10.1017/S095927090800035x

Cocker M (2000) African birds in traditional magico-medicinal use—a preliminary survey. Bull Afr Bird Club 7(1):60–66

Cocker M, Mikkola H (2001) Magic, myth and misunderstanding: cultural responses to owls in Africa and their implications for conservation. Bull Afr Bird Club 8(1):30–35

Crump CM (2003) Need to assess animals used by traditional healers in south Africa. In: Williams VL (ed) Hawkers of health: an investigation of the Faraday street traditional medicine market in Johannesburg. Report to Gauteng directorate for nature conservation. Appendix 4

Cunningham AB (1990) Vultures and the trade in traditional medicine. Vulture News 24:3–10

Cunningham AB (2001) Applied ethnobotany. Earthscan, London

Cunningham AB, Zondi AS (1991) Use of animal parts for the commercial trade in traditional medicines. Institute of Natural Resources, Pietermaritzburg

de Kock AC, Watson RT (1985) Organochlorine residue levels in bateleur eggs from the transvaal. Ostrich 56:278–280

Derwent S, Mander M (1997) Twitchers bewitched. The use of birds in traditional healing. Afr Birds Bird 2(1):22–25

El-Kamali HH (2000) Folk medicinal use of some animal products in central Sudan. J Ethnopharmacol 72:279–282

Etkin NL, Ross PJ (1982) Food as medicine and medicine as food: an adaptive framework for the interpretation of plant utilization among the Hausa of northern Nigeria. Soc Sci Med 16: 1559–1573

Herholdt JJ, Anderson MD (2006) Observations on the population and breeding status of the African white-backed vulture, the black-chested snake eagle, and the secretarybird in Kgalagadi Transfrontier Park. Ostrich 77(3–4):127–135

Hobbs JJ (2003) Problems in the harvest of edible birds' nests in Sarawak and Sabah, Malaysian Borneo. Biodivers Conserv 13:2209–2226. doi:10.1023/B:BIOC.0000047905.79709.7f

Jepson P, Ladle RJ (2005) Bird-keeping in Indonesia: conservation impacts and the potential for substitution based conservation responses. Oryx 39:442–448. doi:10.1017/S0030605305001110

Julliard R, Jiguet F, Couvet D (2003) Common birds facing global changes: what makes a species at risk? Glob Change Biol 10:148–154. doi:10.1046/j.1529-8817.2003.00723.x

Kemp A (2000) The sustainable utilisation of birds. Emu 100:355–365. doi:10.1071/MU0003S

Kizungu B, Ntabaza M, Mburinge M (1998) Ethno-ornithology of the tembo in the eastern DRC (former Zaire): part one, Kalehe zone. Afr Study Monogr 19(2):102–113

Lacaille J (1951) Mai Mai, Harley Street of Jo'burg's "muti" men. Outspan 9(47–49):81

Loeb EM, Koch C, Loeb E-MK (1956) Kuanyama ambo magic. 6. Medicinal, cosmetical, and charm flora and fauna. J Am Folklore 69(272):147–174

Macdonald I (1985) Witch-doctors versus wildlife in southern Africa. Afr Wildl 38(1):4–9

Mahawar MM, Jaroli DP (2008) Traditional zootherapeutic studies in India: a review. J Ethnobiol Ethnomed 4:17. doi:10.1186/1746-4269-4-17

Mander M, Diederichs N, Ntuli L, Mavundla K, Williams V, McKean S (2007) Survey of the trade in vultures for the traditional health industry in south Africa. Future Works, Durban

Marshall NT (1998) Searching for a cure: conservation of medicinal wildlife resources in east and southern Africa. Traffic International, Cambridge

Monadjem A, Garcelon DK (2005) Nesting distribution of vultures in relation to land use in Swaziland. Biodivers Conserv 14:2079–2093. doi:10.1007/s10531-004-4358-9

Msimanga A (2004) Breeding biology of southern ground hornbill *Bucorvus leadbeateri* in Zimbabwe: impacts of human activities. Bird Conserv Int 14:563–568. doi:10.1017/S0959270905000237

Nevadomsky J (1988) The apothecary shop in Benin city. Afr Arts 22(1): 72–83,100

Ngwenya MP (2001) Implications for the medicinal animal trade for nature conservation in KwaZulu-Natal. Unpublished Ezimvelo KZN Wildlife report no. NA/124/04

Nikolaus G (2000) Bird exploitation for traditional medicine in Nigeria. Malimbus 23:45–55

Ntiamoa-Baidu Y (1987a) West African wildlife: a resource in jeopardy. Unasylva 156:27–35

Ntiamoa-Baidu Y (1987b) Tropical rain forest management a status report. West African wildlife: a resource in jeopardy. Unasylva 156:1–10

Ogada DL (2007) The ecology and conservation of mackinder's eagle owls (bubu capensis mackinderi) in central Kenya in relation to agricultural land-use and cultural attitudes. Thesis, Rhodes University, Grahamstown

Otieno PO, Lalah JO, Virani M, Jondiko IO, Schramm K-W (2010) Carbofuran and its toxic metabolites provide forensic evidence for furadan exposure in vultures (*Gyps africanus*) in Kenya. Bull Environ Contam Toxicol 84(5):536–544. doi:10.1007/s00128-010-9956-5

Owens IPF, Bennett PM (2000) Ecological basis of extinction risk in birds: habitat loss versus human persecution and introduced predators. Proc Nat Acad Sci 97:12144–12148. doi:10.1073/pnas.200223397

Rabinowitz D (1981) Seven forms of rarity. In: Synge H (ed) The biological aspects of rare plant conservation, 1st edn. Wiley, Chichester

Rabinowitz D, Cairns S, Dillon T (1986) Seven forms of rarity and their frequency in flora of the British Isles. In: Soule ME (ed) Conservation biology, the science and scarcity of diversity, 1st edn. Sinauer, Sunderland

Sankaran R (2001) The status and conservation of the edible-nest swiftlet (*Collocalia fuciphaga*) in the Andaman and Nicobar islands. Biol Conserv 97:283–294

Serge BK (2010) Birds, the Banso people of northwestern Cameroon and their culture. In: Abstracts from the 12th ISE—recent developments in ethnoornithology. Torfino, Canada. http://uk.groups.yahoo.com/group/Ethnoornithology/message/579

Serle W, Morel JG (1979) Les Oiseaux de l'Ouest Africain. Delachaux et Niestlé, Neuchâtel

Simelane TS, Kerley GIH (1998) Conservation implications of the use of vertebrates by Xhosa traditional healers in South Africa. S Afr J Wildl Res 28:121–126

Sinclair I, Ryan P (2010) Chamberlain's birds of Africa south of the Sahara. Struik, Cape Town

Sodeinde OA, Soewu DA (1999) Pilot study of the traditional medicine trade in Nigeria. Traffic Bull 18:35–40

Soewu DA (2008) Wild animals in ethnozoological practices among the Yorubas of southwestern Nigeria and the implications for biodiversity conservation. Afr J Agric Res 3(6):421–427

Taylor EM, Fox J (1992) The fetish market, Lomé, Togo. Niger Field 57:119–125

Thomsen J, Edwards SR, Mulliken TA (1992) Perceptions, conservation and management of wild birds in trade. TRAFFIC, Cambridge

Vernon CJ (1986) The ground hornbill at the southern extremity of its range. Ostrich 57:16–24

White RM, Cocks M, Herbert DG, Hamer ML (2004) Traditional medicines from forest animals. In: Lawes MJ, Eeley HAC, Shackleton CM, Geach BGS (eds) Indigenous forests and woodlands in south Africa: people, policy and practice. University of Natal Press, Pietermaritzburg

Whiting MJ, Williams VL, Hibbitts TJ (2011) Animals traded for traditional medicine at the Faraday market in south Africa: species diversity and conservation implications. J Zool 284(2):84–96. doi:10.1111/j.1469-7998.2010.00784.x

Yu J, Dobson FS (2000) Seven forms of rarity in mammals. J Biogeogr 27:131–139. doi:10.1046/j.1365-2699.2000.00366.x

Zhao Z (2004) An illustrated Chinese material medica in Hong Kong. Chun Hwa Book Company, Hong Kong

# Chapter 19
# Animals Traded for Traditional Medicine at the Faraday Market in South Africa: Species Diversity and Conservation Implications

**Martin J. Whiting, Vivienne L. Williams and Toby J. Hibbitts**

**Abstract** In South Africa, animals and plants are commonly used as traditional medicine for both the healing of ailments and for symbolic purposes such as improving relationships and attaining good fortune. The aim of this study was twofold: to quantify the species richness and diversity of traded animal species and to assess the trade in species of conservation concern. We surveyed the Faraday traditional medicine market in Johannesburg and conducted 45 interviews with 32 traders during 23 visits. We identified 147 vertebrates representing about 9% of the total number of vertebrates in South Africa and about 63% of the total number of documented species (excluding domestic animals) traded in all South African traditional medicine markets. The vertebrates included 60 mammal species, 33 reptiles, 53 birds and one amphibian. Overall, species diversity in the Faraday market was moderately high and highest for mammals and birds, respectively. Evenness values indicated that relatively few species were dominant. Mammal body parts and bones were the most commonly sold items ($n = 453$, excluding porcupine quills and pangolin scales), followed by reptiles ($n = 394$, excluding

Published in Journal of Zoology, 2011, 284: 84–96.

M. J. Whiting (✉)
Department of Biological Sciences, Macquarie University,
Sydney, NSW 2109, Australia
e-mail: martin.whiting@mq.edu.au

M. J. Whiting · V. L. Williams · T. J. Hibbitts
School of Animal, Plant and Environmental Sciences,
University of the Witwatersrand, Private Bag 3,
2050 Wits, South Africa

T. J. Hibbitts
Department of Wildlife and Fisheries Sciences,
TexasA&M University, College Station,
Texas, TX 77843-2258, USA

R. R. N. Alves and I. L. Rosa (eds.), *Animals in Traditional Folk Medicine*,
DOI: 10.1007/978-3-642-29026-8_19, © Springer-Verlag Berlin Heidelberg 2013

osteoderms), birds ($n = 193$, excluding feathers and ostrich eggs), and amphibians ($n = 6$). Most (87.5%) species traded were of Least Concern using IUCN criteria, although 17 species were of conservation concern. However, a higher than expected proportion of traders (62.5%) were selling listed species, which is a matter for concern and should be monitored in the future.

**Keywords** Biodiversity · Threatened species · Ethnozoology · Mammal · Bird · Reptile

## 19.1 Introduction

Burgeoning human populations not only put pressure on biodiversity through competition for space and limited resources (Ehrlich 2009), but also through direct harvest for human consumption (Wilson 1988). Much of the focus relating to the current global biodiversity crisis is on habitat destruction and the unsustainable use of resources. Conservation practices tend to focus on arresting or ameliorating habitat destruction because biodiversity is conserved as a by-product. What has received far less attention from ecologists and conservation biologists is the harvest of animals for use in traditional medicine. Where rare and endangered species are concerned, the use of biodiversity for traditional medicine can have potentially significant impacts on local populations that are already under pressure (Simelane and Kerley 1998; Still 2003; White et al. 2004; Mander et al. 2007; Williams et al. 2007a). Some of the more notable examples of harvest for traditional medicine include rhino horns, bear gall bladders and tiger penises for the Asian market (But et al. 1990; Li et al. 1995; Still 2003). Many species of high value in traditional medicine may have low reproductive rates, be long-lived, and occur at relatively low densities in the wild. Species with these life history traits are considered more prone to extinction (McKinney 1997) and may therefore be less resilient to harvest.

The use of animal parts for the treatment of ailments affecting both humans and livestock has a long and rich history (Lev 2003). For example, bear gall bladders have been used to treat a variety of ailments in China for over 1,300 years (Li et al. 1995) while rhino horn has similarly been used in China for over 2,000 years (But et al. 1990). Traditional medicine in southern Africa falls into two categories: treatment of medical afflictions "white medicine" and dealing with ancestral conflict or "black medicine" (Bye and Dutton 1991). Traditional healers in southern Africa view health and welfare issues as being tightly linked to supernatural forces, social relationships and an individual's relationship with their ancestors (Bye and Dutton 1991; Simelane 1996). As such, a significant component of traditional healing makes use of the "magical" properties of plants or animal parts. For example, skins and parts from lions, leopards (Fig. 19.1a) and cheetah confer strength to the bearer, while other animal parts may be used to

**Fig. 19.1** Images of trader's stalls from the Faraday market in Johannesburg. **a** leopard paws; **b** a typical stall, selling a variety of animal parts including southern African python and southern ground hornbill; **c** a stall with mainly cowrie shells, crocodile osteoderms, porcupine quills, assorted bones and teeth; **d** a stall with an assortment of marine fauna including star fish, coral and a variety of fish; **e** a stall selling mainly tortoise shells (*Kinixys* sp.), pieces of elephant skin, giant land snails (*Achatina* sp.) and assorted bones; **f** assorted animals, including pangolin scales and an aardvark foot. Photo credits: MJ Whiting—**a**, **c**, **d**, **e** and VL Williams—**b**, **f**)

provide protection against enemies, as a charm in a court case, for intelligence in school children, prosperity and good fortune, to strengthen a relationship, or even to aid an individual committing a crime (Simelane 1996; Cocks and Dold 2000;

White et al. 2004; Mander et al. 2007). A large proportion of South Africans believe in the efficacy of traditional medicine and have at some time purchased traditional medicine or consulted a traditional healer (Cunningham and Zondi 1991; Mander et al. 2007). Furthermore, South Africa has a very low ratio of western doctors to patients (Williams 2007), particularly in rural areas, which leaves very little opportunity for consultation with university trained medical doctors. By comparison, traditional healers are far more accessible to most of the population (Bye and Dutton 1991).

The trade in animal parts in southern Africa is thought to be extensive, but is currently poorly understood and only baseline data has been collected for select areas. Furthermore, many of these studies are internal, unpublished reports. Herbert et al. (2003) report a comprehensive assessment of the invertebrate trade at the Warwick Triangle traditional medicine market in Durban and also briefly review the trade in animal parts for traditional medicine. Specifically, they report on diversity, monetary values and rough quantities of taxonomic groups that were traded and identify a few select marine invertebrates of conservation concern. Simelane and Kerley (1998) interviewed traditional healers in the Eastern Cape Province of South Africa and found that 31% of vertebrates traded were listed in South African Red Data books. Mander et al. (2007) focused on the trade in vultures and identified the demand for traditional medicine as a significant threat to the future viability of several species of vulture. All these studies point to the growing need for baseline data and proper quantification of the trade in animal parts for traditional medicine and whether this trade makes significant use of species of conservation concern.

The purpose of this study was to investigate the trade in animals for traditional medicine at the Faraday traditional medicine market (hereafter Faraday) in Johannesburg, South Africa. We quantified species richness, diversity and quantity of vertebrate and marine invertebrate fauna sold by traders. Finally, we examined the trade of animals in relation to their conservation assessment using IUCN criteria.

## 19.2 Methods

Faraday is the largest informal wholesale and retail market for traditional medicine within the province of Gauteng (Williams 2003), and the second largest outlet for traditional medicine in South Africa after the Warwick Junction market in Durban (Herbert et al. 2003). Previous studies at Faraday have focused on the trade in plant material (Williams 2003; Williams et al. 2005, 2007a b, c), but until now, the trade in animal material has not been assessed. A Faraday survey in 2001 revealed that 5% of traders sold only animal parts while 10% sold a combination of plant and animal material (Williams 2003).

### 19.2.1 Market Survey

We used undergraduate students proficient in local languages (isiZulu, Sesotho) to conduct 45 interviews with 32 traders during 23 visits to Faraday to compile an inventory of animal species available for sale. The survey was conducted between June 2004 and November 2005. Animal identifications were made at the market, although photographs were also taken at most of the stalls as documentary evidence and for identifying some species. Identification to species was further aided by field guides for the major vertebrate groups (birds: Sinclair et al. 1997; reptiles: Branch 1998; mammals: Stuart and Stuart 2001). When we totalled species in a particular taxonomic group, we conservatively counted the minimum number of potential species. For example, in the case of "scrub hare, rock rabbit and unidentified rabbit." we would only count two species. We recorded all domestic animals for sale, but do not include them in any taxonomic counts or in any of the analyses. We treated marine fishes and invertebrate species separately to mammals, birds, reptiles and amphibians, primarily because of the difficulty in identifying the individual species of molluscs, echinoderms, corals and dried fishes. These species were also not included in the diversity analyses, but are discussed separately. Our primary approach was to record observable data and to limit the questioning of the traders due to the difficulties expected with obtaining honest/reliable information, especially concerning the origin of the material. We designed a survey form to list the species, quantities (number of individual organisms) and carcass parts sold. We did not record data on the origin or monetary value of the material. The animal fats and mixtures separately sold in bottles were not recorded since there was no way to verify the identity of the material (Fig. 19.2a). Furthermore, some wholesalers of traditional medicine sell "imitation" fat (often domestic animal fat) to consumers (Cunningham and Zondi 1991).

### 19.2.2 Sampling Performance

Complete enumeration of species within a study area is generally not feasible and consequently a number of methods have been devised for estimating total species richness from a sample (Chiarucci et al. 2003). A challenge for ethno-ecological surveys is establishing the completeness of an inventory and how many more species might be recorded with further sampling of the market (Williams et al. 2007c). We used incidence-based species richness estimators calculated by the public-domain software *EstimateS* (Version 7.5.1, Colwell 2006; viz. ICE, Chao 2, first-order jackknife, second-order jackknife, bootstrap and Michaelis–Menton Means) to estimate the number of species that may have been recorded with further sampling. Incidence based rather than abundance-based estimators are more suitable for market data because inventories mostly record the presence or absence of species rather than the abundance or quantity present. Furthermore, certain body parts such as porcupine quills, pieces of skin and feathers may be highly abundant, but at the

**Fig. 19.2** Examples of raw materials derived from medicinal animals sold in the Faraday market in Johannesburg. **a** assorted animal fats sold in bottles; **b** assorted birds: **c** CMR Bean Beetles (*Mylabris oculata*); **d** assorted animals, including pangolin scales and an aardvark foot; d) trader with a peacock. Photo credits: VL Williams—**a**, **b**, **c** and MJ Whiting—**d**

same time may be harvested from a few individual animals, making quantity an unsuitable abundance variable for calculating the estimators.

We assessed the "best" estimator based on its ability to reach a horizontal asymptote (Toti et al. 2000; Williams et al. 2007c). Furthermore, the richness estimates were compared with a list of species compiled from other studies of animals traded for traditional medicine. The list indicated how many more species have been recorded for sale in other markets in South Africa compared to Faraday. A good species richness estimator would therefore not under estimate the total number of potential species in trade and the richness estimate should be at least greater than or equal to the total number of taxa recorded in all current and previous studies. The literature examined to compile the extended list of animals traded was based on Cunningham and Zondi (1991), Simelane (1996), Derwent and Mander (1997), Marshall (1998), Simelane and Kerley (1998), Ngwenya (2001), Crump (2003), Herbert et al. (2003), White et al. (2004), Mander et al. (2007). The study by Crump (2003) was a rapid assessment of nine traders selling animal parts in Faraday in 2001.

### *19.2.3 Species Richness, Similarity and Diversity*

We calculated species richness, the percentage similarity of species sold by different traders (using the Sørenson index for incidence-based data), species accumulation functions and species diversity indices using EstimateS. These techniques have previously been effective in analysing and interpreting ethnobotanical inventories derived from assessments of resource use in South Africa and South America (e.g. Williams et al. 2005, 2007c; Begossi 1996; Hanazaki et al. 2000). We randomised the sample order (i.e. trader order) 50 times to compute the mean statistic at each sample accumulation level and thereby generated smoothed accumulation curves. EstimateS directly computes the Shannon ($H'$), Simpson ($1/\lambda$) and Fisher's $\alpha$ diversity indices. We used the $-\ln \lambda$ form of Simpson's index (see Williams et al. 2005), hence the data were transformed accordingly. The software does not directly compute Hill's numbers or evenness values; however, the appropriate variables for calculating these indices are an output of EstimateS and these values were subsequently derived using the appropriate formulae indicated in Table 19.1. Because the values for the diversity indices are computed at each sample accumulation level, it was possible to plot cumulative diversity curves that indicate how the indices perform as more traders were sampled.

Diversity measures take into account two factors: species richness (the number of species, $S$, in a sample of a specified size) and evenness/equitability (i.e. how uniformly abundant species are in a sample) (Magurran 1988). $S$ is related to the total number of individuals ($n$) summed over all $S$ species recorded (Williams et al. 2005). As sampling effort increases (e.g. more traders, $n$, are sampled) more individuals are encountered and more species are likely to be recorded (Hayek and Buzas 1997). An "index" of diversity (also called an index of heterogeneity, e.g. Simpson's index) incorporates both richness and evenness into a single value, and is based on the proportional abundance of species in a sample (Ludwig and Reynolds 1988; Magurran 1988). Part of the rationale behind calculating species diversity is that the more singletons (species occurring once) there are in a sample, the more one would expect to find at a site and therefore the greater the expected species diversity. The Shannon ($H'$) diversity index measures the average degree

**Table 19.1** Comparisons of selected measures of diversity between animals sold in the Faraday market

| Index/Measure | Animals ($n$ = 32 traders) $n$ = 608 | Mammals ($n$ = 32 traders) $n$ = 305 | Reptiles ($n$ = 31 traders) $n$ = 178 | Birds ($n$ = 22 traders) $n$ = 123 |
|---|---|---|---|---|
| Species richness ($S$ or $N_0$ or $e^{\mathrm{Hmax}}$) | 147 | 60 | 33 | 53 |
| Mean $S$ per trader ± SD | 25.1 ± 14.7 | 13.3 ± 7.6 | 7.1 ± 4.0 | 4.7 ± 5.2 |
| Shannon ($H'$) | 4.49 | 3.67 | 2.92 | 3.67 |
| Simpson (–ln $\lambda$) | 4.22 | 3.49 | 2.68 | 3.60 |
| Fisher's $\alpha$ | 61.6 | 22.4 | 11.9 | 35.3 |
| Evenness $E_1$(Shannon $J'$) ($H'/H'_{\mathrm{max}}$)[a] | 0.90 | 0.90 | 0.84 | 0.92 |
| Evenness $E_5$ ($N_2-1/N_1-1$) | 0.76 | 0.83 | 0.77 | 0.93 |
| Hill's $N_1$ ($e^{H'}$) | 89.1 | 39.3 | 18.5 | 39.3 |
| Hill's $N_2$ ($1/\lambda$) | 68.3 | 32.9 | 14.5 | 36.4 |
| Hill's $N$ ($N/N_{\mathrm{max}}$)[b] | 26.4 | 13.9 | 7.7 | 8.8 |
| Singletons (no. of species occurring once) | 57 | 18 | 13 | 26 |
| Mean number of shared species | 7.1 ± 4.8 (range = 0–23) $n$ = 496 | 4.1 ± 3.1 (range = 0–15) $n$ = 496 | 2.5 ± 1.8 (range = 0–6) $n$ = 496 | 0.8 ± 1.0 (range = 0–6) $n$ = 300 |
| Mean percentage Sørenson similarity of species sold by traders | 26.7 ± 11.7% (0–62.5%) $n$ = 496 | 28.0 ± 15.5% (0–80%) $n$ = 496 | 33.4 ± 17.8% (0–85.7%) $n$ = 496 | 13.1 ± 14.8% (0–65.7%) $n$ = 300 |

[a] $H_{max}$ = ln $S$ (maximum value of the Shannon index)
[b] $N_{max}$ = the number of individuals of the most abundant species

of "uncertainty" in predicting the identity of a species chosen at random from a sample (Ludwig and Reynolds 1988). The greater the uncertainty, the more difficult it is to predict the identity of a species and therefore the higher the diversity of the sample. The index is sensitive to the abundance of the rarest or least recorded species (Magurran 1988). Simpson's diversity index ($-\ln \lambda$) also increases as diversity increases and indicates the likelihood that two species chosen at random are the same species. The higher the diversity, the less likely two species chosen at random will be the same species. Simpson's index is sensitive to the abundances of the commonest or most recorded species (Magurran 1988). Fisher's $\alpha$ is a diversity index sensitive to sample size, the number of species and the number of species of intermediate abundance. When the number of species is low, alpha is lower and therefore smaller samples with fewer species usually have smaller values of $\alpha$ (Williams et al. 2005). Fisher's $\alpha$ is also a number close to the number of species expected to be represented by one individual (Hayek and Buzas 1997). Hill's numbers represent the number of species that are abundant ($N_1$), very abundant ($N_2$) and most abundant ($N_\infty$) in a sample. These numbers are derived from the Shannon, Simpson and Berger–Parker indices, respectively. Hill's numbers, especially $N$, can help indicate which species may be dominant in the market. To objectively determine the number of species which are of rare, intermediate or common abundance in the market, Williams et al. (2005) recommended transforming Hill's numbers in the following way: the number of common species $= N_\infty$; the number of species of intermediate abundance in the market $= N_1 - N_\infty$; the number of "rare" species (i.e. of low incidence) $= S - N_1$.

Evenness (or equitability) measures are another way of quantifying species dominance in a market. If all species are equally abundant throughout the market, then evenness values would be at a maximum of 1. The evenness value would decrease towards zero if the relative abundances of some species increased and they dominated the stalls in the market. The overall relative abundances of species thus determine the value of an evenness index. We used two evenness indices primarily to better differentiate between data sets if the resultant values from one index were the same. $E_1$ (also called the Shannon $J'$) is the most commonly used index but is sensitive to species richness and singletons (Ludwig and Reynolds 1988). $E_5$, however, tends to remain constant with sampling variations and tends to be independent of sample size (Ludwig and Reynolds 1988). The dispersion of species throughout the market (i.e. uniform, aggregated or random) was calculated using software called "Species diversity and richness" (version 3.02, 2002; Pisces Conservation Ltd., New Milton, UK).

### *19.2.4 Species of Conservation Concern*

All vertebrates were checked against 2001 IUCN Red List Categories and Criteria version 3.1 and used in conjunction with the following sources: Minter et al. (2004) for amphibians; IUCN (2009, Version 2009.1) for reptiles; Barnes (2000) and Hockey et al. (2005) for birds; and Friedmann and Daly (2004) for mammals. We tested whether species of conservation concern (IUCN categories: Critically Endangered [CR], Endangered [EN], Vulnerable [VU] or Near-Threatened [NT]) were proportionally as prevalent among traders as species of Least Concern (LC) (IUCN category) using $\chi^2$ tests (two-tailed). Because of low sample sizes we combined all species of conservation concern to meet the assumptions of the $\chi^2$ tests. A total of 136 species were scored for conservation status and of these, 119 were of LC while 17 were of conservation concern. Our expected values for the $\chi^2$ test were therefore 0.875 and 0.125, respectively. We also tested for a significant difference in the abundance of body parts (including entire animals) that were being sold, between species of conservation concern and species of LC. For this test, a total of 922 body parts were assigned to 136 species of which 17 were of conservation concern (64 items) while the remaining 119 species (865 items) were scored as LC. Therefore, we used expected values of 6.78 items/species under the null hypothesis that species were equally abundant, regardless of their conservation status. For this latter test, we excluded porcupine quills, eggs, feathers, crocodile osteoderms, pangolin scales and teeth, all of which could inflate values for a particular species. In the case of antelope horns we used the minimum number of individuals necessary to constitute the number of horns (i.e. we divided by two or used half the number plus one if it was an odd number of horns). Because of these measures, the total number of species was less than what was used for the first $\chi^2$ test. All means are reported $\pm$ 1 SD.

## 19.3 Results

### *19.3.1 Trade in Vertebrates*

Excluding domestic animals, we identified 147 vertebrate species traded at Faraday, representing one species of frog, 33 species of reptile, 53 species of bird and 60 species of mammal (Table 19.2). Seven domestic mammals were sold by traders: goat, cattle, sheep, horse, donkey, pig and cat (Table 19.2). Of the species identified at Faraday, 41% were mammals (excluding domestic animals), 36% were birds and 22% were reptiles. For South Africa alone, these species counts represent 8% of the reptile fauna (417 taxa, WR Branch, pers. comm.), 6% of the bird fauna (841 taxa, Birdlife International 2009) and 20% of the mammal fauna (299 taxa, Skinner and Chimimba 2005).

The most taxonomically widespread groups were birds (15 orders, 35 families) (Fig. 19.2b) and mammals (15 orders, 24 families). Perching birds (order Passeriformes) had the highest number of recorded bird families and species (nine families, 14 species), with each family within this order only represented by one to three species. Among raptors, members of the family Accipitridae were the most frequently recorded in the market (>5 species). The most common mammals identified in the market were carnivores (seven families and 24 species), of which, cats were the

**Table 19.2** Checklist of Vertebrate Species: Threat Status and Market Frequency

| Classification (*CLASS*, **ORDER**, Family) | Species | Common name | IUCN category | Number of traders |
|---|---|---|---|---|
| ***CLASS: AMPHIBIA*** | | | | |
| **ANURA** | | | | |
| Bufonidae | *Schismaderma carens* | Toad, Red | LC | 2 |
| | - | Frog, unidentified | - | 1 |
| ***CLASS: REPTILIA*** | | | | |
| **CROCODYLIA** | | | | |
| Crocodylidae | *Crocodylus niloticus* | Nile crocodile | LC | 22 |
| **SQUAMATA** | | | | |
| Pythonidae | *Python natalensis* | Python, Southern African | LC | 23 |
| Colubridae | *Lamprophis aurora* | House Snake, Aurora | LC | 1 |
| | *Dispholidus typus* | Boomslang | LC | 6 |
| | *Philothamnus* sp. | Green Snake, unidentified | - | 1 |
| | *Pseudaspis cana* | Mole Snake | LC | 2 |
| | *Psammophis phillipsii* | Grass Snake, Olive | LC | 2 |
| | *Psammophis* sp. | Sand Snake, unidentified | - | 1 |
| | *Amblyodipsas* sp. | Purple-glossed Snake, unidentified | - | 1 |
| | *Psammophylax rhombeatus* | Skaapsteker, Spotted | LC | 1 |
| | *P. tritaeniatus* | Skaapsteker, Striped | LC | 1 |
| Elapidae | *Dendroaspis polylepis* | Mamba, Black | LC | 7 |
| | *Dendroaspis angusticeps* | Mamba, Green | LC | 5 |
| | *Dendroaspis* sp. | Mamba, unidentified | - | 2 |
| | *Naja mossambica* | Spitting Cobra, Mozambique | LC | 5 |
| | *Naja* sp. | Cobra, unidentified | - | 1 |
| | *Hemachatus haemachatus* | Rinkhals | LC | 1 |
| Viperidae | *Bitis arietans* | Puff Adder | LC | 18 |
| | *Bitis* sp. | Adder, unidentified | - | 1 |
| Agamidae | *Acanthocercus atricollis* | Agama, Southern Tree | LC | 9 |

(continued)

**Table 19.2** (continued)

| Classification (*CLASS*, **ORDER**, Family) | Species | Common name | IUCN category | Number of traders |
|---|---|---|---|---|
| Chamaeleonidae | *Chamaeleo dilepis* | Chameleon, Flap-necked | LC | 2 |
| | | Chameleon, unidentified | - | 3 |
| Varanidae | *Varanus albigularis* | Monitor, Rock | LC | 16 |
| | *Varanus niloticus* | Monitor, Water | LC | 19 |
| | *Varanus* sp. | Monitor, unidentified | - | 7 |
| Scincidae | *Acontias plumbeus* | Skink, Giant Legless | LC | 4 |
| Cordylidae | *Cordylus* cf. *vittifer* | Girdled Lizard, Transvaal | LC | 1 |
| | *Cordylus giganteus* | Giant Sungazer | VU | 5 |
| | *Cordylus tropidosternum* | Girdled Lizard, Tropical | LC | 1 |
| | *Cordylus warreni* | Girdled Lizard, Warren's | LC | 2 |
| Gerrhosauridae | *Gerrhosaurus major* | Plated Lizard, Rough-scaled | LC | 1 |
| | *Gerrhosaurus flavigularis* | Plated Lizard, Yellow-throated | LC | 1 |
| - | - | lizard, unidentified | - | 3 |
| - | - | snake, unidentified | - | 13 |
| Unidentified squamates | - | - | - | 1 |
| **TESTUDINES** | | | | |
| Pelomedusidae | - | Terrapin, unidentified | - | 6 |
| Cheloniidae | *Eretmochelys imbricata* | Turtle, Hawksbill | CR | 1 |
| | - | Turtle, unidentified | - | 1 |
| Testudinidae | *Chersina angulata* | Tortoise, Angulate | LC | 1 |
| | *Kinixys belliana* | Hinged Tortoise, Bell's | LC | 2 |
| | *Kinixys speckii* | Hinged Tortoise, Speke's | LC | 1 |
| | *Kinixys* sp. | Hinged Tortoise, unidentified | - | 1 |
| | *Stigmochelys pardalis* | Leopard Tortoise | LC | 8 |
| | *Homopus* sp. | Padloper, unidentified | - | 2 |
| | - | Tortoise, unidentified | - | 10 |

(continued)

**Table 19.2** (continued)

| Classification (*CLASS*, **ORDER**, Family) | Species | Common name | IUCN category | Number of traders |
|---|---|---|---|---|
| ***CLASS: AVES*** | | | | |
| **ANSERIFORMES** | | | | |
| Anatidae | *Thalassornis leuconotus* | Duck, White-backed | LC | 1 |
| - | - | Duck, unidentified | - | 3 |
| **BUCEROTIFORMES** | | | | |
| Bucerotidae | *Bycanistes bucinator* | Hornbill, Trumpeter | LC | 1 |
| | - | Hornbill, unidentified | - | 1 |
| Bucorvidae | *Bucorvus cafer* | Ground-hornbill, Southern | VU | 3 |
| Musophagidae | *Tauraco corythaix* | Turaco, Knysna | LC | 1 |
| **CAPRIMULGIFORMES** | | | | |
| Caprimulgidae | *Caprimulgus* sp. | Nightjar, unidentified | - | 1 |
| **CHARADRIIFORMES** | | | | |
| Burhinidae | *Burhinus capensis* | Thick-knee, Spotted | LC | 4 |
| | *Burhinus* sp. | Thick-knee, unidentified. | - | 1 |
| Charadriidae | *Vanellus armatus* | Lapwing, Blacksmith | LC | 1 |
| Laridae | *Sterna caspia* | Tern, Caspian | NT | 1 |
| **CICONIIFORMES** | | | | |
| Ardeidae | *Bubulcus ibis* | Egret, Cattle | LC | 7 |
| | *Casmerodius albus* | Egret, Great | LC | 1 |
| Scopidae | *Scopus umbretta* | Hamerkop | LC | 2 |
| Ciconiidae | *Ciconia ciconia* | Stork, White | LC | 3 |
| | - | Stork, unidentified (red bill) | - | 2 |
| **COLIIFORMES** | | | | |
| Coliidae | *Colius striatus* | Mousebird, Speckled | LC | 2 |
| | - | Mousebird, unidentified | - | 1 |

(continued)

**Table 19.2** (continued)

| Classification (*CLASS*, **ORDER**, Family) | Species | Common name | IUCN category | Number of traders |
|---|---|---|---|---|
| **COLUMBIIFORMES** | | | | |
| Columbidae | *Stigmatopelia senegalensis* | Dove, Laughing | LC | 6 |
| | *Streptopelia semitorquata* | Dove, Red-eyed | LC | 1 |
| | *Columba livia* | Dove, Rock | LC | 3 |
| **CORACIIFORMES** | | | | |
| Alcedinidae | *Alcedo cristata* | Kingfisher, Malachite | LC | 1 |
| | *Ceryle rudis* | Kingfisher, Pied | LC | 2 |
| | - | Kingfisher, unidentified | - | 1 |
| **CUCULIFORMES** | | | | |
| Cuculidae | *Centropus superciliosus* | Coucal, White-browed | LC | 5 |
| **FALCONIFORMES** | | | | |
| Accipitridae | *Buteo rufofuscus* | Buzzard, Jackal | LC | 1 |
| | *Haliaeetus vocifer* | Fish-eagle, African | LC | 2 |
| | *Polyboroides typus* | Harrier-hawk, African | LC | 1 |
| | *Elanus caeruleus* | Kite, Black-shouldered | LC | 1 |
| | - | Eagle, unidentified | - | 2 |
| | *Gyps africanus* | Vulture, White-backed | VU | 3 |
| | - | Vulture, unidentified | - | 3 |
| Falconidae | - | Kestrel, unidentified | - | 1 |
| | - | hawk/eagle, unidentified | - | 2 |
| **GALLIFORMES** | | | | |
| Phasianidae | *Coturnix coturnix* | Quail, Common | LC | 2 |
| Numididae | *Guttera pucherani* | Guineafowl, Crested | LC | 2 |
| | *Numida meleagris* | Guineafowl, Helmeted | LC | 2 |

(continued)

**Table 19.2** (continued)

| Classification (*CLASS*, **ORDER**, Family) | Species | Common name | IUCN category | Number of traders |
|---|---|---|---|---|
| **GRUIFORMES** | | | | |
| Rallidae | *Amaurornis flavirostra* | Crake, Black | LC | 1 |
| | *Gallinula chloropus* | Moorhen, Common | LC | 1 |
| | *Gallinula* sp. | Moorhen, unidentified | - | 1 |
| | *Porphyrio porphyrio* | Swamphen, Purple | LC | 2 |
| **PASSERIFORMES** | | | | |
| Malaconotidae | *Laniarius ferrugineus* | Boubou, Southern | LC | 1 |
| | *Prionops plumatus* | Helmet-shrike, White | LC | 1 |
| Corvidae | *Corvus capensis* | Crow, Cape | LC | 1 |
| | *Corvus albus* | Crow, Pied | LC | 3 |
| Laniidae | *Lanius collaris* | Fiscal, Common | LC | 3 |
| Pycnonotidae | *Pycnonotus tricolor* | Bulbul, Dark-capped | LC | 1 |
| Zosteropidae | *Zosterops virens* | White-eye, Cape | LC | 2 |
| Cisticolidae | *Cisticola* sp. | Cisticola, unidentified | - | 1 |
| | *Prinia subflava* | Prinia, tawny-flanked | LC | 1 |
| Muscicapidae | *Melaenornis pammelaina* | Flycatcher, Southern Black | LC | 2 |
| | *Cossypha natalensis* | Robin-chat, Red-capped | LC | 1 |
| Turdidae | *Turdus olivaceus* | Thrush, Olive | LC | 1 |
| Sturnidae | *Lamprotornis nitens* | Starling, Cape Glossy | LC | 3 |
| | *Lamprotornis* sp. | Starling, unidentified | - | 1 |
| Passeridae | *Passer domesticus* | Sparrow, House | LC | 1 |
| **PELICANIFORMES** | | | | |
| Threskiornithidae | *Threskiornis aethiopicus* | Ibis, African Sacred | LC | 3 |
| | *Bostrychia hagedash* | Ibis, Hadeda | LC | 5 |

(continued)

**Table 19.2** (continued)

| Classification (*CLASS*, **ORDER**, Family) | Species | Common name | IUCN category | Number of traders |
|---|---|---|---|---|
| Pelecanidae | *Pelecanus onocrotalus* | Pelican, Great White | NT | 1 |
| | - | Pelican, unidentified | - | 1 |
| **STRIGIFORMES** | | | | |
| Tytonidae | *Tyto alba* | Owl, Barn | LC | 4 |
| Strigidae | *Bubo africanus* | Eagle-owl, Spotted | LC | 5 |
| | *Asio capensis* | Owl, Marsh | LC | 3 |
| - | - | Owl, unidentified | - | 3 |
| **STRUTHIONIFORMES** | | | | |
| Struthionidae | *Struthio camelus* | Ostrich | LC | 14 |
| Unidentified birds | - | - | - | **9** |
| ***CLASS: MAMMALIA*** | | | | |
| **CARNIVORA** | | | | |
| Hyaenidae | *Proteles cristatus* | Aardwolf | LC | 3 |
| | *Hyaena brunnea* | Hyaena, Brown | NT | 1 |
| | *Crocuta crocuta* | Hyaena, Spotted | NT | 4 |
| | - | Hyaena, unidentified | - | 5 |
| Felidae | *Felis silvestris* | Cat, African Wild | LC | 1 |
| | *Felis catus* | Cat, Domestic | - | 3 |
| | *Caracal caracal* | Caracal | LC | 1 |
| | *Panthera pardus* | Leopard | LC | 8 |
| | *Panthera leo* | Lion | VU | 3 |
| | *Leptailurus serval* | Serval | NT | 2 |
| Viverridae | *Civettictis civetta* | Civet, African | LC | 1 |
| | *Genetta tigrina* | Genet, Large-spotted | LC | 6 |
| | *Genetta genetta* | Genet, Small-spotted | LC | 4 |
| | *Genette* sp. | Genet, unidentified | - | 4 |

(continued)

**Table 19.2** (continued)

| Classification (*CLASS*, **ORDER**, Family) | Species | Common name | IUCN category | Number of traders |
|---|---|---|---|---|
| Herpestidae | *Mungos mungo* | Mongoose, Banded | LC | 2 |
| | *Herpestes ichneumon* | Mongoose, Large Grey | LC | 1 |
| | *Galerella sanguinea* | Mongoose, Slender | LC | 3 |
| | *Ichneumia albicauda* | Mongoose, White-tailed | LC | 2 |
| | *Suricata suricatta* | Meerkat | LC | 1 |
| | - | Mongoose, unidentified | - | 8 |
| Canidae | *Otoycyon megalotis* | Fox, Bat-eared | LC | 1 |
| | *Vulpes chama* | Fox, Cape | LC | 1 |
| | *Canis mesomelas* | Jackal, Black-backed | LC | 1 |
| | *Canis* sp. | Jackal, unidentified | - | 8 |
| | *Lycaon pictus* | Wild Dog, African | EN | 1 |
| Mustelidae | *Mellivora capensis* | Badger, Honey (Ratel) | NT | 1 |
| | *Aonyx capensis* | Otter, Cape Clawless | LC | 3 |
| | - | Otter, sp. | - | 3 |
| | *Ictonyx striatus* | Polecat, Striped | LC | 14 |
| Otariidae | *Arctocephalus pusillus* | Seal, Cape Fur | LC | 1 |
| - | - | unidentified, small carnivore | - | 1 |
| **CHIROPTERA** | | | | |
| - | - | bat, unidentified | - | 7 |
| **ERINACEOMORPHA** | | | | |
| Erinaceidae | *Atelerix frontalis* | Hedgehog, South African | NT | 8 |
| **HYRACOIDEA** | | | | |
| Procaviidae | *Procavia capensis* | Rock Hyrax | LC | 11 |

(continued)

**Table 19.2** (continued)

| Classification (*CLASS*, **ORDER**, Family) | Species | Common name | IUCN category | Number of traders |
|---|---|---|---|---|
| **LAGOMORPHA** | | | | |
| Leporidae | *Lepus saxatilis* | Hare, Scrub | LC | 7 |
| | *Pronolagus* sp. | Rabbit, Rock | LC | 6 |
| | - | hare/rabbit, unidentified | - | 1 |
| **MACROSCELIDEA** | | | | |
| Macroscelididae | *Elephantulus* sp. | Elephant Shrew, unidentified | - | 2 |
| **PERISSODACTYLA** | | | | |
| Equidae | *Equus asinus* | Donkey | - | 3 |
| | *Equus caballus* | Horse | - | 13 |
| | *Equus burchellii* | Zebra, Plains | LC | 6 |
| **PHOLIDOTA** | | | | |
| Manidae | *Smutsia temminckii* | Pangolin, Ground | VU | 2 |
| **PRIMATES** | | | | |
| Galagidae | *Otolemur crassicaudatus* | Bushbaby, thick-tailed (Greater Galago) | LC | 6 |
| | - | Bushbaby, unidentified | - | 9 |
| Cercopithecidae | *Papio ursinus* | Baboon, Chacma | LC | 22 |
| | *Cercopithecus mitis* ssp. | Monkey, Samango | VU/LC[a] | 2 |
| | *Chlorocebus pygerythrus* | Monkey, Vervet | LC | 16 |
| **PROBOSCIDEA** | | | | |
| Elephantidae | *Loxodonta africana* | Elephant, African | LC | 15 |
| **RODENTIA** | | | | |
| Bathyergidae | - | Molerat, unidentified | - | 2 |
| Hystricidae | *Hystrix africaeaustralis* | Porcupine, Cape | LC | 22 |
| - | - | Rodent, unidentified | - | 1 |

(continued)

**Table 19.2** (continued)

| Classification (*CLASS*, **ORDER**, Family) | Species | Common name | IUCN category | Number of traders |
|---|---|---|---|---|
| **RUMINANTIA** | | | | |
| Giraffidae | *Giraffa camelopardalis* | Giraffe | LC | 4 |
| Bovidae | *Damaliscus pygargus* | Blesbok/Bontebok | VU (Bontebok) | 1 |
| | *Syncerus caffer* | Buffalo, African | LC | 12 |
| | *Tragelaphus scriptus* | Bushbuck | LC | 4 |
| | *Sylvicapra grimmia* | Duiker, Common | LC | 2 |
| | *Cephalophus natalensis* | Duiker, Red | LC | 9 |
| | - | Duiker, unidentified | - | 8 |
| | *Taurotragus oryx* | Eland | LC | 7 |
| | *Oryx gazella* | Gemsbok | LC | 3 |
| | *Capra hircus* | Goat, Domestic | - | 1 |
| | *Aepyceros melampus* | Impala | LC | 7 |
| | *Oreotragus oreotragus* | Klipspringer | LC | 1 |
| | *Tragelaphus strepsiceros* | Kudu, Greater | LC | 7 |
| | *T. angasii* | Nyala | LC | 3 |
| | *Bos taurus* | Cattle | - | 2 |
| | *Redunca arundinum* | Reedbuck | LC | 1 |
| | *Ovis aries* | Sheep | - | 3 |
| | *Antidorcas marsupialis* | Springbok | LC | 3 |
| | *Kobus ellipsiprymnus* | Waterbuck | LC | 4 |
| | *Connochaetes taurinus* | Wildebeest, Blue | LC | 1 |
| | *Connochaetes* sp. | Wildebeest, unidentified | - | 9 |
| | - | ungulate, unidentified | - | 11 |

(continued)

**Table 19.2** (continued)

| Classification (*CLASS*, **ORDER**, Family) | Species | Common name | IUCN category | Number of traders |
|---|---|---|---|---|
| **SUIFORMES** | | | | |
| Suidae | *Potamochoerus larvatus* | Bushpig | LC | 2 |
| | *Phacochoerus africanus* | Warthog, Common | LC | 16 |
| | *Sus domesticus* | Pig | - | 2 |
| **TUBULIDENTATA** | | | | |
| Orycteropodidae | *Orycteropus afer* | Aardvark | LC | 7 |
| **WHIPPOMORPHA** | | | | |
| Hippopotamidae | *Hippopotamus amphibius* | Hippopotamus | LC | 11 |
| Unidentified mammals | | | | 25 |

[a] Depending on species

Check list of vertebrate species, the number of traders recorded selling each species at the Faraday market, and each species' conservation status. The classification follows Minter et al. (2004; frog); Zug et al. (2001); Alexander and Marais (2007) (reptiles); Hockey et al. (2005; birds); and Skinner and Chimimba (2005; mammals). See text for details of conservation assessment (2001 IUCN Red Data Lists v 3.1). *LC* least concern, *NT* Near Threatened, *VU* Vulnerable, *EN* Endangered, *CR* Critically Endangered. We did not identify marine organisms and invertebrates to a sufficient level to assign IUCN categories, but see Table 19.4 for the quantities of these organisms at Faraday most prevalent (five species). The bovids (antelopes and buffalo) were the next most abundant group of mammals (15 species) (Table 19.2). Among reptiles, the squamates (snakes and lizards) were the most common (10 families; 25 species), of which the colubrids (typical snakes) were recorded the most frequently.

The mean number of vertebrate species sold per trader was 25.1 ± 14.7, and ranged from 4.7 ± 5.2 bird species sold per trader to 13.3 ± 7.6 mammal species sold per trader (Table 19.1). The mean number of "shared" species (i.e. species that two traders have in common at their stalls) was 7.1 ± 4.8 species per trader (Table 19.1); hence, the similarity of species sold by traders was relatively low. The Sørenson similarity measure confirms this finding and indicates that species composition at the different traders' stalls is on average only 26.7 ± 1.2% similar for all vertebrates recorded (Table 19.1). However, there is greater similarity of reptiles sold (33.4%) between traders compared to birds (13.1%; Table 19.1). Hence, one is unlikely to find the same bird species being sold by the traders in the market, except for ostriches, owls and a broad spectrum of species from the order Falconiformes. Twelve of the most commonly occurring species were uniformly present throughout the market (Table 19.3, shaded species; including monitors, python, crocodile and baboon), whereas the remaining species occurred randomly at traders' stalls.

### *19.3.2 Commonly Traded Species*

The following species were sold by more than 50% of traders: reptiles: rock (50%), and water monitor (59%); Nile crocodile (69%); southern African python (72%) (Fig. 1b); puff adder (56%); mammals: chacma baboon (69%); Cape porcupine (69%); vervet monkey (50%); warthog (50%) (Table 19.3). African elephants were also commonly traded (47%, 15 traders). Bird species were not as prevalent as mammals and reptiles, but ostriches and owls were the most commonly recorded avian species (44% and 28% of traders respectively).

### *19.3.3 Marine Fauna and Invertebrates*

We recorded an array of invertebrates and fishes that we could only identify at a much higher taxonomic level (Table 19.4). Most invertebrates recorded were marine, representing at least four phyla (Table 19.5). The only non-marine invertebrates were two species of insect and the giant land snail (Table 19.4). Among the marine invertebrates, most were unidentified marine molluscs. Among the fishes, only two were freshwater species (barbel, catfish) while the remainder were marine. The most common of these were sole (seven traders, 70 individuals). We also recorded relatively low numbers of sharks, rays, skates and eel, and 46 unidentified ray-finned fishes (Table 19.4). Of the marine taxa, ray-finned fish, mollusc and echinoderm shells were sold by 56% of traders (Table 19.5). Ninety-one percent of traders sold some marine fauna.

**Table 19.3** Percentage of 32 traders recorded selling species of vertebrates in Faraday

| Mammals | | Reptiles | | Birds | |
|---|---|---|---|---|---|
| Common name | % traders (>20%) | Common name | % traders (>10%) | Common name | % traders (>10%) |
| Chacma baboon | 68.8 | Monitor spp. (rock and water) | 84.4 | Common ostrich | 43.8 |
| Cape porcupine | 68.8 | Nile crocodile | 71.9 | Owl spp. | 37.5 |
| Vervet monkey | 50.0 | Southern African python | 71.9 | Dove spp. | 25.0 |
| Common warthog | 50.0 | Tortoise spp. | 62.5 | Egret spp. | 25.0 |
| Duiker spp. | 46.9 | Puff adder (snake) | 56.3 | Vulture spp. | 18.8 |
| African elephant | 46.9 | Elapids (snakes) | 43.8 | Coucal, Burchell's | 15.6 |
| Bush baby spp. | 43.8 | Southern tree agama (lizard) | 28.1 | Ibis spp. | 15.6 |
| Mongoose spp. | 43.8 | Colubrids (snakes) | 28.1 | Duck spp. | 12.5 |
| Striped polecat | 43.8 | *Cordylus* spp. (girdled lizards) | 21.9 | Starling spp. | 12.5 |
| Horse | 40.6 | Terrapin spp. | 18.8 | Thick-knee spp. | 12.5 |
| African buffalo | 37.5 | Chameleon spp. | 15.6 | | |
| Rock hyrax | 34.4 | Giant legless skink (lizard) | 12.5 | | |
| Hippopotamus | 34.4 | | | | |
| Wildebeest spp. | 31.3 | | | | |
| Genet spp. | 28.1 | | | | |
| Hyaena spp. | 28.1 | | | | |
| Jackal spp. | 28.1 | | | | |
| Southern African hedgehog | 25.0 | | | | |
| Leopard | 25.0 | | | | |
| Aardvark | 21.9 | | | | |
| Bat spp. | 21.9 | | | | |
| Eland | 21.9 | | | | |
| Scrub hare | 21.9 | | | | |
| Impala | 21.9 | | | | |
| Greater kudu | 21.9 | | | | |

Table 19.2 lists the incidence of all individual vertebrate species. Species in grey shading were uniformly distributed throughout the market; the remaining species were randomly dispersed

## 19.3.4 Species Richness

The species accumulation curves for mammals, birds and reptiles approached an asymptote and indicate that further sampling of traders would not yield many more new species for the individual vertebrate classes (Fig. 19.3), hence sampling effort was sufficient. The rate of accumulation of new species was 0.4 new species per trader for reptiles, 0.6 new species per trader for mammals and 1.2 new species per

**Table 19.4** Checklist of Vertebrate Species: Animal Part and Quantity in the Market: Check list of vertebrate species according to the animal part traded and the number of body parts sold by all traders combined, at the Faraday market

**INVERTEBRATES**

| Classification (**CLASS**, Order) | Family | Common name | Animal part | Number of parts |
|---|---|---|---|---|
| **ANTHOZOA** | | | | |
| Scleractinia | - | Coral | - | 3 |
| **MALACOSTRACA** | | | | |
| Decapoda | - | Marine crab | Whole | 2 |
| **CEPHALOPODA** | | | | |
| Sepiida | Sepiidae | Cuttlefish (*Sepia* spp.) | Whole | 45 |
| Octopoda | Octopodidae | Common octopus (*Octopus vulgaris*) | Whole | 2 |
| | | | Tentacle | 2 |
| **BIVALVIA** | - | Clam | Shell | 5 |
| **GASTROPODA** | | | | |
| Sorbeoconcha | Cypraeidae | Cowrie (*Cypraea*) | Shell | 84 |
| Archaeogastropoda | - | Limpit | Shell | 16 |
| - | - | Mollusc (various marine) | Shell | 21 |
| Pulmonata | - | Ocean slugs | Whole | 20 |
| Clade: Stylommatophora | Achatinidae | Giant land snail (*Achatina*) | Shell | 72 |
| - | - | Various unidentified Mollusca | Shell | 934 |
| **ECHINOIDEA** | | | | |
| Cidaroida | Cidaridae | Pencil urchin (*Prionocidaris pistillaris*) | Shell | 75 |
| Echinoida | - | Sea urchin | Shell | 65 |
| **ASTEROIDEA** | | | | |
| Valvatida | - | Star fish | Whole | 34 |
| **INSECTA** | | | | |
| Coleoptera | Meloidae | CMR Bean Beetle (*Mylabris oculata*) | Whole | 200 |
| Orthoptera | Pyrgomorphidae | Grasshopper (*Taphronota* sp.) | Whole | 25 |

**Table 19.4** (continued)

| **FISHES** | | | | |
|---|---|---|---|---|
| Actinopterygii | | | | |
| Anguilliformes | Muraenidae | Eel, Moray | Whole | 2 |
| | | Eel, unidentified | Whole | 2 |
| Cypriniformes | Cyprinidae | Barbel | Head | 1 |
| Tetraodontiformes | Ostraciidae | Box fish | Whole | 30 |
| Tetraodontiformes | - | Puffer fish/porcupine fish | Whole | 14 |
| Siluriformes | - | Catfish | Head | 1 |
| Pleuronectiformes | - | Sole (*Austroglossus pectoralis*) | Whole | 70 |
| - | - | Fish, unidentified | Whole | 46 |
| Chondrichthyes | | | | |
| Rajiformes | - | Rays | Whole | 3 |
| | Rajidae | Skate | Whole | 1 |
| Selachimorpha | - | Shark | Jaw | 1 |
| | | | Skin | 1 |
| **AMPHIBIANS, REPTILES, BIRDS and MAMMALS** | | | | |
| Classification (**CLASS**, **ORDER**, **Family**) | Species | Common name | Animal part | Number of parts |
| **CLASS: AMPHIBIA** | | | | |
| **ANURA** | | | | |
| **Bufonidae** | *Schismaderma carens* | Toad, Red | Whole | 2 |
| | - | Frog, unidentified | Whole | 5 |

(continued)

**Table 19.4** (continued)

| AMPHIBIANS, REPTILES, BIRDS and MAMMALS | | | | |
|---|---|---|---|---|
| Classification (**CLASS**, **ORDER**, **Family**) | Species | Common name | Animal part | Number of parts |
| **CLASS: REPTILIA** | | | | |
| **Crocodylia** | | | | |
| **Crocodylidae** | *Crocodylus niloticus* | Nile crocodile | Skin | 24 |
| | | | Skull | 9 |
| | | | Osteoderms | 65 |
| | | | Lower jaws | 4 |
| | | | Egg | 20 |
| | | | Bone | 1 |
| **Squamata** | | | | |
| **Pythonidae** | *Python natalensis* | Python, Southern African | Skin | 22 |
| | | | Body parts | 16 |
| | | | Skull | 1 |
| **Colubridae** | *Lamprophis aurora* | House Snake, Aurora | Skin | 1 |
| | *Dispholidus typus* | Boomslang | Whole | 5 |
| | | | Skin | 1 |
| | *Philothamnus* sp. | Green Snake, unidentified | Whole | 1 |
| | *Pseudaspis cana* | Mole Snake | Whole | 3 |
| | *Psammophis phillipsii* | Grass Snake, Olive | Whole | 2 |
| | *Psammophis* sp. | Sand Snake, unidentified | Whole | 1 |
| | *Amblyodipsas* sp. | Purple-glossed Snake, unidentified | Whole | 1 |
| | *Psammophylax rhombeatus* | Skaapsteker, Spotted | Whole | 1 |
| | *Psammophylax tritaeniatus* | Skaapsteker, Striped | Whole | 1 |

(continued)

**Table 19.4** (continued)

**AMPHIBIANS, REPTILES, BIRDS and MAMMALS**

| Classification (**CLASS**, **ORDER**, **Family**) | Species | Common name | Animal part | Number of parts |
|---|---|---|---|---|
| **Elapidae** | *Dendroaspis polylepis* | Mamba, Black | Whole | 10 |
| | | | Skin | 1 |
| | *Dendroaspis angusticeps* | Mamba, Green | Whole | 5 |
| | | | Skin | 2 |
| | *Dendroaspis* sp. | Mamba, unidentified | Skin | 2 |
| | *Naja mossambica* | Spitting Cobra, Mozambique | Whole | 10 |
| | *Naja* sp. | Cobra, unidentified | Skin | 1 |
| | *Hemachatus haemachatus* | Rinkhals | Whole | 1 |
| **Viperidae** | *Bitis arietans* | Puff Adder | Skin | 11 |
| | | | Whole | 28 |
| | *Bitis* sp. | Adder, unidentified | Whole | 1 |
| **Agamidae** | *Acanthocercus atricollis* | Agama, Southern Tree | Whole | 20 |
| **Chamaeleonidae** | *Chamaeleo dilepis* | Chameleon, Flap-necked | Whole | 2 |
| | | Chameleon, unidentified | Tail | 1 |
| | | | Whole | 3 |
| **Varanidae** | *Varanus albigularis* | Monitor, Rock | Head | 3 |
| | | | Skin | 10 |
| | | | Whole | 9 |
| | *Varanus niloticus* | Monitor, Water | Foot | 2 |
| | | | Skin | 9 |
| | | | Whole | 18 |
| | *Varanus* sp. | Monitor, unidentified | Head | 2 |

(continued)

**Table 19.4** (continued)

**AMPHIBIANS, REPTILES, BIRDS and MAMMALS**

| Classification (**CLASS**, **ORDER**, **Family**) | Species | Common name | Animal part | Number of parts |
|---|---|---|---|---|
| | | | Skin | 8 |
| | | | Whole | 3 |
| **Scincidae** | *Acontias plumbeus* | Skink, Giant Legless | Whole | 6 |
| **Cordylidae** | *Cordylus* cf. *vittifer* | Girdled Lizard, Transvaal | Whole | 7 |
| | *Cordylus giganteus* | Giant Sungazer | Whole | 11 |
| | *Cordylus tropidosternum* | Girdled Lizard, Tropical | Whole | 1 |
| | *Cordylus warreni* | Girdled Lizard, Warren's | Whole | 10 |
| **Gerrhosauridae** | *Gerrhosaurus major* | Plated Lizard, Rough-scaled | Whole | 2 |
| | *Gerrhosaurus flavigularis* | Plated Lizard, Yellow-throated | Whole | 1 |
| - | - | Lizard, unidentified | Whole | 7 |
| - | - | Snake, unidentified | Head | 2 |
| | | | Skin | 15 |
| | | | Whole | 42 |
| **Testudines** | | | | |
| **Pelomedusidae** | - | Terrapin, unidentified | Shell | 8 |
| | | | Head | 1 |
| | | | Plastron | 1 |
| **Cheloniidae** | *Eretmochelys imbricata* | Turtle, Hawksbill | Head | 1 |
| | - | Turtle, unidentified | Shell | 2 |
| | - | Terrapin/Turtle/Tortoise | Shell | 1 |
| **Testudinidae** | *Chersina angulata* | Tortoise, Angulate | Shell | 1 |
| | *Kinixys belliana* | Hinged Tortoise, Bell's | Carapace | 2 |
| | | | Shell | 2 |

(continued)

**Table 19.4** (continued)

**AMPHIBIANS, REPTILES, BIRDS and MAMMALS**

| Classification (**CLASS**, **ORDER**, **Family**) | Species | Common name | Animal part | Number of parts |
|---|---|---|---|---|
| | *Kinixys speckii* | Hinged Tortoise, Speke's | Shell | 1 |
| | *Kinixys* sp. | Hinged Tortoise, unidentified | Shell | 1 |
| | *Stigmochelys pardalis* | Leopard Tortoise | Foot | 2 |
| | | | Plastron | 3 |
| | | | Shell | 11 |
| | *Homopus* sp. | Padloper, unidentified | Shell | 1 |
| | - | Tortoise, unidentified | Shell | 14 |
| | | | Carapace | 3 |
| | | | Egg | 2 |
| | | | Foot | 2 |
| | | | Neck | 1 |
| | | | Plastron | 4 |
| **CLASS: AVES** | | | | |
| **Anseriformes** | | | | |
| **Anatidae** | *Thalassornis leuconotus* | Duck, White-backed | Head | 1 |
| **-** | - | Duck, unidentified | Foot | 2 |
| | | | Whole | 2 |
| **Bucerotiformes** | | | | |
| **Bucerotidae** | *Bycanistes bucinator* | Hornbill, Trumpeter | Whole | 1 |
| | - | Hornbill, unidentified | Beak | 1 |
| **Bucorvidae** | *Bucorvus cafer* | Ground-Hornbill, Southern | Beak | 1 |
| | | | Skull | 1 |
| | | | Whole | 1 |

(continued)

**Table 19.4** (continued)

**AMPHIBIANS, REPTILES, BIRDS and MAMMALS**

| Classification (**CLASS**, **ORDER**, **Family**) | Species | Common name | Animal part | Number of parts |
|---|---|---|---|---|
| **Musophagidae** | *Tauraco corythaix* | Turaco, Knysna | Whole | 1 |
| **CAPRIMULGIFORMES** | | | | |
| **Caprimulgidae** | *Caprimulgus* sp. | Nightjar, unidentified | Wing | 1 |
| **Charadriiformes** | | | | |
| **Burhinidae** | *Burhinus capensis* | Thick-knee, Spotted | Whole | 4 |
| | *Burhinus* sp. | Thick-knee, unidentified. | Whole | 1 |
| **Charadriidae** | *Vanellus armatus* | Lapwing, Blacksmith | Whole | 1 |
| **Laridae** | *Sterna caspia* | Tern, Caspian | Whole | 1 |
| **CICONIIFORMES** | | | | |
| Ardeidae | *Bubulcus ibis* | Egret, Cattle | Head | 1 |
| | | | Whole | 9 |
| | *Egretta alba* | Egret, Great | Whole | 1 |
| Scopidae | *Scopus umbretta* | Hamerkop | Whole | 2 |
| Ciconiidae | *Ciconia ciconia* | Stork, White | Head-neck | 1 |
| | | | Skull | 1 |
| | | | Whole | 4 |
| | - | Stork, unidentified (red bill) | Head-neck | 2 |
| **COLIIFORMES** | | | | |
| Coliidae | *Colius striatus* | Mousebird, Speckled | Whole | 6 |
| | - | Mousebird, unidentified | Whole | 1 |
| **COLUMBIFORMES** | | | | |
| Columbidae | *Streptopelia senegalensis* | Dove, Laughing | Whole | 4 |
| | *Streptopelia semitorquata* | Dove, Redeyed | Whole | 2 |

(continued)

**Table 19.4** (continued)

| AMPHIBIANS, REPTILES, BIRDS and MAMMALS | | | | |
|---|---|---|---|---|
| Classification (**CLASS**, **ORDER**, **Family**) | Species | Common name | Animal part | Number of parts |
| | *Columba livia* | Dove, Rock (Feral Pigeon) | Whole | 3 |
| | | | Head | 1 |
| **CORACIIFORMES** | | | | |
| Alcedinidae | *Alcedo cristata* | Kingfisher, Malachite | Whole | 1 |
| Cerylidae | *Ceryle rudis* | Kingfisher, Pied | Whole | 2 |
| - | - | Kingfisher, unidentified | Whole | 1 |
| **CUCULIFORMES** | | | | |
| Centropodidae | *Centropus burchellii* | Coucal, Burchell's | Whole | 7 |
| **FALCONIFORMES** | | | | |
| Accipitridae | *Buteo rufofuscus* | Buzzard, Jackal | Whole | 1 |
| | *Haliaeetus vocifer* | Fish-Eagle, African | Whole | 4 |
| | *Polyboroides typus* | Harrier-Hawk, African | Whole | 1 |
| | *Elanus caeruleus* | Kite, Black-shouldered | Whole | 1 |
| | - | Eagle, unidentified | Whole | 2 |
| | *Gyps africanus* | Vulture, White-backed | Whole | 5 |
| | - | Vulture, unidentified | Foot | 2 |
| | | | Head | 1 |
| | | | Whole | 1 |
| Falconidae | - | Kestrel, unidentified | Whole | 1 |
| | - | hawk/eagle, unidentified | Claw | 1 |
| | | | Skull | 1 |
| **GALLIFORMES** | | | | |
| Phasianidae | *Coturnix coturnix* | Quail, Common | Whole | 2 |

(continued)

**Table 19.4** (continued)

| Classification (**CLASS**, **ORDER**, **Family**) | Species | Common name | Animal part | Number of parts |
|---|---|---|---|---|
| **AMPHIBIANS, REPTILES, BIRDS and MAMMALS** | | | | |
| Numididae | *Guttera edouardi* | Guineafowl, Crested | Whole | 7 |
| | *Numida meleagris* | Guineafowl, Helmeted | Whole | 2 |
| **GRUIFORMES** | | | | |
| Rallidae | *Amaurornis flavirostra* | Crake, Black | Whole | 2 |
| | *Gallinula chloropus* | Moorhen, Common | Whole | 1 |
| | *Gallinula* sp. | Moorhen, unidentified | Whole | 2 |
| | *Porphyrio madagascariensis* | Swamphen, African Purple | Whole | 2 |
| **PASSERIFORMES** | | | | |
| Malaconotidae | *Laniarius ferrugineus* | Boubou, Southern | Whole | 2 |
| | *Prionops plumatus* | Helmet-Shrike, White-crested | Whole | 1 |
| Corvidae | *Corvus capensis* | Crow, Cape | Whole | 1 |
| | *Corvus albus* | Crow, Pied | Whole | 7 |
| Laniidae | *Lanius collaris* | Fiscal, Common | Whole | 3 |
| Pycnonotidae | *Pycnonotus tricolor* | Bulbul, Dark-capped | Whole | 1 |
| Zosteropidae | *Zosterops virens* | White-eye, Cape | Whole | 2 |
| Cisticolidae | *Cisticola* sp. | Cisticola, unidentified | Whole | 1 |
| | *Prinia subflava* | Prinia, tawny-flanked | Whole | 1 |
| Muscicapidae | *Melaenornis pammelaina* | Flycatcher, Southern Black | Whole | 2 |
| | *Cossypha natalensis* | Robin-Chat, Red-capped | Whole | 2 |
| | *Turdus olivaceus* | Thrush, Olive | Whole | 1 |
| Sturnidae | *Lamprotornis nitens* | Starling, Cape Glossy | Whole | 4 |
| | *Lamprotornis* sp. | Starling, unidentified | Whole | 1 |
| Passeridae | *Passer domesticus* | Sparrow, House | Whole | 2 |

(continued)

**Table 19.4** (continued)

**AMPHIBIANS, REPTILES, BIRDS and MAMMALS**

| Classification (**CLASS**, **ORDER**, **Family**) | Species | Common name | Animal part | Number of parts |
|---|---|---|---|---|
| **PELICANIFORMES** | | | | |
| Pelecanidae | *Pelecanus onocrotalus* | Pelican, Great White | Whole | 1 |
| | - | Pelican, unidentified | Head | 1 |
| Threskiornithidae | *Threskiornis aethiopicus* | Ibis, African Sacred | Whole | 3 |
| | *Bostrychia hagedash* | Ibis, Hadeda | Whole | 6 |
| | | | Head | 1 |
| Strigiformes | | | | |
| Tytonidae | *Tyto alba* | Owl, Barn | Whole | 4 |
| Strigidae | *Bubo africanus* | Eagle-Owl, Spotted | Whole | 9 |
| | *Asio capensis* | Owl, Marsh | Whole | 3 |
| - | - | Owl, unidentified | Foot | 1 |
| | | | Leg | 1 |
| | | | Whole | 1 |
| **STRUTHIONIFORMES** | | | | |
| Struthionidae | *Struthio camelus* | Ostrich, Common | Beak | 1 |
| | | | Egg | 28 |
| | | | Feather | 30 |
| | | | Leg | 2 |
| | | | Skin | 2 |
| | | | Skull | 2 |
| | | | Toe | 2 |
| Unidentified birds | - | - | Beak | 1 |
| | - | - | Feather | 21 |

(continued)

**Table 19.4** (continued)

**AMPHIBIANS, REPTILES, BIRDS and MAMMALS**

| Classification (**CLASS**, **ORDER**, **Family**) | Species | Common name | Animal part | Number of parts |
|---|---|---|---|---|
| | - | - | Egg shells | 2 |
| | - | - | Head-skull | 2 |
| | - | - | Leg | 7 |
| | - | - | Whole | 5 |
| **CLASS: MAMMALIA** | | | | |
| **CARNIVORA** | | | | |
| Hyaenidae | *Proteles cristatus* | Aardwolf | Skin | 3 |
| | | | Whole | 1 |
| | *Hyaena brunnea* | Hyaena, Brown | Skin | 1 |
| | *Crocuta crocuta* | Hyaena, Spotted | Skin | 3 |
| | | | Skull | 1 |
| | - | Hyaena, unidentified | Skin | 2 |
| | | | Skull | 3 |
| Felidae | *Felis silvestris* | Cat, African Wild | Skin | 1 |
| | *Caracal caracal* | Caracal | Skin | 1 |
| | *Felis catus* | Cat, Domestic | Skin | 3 |
| | *Panthera pardus* | Leopard | Bone | 1 |
| | | | Foot/paw | 6 |
| | | | Skin | 5 |
| | | | Skull | 3 |
| | *Panthera leo* | Lion | Bone | 1 |
| | | | Skin | 1 |
| | | | Skull | 1 |

(continued)

**Table 19.4** (continued)

| AMPHIBIANS, REPTILES, BIRDS and MAMMALS | | | | |
|---|---|---|---|---|
| Classification (**CLASS**, **ORDER**, **Family**) | Species | Common name | Animal part | Number of parts |
| | *Leptailurus serval* | Serval | Skin | 2 |
| Viverridae | *Civettictis civetta* | Civet, African | Skull | 1 |
| | *Genetta tigrina* | Genet, Large-spotted | Skin | 3 |
| | | | Whole | 9 |
| | *Genetta genetta* | Genet, Small-spotted | Skin | 4 |
| | | | Whole | 1 |
| | *Genetta* sp. | Genet, unidentified | Skin | 3 |
| | | | Whole | 2 |
| Herpestidae | *Mungos mungo* | Mongoose, Banded | Skin | 2 |
| | | | Whole | 4 |
| | *Herpestes ichneumon* | Mongoose, Large Grey | Skin | 1 |
| | *Galerella sanguinea* | Mongoose, Slender | Skin | 5 |
| | | | Whole | 1 |
| | *Ichneumia albicauda* | Mongoose, White-tailed | Skin | 3 |
| | *Suricata suricatta* | Meerkat | Whole | 2 |
| | - | Mongoose, unidentified | Head | 1 |
| | | | Skin | 6 |
| | | | Whole | 5 |
| Canidae | *Otoycyon megalotis* | Fox, Bat-eared | Whole | 1 |
| | *Vulpes chama* | Fox, Cape | Whole | 2 |
| | *Canis mesomelas* | Jackal, Black-backed | Skin | 2 |
| | *Canis* sp. | Jackal, unidentified | Skin | 6 |
| | | | Skull | 1 |

(continued)

**Table 19.4** (continued)

| AMPHIBIANS, REPTILES, BIRDS and MAMMALS | | | | |
|---|---|---|---|---|
| Classification (**CLASS**, **ORDER**, **Family**) | Species | Common name | Animal part | Number of parts |
| | | | Whole | 1 |
| | *Lycaon pictus* | Wild Dog, African | Skull | 1 |
| Mustelidae | *Mellivora capensis* | Badger, Honey (Ratel) | Skin | 3 |
| | *Aonyx capensis* | Otter, Cape Clawless | Whole | 1 |
| | - | Otter, sp. | Skins | 4 |
| | *Ictonyx striatus* | Polecat, Striped | Skin | 11 |
| | | | Whole | 15 |
| Otariidae | *Arctocephalus pusillus* | Seal, Cape Fur | Skin | 1 |
| - | - | Unidentified, small carnivore | Skull | 2 |
| **CHIROPTERA** | | | | |
| - | - | Bat, unidentified | Whole | 13 |
| **ERINACEOMORPHA** | | | | |
| Erinaceidae | *Atelerix frontalis* | Hedgehog, South African | Skin | 2 |
| | | | Whole | 7 |
| **HYRACOIDEA** | | | | |
| Procaviidae | *Procavia capensis* | Rock Hyrax | Skull | 1 |
| | | | Whole | 8 |
| **LAGOMORPHA** | | | | |
| Leporidae | *Lepus saxatilis* | Hare, Scrub | Head | 1 |
| | | | Skin | 1 |
| | | | Whole | 7 |
| | *Pronolagus* sp. | Rabbit, Rock | Foot | 1 |
| | | | Leg | 2 |

(continued)

**Table 19.4** (continued)

**AMPHIBIANS, REPTILES, BIRDS and MAMMALS**

| Classification (**CLASS**, **ORDER**, **Family**) | Species | Common name | Animal part | Number of parts |
|---|---|---|---|---|
| | | | Skin | 4 |
| | | | Whole | 1 |
| | - | Unidentified | Whole | 1 |
| **MACROSCELIDEA** | | | | |
| Macroscelididae | *Elephantulus* sp. | Elephant Shrew, unidentified | Whole | 4 |
| Perissodactyla | | | | |
| Equidae | *Equus asinus* | Donkey | Hoof | 2 |
| | | | Skull | 2 |
| | *Equus caballus* | Horse | Hoof | 2 |
| | | | Leg | 26 |
| | | | Penis | 1 |
| | | | Skull | 3 |
| | | | Tail | 1 |
| | *Equus burchellii* | Zebra, Plains | Hoof | 1 |
| | | | Skin | 4 |
| | | | Skull | 3 |
| **PHOLIDOTA** | | | | |
| Manidae | *Smutsia temminckii* | Pangolin, Ground | Foot | 2 |
| | | | Scale | 72 |
| **PRIMATES** | | | | |
| Galagidae | *Otolemur crassicaudatus* | Bushbaby, thick-tailed (Greater Galago) | Skin | 1 |
| | | | Whole | 8 |
| | - | Bushbaby, unidentified | Skin | 1 |

(continued)

**Table 19.4** (continued)

**AMPHIBIANS, REPTILES, BIRDS and MAMMALS**

| Classification (**CLASS**, **ORDER**, **Family**) | Species | Common name | Animal part | Number of parts |
|---|---|---|---|---|
| | | | Skull | 3 |
| | | | Whole | 13 |
| Cercopithecidae | *Papio ursinus* | Baboon, Chacma | Bone | 4 |
| | | | Foot/hand | 8 |
| | | | Skin | 10 |
| | | | Skull | 20 |
| | | | Whole | 14 |
| | *Cercopithecus mitis* ssp. | Monkey, Samango | Skin | 2 |
| | *Chlorocebus pygerythrus* | Monkey, Vervet | Skin | 3 |
| | | | Skull | 12 |
| | | | Whole | 10 |
| **PROBOSCIDEA** | | | | |
| Elephantidae | *Loxodonta africana* | Elephant, African | Bone | 3 |
| | | | Foot | 1 |
| | | | Penis | 1 |
| | | | Skin | 25 |
| | | | Tooth | 1 |
| **RODENTIA** | | | | |
| Bathyergidae | | Molerat, unidentified | Whole | 2 |
| Hystricidae | *Hystrix africaeaustralis* | Porcupine, Cape | Foot | 4 |
| | | | Intestine | 1 |
| | | | Nose | 1 |
| | | | Quill | 388 |

(continued)

**Table 19.4** (continued)

| **AMPHIBIANS, REPTILES, BIRDS and MAMMALS** | | | | |
|---|---|---|---|---|
| Classification (**CLASS**, **ORDER**, **Family**) | Species | Common name | Animal part | Number of parts |
| | | | Skin | 11 |
| | | | Skull | 1 |
| | | | Whole | 1 |
| - | - | Rodent, unidentified | Whole | 1 |
| **RUMINANTIA** | | | | |
| Giraffidae | *Giraffa camelopardalis* | Giraffe | Bone | 15 |
| | | | Skin | 1 |
| | | | Skull | 1 |
| | | | Tail | 1 |
| Bovidae | *Damaliscus pygargus* | Blesbok/Bontebok | Skulls | 6 |
| | *Syncerus caffer* | Buffalo, African | Horn | 10 |
| | | | Skin | 3 |
| | | | Skull | 2 |
| | *Tragelaphus scriptus* | Bushbuck | Horn | 4 |
| | | | Skull | 2 |
| | *Sylvicapra grimmia* | Duiker, Common | Horn | 30 |
| | | | Skin | 1 |
| | *Cephalophus natalensis* | Duiker, Red | Baby/ whole | 1 |
| | | | Horn | 47 |
| | | | Skin | 11 |
| | | | Skull | 3 |
| | - | Duiker, unidentified | Horn | 30 |

(continued)

**Table 19.4** (continued)

| AMPHIBIANS, REPTILES, BIRDS and MAMMALS | | | | |
|---|---|---|---|---|
| Classification (**CLASS**, **ORDER**, **Family**) | Species | Common name | Animal part | Number of parts |
| | | | Leg | 1 |
| | | | Skin | 1 |
| | | | Skull | 3 |
| | *Taurotragus oryx* | Eland | Horn | 5 |
| | | | Scrotum | 1 |
| | | | Skin | 1 |
| | *Oryx gazella* | Gemsbok | Horn | 3 |
| | *Capra hircus* | Goat, Domestic | Skin | 1 |
| | *Aepyceros melampus* | Impala | Horn | 5 |
| | | | Skin | 7 |
| | | | Skull | 1 |
| | *Oreotragus oreotragus* | Klipspringer | Skin | 1 |
| | *Tragelaphus strepsiceros* | Kudu, Greater | Horn | 8 |
| | | | Skull | 2 |
| | *Tragelaphus angasii* | Nyala | Horn | 2 |
| | | | Skin | 2 |
| | *Bos taurus* | Cattle | Belly | 1 |
| | | | Horn | 1 |
| | *Redunca arundinum* | Reedbuck | Horn | 1 |
| | *Ovis aries* | Sheep | Jaw | 2 |
| | | | Skin | 2 |
| | *Antidorcas marsupialis* | Springbok | Hoof | 1 |
| | | | Horn | 4 |
| | *Kobus ellipsiprymnus* | Waterbuck | Hoof | 1 |

(continued)

**Table 19.4** (continued)

**AMPHIBIANS, REPTILES, BIRDS and MAMMALS**

| Classification (**CLASS**, **ORDER**, **Family**) | Species | Common name | Animal part | Number of parts |
|---|---|---|---|---|
| | | | Horn | 3 |
| | | | Skin | 1 |
| | *Connochaetes taurinus* | Wildebeest, Blue | Horn | 2 |
| | *Connochaetes* sp. | Wildebeest, unidentified | Horn | 6 |
| | | | Skin | 3 |
| | | | Skull | 2 |
| | | | Tail | 1 |
| Unidentified bovids | | | | |
| | - | - | Bone | 20 |
| | - | - | Hoof | 29 |
| | - | - | Horn | 105 |
| | - | - | Leg | 5 |
| | - | - | Skin | 8 |
| | - | - | Skull | 9 |
| **SUIFORMES** | | | | |
| Suidae | *Potamochoerus larvatus* | Bushpig | Skin | 2 |
| | *Phacochoerus africanus* | Warthog, Common | Bone | 21 |
| | | | Skin | 5 |
| | | | Skull | 16 |
| | | | Tooth/tusk | 26 |
| | *Sus domesticus* | Pig | Skin | 3 |
| **TUBULIDENTATA** | | | | |
| Orycteropodidae | *Orycteropus afer* | Aardvark | Foot | 3 |

(continued)

**Table 19.4** (continued)

| **AMPHIBIANS, REPTILES, BIRDS and MAMMALS** | | | | |
|---|---|---|---|---|
| Classification (**CLASS**, **ORDER**, **Family**) | Species | Common name | Animal part | Number of parts |
| | | | Skin | 3 |
| | | | Skull | 1 |
| | | | Whole | 1 |
| **WHIPPOMORPHA** | | | | |
| Hippopotamidae | *Hippopotamus amphibius* | Hippopotamus | Bone | 2 |
| | | | Skin | 3 |
| | | | Skull | 1 |
| | | | Tooth/tusk | 19 |
| Unidentified mammals | | | | |
| | - | - | Bone | 1222 |
| | - | - | Carcass | 5 |
| | - | - | Foot | 2 |
| | - | - | Intestine | 1 |
| | - | - | Skin | 15 |
| | - | - | Skull | 4 |
| | - | - | Tooth | 50 |
| | - | - | Vertebra | 239 |

The classification based on Hockey et al. (2005; birds); Skinner and Chimimba (2005, mammals); Zug et al. (2001) and Alexander and Marais (2007; for reptiles)

**Table 19.5** Quantity of material and percentage of 32 traders recorded selling marine fish and invertebrate taxa

| Phylum | Class | Common name | Part | % of traders | No. of pieces |
|---|---|---|---|---|---|
| Chordata | Actinopterygii (ray-finned fish) | Fish spp. (ray-finned fishes, eels) | Whole | 56.3 | 166 |
| | Chondrichthyes (cartilaginous fish) | Fish spp. (sharks, rays and skates) | Whole | 12.5 | 6 |
| Arthropoda | Malacostraca | Crab: marine | Whole | 6.3 | 4 |
| | Insecta | Beetles, grasshoppers | Whole | 9.0 | 225 |
| Cnidaria | Anthozoa | Coral | Coral | 6.3 | 3 |
| Echinodermata | Asteroidea | Starfish | Shell | 40.6 | 34 |
| | Echinoidea | Shell: urchins | Shell | 18.8 | 140 |
| Mollusca | | Shell: molluscs | Shell | 31.3 | 955 |
| | Bivalvia | Shell: clams | Shell | 3.1 | 5 |
| | Cephalapoda | Octopus | Whole | 9.4 | 3 |
| | | Cuttlefish | Whole | 9.4 | 45 |
| | Gastropoda | Shells: cowries, limpets and snails | Shell | 21.9 | 120 |
| | | Snail: giant land | Shell | 43.8 | 72 |

trader for birds. When the vertebrate classes were combined (All), the curve was less asymptotic and indicates that 1.8 new species were recorded per trader sampled. Avian species richness was higher than that of reptiles, despite birds having been recorded at the stalls of fewer traders (Table 19.1, Fig. 19.3).

At least 232 species of vertebrates (excluding domestic animals) have been recorded as being used or traded for traditional medicine from the Faraday survey ($S_{obs}$) and the other surveys conducted in South Africa ($S_{literature}$) combined (Table 19.6). The species identified in Faraday hence represent 63% of the total number of species identified in use or trade in South Africa to date.

The richness estimates generated for all (All) vertebrates ranged from 172 species for the bootstrap estimator to 233 species for the second-order jackknife estimator (Jack 2) (Fig. 19.4, Table 19.6). Both bootstrap and Jack 2 were consistently the lowest and highest estimators respectively of species richness for all data sets. Because ICE and Chao 2 gave unrealistically high estimates for a smaller number of traders (>300 species for "All" after two traders) and MMRuns predicted >6,000 species after 10 traders, the curves of these estimators were not presented and are not considered to be good predictors of species richness. Only the Jack 2 estimator consistently predicted within 0 to +3 species the total number of species recorded to date ($S_{total\ recorded}$) (Table 19.6); the remainder of the estimators underestimated $S_{total\ recorded}$ by 30–50 species. The Jack 2 estimator thus predicted an "upper-bound" estimate for the total number of species that might be recorded in the Faraday market over time (Table 19.6), including opportunistically harvested species that have low use and commercial values.

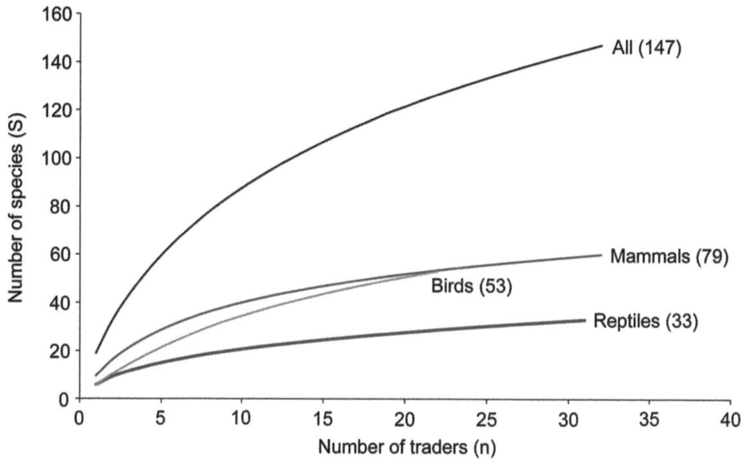


**Fig. 19.3** Species accumulation curves for vertebrate animals traded at the Faraday market. Parentheses indicate sample sizes

**Table 19.6** Comparison of observed species richness in the Faraday market ($S_{obs}$), the number of additional species observed/recorded in other surveys ($S_{literature}$) and the estimated species richness predicted by the second-order jackknife (Jack 2) estimator from EstimateS

| | Observed species richness | | | Estimated species richness |
|---|---|---|---|---|
| | $S_{obs}$ | $S_{literature}$ | $S_{total\ recorded} = S_{obs} + S_{literature}$ | Jack 2 |
| All | 147 | >85 | ≥232 | 233 |
| Mammals | 60 | >23 | ≥84 | 87 |
| Reptiles | 33 | >17 | ≥50 | 52 |
| Birds | 53 | >40 | ≥93 | 93 |
| Amphibians | 1 | >4 | ≥5 | – |

## 19.3.5 Species Diversity

The overall diversity of the species identified in the Faraday market is medium–high (Shannon $H' = 4.49$; Simpson's $-\ln \lambda = 4.23$) (Table 19.1, Figs. 19.5, 19.6). The cumulative diversity curves have reached asymptotes, indicating that the diversity index values would change very little with additional sampling effort (Figs. 19.5, 19.6). Within the vertebrate groups, diversity values indicate that there is a greater diversity of birds traded in the markets compared to mammals and reptiles, even though the overall species richness of birds is lower than that for mammals (Table 19.1, Figs. 19.5, 19.6). The higher bird diversity is partly indicative of the higher number of singletons recorded (Table 19.1).

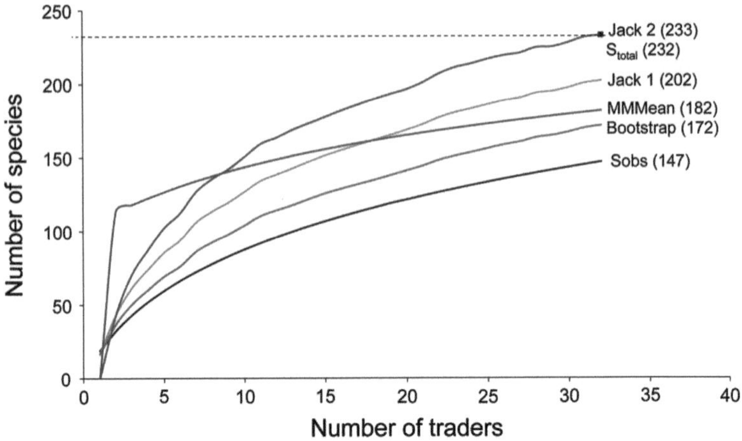


**Fig. 19.4** The performance of four incidence-based species richness estimators compared with the observed species accumulation curve ($S_{obs}$) for "All" animals identified in the Faraday market (mammals, reptiles, birds and amphibian). The Jack 2 estimator predicts within one species the total number of species recorded from 10 other surveys ($S_{total}$) and also the total number of species likely to be recorded in Faraday over time

Overall, evenness values are high, indicating that most species were evenly dispersed throughout the market and that relatively few species were very dominant (Table 19.1, Table19.7). The predominance of crocodile, python and monitor parts within the market accounts for the lower evenness values for reptiles compared to mammals and birds. Of the 33 reptile species identified, eight species (24%) were dominant and were frequently sold at traders' stalls (Table 19.7). Bird fauna exhibited the least dominance of all the vertebrates with only 15% of the species being of very common occurrence. Forty-nine percent of bird species were identified only once in the market (Table 19.7), further confirming the reason for the high diversity values for avian fauna despite the slightly lower species richness values compared to mammals.

### *19.3.6 Quantity Traded and Consumption Levels*

The number of traders selling a species and the quantity of that species in Faraday was significantly positively correlated ($r = 0.91$; $n = 138$; $P < 0.00001$) (Fig. 19.7). Hence, the more traders that sold a species, the more pieces of it there were likely to be. The correlation excluded 30 eggs, 51 ostrich feathers, 388 porcupine quills, 76 pangolin scales, 266 bovine horns and 84 teeth, as well as 65 crocodile osteoderms. When the aforementioned animal parts were included, the correlation was lower, but still significant ($r = 0.56$; $n = 146$; $P < 0.00001$). Table 19.4 lists the quantities of each animal part sold per species.

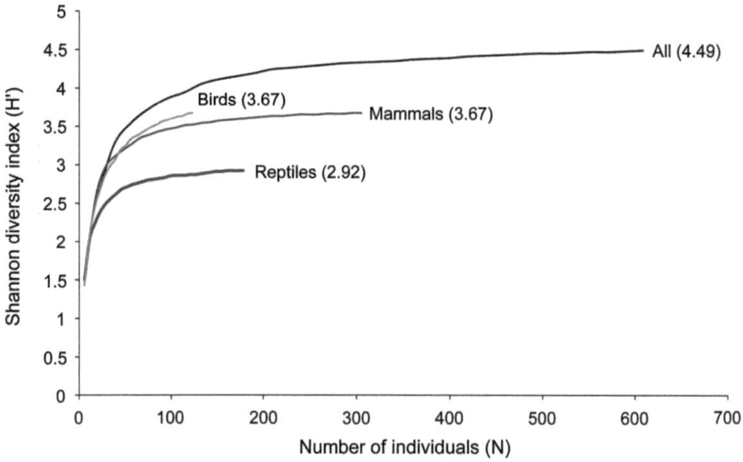


**Fig. 19.5** Cumulative diversity curve for the Shannon diversity index ($H'$)

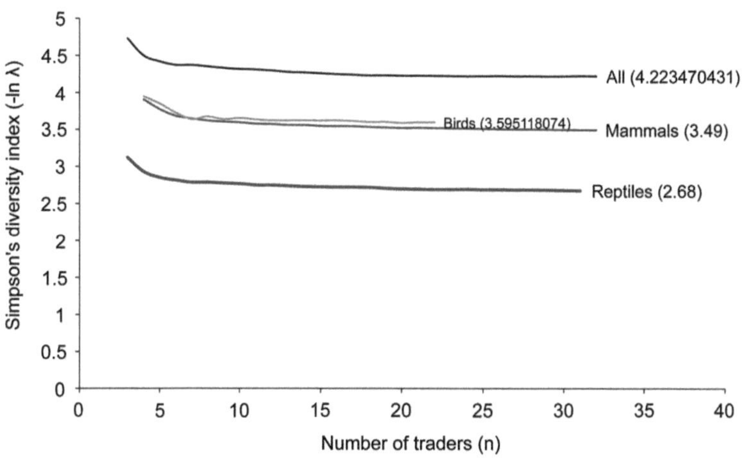


**Fig. 19.6** Cumulative diversity curve for the Simpson's diversity index ($-\ln \lambda$)

**Table 19.7** Number and percentage of species of rare, intermediate and common occurrence within the Faraday market

| | All | Mammals | Reptiles | Birds |
|---|---|---|---|---|
| Total $S$ | 147 | 60 | 33 | 53 |
| Very common occurrence | 28 (19%) | 12 (20%) | 8 (24%) | 8 (15%) |
| Intermediate occurrence | 62 (42%) | 30 (50%) | 12 (36%) | 19 (36%) |
| Rare occurrence | 57 (39%) | 18 (30%) | 13 (39%) | 26 (49%) |

The categories are derived from Hill's numbers $N_0$, $N_1$ and $N_\infty$

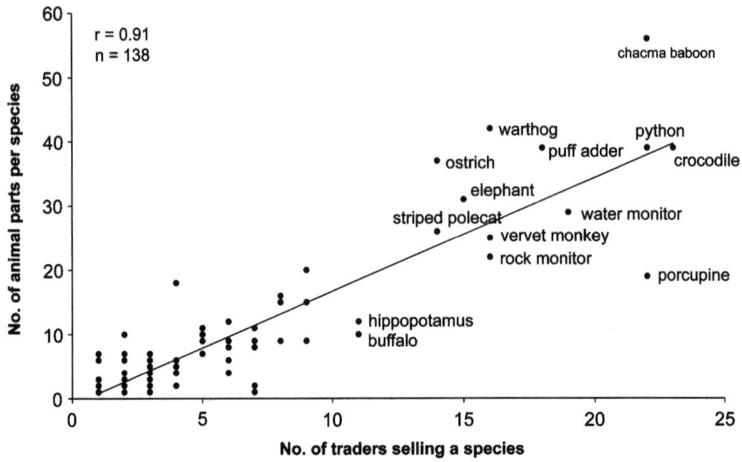


**Fig. 19.7** Relationship between the number of traders selling a vertebrate species and the total number of animal parts per species (excluding ostrich feathers, porcupine quills, pangolin scales, bovine teeth and horns, and crocodile osteoderms). Species sold by more than 10 traders are labelled

Mammal body parts and bones were present in the largest quantities (2453 pieces, excluding porcupine quills and pangolin scales), followed by reptiles (394 pieces, excluding osteoderms), birds (193 pieces, excluding feathers and ostrich eggs) and amphibians (6 parts) (Table 19.8). Mammal bones were the most prevalent body part recorded. Of the vertebrates, fishes were the most likely to be sold as an intact carcass ($n = 172$, 100%) followed by birds ($n = 152$, 55.7% of all body parts), reptiles ($n = 163$, 35.5%) and mammals ($n = 140$, 4.8%). Conversely, mammals were the most likely to be sold as individual body parts and mammal bones were the most frequently documented item ($n = 1528$), followed by porcupine quills ($n = 388$), horns ($n = 266$), pieces of skin ($n = 214$) and skulls/heads ($n = 133$). For the reptiles, body parts that were traded in significant numbers included skins ($n = 107$), crocodile osteoderms ($n = 65$), and pieces of chelonian shells ($n = 55$). Feathers ($n = 51$) and eggs ($n = 30$) were the most frequently traded items of bird origin.

On average, traders ($n = 10$) sold to 5–10 customers a day, but this ranged from 2 to 15 customers and went as high as 25 customers on a very busy day.

### *19.3.7 Conservation Status*

Most ($n = 119$, 87.5%) species traded ($n = 136$) in Faraday were of LC (Table 19.2, Table19.9). Of the 17 taxa of conservation concern, we recorded a single individual (skull) of a CR species (Hawksbill Turtle), and a single individual (skull) from an EN species (Wild Dog). Two traders had parts of a Samango

**Table 19.8** Number of body parts recorded for vertebrates in the Faraday survey

| Mammals | | Reptiles | | Birds | |
|---|---|---|---|---|---|
| Body parts | No. of pieces | Body parts | No. of pieces | Body parts | No. of pieces |
| Bones: unidentified | 1528 | Whole (lizards, monitors and snakes) | 163 | Whole | 152 |
| Porcupine quills | 388 | Skin (Squamata) | 107 | Feathers | 51 |
| Horns | 266 | Osteoderms (crocodile) | 65 | Eggs | 30 |
| Skin (whole or pieces) | 214 | Shell/plastron/carapace (Testudines) | 55 | Skull | 13 |
| Whole body/carcass | 140 | Eggs | 22 | Leg | 11 |
| Skull/head | 113 | Skull/head | 23 | Foot | 6 |
| Teeth/tusks | 84 | Python body parts | 16 | Beak | 4 |
| Scales (pangolin) | 72 | Foot | 6 | Head and neck | 3 |
| Hooves | 35 | Neck | 1 | Skin | 2 |
| Legs | 34 | Tail | 1 | Wing | 1 |
| Foot/paw | 27 | | | | |
| Intestine | 3 | | | | |
| Penis/scrotum | 3 | | | | |
| Tails | 3 | | | | |
| Jaw | 2 | | | | |
| Nose | 1 | | | | |
| TOTAL | 2913 | | 459 | | 273 |

Monkey (Table 19.4), which, depending on the subspecies, is considered either VU or LC (Kingdon et al. 2008). Of the remaining 14 taxa, six were VU (1 lizard, 2 birds, 3 mammals) and eight (2 birds, 6 mammals) were NT (Table 19.9).

Twenty (62.5%) traders sold at least one species of conservation concern (mean = $1.59 \pm 1.79$, range: 0–6, $n = 32$ traders), which was significantly more than expected (Fisher's exact test $P = 0.067$). However, the proportion of traders selling a particular species was unrelated to its conservation status ($\chi^2 = 0.63$, d.f. = 1, $P > 0.1$; Table 19.2, Table19.9). Furthermore, species of conservation concern were not significantly more abundant ($\chi^2 = 1.37$, d.f. = 1, $P > 0.1$; Table 19.4, Table19.9) than species of LC.

## 19.4 Discussion

We provide the first quantification of the trade of animals for use in traditional medicine at the Faraday market in South Africa. Most animals traded were vertebrates, although significant quantities of marine molluscs were also on sale. We identified 147 species of vertebrate, most of which were mammals (41%, 60 taxa), followed by birds (36%, 53 taxa), reptiles (22%, 33 taxa) and a single species of frog. All together, this species richness constitutes *c.* 8.7% of the total

**Table 19.9** Vertebrate species of conservation concern according to 2001 IUCN Red List Categories and Criteria version 3.1 that were traded at the Faraday market

| Common name | Species | IUCN category | Number of traders | Number of parts |
|---|---|---|---|---|
| **Reptiles** | | | | |
| Hawksbill turtle | *Eretmochelys imbricata* | CR | 1 | 1 |
| Sungazer (lizard) | *Cordylus giganteus* | VU | 5 | 11 |
| **Birds** | | | | |
| Southern ground-hornbill | *Bucorvus cafer* | VU | 3 | 3 |
| Caspian tern | *Sterna caspia* | NT | 1 | 1 |
| White-backed vulture | *Gyps africanus* | VU | 3 | 5 |
| Great white pelican | *Pelecanus onocrotalus* | NT | 1 | 1 |
| **Mammals** | | | | |
| Samango monkey | *Cercopithecus mitis* ssp. | VU/LC | 2 | 2 |
| South African hedgehog | *Atelerix frontalis* | NT | 8 | 9 |
| Ground pangolin | *Smutsia temminckii* | VU | 2 | 74 |
| Brown hyaena | *Hyaena brunnea* | NT | 1 | 1 |
| Spotted hyaena | *Crocuta crocuta* | NT | 4 | 4 |
| Unidentified hyaena | – | NT | 5 | 5 |
| Lion | *Panthera leo* | VU | 3 | 3 |
| Serval | *Leptailurus serval* | NT | 2 | 2 |
| African wild dog | *Lycaon pictus* | EN | 1 | 1 |
| Honey badger (Ratel) | *Mellivora capensis* | NT | 1 | 3 |
| Blesbok/Bontebok | *Damaliscus pygargus* | VU (Bontebok) | 1 | 6 |

We did not identify marine organisms and invertebrates to a sufficient level to assign IUCN categories. See Table 19.2 for detail on the specific parts of the animal that were for sale. Note that the number of parts sold does not equate to the number of animals sold. In the case of the Samango monkey, there are two southern African subspecies, one of which is Vulnerable while the other is Least Concern. We could not distinguish between Bontebok (VU) and Blesbok (LC) skulls. We also include unidentified hyaena because both species in southern Africa are NT. *LC* Least Concern, *NT* Near Threatened, *VU* Vulnerable, *EN* Endangered, *CR* Critically Endangered

frog, reptile, bird and mammal fauna of South Africa (1,685 + species total). If frogs (128 species) are excluded, this percentage increases marginally to 9.4%. We separately quantified domestic animals, invertebrates and marine fishes. Parts of seven domestic animals were for sale, but generally only a few individuals of each species and from only a few traders. Therefore, they are likely to be relatively unimportant as a source of medicine. Of the fishes, sole (*Austroglossus pectoralis*) were the most abundant, followed by box fish and an assortment of dried marine ray-finned fishes that we were unable to identify. Marine molluscs, chiefly gastropods, were sold by about a third of traders and were abundant in the market ($n = 955$). We only documented two species of insect. One trader had a large (> 200) batch of CMR Bean Beetles (*Mylabris oculata*) (Table 19.4) while three traders had grasshoppers (*Taphronota*) for sale.

Species richness at Faraday was relatively high for a single source for traditional medicine in South Africa. In comparison to the 147 species of vertebrate that were for sale at Faraday, Simelane and Kerley (1998) reported 44 species (eight reptiles, six birds, 30 mammals) being sold in 19 herbalist shops in the Eastern Cape Province of South Africa. Cunningham and Zondi (1991) examined the trade in animals for traditional medicine in KwaZulu-Natal Province and also review literature reports for South Africa. They report at least 79 species of vertebrate (18 reptiles, 16 birds, 45 mammals), excluding domestic mammals and various marine invertebrates and fishes. More recently, Ngwenya (2001) recorded 132 species of vertebrate (21 reptiles, 32 birds, 79 mammals) in trade across KwaZulu-Natal Province, of which 50 species were in high demand, especially vultures, chacma baboon, green mamba, southern African python, Nile crocodile, puff adder, striped weasel and black mamba. In these studies, mammals are the most commonly sold group, followed by similar numbers of birds and reptiles. At Faraday, mammals were also the most commonly traded group, but we found a higher proportion of bird species than reptiles. In contrast to these studies, Herbert et al. (2003) focused on the invertebrate trade at a large traditional medicine market in Durban. They report a much greater diversity of marine invertebrates (seven phyla compared to four phyla in our study), which can be explained in large part by Durban's coastal location.

The second-order jackknife estimator (Jack 2) predicted that 233 species (an additional 86 species) could be identified with further sampling in the Faraday market over time (Fig. 19.4, Table 19.6). Based on the Jack 2 estimate, the Faraday survey has identified 63% of the total number of species recorded in South Africa to date. Given that samples should aim to record 50–75% of the total richness in a region (Heck et al. 1975), we believe the sampling strategy and the number of traders interviewed to be sufficient and representative. However, estimates of species richness at traditional medicine markets are always conservative because of the large proportion of unidentified material. Of the 3,716 animal parts documented at Faraday, 42% were not identifiable at the level of order, 45% not identifiable to family and 53% not identifiable to species. Most (41%) of the unidentified animal parts were various mammal bones and teeth, while 72% of the fish and invertebrates could not be identified at the level of order, 79% were not identified to family and 87% could not be identified to genus. Therefore, species richness is likely to be higher than what we report here since we took a conservative approach to estimating species richness and diversity by not including "morphospecies" in the analyses (i.e. typological species that could only be identified as mongoose sp., monitor sp., owl sp., etc.). A consequence of this action was a reduction in the total number of singletons and doubletons, variables that are usually positively correlated with diversity and estimates of species richness. The more singletons there are in a sample, the higher the diversity and the greater the total estimated species richness is likely to be. When we included the morphospecies in the analyses, the richness estimate generated by Jack 2 for all (All) vertebrates increased from 233 species (Table 19.6) to 289 species and the Jack 1 estimator predicted 247 species in trade. Hence, Jack 1 and Jack 2 can be viewed as good lower- and upper-bound estimators, respectively, of vertebrate species richness traded commercially in the Faraday market over time, including opportunistically harvested species. In the absence of morphospecies in the analyses, however, Jack 2 was the only estimator that predicted (within three species) the total number of species that have been identified in South African markets to date. Williams et al. (2007c) also found that the Jack 1 and Jack 2 estimators were the best lower- and upper-bound estimators, respectively, of plant species richness at Faraday. Furthermore, the diversity and species accumulation curves were all asymptotic or near asymptotic respectively, indicating that further sampling would not significantly change the diversity index nor the species richness results with further sampling; hence, a sufficient number of traders were interviewed overall.

In general, there was little overlap in what animal species traders had for sale. Any two traders would generally have less than a third of their species in common and this number was also dependent on taxonomic group. For example, traders had about 33% of reptile species in common but only about 13% of bird species in common. Nine species of vertebrate were traded by more than 50% of traders, five of which were reptiles (rock and water monitor, Nile crocodile, southern African python, puff adder) and four of which were mammals (chacma baboon, Cape porcupine, vervet monkey, warthog). In the case of birds, there were relatively low numbers of any one species, with the exception of ostrich (44% of traders) and to a lesser extent, owls. A low abundance of any particular bird species coupled with relatively high species richness meant higher species diversity and evenness. By comparison, mammal and reptile diversity indices were lower because of the relative abundance of certain species.

Establishing the impact of traditional medicine on wildlife is notoriously difficult because traders are reluctant to reveal the source of their stock. At Faraday, we were unable to explore this issue and we only obtained a very rough estimate of the number of customers that bought animal parts per day, from 10 traders. Therefore, we had no data on the actual turn-over of specific species and the rate at which stock was replaced. In addition, many traders sell individual bones or pieces of skin making it impossible to determine how many individual animals are being traded in a particular market. For example, at Faraday, traders frequently sold small pieces of elephant skin. Only prohibitively expensive DNA analysis would allow an estimation of how many individual elephants were present in the market and such an undertaking might only be valuable for the most critically endangered species. Another confounding issue is that because traders are also willing to use animals recovered dead from the wild (death by natural causes or for example, by a vehicle) we also had no data on the proportion of live animals that were harvested from the wild specifically for traditional medicine. However, in one instance we observed a live hedgehog (IUCN NT) and a batch of recently killed Sungazer lizards (VU). Regardless of these constraints, we were still able to provide a crude assessment of the potential impact of the Faraday market on species of conservation concern by enumerating all parts belonging to threatened species. We documented a single CR species (Hawksbill Turtle) and one EN species (Wild Dog), both consisting of a single skull. The remaining 15 species of conservation concern consisted of a single reptile (VU), four birds (2 VU, 2 NT) and 10 mammals. Of the mammals, six horns were identified as belonging to either

Blesbok (LC) or Bontebok (Vulnerable). While the widespread distribution and greater abundance of Blesbok make them a likely candidate, we cannot exclude the possibility that one or more might be Bontebok. In the case of the Samango Monkey, we were unable to establish the subspecies, one of which is a threatened species. Of the remaining mammals, six were NT and two were VU. The 17 species of conservation concern all occurred at relatively low frequency (excluding pangolin, <4 parts/species). In the case of Pangolin, 74 scales were recorded which could potentially come from a single individual. Therefore, the trade of species of conservation concern at Faraday is unlikely to pose a significant threat to the viability of any one species. However, our study is a snap-shot in time and given the extensive country-wide trade in animals for traditional medicine, future monitoring is necessary to prevent over-exploitation of threatened species. This is particularly true for animals such as vultures, which are highly prized for traditional medicine, and which can and have been killed in significant numbers during a single event such as poisoning (Cunningham and Zondi 1991; Mander et al. 2007). While we need to respect the individual's need to access traditional medicine, it is in everyone's interest to ensure that these age-old practices are sustainable.

**Acknowledgments** This study was funded by a National Research Foundation grant to M.J.W. In addition, V.L.W. was funded by a Postdoctoral Fellowship from the University of the Witwatersrand during the write-up of this paper. We are very grateful to Tebogo Mothupi, Zuko Mafani, Mdu Seakamela and Simphiwe Khuluse for collecting a significant portion of the data presented here. We also thank Solomon Mvubu, chairman of the Faraday Traders Committee, and all the Faraday traders for their co-operation, without whom, this study would not have been possible. We are grateful for updated species counts for South Africa from Bill Branch (reptiles), Louis du Preez (frogs), Trevor Hardaker (birds) and Chris Chimimba (mammals). The protocol for this research was approved by the University of the Witwatersrand Human Research Ethics Committee (Medical) (Protocol M0500945).

## References

Alexander G, Marais J (2007) A guide to the reptiles of Southern Africa, 1st edn. Struik, Cape Town

Barnes KN (ed) (2000) The Eskom red data book of birds of South Africa, Lesotho and Swaziland, 1st edn. BirdLife, South Africa

Begossi A (1996) Use of ecological methods in ethnobotany: diversity indices. Econ Bot 50(3):280–289

BirdLife International (2009) World bird database (WBDB). Cambridge. http://www.birdlife.org/datazone/index.html. Accessed 10 July 2009

Branch B (1998) Field guide to snakes and other reptiles of southern Africa, 3rd edn. Struik, Cape Town

But PPH, Lung LC, Tam YK (1990) Ethnopharmacology of rhinoceros horn. I: antipyretic effects of rhinoceros horn and other animal horns. J Ethnopharmacol 30:157–168

Bye SN, Dutton MF (1991) The inappropriate use of traditional medicines in South Africa. J Ethnopharmacol 34:253–259

Chiarucci A, Enright NJ, Perry GLW, Miller BP, Lamont BB (2003) Performance of nonparametric species richness estimators in a high diversity plant community. Divers Distrib 9:283–295

Cocks M, Dold A (2000) The role of 'African Chemists' in the health care system of the Eastern Cape Province of South Africa. Soc Sci Med 51(10):1505–1515
Colwell R (2006) EstimateS: statistical estimation of species richness and shared species from samples, version 7.5.1. http://viceroy.eeb.uconn.edu/estimates. Accessed 10 July 2009
Crump CM (2003) Need to assess animals used by traditional healers in South Africa. In: Williams VL (ed) Hawkers of health: an investigation of the Faraday Street traditional medicine market in Johannesburg, Appendix 4. Unpublished Report to Gauteng Directorate for Nature Conservation, DACEL
Cunningham AB, Zondi AS (1991) Use of animal parts for the commercial trade in traditional medicines, 1st edn. Institute of Natural Resources, Pietermaritzburg
Derwent S, Mander M (1997) Twitchers bewitched. The use of birds in traditional healing. *Africa: Birds & Birding* 2(1):22–25
Ehrlich PR (2009) Cultural evolution and the human predicament. TREE 24:409–412
Friedmann Y, Daly B (eds) (2004) Red data book of the mammals of South Africa: a conservation assessment. CBSG Southern Africa, conservation breeding specialist group (SSC/IUCN). Endangered Wildlife Trust, South Africa
Hanazaki N, Tamashiro JY, Leitão-Filho HF, Begossi A (2000) Diversity of plant uses in two Caiçara communities from the Atlantic forest coast, Brazil. Biodivers Conserv 9:597–615
Hayek LC, Buzas MA (1997) Surveying natural populations, 1st edn. Columbia University Press, New York
Heck KL, van Belle G, Simberloff D (1975) Explicit calculation of the rarefaction diversity measurement and the determination of sufficient sample size. Ecology 56(6):1459–1461
Herbert DG, Hamer ML, Mander M, Mkhize N, Prins F (2003) Invertebrate animals as a component of traditional medicine trade in KwaZulu-Natal South Africa. Afr Invertebr 44(2):327–344
Hockey PAR, Dean WRJ, Ryan PG (eds) (2005) Roberts birds of southern Africa, 7th edn. The trustees of the John Voelcker Bird Book Fund, Cape Town
IUCN (2009) 2009 IUCN Red list of threatened species. IUCN, Gland, Switzerland. http://www.iucnredlist.org. Accessed 24 Aug 2009
Kingdon J, Butynski TM, de Jong Y (2008) *Cercopithecus mitis* ssp. *albogularis*. In: IUCN 2010. IUCN Red List of Threatened Species. Version 2010.4. http://iucnredlist.org/. Accessed 10 July 2009
Lev E (2003) Traditional healing with animals (zootherapy): medieval to present-day Levantine practice. J Ethnopharmacol 85:107–118
Li YW, Zhu X, But PPH, Yeung HW (1995) Ethnopharmacology of bear gall bladder: I. J Ethnopharmacol 47:27–31
Ludwig JA, Reynolds JF (1988) Statistical ecology: a primer on methods and computing. Wiley, Toronto
McKinney ML (1997) Extinction vulnerability and selectivity: combining ecological and paleontological views. Annu Rev Ecol Syst 28:495–516
Magurran AE (1988) Ecological diversity and its measurement, 1st edn. Princeton University Press, Princeton
Mander M, Diederichs N, Ntuli L, Mavundla K, Williams V, McKean S (2007) Survey of the trade in vultures for the traditional health industry in South Africa, 1st edn. FutureWorks, Durban
Marshall NT (1998) Searching for a cure: conservation of medicinal wildlife resources in east and southern Africa, 1st edn. TRAFFIC International, Cambridge
Minter LR, Burger M, Harrison JA, Braack HH, Bishop PJ, Kloepfer D (eds) (2004) Atlas and red data book of the frogs of South Africa, Lesotho and Swaziland. SI/MAB Series 9. Smithsonian Institution, Washington
Ngwenya MP (2001) Implications for the medicinal animal trade for nature conservation in Kwazulu-Natal. Ezemvelo KZN Wildlife Report No. NA/124/04
Simelane TS (1996) The traditional use of indigenous vertebrates. PhD. Thesis, University of Port Elizabeth, Port Elizabeth

Simelane TS, Kerley GIH (1998) Conservation implications for the use of vertebrates by Xhosa traditional healers in South Africa. S Afr J Wildl Res 28(4):121–126
Sinclair I, Hockey P, Tarboton W (1997) Sasol birds of southern Africa, 2nd edn. Struik, Cape Town
Still J (2003) Use of animal products in traditional Chinese medicine: environmental impact and health hazards. Complement Therap Med 11:118–122
Skinner JD, Chimimba CT (2005) The mammals of the southern African subregion, 3rd edn. Cambridge University Press, Cambridge
Stuart C, Stuart T (2001) Field guide to mammals of southern Africa, 3rd edn. Struik, Cape Town
Toti DS, Coyle FA, Miller JA (2000) A structured inventory of Appalachian grass bald and heath bald spider assemblages and a test of species richness estimator performance. J Arachnol 28:329–345
White RM, Cocks M, Herbert DG, Hamer ML (2004) Traditional medicines from forest animals. In: Lawes MJ, Eeley HAC, Shackleton CM, Geach BGS (eds) Indigenous forests and woodlands in South Africa: people, policy and practice, 1st edn. University of Natal Press, Pietermaritzburg
Williams VL (2003) Hawkers of health: an investigation of the Faraday street traditional medicine market in Johannesburg. Unpublished Report to Gauteng Directorate for Nature Conservation, DACEL
Williams VL (2007) The design of a risk assessment model to determine the impact of the herbal medicine trade on the Witwatersrand on resources of indigenous plant species. PhD thesis, University of the Witwatersrand, Johannesburg
Williams VL, Witkowski ETF, Balkwill K (2005) Application of diversity indices to appraise plant availability in traditional medicine markets in Johannesburg, South Africa. Biodivers Conserv 14:2971–3001
Williams VL, Balkwill K, Witkowski ETF (2007a) Size-class prevalence of bulbous and perennial herbs sold in the Johannesburg medicinal plant markets between 1995 and 2001. S Afr J Bot 73:144–155
Williams VL, Witkowski ETF, Balkwill K (2007b) Volume and financial value of species traded in the medicinal plant markets in Gauteng, South Africa. Int J Sustain Dev World Ecol 14:584–603
Williams VL, Witkowski ETF, Balkwill K (2007c) The use of incidence-based species richness estimators, species accumulation curves and similarity measures to appraise ethnobotanical inventories from South Africa. Biodivers Conserv 16:2495–2513
Wilson EO (ed) (1988) Biodiversity. National Academy Press, Washington
Zug GR, Vitt LJ, Caldwell JP (2001) Herpetology: an introductory biology of amphibians and reptiles, 2nd edn. Academic Press, California

# Chapter 20
# Finely Ground-Hornbill: The Sale of *Bucorvus Cafer* in a Traditional Medicine Market in Bulawayo, Zimbabwe

**Robin K. Bruyns, Vivienne L. Williams and Anthony B. Cunningham**

**Abstract** Southern Ground-hornbill (*Bucorvus cafer*) is used in traditional medicine in Zimbabwe, and the trade of this species was investigated in an informal sector market in Bulawayo. The frequency of hunting for *B. cafer* was found to be currently insignificant, and could neither be causally linked to population decline nor proved to be a driver of the low abundances of ground-hornbills in the surrounding areas where hunting occurs. Cultural perceptions and the preferences of users in western Zimbabwe, primarily Matabeleland, appear to be a factor in the low levels of utilization. Hunting of ground-hornbills for traditional medical practices in this area is therefore not a significant threat to the species, since hunting practices appear to be rare and opportunistic rather than directly intentional. Habitat loss and degradation (largely due to agricultural expansion, charcoal and fuelwood demand) is likely to be the main threat to *B. cafer* populations.

R. K. Bruyns · V. L. Williams (✉)
School of Animal, Plant and Environmental Sciences,
University of the Witwatersrand, Private Bag 3, Wits 2050, South Africa
e-mail: vivwill@netdial.co.za

A. B. Cunningham
School of Plant Biology, University of Western Australia,
35 Stirling Highway, Crawley 6009, Australia
e-mail: tonyc05@bigpond.net.au

A. B. Cunningham
People and Plants International, 84 Watkins Street,
White Gum Valley, Fremantle, Australia

R. R. N. Alves and I. L. Rosa (eds.), *Animals in Traditional Folk Medicine*,
DOI: 10.1007/978-3-642-29026-8_20, 

## 20.1 Introduction

In 1901, Hollis (1901) described the respect that the Taveta people in southern Kenya had for the "turkey-buzzard": "*Woe betide the native who either on purpose or by accident kills one of these birds; he will be struck down by a strange and mysterious illness, which, in a few days, will terminate in his death*". From his description, the alleged "turkey-buzzard" was more likely a Southern Ground-hornbill (*Bucorvus cafer*, syn. *B. leadbeateri*). Similar reverence for the ground-hornbill, or *intsingizi*, is held by the amaXhosa people in the Eastern Cape Province of South Africa (Godfrey 1941). The bird is considered sacred, should not be killed, can be sent as a messenger of death by a witch-doctor, and can be used to bring rain during times of drought (Godfrey 1941). Such records of the reverence and role of birds in African cultures are numerous (e.g., Hollis 1901; France 1905; Garbutt 1909; Parrinder 1956; Conco 1972; Barnard 1979; Msimanga 2000; Abrahams et al. 2002); however, the actual documentation of taxa used is fairly recent (e.g., Ntiamoa-Baidu 1987; Cunningham and Zondi 1991; Marshall 1998; Simelane and Kerley 1998; Nikolaus 2000; Cocker 2000; Ngwenya 2001; Whiting et al. 2011). The tendency to document the species used appears to stem from an increasing interest in the use of avifauna for traditional medicine, and, moreover, a growing concern over the conservation implications of unchecked use.

Vultures, Bateleur eagles (*Terathopius ecaudatus*), and Southern Ground-hornbills have been identified as priority conservation species in the commercialized animal traditional medicine trade (Cunningham and Zondi 1991). In southern Africa, the monitoring of ground-hornbill sales in informal sector markets is important since the species is threatened and is being hunted from areas specifically set aside for conservation. Additionally, the species is threatened by habitat destruction, low clutch sizes, and sexual maturity only being reached after 5 years. Hence, there is a need to conserve and protect areas that act as a population source for priority species such as the ground-hornbill (Cunningham and Zondi 1991). This not only places emphasis on core areas linked to the persistence of species vulnerable to human impacts, but also facilitates stricter monitoring of an area for any traditional medicine harvesting activities. To this end, Cunningham and Zondi (1991) firmly state: "*Although sustainable use is possible in theory, the margin between sustainable use and over-exploitation for species with a low fecundity and long period before sexual maturity is so small that with the constraints of money and manpower on conservation agencies, for practical purposes it is not a realistic option, and certainly not when populations of Cape vulture, Lappet faced vulture, Bateleur Eagle, Ground Hornbill and other species are in decline*". Conservation concerns about the vulnerability of large-bodied, slow-breeding bird species that become threatened through habitat loss or when the fecundity-mortality balance is disrupted is widespread. In a global review of the main causes of species extinction risk in birds, Owens and Bennett (2000) pointed out that different bird lineages may follow different paths to extinction, with some bird families [viz. parrots (Psittacidae), rails (Rallidae), pheasants, partridges and francolins (Phasianidae), pigeons (Columbidae), cranes (Gruidae) and white-eyes (Zosteropidae)] threatened both by habitat loss and impacts on the fecundity-mortality balance.

The Southern Ground-hornbill (*B. cafer*) has a southern African distribution ranging from eastern South Africa to south-west Angola and southern Kenya (BirdLife International 2010). The actual global population size has not been quantified (BirdLife International 2010), and no data on population numbers are available for Zimbabwe. In South Africa, however, Kemp (2000) estimated that there were 1,500–2,000 birds in 2000, and Spear et al. (2005) estimated the species had experienced a 50% decline in range and a > 10% decline in numbers of mature individuals over the past three decades. Since 1998, regular aerial surveys of the Niassa National Reserve in northern Mozambique showed a population of 3,662–5,122 Southern Ground-hornbills in 2009 (Craig 2009). This is double the total South African population in an area of 42,713 km$^2$. The species was first assessed according to IUCN Red List criteria in 1988 and classified as Lower Risk/ Least Concern (Barnes 2000). In 2010, however, the species' global status was upgraded to Vulnerable (A4bcd) due to rapid population declines resulting from "habitat destruction and persecution" (BirdLife International 2010).

In the past few years, conservation-based projects aimed at *B. cafer* have increased. These include programs to collect second chicks from the Kruger National Park for rearing and release into protected areas, the use of artificial nest boxes, and research into the breeding habits and general monitoring of the birds (Spear et al. 2005). In the Niassa National Reserve, where habitats are largely intact, there has been a marked increase of *B. cafer* since 1998 (Craig 2009).

In Zimbabwe, ground-hornbill numbers have declined significantly outside most protected areas (Chiweshe 2007). The five major factors contributing to this trend were considered to be competition for space with the expanding human population, the loss of large hollow tree trunks used for nesting, birds eating poisoned baits, snaring, and persecution (e.g., residents shooting the birds when they smashed window panes that reflected their images). Similarly in Malawi, rapid deforestation for agriculture has restricted ground-hornbills to national parks, wildlife, and forest reserves (Kalamira 2007). Here, low number of birds are attributed to habitat loss, and the hunting of birds for bushmeat, and feathers for traditional dress (Kalamira 2007).

The presence of *B. cafer* in South African traditional medicine markets, and the need for further research, was reported by Cunningham and Zondi (1991), Ngwenya (2001) and Whiting et al. (2011). However, only Whiting et al. (2011) give quantifiable values of what was present, and none state what the bird is used for. Hence, in the context of threats posed by the traditional medicine trade, the aim of this study was to evaluate the trade and throughput of *B. cafer* in an informal sector market in Bulawayo, Zimbabwe, where no known market survey work had been conducted. The study also aimed at establishing the manner by which the birds were obtained (hunted), and to note the traders' perceptions on use and decline.

## 20.2 Methods

Interviews with 18 traders selling traditional medicines were conducted in the Makokoba informal sector market in Bulawayo, Zimbabwe (20.1437S; 28.5755E), in August 2010, aided by an interpreter. Traders in the market primarily sell traditional medicines, but vendors selling fresh produce are also present. Prior permission to conduct the survey was obtained from the Chairman of the market and the local authorities. Traders were first asked whether they used and/or sold *B. cafer* (commonly called *iNsingizi* in isiNdebele or *Dendera* in chiShona) and, if the answer was affirmative, then the interview proceeded using a questionnaire.

By making use of semi-structured interviews and questionnaires containing focused questions, surveys can facilitate a general understanding of the structure and dynamics of informal markets. The questionnaires designed for the Makokoba survey are listed in Appendix 1 and 2, and were adapted from the questionnaires used for the study on vultures sold for traditional medicine by Mander et al. (2007) in South Africa. Once the survey was completed, a scientist working at the Bulawayo Natural History Museum visited Makokoba at least twice a month for seven months to monitor the sales of *B. cafer* and to note the arrival of any new specimens.

## 20.3 Results

### *20.3.1 Trader Demographics*

Of the 18 traders interviewed in the market, 12 (67%) indicated that they occasionally sold ground-hornbill parts. While 67% of all the vendors in Makokoba are male, 75% of the 12 traders who sold ground-hornbill were male; hence, compared to the entire market, a slightly lower proportion of the female traders (25%) sold ground-hornbill parts. Regarding trader ethnicity, the majority (67%) were isiNdebele-speaking, with some chiShona (17%), VaChangana ('Shangaan'), and chiKalanga speakers present as well. However, 83% of the 12 vendors selling ground-hornbill were isiNdebele, and no chiKalanga-speakers sold these birds.

### *20.3.2 Southern Ground-Hornbill Trade and Acquisition*

Most traders in Makokoba sold plant material for traditional medicine interspersed with animal parts (Fig. 20.1). Of the 18 vendors interviewed, six sold no animal material. On average, each trader had been "in business" for 23 ± 17 years had some local long-term knowledge concerning the change in the availability of ground-hornbill as a resource. All the traders who sold ground-hornbill said that it had become more difficult to acquire them. Reasons given for the decline in availability included: drought; too many hunters; people not following their traditional ways; and, that the birds protect themselves through the spirits and 'seeing' who intends to kill them. Fifty-five percent of traders noted that it could take them 2 years or more to acquire a single bird. In addition, once acquired, the parts could remain at their stalls for between 3 months and several years. The monthly monitoring of Makokoba from August 2010 to February 2011 revealed that no new birds were acquired by the traders in the 7-month period.

**Fig. 20.1** Traders in the Makokoba informal sector market in Bulawayo, Zimbabwe. Stalls sell mostly plant material interspersed with animal parts. **a** typical stalls; **b** a crow and the paw of a hyena are visible in the foreground (Photos: VL Williams)

Despite an eye and socket of *B. cafer* having been observed in Makokoba 3 months before we conducted the survey (VL Williams pers. obs.), no birds (body parts or whole) were present in the market at the time of the study in August 2010. Hence, we queried the traders instead regarding their *preference* for various body parts should they be able to acquire them and also what they tended to sell when they did acquire the parts. There were some differences in preferences versus what they usually sold. For example, 67% of traders said they tended to sell any *B. cafer* body parts when they became available—however, 29% of these traders would have preferred to have acquired the head, 29% the heart, 21% the brain, 14% the intestine, and 7% would have chosen only the feathers instead of what they were offered by the hunters/suppliers. Overall, however, the preferences were for the brains, followed by the head and heart, the feathers, the intestines, and the bones (Fig. 20.2). Interestingly, 50% of traders credited sales of ground-hornbill to a combination of specific customer requests and their recommendation of the species. However, 42% said their sales were based solely on the customer asking for *B. cafer* parts, and 8% said that sales were made only on their recommendation to the customer.

In terms of the acquisition of the birds, 18% of traders hunted for ground-hornbills themselves, 18% acquired them opportunistically, and 64% were supplied by hunters who either came to the market or they specifically placed orders with them. The hunting localities cited as a source of *B. cafer* are shown in Fig. 20.3. The most cited areas were the Matobo National Park (±50 km from Bulawayo, or a 45 min drive) and the Nkayi area (±150 km from Bulawayo, or 2 h drive). The Hwange National Park is furthest away from Bulawayo and at least 5.5 h drive.

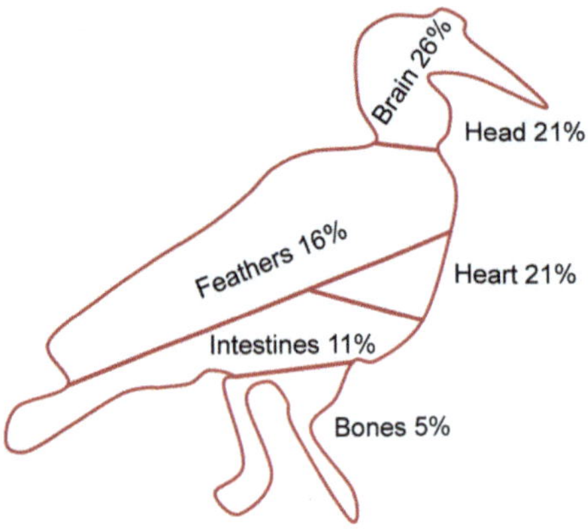


**Fig. 20.2** Trader preferences for Southern Ground-hornbill body parts

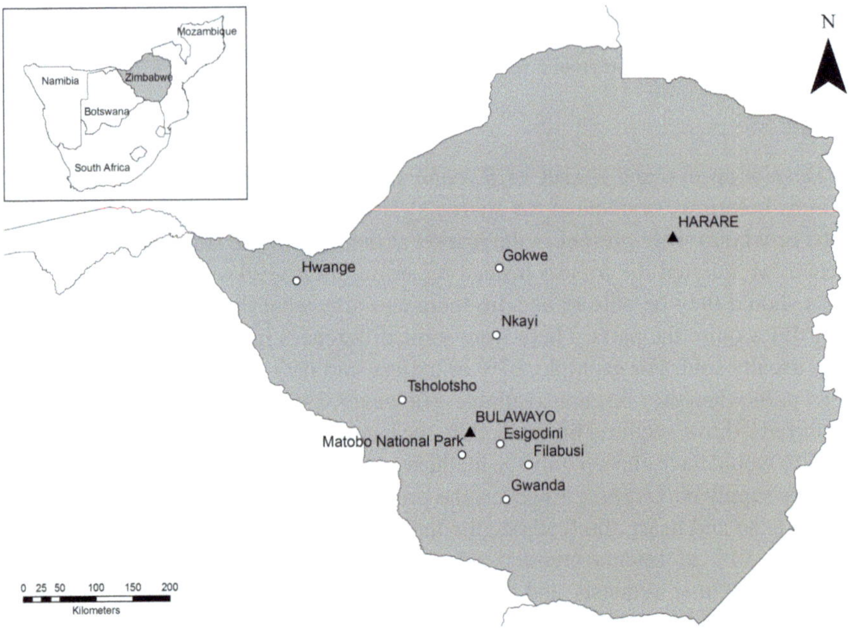


**Fig. 20.3** Map of Zimbabwe and the "best" hunting areas for Southern Ground-hornbill

The areas cited by the traders as good sources for *B. cafer* are, on average, 108 ± 82 km from Bulawayo and encompass a minimum area of 53,993 km$^2$. The survey results were inconclusive about the most common means to catch ground-hornbills, and traps (snares), dogs, poisons, and catapults were mentioned equally.

### 20.3.3 Uses of Southern Ground-Hornbill

When asked what customers wanted or needed *B. cafer* parts for, seven different uses were mentioned. Five (42%) traders said that it was used for starting fights, especially to "get back" at someone and to get revenge. Two traders said the parts were burned as incense to get rid of bad spirits. The remaining uses mentioned were: protection of the homestead; making a person strong; making dreams come true; stopping theft; and, to bring ancestral spirits to young *inyangas* (herbalists).

## 20.4 Discussion

In studying relations between cultures and their respective environments, Anyinam (1995) observed that traditional medicine has very close ties with local ecosystems and, unless regulated, traditional medical practices may have adverse effects on local natural environments as they do not contribute to the preservation or conservation of these areas. Anyinam (1995) also makes special mention of the conservation of plants and animals with respect to the level of harvest shown by the traditional medicine trade as many of the targeted species are habitat specific, which raises questions of the overexploitation of rare, endemic species or those with low reproductive rates. Simelane and Kerley (1998) propose that such dependency on natural resources for traditional medicine may also be used as a tool to protect and conserve these same resources. Such an objective would seek to equate the cultural value of species to local contextual conservation values, thereby, ideally, creating sustainable species use both for cultural and for conservation purposes.

In Bulawayo, we noted that *B. cafer* is very well-known by healers in the traditional medicine market of Makokoba, but it is not used very often—partially due to the ritual process and traditional laws healers and hunters need to follow to be able to obtain a bird. In short, all Ndebele traders described how they first have to dream about the bird, then use a specific medicine on themselves before killing the bird to avoid becoming insane. Then, they have to find a bird, and, once they have killed it, sprinkle seeds of *rappoko* (probably from finger millet, *Eleusine coracana*) on the carcass. The bird must then be left where it was killed for 2 days before any parts are removed, and the carcass must remain behind. Sprinkling *rappoko* seeds (a traditional African crop widely used for ritual purposes linked to the ancestors) probably occurs because of the link between rain-making rituals (using ground-hornbill parts) and agricultural production (since finger millet is a traditional crop).

Although this ritual process may take place over a few months or even years, this does not preclude *B. cafer* from being sought after, as can be seen by the large geographical range that hunters and healers are prepared to cover in order to find them. Traders openly discussed the best areas to obtain the birds, most of which occurred within or near nature reserves. While a large proportion of traders asked hunters to acquire the birds for them, the chances of one such hunter finding a bird appears near opportunistic and rare, and the parts sold in the market coincide with the type of parts that would be taken from a ground-hornbill carcass opportunistically found in the wild.

The monitoring of the market after the initial survey revealed that no new ground-hornbill specimens entered the market over a period of 7 months. The findings of this study appear to *suggest* that *B. cafer* is not commonly used or traded for traditional medicine in western Zimbabwe and hence the impact of use is negligible; however, this inference is not necessarily broadly valid for the entire country. Beliefs, practices, and taboos relating to the use of birds may either favor birds or cause them to be persecuted (Msimanga 2000). Hence, species-specific persecution may depend on the unique belief system of a particular ethnic group, and, may be exclusive to that group (Msimanga 2000). Bulawayo is located on the western side of Zimbabwe in the province of Matebeleland, a region that is largely isiNdebele-speaking. Most traders in the market were noted to be Ndebele, hence subscribing to the associated cultural values and belief systems. On the eastern side of Zimbabwe, however, is the province of Mashonaland where people are predominantly Shona and have a different set of cultural practices relating to birds. It is considered taboo to kill a ground-hornbill by the Ndebele since a certain amount of 'bad luck' is said to follow the person who kills it, especially if they take it home (as reported by many traders we interviewed). These same cultural restrictions when it comes to hunting birds do not necessarily apply in other parts of Zimbabwe. Hence, one cannot presume that hunting has no impact on subpopulation abundance in specific areas of *B. cafer* occurrence, or that the decline in the Zimbabwean *B. cafer* subpopulation is entirely due to habitat degradation. Hence, further contextual cultural studies would be beneficial to ascertain 'hotspots' where the local culture may or may not have an impact on local subpopulations of *B. cafer*.

## 20.5 Conclusion

Currently, traditional medicinal practices in western Zimbabwe appear to pose no immediate threat to *B. cafer* populations as long as large areas of suitable habitat are conserved. The traditional beliefs of the local Ndebele population, combined with the rarity of the species, appear to limit its persecution by people for traditional medicine in Matabeleland. It is not known whether an increase in the *B. cafer* population would give rise to an increase in the species' use for traditional medicine. Hence, it is proposed that for ground-hornbills, traditional medicinal use outside of high population areas is influenced by cultural value and not by economic value, as found in larger traditional medicine markets in major cities (Alves and Rosa 2007).

**Acknowledgments** The study was funded by a National Research Foundation grant from Dr CT Symes (NRF 70984), and the University of the Witwatersrand. We would like to thank the traders at the Makokoba Market, Bulawayo, for their cooperation, Dr. Peter Mundy for his valued cultural input, and Dr. Don Broadley and Sheila Broadley for their exceptional hospitality.

## 20.6 Appendix 1: Market Information Sheet

| | | | | | |
|---|---|---|---|---|---|
| Date | | Surveyor | | Market No. | |
| Country | | City/Town | | Market name | |

| | |
|---|---|
| Total number of traders in the market | |
| Number of traders selling animals | |
| Number of traders selling birds | |

| | | |
|---|---|---|
| Camera Type (Tick appropriate) | Digital | Disposable |

| | |
|---|---|
| Problems encountered / Notes on the market | |

Does the trader recognize any of the following birds?

Male Southern Ground-Hornbill

Female Southern Ground-Hornbill

Juvenile Southern Ground-Hornbill

## 20.7 Appendix 2: Market Questionnaire for Ground-Hornbills

**<u>Ground Hornbill Market Questionnaire</u>**

| | | | |
|---|---|---|---|
| Interview Number | | Interviewer | |
| Date | | Location | Country City Market |

**Trader Information**

| | | | |
|---|---|---|---|
| Home language | | Gender | |
| Local name for Ground-Hornbill | Name Language | | |

| | | | | |
|---|---|---|---|---|
| 1. Do customers ask for Ground-Hornbill, or do you recommend it to them? | ASK | RECOMMEND | | |
| 2. What do customers use Ground-Hornbill for? | | | | |
| 3. Where are your customers from? | Rural | City | Other countries | Other(Specify) |
| 4. Do you sell parts or do you sell as a mixed medicine? | Parts | Medicine | Both | |

5. What parts do you sell?

| | | | | | |
|---|---|---|---|---|---|
| Head | Beak | Feathers | Intestine | Throat | Eyes |
| Feet | Bones | Wings | Neck | Leg | Other |

| | |
|---|---|
| 6. Do you prefer to use a specific part of the bird? | |
| 7. Where is the best area to get Ground-Hornbill? | |

| | | | |
|---|---|---|---|
| 8. How do you get Ground Hornbill? | I hunt it myself | Someone hunts it for me | Opportunistic |

| | |
|---|---|
| 9. How often do you get a Ground Hornbill? (Whole or parts) | |
| 10. How long does it take to get one? | |
| 11. How long does one whole bird last you? | |

| | | |
|---|---|---|
| 12. Is it becoming easier or more difficult to find this bird? | Easier | More difficult |

| | |
|---|---|
| 13. Why do you think this is? | |

| | | | |
|---|---|---|---|
| 14. Does it make a difference if you get a male Ground Hornbill or a female Ground Hornbill? | Yes | No | Preference<br>Male Female Juvenile |

| | |
|---|---|
| 15. What traditional laws are there about trading or using Ground Hornbill? | |

## References


Abrahams N, Jewkes R, Mvo Z (2002) Indigenous healing practices and self-medication amongst pregnant women in Cape Town South Africa. Afr J Reprod Health 6(2):79–86

Alves RRN, Rosa IL (2007) Zootherapy goes to town: the use of animal-based remedies in urban areas of NE and N Brazil. J Ethnopharm 113:541–555

Anyinam C (1995) Ecology and ethnomedicine: exploring links between current environmental crisis and indigenous medical practices. Soc Sci Med 40:321–329

Barnard A (1979) Nharo Bushman medicine and medicine men. Afr J Int Afr Inst 49(1):68–80

Barnes KN (ed) (2000) The Eskom Red Data book of birds of South Africa Lesotho and Swaziland. BirdLife, South Africa

BirdLife International (2010) Species factsheet: *Bucorvus cafer*. Downloaded from http://www.birdlife.org on 9/7/2010 and 01/12/2011

Chiweshe N (2007) The current conservation status of the Southern ground hornbill Bucorvus leadbeateri in Zimbabwe. In: Kemp AC, Kemp MI (eds) The active management of hornbills and their habitats for conservation, pp 252–266

Cocker M (2000) African birds in traditional magico-medicinal use—a preliminary survey. Bull Afr Bird Club 7(1):60–66

Conco WZ (1972) The African Bantu traditional practice of medicine: some preliminary observations. Soc Sci Med 6:283–322

Craig GC (2009) Aerial survey of wildlife in the Niassa Reserve and adjacent areas. October 2009 Sociedade para a Gestão e Desenvolvimento da Reserva do Niassa, Maputo

Cunningham AB, Zondi AS (1991) Use of animal parts for the commercial trade in traditional medicines. Institute of Natural Resources, Pietermaritzburg

France H (1905) Customs of the Awuna tribes. J Roy Afr Soc 5(17):38–40

Garbutt HW (1909) Native witchcraft and superstition in South Africa. J Roy Anthropol Inst G B Irel 39:530–558

Godfrey R (1941) Bird-lore of the Eastern Cape Province. Witwatersrand University Press, Johannesburg

Hollis C (1901) Notes on the history and customs of the people of Taveta. East Africa. J Roy Afr Soc 1(1):98–125

Kalamira N (2007) The status of the Southern Ground Hornbill Bucorvus leadbeateri in some protected areas of Malawi. In: Kemp AC, Kemp MI (eds) The active management of hornbills and their habitats for conservation, pp 245–251

Kemp AC (2000) Southern ground hornbill. In: The Eskom Red Data book of birds of South Africa, Lesotho and Swaziland. Birdlife, South Africa

Mander M, Diederichs N, Ntuli L, Mavundla K, Williams V, McKean S (2007) Survey of the trade in vultures for the traditional health industry in South Africa. FutureWorks, Durban

Marshall NT (1998) Searching for a cure: conservation of medicinal wildlife resources in East and Southern Africa. Traffic International, Cambridge

Msimanga A (2000) The role of birds in the culture of the Ndebele people of Zimbabwe. Ostrich 71(1&2):2–24

Ngwenya MP (2001) Implications for the medicinal animal trade for nature conservation in KwaZulu-Natal. Unpublished Ezimvelo KZN wildlife report no. NA/124/04

Nikolaus G (2000) Bird exploitation for traditional medicine in Nigeria. Malimbus 23:45–55 http://malimbus.free.fr/articles/V23/23045055.pdf

Ntiamoa-Baidu Y (1987) Tropical rain forest management a status report. West African wildlife: a resource in jeopardy. Unasylva 156:1–10

Owens IPF, Bennett PM (2000) Ecological basis of extinction risk in birds: habitat loss versus human persecution and introduced predators. Proc Nat Acad Sci 97:12144-12148. doi: 10.1073/pnas.200223397

Parrinder EG (1956) African ideas of witchcraft. Folklore 67(3):142–150

Simelane TS, Kerley GIH (1998) Conservation implications of the use of vertebrates by Xhosa traditional healers in South Africa. S Afr J Wildl Resour 28:121–126

Spear D, Morrison K, Daly B, du Plessis M, Turner A, Friedmann Y (2005) Southern ground hornbill, *Bucorvus leadbeateri*, population and habitat viability assessment (PHVA) briefing book. The conservation breeding specialist group (CBSG Southern Africa). Endangered Wildlife Trust, Johannesburg

Whiting MJ, Williams VL, Hibbitts TJ (2011) Animals traded for traditional medicine at the Faraday market in South Africa: species diversity and conservation implications. J Zool 284(2):84–96. doi:10.1111/j.1469-7998.2010.00784

# Index



R. R. N. Alves and I. L. Rosa (eds.), *Animals in Traditional Folk Medicine*,
DOI: 10.1007/978-3-642-29026-8, © Springer-Verlag Berlin Heidelberg 2013

MIX
Papier aus verantwortungsvollen Quellen
Paper from responsible sources
FSC® C105338

If you have any concerns about our products,
you can contact us on
**ProductSafety@springernature.com**

In case Publisher is established outside the EU,
the EU authorized representative is:
**Springer Nature Customer Service Center GmbH**
**Europaplatz 3, 69115 Heidelberg, Germany**

Printed by Libri Plureos GmbH
in Hamburg, Germany